An Invitation to Social Research

How It's Done

FIFTH EDITION

An Invitation to Social Research

How It's Done

Emily Stier Adler

Rhode Island College

Roger Clark

Rhode Island College

Australia • Brazil • Mexico • Singapore • United Kingdom • United States

CENGAGE
Learning

An Invitation to Social Research:
How It's Done, Fifth Edition
Emily Stier Adler and Roger Clark

Product Manager: Seth Dobrin

Content Developer: Carolyn Lewis

Content Coordinator: Jessica M Alderman

Product Assistant: Nicole Bator

Media Developer: John Chell

Marketing Manager: Michelle Williams

Art and Cover Direction: Carolyn Deacy,
MPS Limited

Production Management and Composition:
Jitendra Kumar, MPS Limited

Manufacturing Planner: Judy Inouye

Rights Acquisitions Specialist: Thomas
McDonough

Photo Researcher: PMG/Hemalatha
Dhanapal

Text Researcher: PMG/Sachin Das

Cover Designer: Lisa Buckley

Cover Image: imagebroker.net/SuperStock,
Voisin/Phanie/SuperStock, and Photodisc

For product information and technology assistance, contact us at
Cengage Learning Customer & Sales Support, 1-800-354-9706.

For permission to use material from this text or product,
submit all requests online at **www.cengage.com/permissions**.

Further permissions questions can be e-mailed to
permissionrequest@cengage.com.

Library of Congress Control Number: 2013950268

Student Edition:

ISBN-13: 978-1-285-74642-5

ISBN-10: 1-285-74642-2

Cengage Learning
200 First Stamford Place, 4th Floor
Stamford, CT 06902
USA

Cengage Learning is a leading provider of customized learning solutions with office locations around the globe, including Singapore, the United Kingdom, Australia, Mexico, Brazil, and Japan. Locate your local office at **www.cengage.com/global**.

Cengage Learning products are represented in Canada by
Nelson Education, Ltd.

To learn more about Cengage Learning Solutions, visit **www.cengage.com**.

Purchase any of our products at your local college store or at our preferred online store **www.cengagebrain.com**.

Printed in the United States of America
1 2 3 4 5 6 7 17 16 15 14

We dedicate this book to David, Evan, and Matthew

BRIEF CONTENTS

CONTENTS

PREFACE

We'd like to invite you to participate in one of the most exciting, exhilarating, and sometimes exasperating activities we know: social science research. We extend the invitation not only because we know, from personal experience, how rewarding and useful research can be, but also because we've seen what pleasure it can bring other students of the social world. Our invitation comes with some words of reassurance, especially for those of you who entertain a little self-doubt about your ability to do research. First, we think you'll be glad to discover, as you read *An Invitation to Social Research: How It's Done*, how much you already know about how social research is done. If you're like most people, native curiosity has been pushing you to do social research for much of your life. This book is meant simply to assist you in this natural activity by showing you some tried-and-true ways to enlightening and plausible insights about the social world.

SPECIAL FEATURES

Active Engagement in Research

Our second word of reassurance is that we've done everything we can to minimize your chances for exasperation and maximize your opportunities for excitement and exhilaration. Our philosophy is simple. We believe that honing one's skill in doing social research is analogous to honing one's skills in other enjoyable and rewarding human endeavors, like sport, art, or dance. The best way isn't simply to read about it. It's to do it and to watch experts do it. So, just as you'd hesitate to teach yourself tennis, ballet, or painting only by reading about them, we won't ask you to try learning the fine points of research methodology by reading alone. We'll encourage you to get out and practice the techniques we describe. We've designed exercises at the end of each chapter to help you work on the "ground strokes," "serve," "volleys," and "overheads" of social research. We don't think you'll need to do all the exercises at home. Your instructor might ask you to do some in class and might want you to ignore some altogether. In any case, we think that, by book's end, you should have enough control of the fundamentals to do the kind of on-the-job research that social science majors are increasingly asked to do, whether they find themselves in social service agencies, the justice system, business and industry, government, or graduate school.

The exercises reflect our conviction that we all learn best when we're actively engaged. Other features of the text also encourage such active engagement, including the "Stop & Think" questions that run through each chapter, which encourage you to actively respond to what you're reading.

Engaging Examples of Actual Research

Moreover, just as you might wish to gain inspiration and technical insight for ballet by studying the work of Anna Pavlova or Mikhail Baryshnikov, we'll encourage you to study the work of some accomplished researchers. Thus, we build most of our chapters around a research essay, what we call focal research, that is intended to make the research process transparent, rather than opaque. We have chosen these essays for their appeal and accessibility, and to tap what we hope are some of your varied interests: for instance, crime, gender, election polls, life in prison, attitudes toward environmentalism, immigrants' lives, and others.

Behind-the-Scenes Glimpses of the Research Process

These focal research pieces are themselves a defining feature of our book. In addition to such exemplary "performances," however, we've included many behind-the-scenes glimpses of the research process, often written by researchers whose work we admire. We're able to provide these glimpses because many researchers have given generously of their time to answer our questions about what they've done, the special problems they've encountered, and the ways they've dealt with these problems. The glimpses should give you an idea of the kinds of choices and situations the researchers faced, where often the "real" is far from the "ideal." You'll see how they handled the choices and situations and hear them present their current thinking about the compromises they made. In short, we think you'll discover that good research is an achievable goal, and a very human enterprise.

Clear and Inviting Writing

We've also tried to minimize your chances for exasperation by writing as clearly as we can. A goal of all social science is to interpret social life, something you've all been doing for quite a while. We want to assist you in this endeavor, and we believe that an understanding of social science research methods can help. But unless we're clear in our presentation of those methods, your chances of gaining that understanding are not great. There are, of course, times when we'll introduce you to concepts that are commonly used in social science research that might be new to you. When we do, however, we will try to provide definitions to make the concepts as clear as possible. The definitions are highlighted in the margin of the text and in the glossary at the end of the text.

Focus on Ethics

Given the importance of doing research that is methodologically correct and practical as well as ethical, we've put a focus on ethical principles in each chapter. The "Thinking about Ethics" section of each chapter applies the ethical principles we cover in depth in Chapter 3 to research projects presented in the subsequent chapters.

Balance between Quantitative and Qualitative Approaches

We think you'll also appreciate the balance between quantitative and qualitative research methods presented here. Quantitative methods focus on things that are measured numerically. ("He glanced at her 42 times during the performance.") Qualitative methods focus on descriptions of the essence of things. ("She appeared annoyed at his constant glances.") We believe both methodological approaches are too useful to ignore. Emblematic of this belief is the inclusion of a chapter (Chapter 15) that devotes about as much space to the discussion of qualitative data analysis as it does to quantitative data analysis. The presence of such a chapter is another defining feature of the book.

Moreover, in addition to more conventional strategies, we will introduce you to some relatively new research strategies, such as using the Internet to refine ideas and collect data and visual methodologies. We cover the link between theory and research, compare research to other ways of knowing, and focus on basic and applied research.

Our aims, then, in writing this book have been (1) to give you firsthand experiences with the research process, (2) to provide you with engaging examples of social science research, (3) to offer behind-the-scenes glimpses of how professional researchers have done their work, (4) to keep our own presentation of the "nuts-and-bolts" of social science research as clear and inviting as possible, (5) to focus on doing research following ethical principles, (6) to give a balanced presentation of qualitative and quantitative research methods, (7) to introduce recent technological innovations, and (8) to present some research done in the "real world" to solve important social problems. Whether we succeed in these goals, and in the more important one of sharing our excitement about social research, remains to be seen. But rest assured, however, of our conviction that there is excitement to be had.

WHAT IS NEW IN THE FIFTH EDITION

The fifth edition represents a substantial revision of the fourth. Once again, we've rewritten major sections of every chapter to clarify the process of social research and to provide up-to-date material from the social research literature. In doing so, we've focused our presentation on the essentials of social research and covered some new material as well as classic sources.

We've added two new features that should underline what's important about each chapter. On the one hand, we provide chapter objectives at the beginning of each chapter, objectives that should help students know, in some detail, what they may expect to take away from the chapter. Second, we've added, in many chapters, a brief overview of social science research that's made the national or international news and thereby shows how critically important the topic of the chapter can be for understanding the real world. For example, in Chapter 14, we focus on evidence-based programs and policies and our "Research in the News" features a legal battle over a drug education program that more than a decade of research has shown to be ineffective. Our data analysis chapter (Chapter 15) also reflects, as does the rest of the current text, our belief that research, as practiced by social (and all other) scientists, is increasingly computer assisted and Internet based. So, for instance, in the data analysis chapter, we introduce students to data that they can analyze online. In other chapters, we also present ways of finding research reports and data that can be accessed quickly online.

Themes from the first three editions have been retained here. This edition has 13 focal research pieces—four of them new. While incorporating the new pieces, we have maintained the balance between qualitative and quantitative research in the book. In Chapter 6, for instance, we present two new focal research pieces on measurement: one explores ways to measure "happiness" and the other considers reliability and validity issues with current measures of "suicide." In Chapter 11, we present a qualitative, observation-based study of Pittsburgh Steelers fans, a study that argues that being a member of a sports team's fan base can be like belonging to a quasi-religion. In Chapter 13, we present a new piece that examines the relationship between the gender, race, ethnicity, and other identity statuses of authors of children's book, on the one hand, and the visibility of female characters in their works, on the other. In all cases, our new contributors have volunteered important "behind-the-scenes" insights into the research process, insights that we gratefully share here.

ACKNOWLEDGMENTS

We cannot possibly thank all those who have contributed to the completion of this edition, but we can try to thank those whose help has been most indispensable and hope that others will forgive our neglect. We'd first like to thank all the students who have taken research methods courses with us at Rhode Island College for their general good-naturedness and patience as we've worked out ideas that are crystallized here, and then worked them out some more. We'd also like to thank our colleagues in the Sociology Department and the administration of the college, for many acts of encouragement and more tangible assistance, including released time.

We'd like to thank colleagues, near and far, who have permitted us to incorporate their writing as focal research or in text boxes, and, in many cases, have then read how we've incorporated it and told us how we might do better. They are Nichole Banton, Susan Chase, Erica Chito Childs, Leah Christian, Desirée Ciambrone, Allyssa Coon, Marci Cottingham, Michael Dimock, Danielle Dirks, Sandra Enos, Rachel Filinson, John Grady, Jill Harrison, Michele Hoffnung, Scott Keeter, Julie Kessler, Brandon Lang, Matthew T. Lee, Kristy Maher, Ramiro Martinez, Donald C. Naylor, Maureen Norton-Hawk, Chris Podeschi, Matthew M. Reavy, Ian Rockett, Jennifer Silva, and Jessica Holden Sherwood. These researchers include those who've given us the behind-the-scenes glimpses of the research process that we think distinguishes our book.

We're very grateful to our editor at Wadsworth, Carolyn Lewis, for help with the fifth edition. We've benefited greatly from the comments of the social science colleagues who have reviewed our manuscripts at various stages of development: Mary Archibald; Cynthia Beall, Case Western Reserve University; Daniel Berg, Humboldt State University; Susan Brinkley, University of Tampa; Jeffrey A. Burr, State University of New York, Buffalo; Richard Butler, Benedict College; Daniel Cervi, University of New Hampshire; Charles Corley, Michigan State University; Norman Dolch, Louisiana State University, Shreveport; Craig Eckert, Eastern Illinois University; Pamela I. Jackson, Rhode Island College; Alice Kemp, University of New Orleans; Michael Kleinman, University of South Florida; James Marquart, Sam Houston State University; James Mathieu, Loyola Marymount University; Steven Meier, University of Idaho; Deborah Merrill, Clark University; Lawrence Rosen, Temple University; Josephine Ruggiero, Providence College; Kevin Thompson, University of North Dakota; Steven Vassar, Mankato State University; and Gennaro Vito, University of Louisville, offered important assistance for the first edition.

For the second edition, Gai Berlage, Iowa College; Susan Chase, University of Tulsa; Dan Cooper, Hardin-Simmons University; William Faulkner, Western Illinois University; Lawrence Hazelrigg, Florida State University; Josephine Ruggiero, Providence College; Jeffrey Will, University of North Florida; and Don Williams, Hudsonville College, have been most helpful.

For the third edition, we would like to thank Ronda Priest, University of Southern Indiana; Dennis Downey, University of Utah; Scott M. Myers, Montana State University; William Wagner, California State University, Bakersfield; and John Mitrano, Central Connecticut State University, for their valuable comments.

For the fourth edition, we would like to thank Allan Hunchuck, Theil College; Aaryn Ward, Louisiana State University; Erin Leahey, University of Arizona; and Kevin Yoder, University of North Texas for their help.

For the fifth edition, we thank Timothy Hayes, North Georgia College & State University, Anna Bellatorre, University of Nebraska—Lincoln, Stacy Keogh, University of Montana, Julie Kmec, Washington State University, Aaron McCutcheon, Florida State University, Summer McWilliams, Florida State University, and Bryan Sykes, DePaul University.

Finally, we'd like to thank our spouses, George L. Adler and Beverly Lyon Clark, for providing many of the resources we've thanked others for: general good-naturedness and patience, editorial assistance, encouragement, and other tangible aids, including released time.

An Invitation to Social Research

How It's Done

CHAPTER 1

The Uses of Social Research

CHAPTER OBJECTIVES

This chapter is designed to help students

- **Recognize the advantages of knowledge based on research over knowledge from authorities and knowledge from personal inquiry**

- **Appreciate the importance of both care and community in science**

- **Distinguish between basic and applied research**

- **Learn about the four purposes of scientific research: exploration, description, explanation and evaluation**

CHAPTER OUTLINE

Research versus Other Ways of Knowing

The Uses and Purposes of Social Research

Summary

Exercises

© Emily Stier Adler

How do you see yourself 10 years from now? Will you be single? Will you be a parent? What about people you know? Does it seem to you that almost everyone with whom you've talked about such things is either planning to marry and have children or is already married with children? Or does it seem to be the other way around? Does everyone seem to be planning to avoid marriage, perhaps to live alone, at least for extended periods of his or her life? What's the reality? Studying something systematically, with method, means not relying on your impressions—or on anyone else's, either. It means checking your impressions against the facts. It also means being sure about what you are looking for. One way researchers achieve this level of definiteness is to begin their research with a **research question**, or a question about one or more topics that can be answered through research. One such focusing question about our initial concern, for instance, might be: Is almost everyone in the country married with children or are they living alone?

research question, a question about one or more topics or concepts that can be answered through research.

The U.S. Census Bureau's *Statistical Abstract* (2012a) provides one apparently relevant set of facts. It shows that in 2010, the number of Americans living alone was about 26.7 percent of all households and was greater than the proportion of American households made up of a married couple with children, about 20.9 percent. Until 2000, the proportion of American households made up of married couples with children had always been greater than the proportion of Americans living alone (Hobbs, 2005). This could be the starting point of your investigation.

STOP & THINK   *This would only be a start, however. What element of our initial interest in the future of you and people like you is not addressed by the "facts" of the Census report?*

Although the Census report does tell us something that's relevant to our main research—whether young people today are planning to marry and have children or planning not to marry—it requires a little interpreting, doesn't it? At first glance, Census facts seem almost to address this question. They show that, in 2010, more households in the United States were made up of people living alone than of married couples with children and so seem to imply that more people will end up alone than in families with children. Or do they?

We've made up the pie chart in Figure 1.1 to help us think about the question of whether the Census data really settle the issue addressed in our research question. The chart indicates one thing that we already knew: More American households are made up of people living alone than of married couples with children. But perhaps looking at the chart will remind you that all households made up of married couples with children had at least *twice as many adults in them* as households made up of people living alone. So, in fact, there actually were more adults living as parts of married couples with children in 2010 than there were adults living alone. It also reminds us that we should consider other kinds of households. One such group is married couples without children, a group that is more numerous than households made up of people living alone and, again, would therefore be households that have at least twice as many adults in them as households made up of people living alone. The presence of

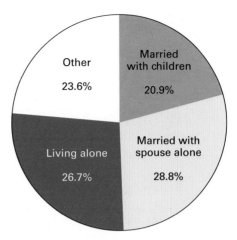

FIGURE **1.1** Total Households in 2010: 117,538,000

Source: Data from U.S. Census (2012).

this large group might make us want to revise our original question to include something about the number of people who plan to live with another adult (or adults) but without children. Moreover, we suspect that more of these people—people in couples without children—"plan" to become couples with children than "plan" to live alone. In effect, then, the chart also leads us to a second research question, a question that, in some ways, might even better embody the concern expressed at the beginning of paragraph one: Are more young people planning to marry and not have children, to marry and have children, or to live alone?

STOP & THINK   *Can you think of a better way to find out how young people plan to live in the future than with the Census data of 2010?*

We suspect that you can think of such a way. Maybe you've noticed that Figure 1.1 is misleading in at least two ways. First, it focuses on the way things are, or more precisely, the way things were in 2010. It doesn't help with the question of what people are planning to do in the future. Second, it focuses on households, not people. We've obviously been able to make some inferences about people from the information about households, but the inferences have been somewhat awkward and would be even more so if we focused more attention on the fourth part of Figure 1.1: the part about "other" households. People, not households, are what we're really interested in, so Figure 1.1 ultimately provides information about the wrong kind of thing, the wrong **unit of analysis,** the unit about which information is collected. One of the first decisions you need to make when planning research is, "What should my unit of analysis be?" Moreover, one of the first questions to ask of any research project you read is, "What is this study's unit of analysis?"

unit of analysis, a unit about which information is collected.

STOP & THINK   *See if you can identify the units of analysis for each of the following studies: (1) Rosenfeld and Thomas' (2012) finding that people are increasingly likely to find prospective partners, often complete strangers, online; (2) Fallon, Swiss and Viterna's (2012) finding that in developing countries that began democratizing after 1975, early elections brought a drop in women's access to the nation's legislature, but later elections brought increased women's access; and (3) Soule and King's (2008) finding that social movement organizations (like women's rights organizations) that develop specialized goals, rather than more general ones, are less likely to survive than organizations that can maintain generalized goals.*

Returning to our question about how you might find out about young people's future plans, perhaps you've thought of doing some kind of questionnaire survey, the nature of which we talk about more in Chapter 9. Perhaps you've also realized that you'd want to survey some kind of representative sample of young people. We talk about such samples in Chapter 5. Perhaps you've thought a little about the kinds of information you might want. We deal with those issues in Chapter 6. Perhaps you've even thought about examining what others have had to say about the issue. You might, for instance, be interested in Michele Hoffnung's research, the focal piece of Chapter 4. Maybe someone's already collected just the information you need and would be willing to share it with you. We examine this possibility in Chapter 12, on finding and using available data.

Whatever you've thought, you've clearly begun to engage the question, as we (Emily and Roger) have done with questions we've studied, as a mystery to be solved. Learning about social research is a lot like learning to solve mysteries. It's about challenge, frustration, excitement, and exhilaration. We, Emily and Roger, are addicted to social research. We've recently enjoyed working together on a study of how people make the transition to retirement (and will tell you about our findings in Chapter 7). We're passionately interested in this mystery, partly because we're thinking about the process of retirement ourselves.

RESEARCH VERSUS OTHER WAYS OF KNOWING

Knowledge from Authorities

The data about households in America—that slightly more of them are made up of people living alone than of people living with a spouse and children—are fascinating. But perhaps they're not as fascinating as what we, Emily and Roger, learned many years ago, literally at our mothers' knees, that "In fourteen hundred and ninety-two, Columbus sailed the ocean blue," despite nearly everyone's belief that he was sailing in a direction that jeopardized his very existence.

Now, why do we think we "know" these things? Basically, it's because some authority told us so. In the first case, we read it in a Census Bureau report. In the second case, we relied on our moms, who were reporting a commonly accepted version of America's "discovery." **Authorities,** such as the

authorities, socially defined sources of knowledge.

Census Bureau and our moms, are among the most common sources of knowledge for most of us. Authorities are socially defined sources of knowledge.

Many social institutions, such as religion, news media, government agencies, and schools, are authorities, and individuals within them are often seen as having superior access to relevant knowledge. In modern societies like ours, we often attribute authority to what newscasters on television say or what's posted in an online newspaper such as the *Huffington Post*. Sometimes authorities use research as their basis for knowledge, but we usually don't evaluate their sources. For most of us, most of the time, learning from authorities is good enough, and it certainly helps keep life simpler than it would be otherwise. Life would be ridiculously difficult and problematic if, for instance, we who live in the Western world had to reinvent a "proper" way of greeting people each time we met them. At some point, we're told about the customs of shaking hands, bumping fists, or saying "Hi," and we move on from there.

 STOP & THINK   *What, do you think, are the major disadvantages of receiving our "knowledge" from authorities?*

Although life is made simpler by "knowledge" from authorities, sometimes such knowledge is inappropriate, misleading, or downright incorrect. Very few people today take seriously the "flat Earth" theories that have been attributed to Columbus's social world. More interesting, perhaps, is an increasingly accepted view that, our mothers' teachings notwithstanding, very few people in Columbus's social world took seriously the flat-Earth view either; that, in fact, this view of the world was wrongly attributed to them by late-nineteenth-century historians to demonstrate how misguided people who accepted religious over scientific authority could be (e.g., Gould, 1995: 38–50). In fact, we mean no offense to our moms when we say they represent a whole category of authorities who can mislead: authorities in one area of expertise (in the case of our moms, us) who try to speak authoritatively on subjects in which they are not experts (in the case of our moms, the worldview of people in Columbus's society).

Knowledge from Personal Inquiry

But if we can't always trust authorities, like our moms, or even experts, like those late-nineteenth-century historians (and, by extension, even our teachers or the books they assign) for a completely truthful view of the world, who or what can we trust? For some of us, the answer is that we can trust the evidence of our own senses. **Personal inquiry**, or inquiry that employs the senses' evidence for arriving at knowledge, is another common way of knowing.

personal inquiry, inquiry that employs the senses' evidence.

 STOP & THINK   *Can you think of any disadvantages in trusting personal inquiry alone as a source of "knowledge"?*

The problem with personal inquiry is that it, like the pronouncements of authorities, can lead to misleading, even false, conclusions. As geometricians like to say, "Seeing is deceiving." This caution is actually as appropriate for students of the social world as it is for students of regular polygons because the evidence of our senses can be distorted. Most of us, for instance, developed our early ideas of what a "family" was by observing our own families closely. By the time we were six or seven, each of us (Emily and Roger) had observed that our own families consisted of two biological parents and their children. As a result, we concluded that all families were made up of two biological parents and their children.[1]

There's obviously nothing wrong with personal inquiry ... except that it frequently leads to "knowledge" that's pretty half-baked. There are many reasons for this problem. One of them is obvious from our example: Humans tend to *overgeneralize* from a limited number of cases. Both of us had experienced one type of family and, in a very human way, assumed that what was true of our families was true of all human families. Another barrier to discovering the truth is the human tendency to *perceive selectively* what we've been conditioned to perceive. Thus, even as 10-year-olds, we might have walked into an intentional community such as a commune, with lots of adults and children, and not entertained the possibility that this group considered itself a family. We just hadn't had the kind of experience that made such an observation possible. A third problem with knowledge from personal inquiry is that it often suffers from *premature closure*—our tendency to stop searching once we think we have an answer. At 10, we (Emily and Roger) thought we knew what a family was, so we simply didn't pursue the issue further.

So, neither relying on authorities nor relying on one's own personal inquiry is a foolproof way to the truth. In fact, there might not be such a way. But authors of research methods books (ourselves included) tend to value a way that's made some pretty astounding contributions to the human condition: the scientific method. In the next two subsections, we'll, first, discuss some relative strengths of the scientific method for knowledge acquisition and, second, give you some idea about what this knowledge is intended to do.

The Scientific Method and Its Strengths

scientific method, a way of conducting empirical research following rules that specify objectivity, logic, and communication among a community of knowledge seekers and the connection between research and theory.

positivist view of science, a view that human knowledge must be based on what can be perceived.

Specifying precise procedures that constitute the **scientific method** is a dicey business at best. Part of what makes it so hard is that there are fundamental philosophical differences in beliefs about what science is. Without getting too bogged down in weighty philosophical matters, we should probably confess that our own philosophical approach to science, and that of most contemporary social scientists, isn't the classical positivist view. A **positivist view of science** suggests we should stick to those things we can observe and measure directly, such as the time it takes a person to solve a Rubik's cube, whether someone lives alone, or how many times a week an individual goes to church. The goals of positivist science are unchallengeable propositions about the world, like, in the physical sciences, "every action has an equal and opposite reaction." Things that can't be directly observed, such as what people think about a

---

[1] Of course, by the time that Roger, in his early forties, had adopted two children from another country, his ideas of "family" had changed many times.

post-positivist view of science, a view that knowledge is not based on irrefutable observable grounds, that it is always somewhat speculative, but that science can provide relatively solid grounds for that speculation.

friend's divorce, how they feel about losing a job, or the economic impact of globalization, are irrelevant. Ours is a **post-positivist view of science**, a view that knowledge is not based on irrefutable observable grounds, that it is always somewhat speculative, but that science can provide relatively solid grounds for that speculation. Moreover, most post-positivists would argue that the pursuit of unchallengeable propositions about the world is unrealistic and, perhaps, unscientific. Most physicists today feel that Newton's laws of motion, of which "every action has an equal and opposite reaction" is but one example, are limited in their scope. Good as they are, they simply don't apply when things get very small (say, subatomic) or very fast (say, close to light speed). They're certainly not unchallengeable.

objectivity, the ability to see the world as it really is.

intersubjectivity, agreements about reality that result from comparing the observations of more than one observer.

Positivists have often claimed that they strive for **objectivity**, or the ability to see the world as clearly as possible, free from personal feelings, opinions, or prejudices about what it is or what it should be. Post-positivists, on the other hand, suggest that the most even scientists can strive for is **intersubjectivity**, or agreements about reality that come from the practice of comparing one's results with those of others and discovering that the results are consistent with one another. Generally, post-positivists, such as Emily and Roger, believe that scientists may be more careful than everyday people, but their work is nonetheless fallible. The "goal of science," according to Trochian (2006), "is to hold steadfastly to the goal of getting it right about reality, even though we can never achieve that goal!"

But how are we to hold steadfastly to that goal? We'd like to suggest four steps that are often involved in doing science and point to the relative emphasis placed on *care* and *community* that distinguishes science from other modes of knowing, whatever one's philosophical views of science are. An early step is to specify the goals or objectives that distinguish a particular inquiry (here care is paramount). In our first example, we eventually specified the goal of answering the research question, "Are more young people planning to marry and have children or to live alone?" A subsequent step involves reviewing literature or reading what's been published about a topic (here, learning what a relevant community thinks is the goal). We could do worse, in the pursuit of that early research question—about whether young people are planning to marry and have children or to live alone, for instance—than read Michele Hoffnung's work (see Chapter 4) on what an earlier generation of "young people" planned to do. At some point, it becomes important to specify what is actually observed (care again). We'd want to define, for instance, what we mean by "young people" and how we plan to measure what their plans are. A later step is to share one's findings with others in a relevant community so that they can scrutinize what's been done (community again). We might want, for instance, to prepare a paper for a conference of family sociologists. These steps or procedures will come up again throughout this book, but we'd now like to stress some of the strengths that accrue to the scientific method because of their use.

The Promotion of Skepticism and Intersubjectivity

One great strength of the scientific method, over modes that rely on authorities and personal inquiry, is that, ideally, it promotes skepticism about its own knowledge claims. Perhaps you looked at our presentation of the data about today's households (e.g., that fewer are made up of married couples with children than are made up of people living alone) and said, "Hey, those facts alone don't tell us much about people's intentions. And they don't even indicate that more people are living alone than living in couples today." If so, we applaud your skepticism. One way in which healthy skepticism is generated is through the communities of knowledge seekers. Each member of these communities has a legitimate claim to being a knowledge producer, as long as he or she conforms to other standards of the method (mentioned later). When intersubjective agreement eludes a community of scientists because various members get substantially different results, it's an important clue that knowledge remains elusive and that knowledge claims should be viewed with skepticism. Until the 1980s, for instance, the medical community believed that stomach ulcers were caused by stress. Then a couple of Australian scientists, Barry Marshall and Robin Warren, found, through biopsies, that people with ulcerous stomachs often had *Helicobacter pylori* bacteria lurking nearby and theorized that the bacteria had caused the ulcers. Marshall and Warren thus became skeptical of the medical community's consensus about the causes of ulcers. Few members of that community took them seriously, however, until Marshall, experimenting on himself, swallowed *H. pylori* and developed pre-ulcerous symptoms. Subsequently, Marshall and Warren found that most stomach ulcers could be successfully treated with antibiotics. For their work in discovering the bacterium that causes stomach inflammation, ulcers, and cancer, Marshall and Warren won the 2005 Nobel Prize for Physiology or Medicine (Altman, 2005: D3; Marshall, 2002).

The Extensive Use of Communication

Another related ideal of the scientific method is *adequate communication* within the community of knowledge seekers, implicit in the scientific procedures of referring to previous published accounts in an area and of sharing findings with others. Unlike insights that come through personal inquiry, scientific insights are supposed to be subjected to the scrutiny of the larger community and, therefore, need to be as broadly publicized as possible. Communication of scientific findings can be done through oral presentations (as at conferences) or written ones (especially through publication of articles and books). For example, Hoffnung has presented her findings about college students' plans for the future at scientific conferences and in scholarly articles. In Chapter 4, she offers you a glimpse of her findings. Increasingly, new technology (discussed in Chapter 12) allows for increased communication about research and the exchange of data. You, for instance, can use a variety of print and online resources that your college library may make available to you to find references to, and even copies of, research articles about your topic. We did and found research by Barr and Simons (2012) and Johnstone and Lee (2009) that supplement Hoffnung's work and our interests in the kinds of family lives that young people envision by examining recent survey data on how young men and women are expecting to deal with work–family conflicts in their future. Once findings are communicated, they then become grist for a critical mill. Others are thereby invited to question (or openly admire) the particular approach that's been reported or to try to reproduce (or *replicate*) the findings using other approaches or other circumstances. Adequate communication thus facilitates the ideal of reaching intersubjective "truths."

Testing Ideas Factually

These communal aspects of the scientific method are complemented by at least three other goals, goals that underscore the care admired by scientists: that "knowledge" be *factually testable*, that it be *logical*, and that it be *explicable through theory*. Factual testability means that scientific knowledge, like personal inquiry, must be supported by observation. Unlike positivists, post-positivists don't insist that observations themselves can't be affected by personal biases or theories. For example, people concerned about environmental warming might be most likely to notice that glaciers are melting. All observations are fallible, but most post-positivists would nonetheless accept the notion that if what is being proposed as knowledge doesn't stand up to many people's observation, it's not really knowledge at all. And rather than simply using evidence to support a particular view, as we sometimes do in personal inquiry, scientific observation also includes trying to imagine the kinds of observations that would undermine the view and then pursuing those observations. Confronted with the Census data about, say, the increase in the number of households with people living alone, we questioned whether those data were in fact the best data for answering our question and imagined other possible sources (e.g., a questionnaire survey of our own). Similarly, when confronted with the idea that people of Columbus's day held a "flat-Earth" view of the world, historians consulted the writings of scientists of that day and earlier and found evidence of a pretty widespread belief that the earth was spherical—surprisingly similar to our beliefs today (Gould, 1995: 38–50). The pursuit of counterexamples, and the parallel belief that one can never fully prove that something is true (but that one can cast serious doubt on the truth of something), is a key element of the scientific method.

The Use of Logic

Scientists are often thought of being especially logical. Television characters from *Star Trek*'s Mr. Spock to *The Big Bang Theory*'s Sheldon Cooper and Amy Farrah Fowler are meant to embody the desirability of logical reasoning in science. Spock, Cooper, and Fowler would all recognize the logical fallacy in appealing to authority we've already mentioned. (Just because someone is an authority in field X doesn't necessarily mean that that person's claim, C, in field Y is true.) Spock, Cooper, and Fowler, like most scientists, approve of logical reasoning. You used logical reasoning, too, if, when confronted with the Census data about the compositions of households in America in 2010, you wondered if those data were adequately related to the question about young people's plans.

Like many scientists, Spock, Cooper, and Fowler are quick to point out illogical reasoning. If young Roger or Emily had proposed the notion that all families consisted of two biological parents and their offspring and presented the reasoning "We belong to such families. Therefore everyone must," then Spock, Cooper, or Fowler might have responded, "That's illogical." One tenet of the scientific method is that one must adhere, as closely as possible, to the rigors of logical thinking. Few practicing scientists invoke the scientific standard of logic as frequently as Spock, or even Cooper and Fowler, did, but fewer still would wish to appear as illogical as Roger and Emily were in their reasoning.

STOP & THINK   *Sam Roberts (2007) broke the news that in 2005, for the first time, more American women were living without a husband than with one. Does it follow logically from this news that more than 50 percent of American women were unmarried?*

Theoretical Explanation

Logically, the fact that more American women are living without a spouse than with one doesn't mean that more than 50 percent of women are unmarried. In fact, in 2005, almost 54 percent of American women *were* married, but almost 5 percent of them were either legally separated or said their spouses weren't living at home for some reason (Roberts, 2007). Nonetheless, by 2005, 51 percent of American women were living without a spouse. Does this single observation, interesting as it is, constitute "knowledge"? Not according to most scientists or, in fact, most other people. In order for it to rise to the level of knowledge, most people (and scientists) would want some kind of explanation of the fact or some kind of **theory**. Roberts (2007) provides such an explanation in terms of two trends: (1) younger women marrying later and (2) older women not remarrying after being widowed or after divorce. The fact that one might logically deduce from these two trends that fewer women would be living with spouses than before really enhances your sense that you know something. Doesn't it?

theory, an explanation about how and why something is as it is.

The relative strengths of the scientific method, then, derive from several attributes of its practice. Ideally, the method involves communities of relatively equal knowledge seekers (scientists) among whom findings are communicated freely for careful scrutiny. Knowledge claims are ideally subjected to factual tests and to tests of logical reasoning. They're also supposed to be explicable in terms of theory.

STOP & THINK   *Suppose I submit a research report to a journal and the journal's editor writes back that the journal won't publish my findings because expert reviewers don't find them persuasive. Which of the strengths of the scientific method is the editor relying on to make his or her judgment?*

THE USES AND PURPOSES OF SOCIAL RESEARCH

By now, you might be saying, "OK, research methods may be useful for finding out about the world, for solving mysteries. But aren't there any more practical reasons for learning them?" We think there are.

We hope, for instance, that quite a few of you will go on and apply your knowledge of research methods as part of your professional lives. We know that many of our students have done so in a variety of ways: as graduate students and professors in the social sciences, as social workers, as police or correctional officers, as analysts in state agencies, as advocates for specific groups or policies, as community organizers, or as family counselors, to name but a few. These students, like us, have tended to direct their research toward two different audiences: toward the scientific community in general, on the one hand, and toward people interested in specific institutions or programs, on the other. When they've engaged the scientific community in general, as Sandra Enos does in the research, reported in Chapter 10, on how women in prison mother their children, they've tended to engage in what is sometimes called **basic research**. Enos reports to other scientists, for instance, that, while incarcerated, white women tend to place their children in the care of their husbands or the state (e.g., foster care), and African American women tend to place their children in the care of their own mothers. Basic research is designed to add to our knowledge and understanding of the social world for the sake of that knowledge and understanding. Much of the focal research in this book, including both Enos's work on women in prison (Chapter 10) and Hoffnung's work on college students' plans about (and achievements in) their futures (Chapter 4), is basic research.

When our former students have addressed a clientele with interests in particular institutions or programs, like the students quoted in the left-hand column of Box 1.1, they've tended to engage in what is called **applied research**, which aims to have practical results and produce work that is intended to be useful in the immediate future. Schools, legislatures, government and social service agencies, health care institutions, corporations, and the like all have specific purposes and ways of "doing business." Applied research, including evaluation research and action-oriented research, is designed to provide information that is immediately useful to those participating in institutions or programs. Such research can be done for or with organizations and communities and can include a focus on the action implications of the research. Evaluation research, for example, can be designed to assess the impact of a specific program, policy, or legal change. It often focuses on whether a program or policy has succeeded in effecting intentional or planned changes. Participatory action research, which we'll discuss in Chapter 14, is done jointly by researchers and community members. It often has an emancipatory purpose. Participatory action research focuses on the differentiated consequences of social oppression and "lifts the multiple stories and counter stories dwelling within any complex institution or social arrangement" (Fine and Torre, 2006: 255). You'll find an example of applied research in Chapter 14, where Harrison and Norton-Hawk assess the needs of incarcerated women in Ecuador.

Even if you don't enter a profession in which you'll do research of the sort we discuss in this book, we still think learning something about research methods can be one of the most useful things you do in college. Why? Oddly, perhaps, our answer implicates another apparently esoteric subject: theory.

basic research, research designed to add to our fundamental understanding and knowledge of the social world regardless of practical or immediate implications.

applied research, research intended to be useful in the immediate future and to suggest action or increase effectiveness in some area.

When we speak of theory, we're referring not only to the kinds of things you study in specialized social theory courses, although we do include those things. We view theories as explanations about how and why things are as they are. In the case of social theories, the explanations are about why people "behave, interact, and organize themselves in certain ways" (Turner, 1991: 1). Such explanations are useful, we feel, not only because they affect how we act as citizens—as when, for instance, we inform, or fail to inform, elected representatives of our feelings about matters such as welfare, joblessness, crime, and domestic violence—but also because we believe that Charles

---

BOX 1.1 Two Kinds of Research

Examples of Applied Research

Here are just a few examples of applied research some of our graduates have done or are planning to do:

"When I worked for the Department of Children, Youth, and Their Families, we conducted a survey of foster parents to see what they thought of foster care and the agency's services. The parents' responses provided us with very useful information about the needs of foster families, their intentions for the future, and the kinds of agency support that they felt would be appropriate." – *Graduate employed by a state Department of Children, Youth, and Their Families*

"As the Department of Corrections was under a court order because of crowding and prison conditions, it was important that we plan for the future. We needed to project inmate populations and did so using a variety of data sources and existing statistics. In fact, we were accurate in our projections." – *Graduate employed by a state Department of Corrections*

"I'm working at a literacy program designed to help children in poverty by providing books for the preschoolers and information and support for their parents. I've realized that while the staff all think this is a great program, we've never really determined how effective it is. It would be wonderful if we could see how well the program is working and what we could do to make it even better. I plan on talking to the director about the possibility of doing evaluation research." – *Graduate employed by a private pediatric early literacy program*

Examples of Basic Research

Here are three more examples of basic research that former students and current colleagues have done:

Paul Khalil Saucier (2008) conducted fieldwork in Greater Boston from 2007 to 2008 to explore ways in which second-generation Cape Verdean youth negotiate their identity as Cape Verdean and as black. He paid particular attention to hip-hop culture to see how it was used as a site where new identities were fashioned and reworked.

Desiree Ciambrone (2001) interviewed women with human immunodeficiency virus (HIV) infection and found that many did not consider HIV to be the most devastating event in their lives. She found, rather, that violence, mother–child separation, and drug use were seen to be more disruptive than HIV infection. You may read more about ethical considerations involved in Ciambrone's work in Chapter 3.

Mikaila Arthur (2007) examined the rise of women's studies, Asian-American studies, and queer studies programs in colleges and universities in the United States. She found, among other things, that while external market forces were weak predictors of curricular change, pressure by students and faculty for such change was the necessary condition for the rise of such programs.

Lemert (1993: 1) is right when he argues that "social theory is a basic survival skill." Individuals survive in society to the extent that they can say plausible and coherent things about that society.

Useful social theory, in our view, concerns itself with those things in our everyday lives that can and do affect us profoundly, even if we are not aware of them. We believe that once we can name and create explanations (or create theories) about these things, we have that much more control over them. At the very least, the inability to name and create such explanations leaves us powerless to do anything. These explanations can be about why some people live alone and some don't, why some are homeless and some aren't, why some commit crimes and some don't, why some do housework and some don't, and why some people live to be adults and some don't. These explanations can come from people who are paid to produce them, like social scientists, or from people who are simply trying to make sense of their lives. Lemert (1993) reminds us that the title for Alex Kotlowitz's (1991) *There Are No Children Here* was first uttered by the mother of a 10-year-old boy, Lafeyette, who lived in one of Chicago's most dangerous public housing projects. This mother observed, "But you know, there are no children here. They've seen too much to be children" (Kotlowitz, 1991: 10). Hers is eloquent social theory, with serious survival implications for those living in a social world where nighttime gunfire is commonplace.

We'll have more to say about theory and its connection to research methods in the next chapter. But, for now, we'll simply say that we believe the most significant value of knowledge of research methods is that it permits a critical evaluation of what others tell us when we and others develop social theory. This critical capacity should, among other things, enable us to interpret and use the research findings produced by others. Our simple answer, then, to the question about the value of a research methods course is not that it adds to your stock of knowledge about the world, but that it adds to your knowledge of *how* you know the things you know, *how* others know what they know, and ultimately, *how* this knowledge can be used to construct and evaluate the theories by which we live our lives.

The major purposes of scientific research, in many ways overlapping with the "uses" (of, say, supplying other scientists with basic information, supplying interested persons with information about programs, or developing theories) we've mentioned, include *exploration*, *description*, *explanation*, and *evaluation*. Although any research project can have more than one purpose, let's look at the purposes individually.

Exploratory Research

exploratory research, groundbreaking research on a relatively unstudied topic or in a new area.

In **exploratory research,** the investigator works on a relatively unstudied topic or in a new area, to become familiar with this area, to develop some general ideas about it, and perhaps even to generate some theoretical perspectives on it. Exploratory research is almost always inductive in nature, as the researcher starts with observations about the subject and tries to develop tentative generalizations about it (see Chapter 2).

An example of exploratory research is Enos's study of how women inmates manage to mother their children when they're in prison, presented in Chapter 10. Enos began by observing not only how much the U.S. prison population has grown in recent decades, but also how much the population of women prisoners had grown—between just 1985 and 2000, it had increased by more than 300 percent. Moreover, she noted that the imprisonment of women created a special problem for their children: where and with whom to live. Enos wondered how these women were dealing with an increasingly urgent issue: the "managing of motherhood from prison." Enos wondered whether she'd be able to spot patterns if, first, she observed the interactions of incarcerated women and their children in a prison's weekend "parenting" program and, then, if she interviewed 25 women inmates intensively. What she noticed was that white and African American women had quite distinctive approaches to placing their children while incarcerated (e.g., African American women were much more likely to place their children with their own mothers or other relatives than white women) and that these patterns had a lot to do with the distinctive ways they'd experienced childhood themselves (e.g., African American women were less likely than white women to blame the people in their childhood families for the behavior that led them to prison). Similarly, Silva (2012) interviewed young working-class men and women to find out how they marked the transition to adulthood, now that access to steady, well-paying jobs has become difficult and an unreliable external marker. She found that both men and women now tend to look inward for feelings that signal that they have overcome the pain of early family trauma. But she finds that they almost always need a witness to this transformation, be it a therapist, a support group, or a family member, to fully believe that it has occurred. Both Enos and Silva, then, did exploratory research—research into a relatively new subject (mothers in prison and working-class men and women in a post-industrial economy that makes well-paying jobs almost unattainable), collected data through observations or in-depth interviews of a relatively few cases, and tried to spot themes that emerged from their data. Although exploratory analyses, with their focus on relatively unexplored areas of research, do not always employ this kind of thematic analysis of data on relatively few cases, when they do they undertake what is called **qualitative data analysis**, or analysis that tends to involve the interpretation of actions or the representations of meanings in words (see Chapter 15).

> **qualitative data analysis,** analysis that results in the interpretation of action or representation of meanings in the researcher's own words.

Descriptive Research

> **descriptive research,** designed to describe groups, activities, situations, or events.

In a **descriptive research**, a researcher describes groups, activities, situations, or events, with a focus on structure, attitudes, or behavior. Researchers who do descriptive studies typically know something about the topic under study before they collect their data, so the intended outcome is a relatively accurate and precise picture. Examples of descriptive studies include the kinds of polls done during political election campaigns, which are intended to describe how voters intend to vote, and the U.S. Census, which is designed to describe the U.S. population on a variety of characteristics. The Census Bureau description of the makeup of households in the United States (2010) is just that a description, in this case of the whole U.S. population. As the pie chart in Figure 1.1 indicates, about 26.7 percent of American households in 2010 were made up of people living alone. Unlike exploratory studies, which might help readers become familiar with a new topic, descriptive studies can provide a very detailed and precise idea of the way things are. They can also provide a sense of how things have changed over time. Thus, a study by Clark and Nunes (2008) of the pictures in introductory sociology textbooks found that, while in textbooks of the 1980s only 34 percent of identifiable individuals shown in pictures were women, in textbooks of the 2000s almost 50 percent of those images were of women. And Armstrong, England and Fogarty's (2012) investigation of a large sample of heterosexual, undergraduate women showed that sex was better in the context of a relationship than in hookups. Descriptive research, in its search for a picture of how the land lies, can be based upon data from surveys (as the Census and the Armstrong, England, and Fogarty reports are) or on content analyses (as the Clark and Nunes study is). Often, as in the case of the Census Bureau's study of American households, descriptive research generates data about a large number of cases: there were actually 117,538,000 households enumerated by the census of 2010. To analyze these data meaningfully, descriptive researchers frequently use **quantitative data analysis**, or analysis that is based on the statistical summary of data (see Chapter 15).

quantitative data analysis, analysis based on the statistical summary of data.

Explanatory Research

Unlike descriptive research, which tends to focus on *how* things are, the goal of **explanatory research** is to explain *why* things are the way they are. Explanatory research looks for causes and reasons. Unlike exploratory research, which tends to be inductive, building theoretical perspectives from data, explanatory research tends to be deductive, moving from more general to less general statements. Thus, for instance, explanatory research often uses preexisting theories to decide what kinds of data should be collected (see Chapter 2). In a study present in Chapter 7, we, Emily and Roger, asked ourselves the explanatory question, "Why are some people more likely than others to do work for pay after they retire from their regular lifetime work?" We knew that some people were likely to do it because they genuinely needed to augment their post-retirement incomes. But we believed that financial hardship couldn't possibly explain all the variation in retired people's actual participation in post-retirement work. Even some professional workers, people with decent savings and good pensions, returned to work after their official retirements. And so we did a study, primarily of professional people, to see if we could figure out why some of them were more likely than others to do post-retirement work.

explanatory research, research designed to explain why subjects vary in one way or another.

We did what many scientists do: We consulted theory. In fact, we consulted a couple of theories about retirement, one of which is "continuity theory," a theory that suggests that some people are more invested in roles, like work roles, than others.

In Chapter 2, you'll read more about continuity theory, but for now we'll just mention a key element of the theory. It expects retired people, when deciding whether to work for pay or not, to make decisions based upon how much of their self-worth was dependent on work during their working lives. From continuity theory, we derived an expectation about the length of a person's pre-retirement career and his or her likelihood of continuing to work after retirement. After surveying people twice, once before they retired and once afterward, we found, in fact, that people who had had longer careers were considerably more likely to work after retirement than people who had had shorter careers. In our work, we describe something (the length of a professional's career) that "goes with" something else (the likelihood of working after formal retirement). In doing so, we explain this connection with a theory (continuity theory), one of the goals of the scientific method.

Explanatory analyses, with their focus on areas upon which a researcher might be able to shed theoretical light in advance of collecting data, may, like descriptive analyses, generate data about relatively large numbers of cases and employ statistical analyses to make sense of these cases. When they do, they, like many descriptive analyses, involve quantitative data analysis (see Chapter 15).

Evaluation Research

evaluation research, research designed to assess the impacts of programs, policies, or legal changes.

Although **evaluation research** can be seen as a special kind of explanatory research, it is distinctive enough that we feel it is worth its own place. Evaluation research is research designed to assess the impacts of programs, policies, or legal changes. It often focuses on whether a program or policy has succeeded in effecting intended or planned change, and when such successes are found, the program or policy explains the change. Thus, when Schenk et al. (2010) found that working directly with women and actively addressing stigma were characteristic of the most successful interventions supporting HIV-infected children in sub-Saharan Africa, they were providing an explanation of those successes. But this is a different kind of explanation than the theoretical explanation that Roger and Emily used in explaining why some people work after retirement and others do not. We will be discussing evaluation and other applied research at much greater length in Chapter 14.

⬡ **STOP & THINK**   *Suppose you've been asked to learn something about the new kinds of communities that have arisen out of people's use of tweets on Twitter. Of the four kinds of research outlined earlier (exploratory, descriptive, explanatory, and evaluation), what kind of study have you been asked to do?*

SUMMARY

We've argued here that, at its best, research is like trying to solve a mystery, using the scientific method as your guide. We've distinguished the scientific approach of social research methods from two other approaches to knowledge about the social world: a reliance on the word of "authorities" and an exclusive dependence on "personal inquiry." We've suggested that the scientific method compensates for the shortcomings of these two other approaches in several ways. First, science emphasizes the value of communities of relatively equal knowledge seekers who are expected to be critical of one another's work. Next, science stresses the simultaneous importance of empirical testing, logical reasoning, and the development or testing of theories that make sense of the social world. Two communities that researchers report to are the community of other scientists (basic research) and communities of people interested in particular institutions or programs (applied research). We've argued that knowing research methods may have both professional and other practical benefits, not the least of which is the creation of usable theories about our social world. We suggest that social research methods can help us explore, describe, and explain aspects of the social world, as well as evaluate whether particular programs or policies actually work.

EXERCISE 1.1

Ways of Knowing about Social Behavior

This exercise compares our recollections of our everyday world with what we find out when we start with a research question and collect data.

Part 1: Our Everyday Ways of Knowing

Pick any *two* of the following questions and answer them based on your recollections of past behavior.

1. What does the "typical" student wear as footwear to class? (Will the majority wear shoes, boots, athletic shoes, sandals, and so on?)
2. While eating in a school cafeteria, do most people sit alone or in groups?
3. Of those sitting in groups in a cafeteria, are most of the groups composed of people of the same gender or are most mixed-gender groups?
4. Of your professors this semester who have regularly scheduled office hours, how many of them will be in their offices and available to meet with you during their next scheduled office hour?

Based on your recollection of prior personal inquiry, describe your expectations of social behavior.

Part 2: Collecting Data Based on a Research Question

Use the two questions you picked for Part 1 as "research questions." With these questions in mind, collect data by carefully making observations that you think are appropriate. Then answer the same two questions, but this time base your answers on the observations you made.

Part 3: Comparing the Ways of Knowing

Write a paragraph comparing the two ways of knowing you used. (For example, was there any difference in accuracy? Was there any difference in ease of collecting data? Which method do you have more confidence in?)

EXERCISE 1.2

Social Science as a Community Endeavor

This exercise is meant to reinforce your appreciation of how important the notion of a community of relatively equal knowledge seekers is to social research. We'd like you to read any one of the "focal research" articles in subsequent chapters and simply analyze that article as a "conversation" between the author and one or two authors who have gone before. In particular, summarize the major point of the article, as you see it, and see how the author uses that point to criticize, support, or amend a point made by some other author in the past.

1. What is the name of the focal research article you chose to read?
2. Who is (are) the author(s) of this article?

3. What is the main point of this article?
4. Does the author seem to be criticizing, supporting, or amending the ideas of any previous authors to whom she or he refers?
5. If so, which author or authors is/are being criticized, supported, or amended in the focal research?
6. Write the title of one of the articles or books with which the author of the focal research seems to be engaged.

7. Describe how you figured out your answer to the previous question.
8. What, according to the author of the focal research, is the central idea of the book or article being discussed?
9. Does the author of the focal research finally take a critical or supportive position in relation to this idea?
10. Explain your previous answer.

EXERCISE 1.3

Ways of Knowing about the Weather

This exercise is designed to compare three ways of knowing about the weather.

Part 1:

Knowledge from Authorities.   The night before, see what the experts have to say about the weather for tomorrow by watching a television report, listening to a radio newscast, or checking online. Write what the experts said about tomorrow's weather (including the temperature, the chances of precipitation, and the amount of wind).

Knowledge from Casual Personal Inquiry.   That day, before you go outside, look through only one window for a few seconds but don't look at a thermometer. After taking a quick glance, turn away, and then write down your perceptions of the weather outside (including the temperature, the amount and kind of precipitation, and the amount of wind).

Knowledge from Research.   Although we're not asking you to approximate the entire scientific method (such as reviewing the literature and sharing your findings with others in a research community), you can use some aspects of the method: specifying the goals of your inquiry and making and recording careful observations.

Your research question is, "What is the weather like outside?" To answer the question, use a method of collecting data (detailed observation of the outside environment) and any tools at your disposal (thermometer, barometer, and so on). Go outside for at least five minutes and make observations. Then come inside and write down your observations of the weather outside (including the temperature, the amount and kind of precipitation, and the amount of wind).

Part 2: Comparing the Methods

Write a paragraph comparing the information you obtained using each of the ways of knowing. (For example, was there any difference in accuracy? Was there any difference in ease of collecting data? Which method do you have the most confidence in?)

CHAPTER 2

Theory and Research

CHAPTER OBJECTIVES

This chapter is designed to help students

- Know what concepts, variables and hypotheses are
- Distinguish between independent and dependent variables
- Learn the criteria for demonstrating a causal relationship
- Identify antecedent, intervening and extraneous variables
- Understand the ways in which research informs theory, particularly:
  1. To test theory, through a process of deduction and observation
  2. To build theory, through a process of induction from observation

CHAPTER OUTLINE

CONCEPTS, VARIABLES, AND HYPOTHESES

We imagine that you've given some thought as to whether abortion is moral or not and therefore whether it should be legal or not. But have you ever considered the ways in which legalized abortion might be associated with legal (or illegal) behaviors of other kinds? In this chapter, on the relationship between theory and research, we will begin with an example that forces us to think about such an association. Let's examine Box 2.1, a summary of a theory by John Donohue and Steven Levitt that connects the legalization of abortion in the early 1970s and a decrease in crime rates in the United States in the early 1990s. You'll recall from Chapter 1 that a theory is an explanation of the hows and whys of something.

Donohue and Levitt's theory about the connection between abortion legalization and crime rates, like all theories, is formulated in terms of **concepts**, which are words or signs that refer to phenomena that share common characteristics. Some, but not all, of these concepts refer to phenomena that are, at least for the purposes of the theory, relatively fixed or invariable in nature. Donohue and Levitt, for example, refer to the "high-crime late adolescent years" and "birth cohort." These concepts require clarification for the purposes of research, or what researchers call **conceptualization**, but once defined, they do not vary. For Donohue and Levitt, the high-crime late adolescent years begin at about 17 years of age, and a birth cohort is those born in a calendar year. The first birth cohort in the United States that could have been affected by the 1973 *Roe v. Wade* decision would have been born in 1974; therefore Donohue and Levitt expect that there should have been a diminution in this cohort's crime rate in comparison with previous cohorts 17 years later, or about 1991.

concepts, words or signs that refer to phenomena that share common characteristics.

conceptualization, the process of clarifying what we mean by a concept.

STOP & THINK *Do you see how the concept of a "high-crime late adolescent years" is relatively fixed? How its definition does not admit much variation? (It begins at about 17 and ends, one guesses, at about 19 years of age.) Can you identify concepts used by Donohue and Levitt that could vary from one person or birth cohort to another?*

BOX 2.1 — A Theoretical Statement: Why the Legalization of Abortion Contributed to the Decline of Crime Rates

In 2001, John Donohue and Steven Levitt published a highly controversial research report, "The Impact of Legalized Abortion on Crime," in which they claimed that the introduction of legalized abortion in the United States in the early 1970s led to a drop in crime in the early 1990s. The key theoretical argument in Donohue and Levitt's paper was that abortion "is more frequent among parents who are least willing or able to provide a nurturing home environment" (2001: 386). Moreover, children from less nurturing home environments are more likely than others to engage in criminal activity, once they reach the "high-crime late adolescent years" (386). Therefore, to the extent that the legalization of abortion in the early 1970s enabled parents who would provide the least nurturing home environments to avoid having children, it also led to a lower crime rate when the affected birth cohort reached its late adolescent years.

In addition to using concepts that identify phenomena that are relatively fixed in nature, which don't vary for the purposes of the research, like "high-crime late adolescent years," Donohue and Levitt use concepts such as "nurturing home environment" and "criminal activity" that might vary from one person to another or from one birth cohort to another. A person might experience a nurturing home or not and might engage in criminal activity later in life or not. A higher percentage of the children in one birth cohort, for instance, might experience a nurturing home environment than those in another birth cohort. A higher percentage of the people in one birth cohort might engage in criminal activity when they are 17 years of age than those in another birth cohort. Whether one focuses on individuals or birth cohorts would, of course, make a big difference for a particular research project. Individuals and birth cohorts are different units of analysis, which are, as you may recall, units about which one collects information. But what do we call the information that is gathered about these units? In the sciences, this information is in the form of categories or values of variables—or categories of characteristics that can vary from one unit of analysis to another or for one unit of analysis over time. One can also think of a **variable** as a concept that varies.

variable, a characteristic that can vary from one unit of analysis to another or for one unit of analysis over time.

 STOP & THINK

The lowest possible value for the percentage of a birth cohort that has experienced a nurturing home environment is zero percent. What's the highest possible value? What are the lowest and highest possible values for the percentage of a birth cohort that engages in criminal activity when it is 17 years old?

 STOP & THINK AGAIN

Suppose, in testing Donohue and Levitt's theory, you wanted to make the individual, not the birth cohort, your unit of analysis. What categories could you use to determine whether someone engaged in criminal activity after he or she reached his or her seventeenth birthday?

To be a variable, a concept must have at least two categories. For the variable "gender," for instance, we can use the categories "female" and "male." Gender is most frequently used to describe individual people. As we've indicated, the variable "criminal activity" used by Donohue and Levitt, however, can be meant, in the theoretical formulation explained earlier, to describe a birth cohort. However, getting information about a birth cohort is very difficult, so Donohue and Levitt use substitutes for birth cohort, focusing instead on the crime rates for the whole U.S. population in different years. They end up measuring the concept of "criminal activity" in a variety of ways. One of these ways is as a characteristic of the whole nation and in terms of the number of crimes committed in a year per 100,000 people (see Table 2.1). This variable could have many more than two categories, and perhaps even an infinite number.

STOP & THINK   *Notice that the crime rate is a variable characteristic of the United States because it can vary over time. Notice also that this variable could take on many, rather than just two, values. What pattern in Table 2.1 do you spot in the U.S. crime rate between 1985 and 2009? In what year did a reversal of the previous pattern of increases begin? When was the first cohort affected by the 1973 Roe v. Wade decision born? How many years was this before the downturn in the national crime rate? How might this be significant for the thesis advanced by Donohue and Levitt?*

TABLE **2.1**
What Happened to the Crime Rate in the United States between 1985 and 2006?

Year	Total Crime Rate (number of crimes per 100,000 people)
1985	5,207.1
1986	5,480.4
1987	5,550.0
1988	5,664.2
1989	5,741.0
1990	5,820.3
1991	5,897.8
1992	5,660.2
1993	5,484.4
1994	5,373.9
1995	5,275.9
1996	5,086.6
1997	4,930.0
1998	4,615.5
1999	4,266.5
2000	4,124.8
2001	4,162.6
2002	4,118.8
2003	4067.0
2004	3977.3
2005	3900.5
2006	3808.1
2007	3739.0
2008	3669.0
2009	3466.0

Source: The Disaster Center (2008) and US Census Bureau, 2012a.

When we think of "crime activity" as the crime rate of the country, then, we see that it does decline at about the time that Donohue and Levitt's theory predicts: about 17 years (in 1991) after *Roe v. Wade* would have had its initial effects (in 1974). This illustrates a frequent use of theory for research: to point to variable characteristics of subjects that can be expected to be associated with each other. Thus, Donohue and Levitt are claiming that, among other things, they expect that a downturn in the crime rate should occur in jurisdictions where abortion became legal sometime earlier. Similarly, a research project that we (Emily and Roger) completed—mentioned in Chapter 1—found that people who do volunteer work before they retire are much more likely to do volunteer work after they retire than those who did not do volunteer work before retirement. Using the categories of two variables, "whether one did volunteer work before retirement" and "whether one does volunteer work after retirement," we found that "those who did volunteer work before retirement" were more likely "to do volunteer work after retirement" and "those who did not do volunteer work before retirement" were more likely "to not do volunteer work after retirement."

hypothesis, a testable statement about how two or more variables are expected to relate to one another.

The finding supported an expectation that we had derived from role theory about the way the two variables would be associated. We stated this expectation in the form of a **hypothesis,** which is a statement about how two or more variables are expected to relate to one another. We developed this hypothesis:

> People who have performed volunteer roles before retirement will be much more likely to perform such roles after retirement than people who have not performed such roles. (Chapter 7)

Note that this hypothesis links two variable characteristics of individuals: their participation in volunteer work before retirement and their participation in such work after retirement. You might also note that, like many other social science hypotheses, this one doesn't refer to absolute certainties (such as, "If one does volunteer work before retirement, one will always engage in volunteer work after retirement, and if one does not do volunteer work before retirement, one will never engage in volunteer work after retirement"). Rather, this hypothesis speaks of tendencies.

The business of research, you've probably surmised, can be complex and can involve some fairly sophisticated ideas (theory, hypotheses, concepts, and variables, to name but a few), all of which we'll discuss again later. Meanwhile, we'd like to introduce you to one more fairly sophisticated idea before moving on to a more general consideration of the relationship between research and theory. This is the idea that explanations involve, at minimum, the notion that change in one variable can affect or influence change in another. Not surprisingly, scientists have special names for the "affecting" and "affected" variables. Variables that are affected by change in other variables are called **dependent variables**—dependent because they depend on change in the first variable. **Independent variables,** on the other hand, are variables that affect change in dependent variables.

dependent variable, a variable that a researcher sees as being affected or influenced by another variable (contrast with independent variable).

independent variable, a variable that a researcher sees as affecting or influencing another variable (contrast with dependent variable).

Thus, when Donohue and Levitt hypothesize that making abortion legal leads to declines in the crime rate, they are implying that a change in the

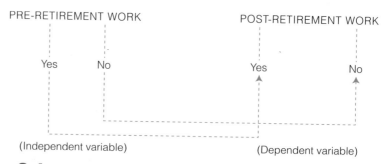

FIGURE **2.1** A Diagram of the Hypothesized Relationship Between Pre-Retirement Volunteer Work and Post-Retirement Volunteer Work

independent variable (whether abortions are legal) will cause change in the dependent variable (the crime rate). And, when we hypothesize that people who do volunteer work before retirement are more likely to do volunteer work after retirement than those who do not do volunteer work before retirement, we are suggesting that a change in the independent variable (doing or not doing volunteer work before retirement) will cause change in the dependent variable (doing or not doing volunteer work after retirement). Having an arrow stand for something like "is associated with," hypotheses can be depicted using diagrams like the one in Figure 2.1. The arrows in this diagram suggest that having done volunteer work before retirement "is associated with" doing volunteer work after retirement, while not having done volunteer work before retirement "is associated with" not doing volunteer work after retirement. And, in general, when one or more categories of one variable are expected to be associated with one or more categories of another variable, the two variables are expected to be associated with each other.

STOP & THINK   *Can you draw a diagram of Donohue and Levitt's hypothesis linking the legalization of abortion to declines in criminal activity? (Hint: It might be easiest to think in terms of individuals, not cohorts or countries. What kind of person, according to Donohue and Levitt's thesis, is more likely to engage in criminal activity: someone born when abortion is legal or someone born when abortion is not legal?)*

STOP & THINK AGAIN   *See if you can identify the independent and dependent variable of hypotheses that guided Burgard and Ailshire's (2013) reexamination of the hypothesis that women sleep more than men.*

We'd like to give you one more example of how to take apart a hypothesis. Let's look at the hypothesis mentioned in the previous "Stop & Think," that women sleep longer than men. Here the units of analysis are people, the independent variable is gender (which can vary from men to women), and the dependent variable is time of sleep (which can also vary from long to short). We would draw a diagram of this hypothesis as shown in Figure 2.2.

FIGURE **2.2** A Diagram of the Hypothesized Relationship Between Gender and Time of Sleep

We want to assure you that at this point we don't necessarily expect you to feel altogether comfortable with all the ideas we've introduced so far. We'll discuss them all again, so you'll gain more familiarity with them. For now, we want to emphasize a major function of theory for research: to help lead us to hypotheses about the relationship between or among variables. We'd like to turn to a more general consideration of the relationship between research and theory. But first let us say a little more, by way of a word of caution, about the difficulty of establishing causality in social science.

SOCIAL SCIENCE AND CAUSALITY: A WORD OF CAUTION

The fact that two variables are associated with each other doesn't necessarily mean that change in one variable causes change in another variable. Table 2.1 shows, generally, what Donohue and Levitt (2001), in their much more sophisticated analysis, contend: When abortion rates (the independent variable) go up, societies are likely to experience a decline in crime rates (the dependent variable) later, when the more "wanted" generation grows up. But does this mean that making abortion legal causes lower crime rates? Donohue and Levitt suggest that it does. Others argue that it doesn't.

In the social sciences, we have difficulty establishing causality for several reasons. One reason is that to show that change in one variable causes change in another, we want to be sure that the "cause" comes before, or at least not after, the "effect." But many of our most cherished methods of collecting information in the social sciences just don't permit us to be sure which variable comes first. Kornrich, Brines, and Leupp (2013), for instance, found that married couples who reported more traditional housework patterns (i.e., women doing more housework than men) also said, in a questionnaire survey, that they had higher sexual frequency than couples who shared housework more. The association of these two variables, however, doesn't prove that the division of household labor causes the higher frequency of sex because those answering the survey reported their behavior about both variables at the same time. The researchers can't tell whether sexual frequency came before the division of household labor or the division of household labor came before sexual frequency. It would seem that Donohue and Levitt's research at least passes the time-order test: that the "cause" (abortions, subsequent to *Roe v. Wade*) comes before the "effect" (a decline in crime rates in the 1990s). In this case, however, there can remain doubt about whether the "right" time "before" has been specified. Donohue and Levitt, in their sophisticated statistical analysis, make very plausible guesses about appropriate time "lags," or periods over which abortion will have its greatest effects on crime, but even they can't be sure that their "lags" are exactly right.

STOP & THINK

Donohue and Levitt (2001: 382) think there was something pretty significant about the 17 years between when Roe v. Wade *would have affected its first birth cohort (1974) and the downturn, beginning in 1991, in the overall crime rate in the United States. Can you see any reasons why 17 years might be a significant length of time? Can you imagine, assuming that abortion has any effect on crime, why one might expect a shorter or longer period of time between the legalization of abortion and the onset of significant declines?*

Another technical problem with establishing causality is that even if we can hypothesize the correct causes of something, we usually can't demonstrate that another factor isn't the reason why the "cause" and "effect" are associated. We might be able to show, for instance, that members of street gangs are more likely to come from single-parent households than nonmembers. However, because we're not likely to want or be able to create street gangs in a laboratory, we can't test whether other things (like family poverty) might create the association between "family structure" (an independent variable) and "street gang membership" (a dependent variable). Similarly, it's possible that some "other thing" (like swings in the economic cycle) might create conditions that make "abortion rates" (people may be more likely to have more abortions during economic downturns) and "subsequent crime rates" (people may be less likely to commit crimes during economic upturns) go together. Doing experiments (discussed in Chapter 8) is most useful in demonstrating causality, but this strategy frequently doesn't lend itself well to ethical and practical reasons to social science investigations.

antecedent variable, a variable that comes before both an independent variable and a dependent variable.

The idea of a third variable, sometimes called an **antecedent variable**, which comes before, and is actually responsible for, the association between an independent variable and a dependent variable, is so important that we'd like to give you another example to help you remember it. Firefighters will tell you of the association between the number of firefighters at a fire (an independent variable) and the damage done at the fire (a dependent variable): that the more firefighters at a fire, the more damage occurs.

STOP & THINK

Can you think of an antecedent variable that explains why fires that draw more firefighters are more likely to do more damage than fires that draw fewer firefighters?

Perhaps it occurred to you that a characteristic of fires that would account for both the number of firefighters drawn to them and the amount of damage done is the size of the fire. Smaller fires tend to draw fewer firefighters and do less damage than larger fires. In this case, the antecedent variable, the size of the fire, explains why the independent variable and the dependent variable are associated. When this happens, when an antecedent variable provides such an explanation, the original association between the independent variable and the dependent variable is said to be **spurious**, or non-causal. Neither one causes the other; their association is due to the presence of an antecedent variable that creates the association. It is only when the association between an independent variable and a dependent variable is non-spurious (not generated by some third variable acting on the two) that we can conclude that it is causal. But this is hard to demonstrate, because it means taking into account *all possible* antecedent variables, of which there are an infinite number.

spurious, non-causal.

The three conditions that must exist before we can say an independent variable "causes" a dependent variable are, then:

1. the two variables are associated in fact. In the case of the fires, this would mean showing that fires that drew many firefighters actually had more damage than other fires. This is the condition of *empirical association*.
2. the independent variable, in fact, precedes, or at least doesn't come after, the dependent variable. In the case of the fires, this would mean showing that fire damage never came before the arrival of the firefighters. This is the condition of *temporal precedence* or *time order*.
3. there is no third variable, antecedent to the independent variable and the dependent variable, that is responsible for their association. In the case of the fires, this would mean showing that there is no variable, like the size of the fire, that led to the association between the number of firefighters at the fire and the damage done at the fire. This is the condition of *elimination of alternative explanations* (or *demonstrating non-spuriousness*).

Most social science research designs make it difficult to establish the condition of temporal precedence and make it impossible to establish the condition of non-spuriousness, so we can rarely say we're sure that one thing causes another. In fact, only the experimental design is meant to establish causality. We often settle for establishing the first condition, the condition of empirical association, knowing that it is a necessary condition for, but not a completely satisfactory demonstration of, causality.

STOP & THINK   *Philip Maymin (reported by Pothier [2009]) reports, after studying the last fifty years of popular music and over 5,000 hit songs, that when the stock market is jumpy, we prefer music with a steady beat (like Kanye West's "Heartless"), but when the stock market is calm, we prefer music with more unpredictable beats (like Sean Paul's "Like Glue"). Assuming Maymin is right, which of the conditions that must exist to show that market behavior affects our musical tastes seems to hold? Which one or ones has not yet been demonstrated?*

Proving causality is, as we've suggested, pretty tough to do, partly because one has to show that no third variable, antecedent to the independent variable and the dependent variable, is responsible for their association. Because there are, in principle, an infinite number of antecedent variables for all combinations of independent and dependent variables, eliminating all of them as possible instigators of the relationship is impossible. While we're thinking about causation, though, we'd like to mention two other kinds of variables—intervening and extraneous variables—whose presence would not challenge the possibility that an independent variable causes variation in a dependent variable. An **intervening variable** is a variable that comes between an independent and a dependent variable. Here, the researcher conceives of the independent variable affecting the intervening variable, which in turn is conceived to affect the dependent variable. Donohue and Levitt, for instance, believe legalizing abortion means that more children experience nurturing homes and that this experience reduces their likelihood of crime. Here, as in the case of all intervening variables, the experience of a nurturing home is posited to be the reason why legalizing abortion reduces the likelihood of crime. An **extraneous variable** is a variable that the researcher sees as having an effect on the dependent variable in addition to the effect of the independent variable. Unlike antecedent variables, whose presence may demonstrate that the relationship between the independent and the dependent variable is non-causal, and intervening variables, whose presence may account for how an independent variable affects a dependent variable, an extraneous variable simply provides a complementary reason for variation in the dependent variable. Social scientists rarely seek a single cause for given effects. They tend, instead, to look for multiple reasons for variation in a dependent variable. Thus, for instance, while Donohue and Levitt were initially interested in the effects of legalizing abortion on rates of crime, Levitt (2004) eventually explained the 1990s downturn in the crime rate in terms of four factors, each extraneous to the others: the legalization of abortion, an increase in the number of police, a rising prison population, and a waning crack epidemic.

intervening variable, a variable that comes between an independent and a dependent variable.

extraneous variable, a variable that has an effect on the dependent variable in addition to the effect of the independent variable.

THE RELATIONSHIP BETWEEN THEORY AND RESEARCH

If theories are explanations of the hows and whys of things, then good *social* theories might be considered to be theories that enable people to articulate and understand something about everyday features of social life that had previously been hidden from their notice. We begin this section with two examples of social theories used by the social-scientist authors of focal research pieces that we use later in this text. Although the complete articles are included in later chapters, here we will excerpt the authors' descriptions of a theory they've employed or developed through their research. The first excerpt is from our research on retirement that we include in Chapter 7.

🛑 STOP & THINK   *As you read each excerpt, see if you can guess whether it appears before or after the authors' presentation of their own research.*

FOCAL RESEARCH

Excerpt from "Moving On? Continuity and Change after Retirement": Role Theory

By Emily Stier Adler and Roger Clark

Role theory (e.g., Ashforth, 2001; Wang, 2007) suggests that to the extent people are highly invested in a particular role, their feelings of self-worth are related to the ability to carry out that role. Role theorists also argue that the loss of roles can lead to depression or anxiety. We posit, then, that after retirement, to the extent possible, people will retain previous roles because doing so enables them to maintain feelings of self-worth and avoid feelings of anxiety and depression. Moreover, we posit that pre-retirement expressed desires to perform new roles (like new volunteer roles) after retirement are less likely to be based upon the same emotional calculus of self-worth, depression, and anxiety that informs the desire to continue performing a pre-retirement role. In fact, we expect that all workers have at least some cognitive access to the emotional and psychological needs that work roles have satisfied for them and that their plans to do some kind of work after retirement will therefore be related to whether they actually work for pay after retirement.

References

Ashforth, B. 2001. *Role transitions in organizational life.* Mahwah, NJ: Erlbaum.

Wang, M. 2007. Profiling retirees in the retirement transition and adjustment process: Examining the longitudinal change patterns of retirees' psychological well-being. *Journal of Applied Psychology* 92(2): 455–474.

Our second excerpt is from Marci Cottingham's research on football fans that is included in Chapter 11.

FOCAL RESEARCH

Excerpt from "The Terrible Towel and Fair-Weather Fans: Steelers Nation as a Quasi-Religion"

By Marci D. Cottingham

Furthermore the study reveals an important feature of this quasi-religion [that of Pittsburgh Steelers football fans]: its emphasis on experience and emotion over cognition and formalized beliefs. Fans do not adhere to an official set of beliefs about the team or the game, but rather hold an informal set of tacit beliefs about appropriate behavior and the knowledge base that signal a truly devoted fan. Their emotional response to profane treatment of the Terrible Towel and Ben Roethlisberger suggests a moral order among fans that is neither codified nor explicitly stated. Lacking a set of formal beliefs, however, should not disqualify the group from consideration as a quasi-religion. To do so would be to ignore the profound meaning that fandom affords members and the important role of sport in contemporary American culture.

Deductive Reasoning

Adler and Clark's description of "role theory" comes at the beginning of their article, before they present their own research; Cottingham's theory that football's quasi-religion privileges experience and emotion over formalized beliefs at the end of her article, after the presentation of her findings. Note, however, that both theories attempt to shed light on what might have been a previously hidden aspect of social life. They do so with concepts (such as "self-worth" and "anxiety" or "tacit beliefs" and "moral order") that are linked, explicitly or implicitly, with one another. These concepts are neither so personal as the one espoused by Lafeyette's mother at the beginning of *There Are No Children Here* (remember the boy mentioned in Chapter 1?) nor as grandiose as some theories you might have come across in other social science courses. But they are theories, nonetheless.

What distinguishes these theories, and many others presented by scientists, from those presented by pseudoscientists (astrologers and the like) is that they are offered in a way that invites skepticism. Neither we nor Cottingham are 100 percent sure about the correctness of the explanation for the social world. For both sets of researchers, a major value of research methods is that they offer a way to test or build social theories that fit the observable social world. We'll spend the rest of this chapter discussing how research can act as a kind of "go-between" for theory and the observable world.

Many social scientists engage in research to see how well existing theory stands up against the real world. That is, they want to *test* at least parts of the theory. We did our research because of an interest in explaining why some people work for pay and do volunteer work after retirement and others don't. We started with a general theory of people's behavior in later life: role theory. This theory, like many others, is made up of very general, though not necessarily testable, propositions. One of these propositions is that people's feeling of self-worth in later life is dependent on their ability to continue to play roles they played earlier in life. From this theory, we deduced, among others, the testable hypothesis: people who did volunteer work before retirement were more likely to do volunteer work after retirement than those who hadn't.

Having made this deduction, we needed to decide what kind of observations would count as support for the hypothesis and what would count as nonsupport. In doing so, we devised a way of measuring the variables for the purpose of the study. **Measurement**, as we will suggest in Chapter 6, involves devising strategies for classifying units of analysis (in this case, people around retirement age) by categories (say, those who did volunteer work and those who didn't) to represent variable concepts (say, volunteering). To measure whether respondents did volunteer work before retirement, Emily, who actually did the interviewing, asked questions, during a pre-retirement interview, about work history, followed by a question like, "So thinking about life outside of work, how would you say you spend your free time?" If respondents clearly indicated that part of their free time was spent as volunteers, they were counted as doing volunteer work. If respondents didn't mention volunteer work during the response, Emily asked, "Now, you've mentioned doing several

measurement, the process of devising strategies for classifying subjects by categories to represent variable concepts.

things with your free time. Do you spend some time doing volunteer work?" If respondents now indicated that they did, they were also counted doing volunteer work. If they indicated that they did not, they were counted as not volunteering. A similar set of questions was used in post-retirement interviews to discover whether respondents were engaged in volunteer work then. Responses to questions, then, were the ways in which Emily and Roger observed the social world.

deductive reasoning, reasoning that moves from more general to less general statements.

Here we'd like to emphasize that in the process of testing role theory, Emily and Roger did what is typical of most such exercises: They engaged in a largely **deductive reasoning** in which more general statements were used as a basis for deriving less general, more specific statements. Figure 2.3 summarizes this process.

empirical generalizations, statements that summarize a set of individual observations.

Theory testing rarely ends with the kinds of observational statements suggested in Figure 2.3. These statements are usually summarized with statistics, some of which we'll introduce in Chapter 15. The resulting summaries are often **empirical generalizations,** or statements that summarize a set of individual observations. We report, for example, the empirical generalization that people who did volunteer work before retirement were indeed more likely to do it after retirement as well.

As with our study, research is frequently used as a means of testing theory: to hold it up against the observable world and see if the theory remains plausible or becomes largely unbelievable. When researchers use their work to test theory in this way, they typically start by deducing hypothetical statements from more theoretical ones and then make observational statements that, when summarized, support or fail to support their hypotheses. In general, they engage in deductive processes, moving from theories to observations.

Type of Statement	Illustrative Statement
Theoretical ↓	After retirement, to the extent possible, people will retain previous roles, including non-work roles, because doing so enables them to maintain feelings of self-worth and avoid feelings of anxiety and depression.
Hypothetical ↓	People who do volunteer work before retirement are more likely to do it after retirement than others.
Observational	Respondent A was a lay leader in his church at the time of his pre-retirement interview. After retirement, he'd continued as a volunteer in his church and found other volunteer opportunities as well, including work at an HIV/AIDS drop-in center at a nearby hospital.

FIGURE **2.3** An Illustration of the Logic of Theory Verification from Adler and Clark

Inductive Reasoning

In contrast with the more or less deductive variety of research we've been discussing is a kind that emphasizes moving from more specific kinds of statements (usually about observations) to more general ones and is, therefore, a process called **inductive reasoning**. Many social scientists engage in research to develop or build theories about some aspect of social life that has previously been inadequately understood. Theory that is derived from data in this fashion is sometimes called **grounded theory** (Glaser, 1993, 2005; Glaser and Strauss, 1967).

inductive reasoning, reasoning that moves from less general to more general statements.

grounded theory, theory derived from data in the course of a study.

The process of building grounded theory from data is illustrated by Cottingham's research reported in Chapter 11. Cottingham observed Pittsburgh Steelers fans at games, tailgate parties, and sports bars. She did not work deductively but rather focused initially on what fans did and said. She observed, for instance, fans at a game played in 15-degree weather excitedly reacting to good Steelers plays by taking off all the outerwear above their waists.

Having immersed herself in many such observations, Cottingham then developed statements at a higher level of generality, statements that summarize a much larger set of observations and therefore can be seen as empirical generalizations. One generalization suggests that fans at Steelers games were even more likely to engage in orchestrated rituals than fans at sports bars or even tailgate parties.

Cottingham might easily have left her study here, providing us with a kind of descriptive knowledge about Pittsburgh Steelers fans. Instead, she offers an elaboration of previous theory about quasi-religions. She argues, in fact, that fans of football teams behave as if in quasi-religions that emphasize rituals and symbols that intensify emotional commitment, but that tend to have few formalized beliefs. Figure 2.4 summarizes this approach.

The Cyclical Model of Science

A division exists among social scientists who prefer to describe the appropriate relationship between theory and research in terms of theory testing and those who prefer theory building. Nonetheless, as Walter Wallace (1971) has suggested, it is probably most useful to think of the real interaction between theory and research as involving a perpetual flow of theory building into theory testing and back again. Wallace captured this circular or cyclical model of how science works in his now famous representation, shown in Figure 2.5.

Type of Statement	Illustrative Statement
Theoretical ↑	Fan quasi-religions put greater emphasis on ritual that heighten emotion than on formalized beliefs.
Empirical Generalization ↑	Ritualized forms of behaviour occur at games.
Observational	Two males fans bare their upper-bodies in response to big plays.

FIGURE **2.4** An Illustration of the Process of Discovery Based on Cottingham

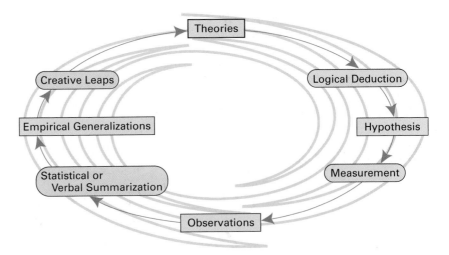

FIGURE **2.5** Wallace's Cyclical Model of Science Theories

Source: Adapted from Walter Wallace (1971). Boxed items represent the kinds of statements generated by researchers during their research. Items in ovals represent the typical ways in which researchers move from one type of statement to the next. Reprinted with permission from: Wallace, Walter L. *The logic of science in sociology* (Chicago: Aldine). Copyright © 1971 by Walter L. Wallace.

The significance of Wallace's representation of science is manifold. First, it implies that if we look carefully at any individual piece of research, we might see more than the merely deductive or inductive approach we've outlined. We've already indicated, for instance, that our research on retirement goes well beyond the purely deductive work implied by the "right half" of the Wallace model. We do use statistical procedures that lead to empirical generalizations about the observations. (One such generalization: people who do volunteer work before retirement are more likely than others to do it after retirement.) This led, in turn, to affirmation of "role theory." Moreover, although much of Cottingham's work is explicitly inductive, involving verbal summaries of observations and creative leaps from empirical generalizations to theoretical statements shown on the "left half" of Wallace's model, she actually did some hypothesis testing as well. Thus, in arriving at the inference that fans held no formalized beliefs, she reexamined her data and found that, while there were many examples of fans knowing considerable amounts of trivia about the team, there really were no formal beliefs expressed by any informant. This kind of cross-checking of hypotheses against observations is actually an important element in the development of grounded theory (Glaser, 1993, 2005; Glaser and Strauss, 1967).

A second implication of Wallace's model is that, although no individual piece of research need touch on all elements of the model, a scientific community as a whole might very well do so. Thus, for instance, some researchers might focus on the problem of measuring key concepts, whereas others use those measurement techniques in testing critical hypotheses.

RESEARCH IN THE NEWS

Another example of how one might use Wallace's model to analyze the workings of a scientific community comes out of the controversy stirred by Donohue and Levitt's work on abortion and crime, a controversy that reached the *New York Times* in 2001. Levitt told how he became interested in the connection (Stille, 2001). It began, Levitt says, with a simple observation: that before *Roe v. Wade* in 1973, about 700,000 abortions were performed each year; the number reached 1.6 million in 1978, remained fairly constant through the 1980s, and began to decline during the 1990s. "When I first learned of the magnitude of abortions performed in the late 1970s, I was staggered by the number and thought, 'How could this not have had some dramatic social impact?'" (Stille, 2001). From this simple fact, Levitt reports, and from the observation that "the five states that legalized abortion … [before *Roe v. Wade*] saw drops in crime before the other 45 states" (Stille, 2001) (which methodology students will recognize as an empirical generalization), he and Donohue built their theory that increased abortion rates lead, via the reduction of unwanted births, to decreased crime rates (general proposition). They then came up with a series of hypotheses, suggesting, among other things, that various crime rates in different states should drop at particular times after the rise of abortion rates in those states (hypotheses). They collected appropriate data to test these hypotheses (observations) and found that the resulting analyses (giving them empirical generalizations) affirmed those hypotheses. In other words, Levitt's backstory for his research involves both induction (from observation to empirical generalization to theory) and deduction (from theory to hypothesis to observation).

Source: http://www.nytimes.com/2001/04/14/arts/new-attention-for-the-idea-that-abortion-averts-crime.html?pagewanted=all&src=pm

Most important, Wallace's model reinforces the basic point that theory and research are integrally related to each other in the sciences. Whether the scientist uses research for the purpose of theory testing, theory building, or both, the practitioner learns to think of theory and research as necessarily intertwined. Perhaps even more than "the chicken and the egg" of the old riddle or the "love and marriage" of the old song, when it comes to theory and research, "you can't have one without the other."

SUMMARY

In this chapter, we added to our discussion of theory by introducing the terminology of concept, variable, hypothesis, and empirical generalization. We introduced the criteria for showing that one variable causes another variable. Using these terms and several examples of actual research, we illustrated the relationships that most scientists, including social scientists, see existing between research and theory. Many of the nuances of the connections between theory and research have been captured by Walter Wallace's cyclical model of how science works. Wallace's model suggests, on the one hand, how research can be used to test theories or explanations of the way the world really is. On the other, the model suggests how research can be used as a tool to discover new theories. Moreover, Wallace's model suggests that an appropriate model of science encompasses not only both of the previous views of the relationship between theory and research but also the view that research and theory "speak" to each other in a never-ending dialogue.

EXERCISE 2.1

Hypotheses and Variables

1. Name a subfield in the social sciences that holds some interest for you (e.g., marriage and family, crime, aging, stratification, gender studies).
2. Name a topic that you associate with this subfield (e.g., divorce, domestic violence, violent crime, medical care, the cycle of poverty, sharing of domestic work).
3. Form a hypothesis about the topic you've named that involves a guess about the relationship between an independent variable and a dependent variable (e.g., and this is just an example—you have to come up with your own hypothesis—"children whose parents have higher status occupations will begin their work lives as adults in higher-status occupations than will children whose parents have low-status occupations").
4. What, in your opinion, is the independent variable in your hypothesis (e.g., "parents' occupation")?
5. What is the dependent variable of your hypothesis (e.g., "child's occupation")?
6. Rewrite your hypothesis, this time in the form of a diagram like the one in Figure 2.1, a diagram that connects the categories of your variables.

EXERCISE 2.2

The Relationship between Theory and Research

Using your campus library (or an online connection through it), find a recent research article in one of the following journals: *Social Science Quarterly, American Sociological Review, American Journal of Sociology, Journal of Marriage and the Family, Criminology, The Gerontologist, Social Forces, Qualitative Sociology,* and *The Journal of Contemporary Ethnography.*

Or find an article in any other social science research journal that interests you. Read the article and determine how the research that is reported is related to theory.

1. What is the article's title?
2. Who is (are) the author(s)?
3. What is the main finding of the research that is reported?
4. Describe how the author(s) relate(s) the research to theory.

   Hint: Do(es) the author(s) begin with a theory, draw hypotheses for testing, and test them? Do(es) the author(s) start with observations and build to a theory or a modification of some theory? Do(es) the author(s) work mostly inductively or deductively? Do(es) the author(s) use theory in any way?
5. Give support for your answer to Question 4 with examples of the use of theory and its connection to observations from the research article.

Ethics and Social Research

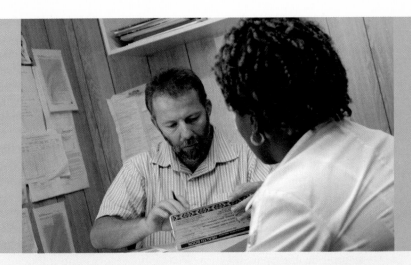

CHAPTER 3

CHAPTER OBJECTIVES

This chapter is designed to help students

- Know the ethical responsibilities of researchers
- Gain a historical perspective on research ethics
- Learn about American Sociological Association's guidelines for ethical research
- Identify ethical issues in research projects
- Understand how to protect participants from harm
- Define informed consent
- Differentiate between anonymity and confidentiality
- Understand researchers' ethical responsibilities to colleagues, the community, and the public
- Know about potential conflicts between ethical principles and research interests
- Be aware of legal issues when doing research
- Identify responsibilities when publishing research

CHAPTER OUTLINE

The Responsibilities of Researchers

A Historical Perspective on Research Ethics

Principles for Doing Ethical Research

Ethical Dilemmas During and After Data Collection

Ethics and Publishing Results

Making Decisions to Maximize Benefit and Minimize Risk

Summary

Exercises

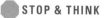**STOP & THINK**   *Have you been the victim of a crime—either a relatively small one, such as having some personal property stolen, or a more serious crime, such as assault? If so, would you be willing to fill out a questionnaire or talk to a researcher who asks about your experiences? Would you have any concerns about participating in such a study?*

THE RESPONSIBILITIES OF RESEARCHERS

Researchers have responsibilities to those who participate in their studies. Criminologist Peter Manning (2002: 546) believes "[w]e owe them, I think, succor, aid in times of trouble, protection against personal loss or embarrassment, some distance that permits their growth and perhaps feedback on the results of research if requested or appropriate." Although not all social scientists offer succor and aid in times of trouble, most aim to act *ethically* as they plan research, conduct studies and disseminate their findings. In the next chapters we'll be discussing practical and methodological issues, but first we'll focus on **ethical principles in research**, the set of standards and principles used to determine appropriate and acceptable research conduct.

ethical principles in research, the set of values, standards, and principles used to determine appropriate and acceptable conduct at all stages of the research process.

Determining appropriate research behavior is a complex topic and a source of heated debate. Social scientists question whether there is a set of universally applicable ethical principles or whether it's better to take a case-by-case approach and consider each project's specific circumstances when deciding what is, and what is not, ethical. We also know that all possibilities cannot be anticipated. Unforeseen circumstances and unintended consequences at every stage of research—from planning to publication—can create ethical problems. Finally, there is the issue of whether the ethical *costs* of the study should be measured against the study's *benefits* for the participants or others.

A HISTORICAL PERSPECTIVE ON RESEARCH ETHICS

Ethics did not become a concern of social science organizations as a whole until the middle of the twentieth century although individual scholars might have considered the issue earlier. The first ad hoc committee of the American Psychological Association (APA) established in the late 1930s to discuss the creation of a code of ethics did not support developing one. It wasn't until the 1950s that the APA adopted a code of ethics (Hamnett, Porter, Singh, and Kumar, 1984: 18). The members of the American Sociological Association (ASA) were so divided about creating a statement on ethics that it took more than eight years from proposal to adoption in 1971 (Hamnett et al., 1984). Today, most professional organizations and many government agencies have a code of ethics for research (see Box 3.1). The most recent version of the ASA Code of Ethics focuses on values for professional and scientific work and presents rules and procedures for investigating complaints of unethical behavior (see Box 3.2).

Several factors encouraged disciplines to consider ethical issues. One was revulsion at the total disregard for human dignity perpetrated during World

BOX **3.1**   **A Timeline of Milestones in Ethics in Research**

1932–1972	The Tuskegee Syphilis Study is conducted.	1991	The Federal Policy for the Protection of Human Subjects or the "Common Rule" is adopted by some agencies.
1948	The Nuremberg Code, a set of ethical directives for human experimentation, is developed.	1998	In Canada, the Tri-Council Policy Statement: Ethical Conduct for Research Involving Humans is adopted.
1953	The APA adopts its first code of ethics.	2000	The Academy of Criminal Justices Sciences adopts its first code of ethics.
1961	Stanley Milgram begins experiments on obedience.	2005	The UK Economic and Social Research Council (ESRC) specifies a social science research ethics framework to help researchers meet certain ethical principles.
1967	The American Anthropological Association adopts its first code of ethics.		
1970	Laud Humphreys's book *Tearoom Trade: Impersonal Sex in Public Places* is published.	2006	The Council on Graduate Education receives a grant to expand ethics education to science and engineering students.
1971	The ASA adopts its first code of ethics.	2007	The UN Declaration on Indigenous Peoples by the UN General Assembly includes statements about the rights of indigenous peoples specifying consideration of whether research participation is in their best interest.
1973	The British Sociological Association adopts its first code of ethics.		
1974	The National Research Act creates the National Commission for the Protection of Human Subjects of Biomedical and Behavioral Research.		
1979	The National Commission drafts the Belmont Report, a statement of basic ethical principles and guidelines in conducting research with human subjects in the United States.	2012	Many federal departments and agencies use the "Common Rule" for human-subjects research; other federal departments and agencies have their own regulations for research with human subjects.

War II by researchers in concentration camps controlled by Nazi Germany. After the Nuremberg trials, the United Nations General Assembly adopted a set of principles about research on human beings (Seidman, 1991: 47). The first principle of the Nuremberg Code is that voluntary consent of human subjects is absolutely essential (U.S. Government Printing Office, 1949).

There has been increasing awareness of and distress among scientists, policy makers, and the general public about problematic biomedical research involving human subjects. For example, government-sponsored research conducted during the 1950s studied the effects of radiation on human subjects. Funded by the Atomic Energy Commission, this research included medical research using radioisotopes to diagnose or cure disease. However, some

BOX **3.2**

The ASA's Code of Ethics sets forth the principles and ethical standards that underlie sociologists' professional responsibilities and conduct. These principles and standards should be used as guidelines when examining everyday professional activities. They constitute normative statements for sociologists and provide guidance on issues that sociologists may encounter in their professional work. The ASA's Code of Ethics consists of an Introduction, a Preamble, five General Principles, specific Ethical Standards, and the Rules and Procedures of the ASA Committee on Professional Ethics.

General Principles

The following General Principles are aspirational and serve as a guide for sociologists in determining ethical courses of action in various contexts. They exemplify the highest ideals of professional conduct.

Principle A: Professional Competence

Sociologists strive to maintain the highest levels of competence in their work; they recognize the limitations of their expertise; and they undertake only those tasks for which they are qualified by education, training, or experience. They recognize the need for ongoing education in order to remain professionally competent; and they utilize the appropriate scientific, professional, technical, and administrative resources needed to ensure competence in their professional activities. They consult with other professionals when necessary for the benefit of their students, research participants, and clients.

Principle B: Integrity

Sociologists are honest, fair, and respectful of others in their professional activities—in research, teaching, practice, and service. Sociologists do not knowingly act in ways that jeopardize either their own or others' professional welfare. Sociologists conduct their affairs in ways that inspire trust and confidence; they do not knowingly make statements that are false, misleading, or deceptive.

Principle C: Professional and Scientific Responsibility

Sociologists adhere to the highest scientific and professional standards and accept responsibility for their work. Sociologists understand that they form a community and show respect for other sociologists even when they disagree on theoretical, methodological, or personal approaches to professional activities. Sociologists value the public trust in sociology and are concerned about their ethical behavior and that of other sociologists that might compromise that trust. While endeavoring always to be collegial, sociologists must never let the desire to be collegial outweigh their shared responsibility for ethical behavior. When appropriate, they consult with colleagues in order to prevent or avoid unethical conduct.

Principle D: Respect for People's Rights, Dignity, and Diversity

Sociologists respect the rights, dignity, and worth of all people. They strive to eliminate bias in their professional activities, and they do not tolerate any forms of discrimination based on age; gender; race; ethnicity; national origin; religion; sexual orientation; disability; health conditions; or marital, domestic, or parental status. They are sensitive to cultural, individual, and role differences in serving, teaching, and studying groups of people with distinctive characteristics. In all of their work-related activities, sociologists acknowledge the rights of others to hold values, attitudes, and opinions that differ from their own.

Principle E: Social Responsibility

Sociologists are aware of their professional and scientific responsibility to the communities and societies in which they live and work. They apply and make public their knowledge in order to contribute to the public good. When undertaking research, they strive to advance the science of sociology and to serve the public good.

Source: *The entire code can be found on the ASA website at* http://www.asanet.org/images/asa/docs/pdf/CodeofEthics.pdf.

experiments were gruesome, such as one where people with "relatively short life expectancies," such as semi-comatose cancer patients, were injected with plutonium to determine how much uranium was needed to produce kidney damage (Budiansky, 1994). In cancer research in 1963, physicians injected live cancer cells into 22 elderly patients without informing them that live cells were being used or that the procedure was unrelated to therapy (Katz, 1972).

One long-lasting research scandal ended as the result of public outrage. The Tuskegee Syphilis Study (labeled as such because the project was conducted in Tuskegee, Alabama), conducted by the U.S. Public Health Service, studied the effects of untreated syphilis to determine the natural history of the disease (Jones, 1981). The 40-year research project began in 1932 with a sample of 399 poor black men with late-stage syphilis. The study ultimately included 600 men, all of whom were offered free medical care in exchange for their medical data, and none of whom were told that they had syphilis. More than 400 of the men were not offered the standard (and often fatal) treatment for syphilis in the 1930s. In later years, they were not provided with penicillin even though it was available by the late 1940s (Jones, 1981). Many of the men died of the disease, and some of them unknowingly transmitted it to their wives and children (*The Economist*, 1997: 27). As late as 1965, an outsider's questioning of the morality of the researchers was dismissed as eccentric (Kirp, 1997: 50). The project ended in 1972, but government response was not offered until 1997 when President Bill Clinton apologized to the surviving victims and their families (Mitchell, 1997: 10).

Recently an even more egregious study on the same topic from the same era and agency has come to light. Between 1946 and 1948, the U.S. Public Health Service conducted a study in Guatemala where doctors actually infected people with syphilis and gonorrhea. After learning about the factors that caused infection, the researchers administered penicillin, expecting to cure the infection, which they did in most, but not all cases (Reverby, 2011: 9). Deception was central to this study as the researchers did not tell most participants what they were being exposed to. Instead, they argued that secrecy and law breaking were sometimes necessary to further research (Reverby, 2011: 16). This research came to light only after historian Susan Reverby came across information about the experiment in the papers of Dr. John Cutler in archives at the University of Pittsburgh while looking for information about the Tuskegee study. When Secretary of State Hillary Clinton and Secretary of Department of Health and Human Service Kathleen Sebelious learned about the U.S. involvement in this research, they formally apologized to Guatemala and President Barack Obama called Guatemalan President Colom to express his deep regret (Reverby, 2011: 22).

The Tuskegee study and others were an impetus for the passage of the National Research Act of 1974. This act created a commission to identify principles for conducting ethical biomedical and behavioral research and prompted the development of regulations to protect the rights of research subjects at institutions receiving federal funds (Singer and Levine, 2003). Throughout the 1960s, National Institutes of Health lawyers were concerned with study participants' rights as well as protecting them from harm (Stark, 2007: 779). Social

BOX **3.3**   **Overall Goals of the Federal Policy on the Ethical Treatment of Research Subjects**

Many federal agencies have adopted the "Common Rule," a set of regulations governing human subjects of research. Some agencies have additional provisions for research with vulnerable populations such as children, prisoners, pregnant women, or handicapped persons.

The policy is designed to ensure minimal standards for the ethical treatment of research subjects. The major goal is to limit harm to participants in research. That means that no one should suffer harm just because they became involved as subjects or respondents in a research project. Institutions engaged in research should foster a culture of ethical research.

ETHICAL RESEARCH rests on three principles:

1. RESPECT for persons' autonomy, meaning the researcher gives adequate and comprehensive

information about the research and any risks likely to occur, understandable to the participant, and allows them to voluntarily decide whether to participate.
2. BENEFICENCE, meaning the research is designed to maximize benefits and minimize risks to subjects and society.
3. JUSTICE, meaning that the research is fair to individual subjects and does not exploit or ignore one group (e.g., the poor) to benefit another group (e.g., the wealthy).

Research produces benefits valued by society. Regulatory oversight seeks to ensure that any potential harm of the research is balanced by its potential benefits.

Source: http://www.nsf.gov/bfa/dias/policy/hsfaqs.jsp#relation

science came under review in 1966 when Surgeon General William Stewart announced that it was to be reviewed by local boards at hospitals and universities (Stark, 2007: 780). By the 1970s, most federal agencies and professional organizations had developed guidelines for conducting ethical research. See, for example, the National Science Foundation (NSF) policy on the ethical treatment of subjects in Box 3.3.

institutional review board (IRB), the committee at a college, university, or research center responsible for evaluating the ethics of proposed research.

An **institutional review board** (IRB or REB in Canada and REC in Britain) or committee on human subjects research is charged with evaluating the ethics of research proposals, and almost every college, university, and research center in the United States, Canada, and Europe has one.[1] The research plan must be approved by the committee, at least one of whose members is unaffiliated with the institution, before any federal, state, or institutional monies can be spent. For the past 30 years, IRBs have attempted to verify that the benefits of proposed research outweigh the risks and to determine whether subjects have been informed of participation risks. Today there are approximately 4,000 IRBs in the U.S. alone charged with the moral and legal tasks of assessing risks and safeguarding participants' rights (Stark, 2012). While the general guidelines are the same, responses can vary. Laura Stark (2012) sent the same

---

[1] In Chapter 14, we'll read about research conducted by Jill Hume Harrison and Maureen Norton-Hawk in a women's prison in Ecuador. Although their research was approved by an IRB at an American university, they discovered that the prison did not require an IRB and prison administrators were expected to protect the privacy of the inmates.

research proposal to IRBs at multiple research universities but got different decisions about the proposal's acceptability.

The federal policy, using the Common Rule (outlined in Box 3.3), allows the exemption of surveys and observations of public behavior when subjects cannot be identified (NSF, 2012). The need to seek approval from an IRB depends on the topic, the source of funding, and the specific institutions involved. Consulting appropriate ethical codes and being informed about the requirements and procedures for submitting proposals to review committees is important as, ultimately, the responsibility for ethical research behavior lies with the individual researcher.

There is some controversy about how IRBs deal with social research. The original impetus for IRB regulation came from abuses in biomedical research and the members of boards are often mostly from the medical and biological sciences. With few social scientists serving on IRBs, some argue that social science research is assessed using a biomedical model and that IRB requirements can cause undue hurdles, censorship, and costly delays, which creates impediments to a socially relevant and critical sociology (Feeley, 2007: 765–766).

In general, researchers support the IRB process with some reservations. In response to a survey of researchers at one university, the overwhelming majority of faculty and graduate students said that IRB approval is an important protection for human subjects (Ferraro, Szigeti, Dawes, and Pan, 1999). A national survey of 40 U.S. health researchers found they supported the goals of IRB regulation but questioned the effectiveness, efficiency, and fairness of regulation implementation (Burris and Moss, 2006). A recent survey of 319 social scientists, mostly ethnographers, found 76 percent reporting that the ethics review committee did not ask for significant modifications of their research plans; however, some respondents critiqued the process as existing mainly to protect universities rather than research subjects (Wynn, 2011).

Desirée Ciambrone, a sociologist whose work we'll read in the following focal research section, found the IRB at Brown University to be very helpful. While taking a graduate seminar in medical sociology at Brown University, Ciambrone became interested in the topic of women with acquired immunodeficiency syndrome (AIDS). Conversations about AIDS in Africa with a fellow graduate student who had worked on public health and policy issues convinced Ciambrone that focusing on American women's experiences with human immunodeficiency virus (HIV)/AIDS was a worthwhile project. In this, her first totally independent research project, she learned a great deal about conducting a study from start to finish. She found out about the impact of illness on people's lives and how an HIV diagnosis affects women's social and familial roles. She also learned that submitting a research proposal to an IRB can bring one's attention to issues not considered previously.

Ethical Concerns and Researcher Responsibilities in Studying Women with HIV/AIDS

By Desirée Ciambrone[2]

Conducting one's first independent research project is a daunting task; [while one is] eager to immerse oneself in the project, there is a great deal of preparatory work to be done. I chose to study women with HIV infection and found that the anxiety and complexity of conducting social science research are heightened when people facing difficult circumstances are involved.

Studying Women with HIV: Getting Ready

Respondent Recruitment

Finding women with HIV to speak with was no easy task. I sought respondents from health and social service agencies located primarily in the northeastern United States. Some agencies requested copies of my interview guide, my research proposal, and/or the Brown University's IRB approval. I discussed the goal of my study with my contacts and urged them to post/distribute a letter describing the study and ways to contact me with questions about the study or to participate. Better yet, I urged that agency contacts speak with women directly and encourage their participation.

Many organizations I contacted were justifiably protective of their clients and appeared unwilling to assist in respondent recruitment despite their verbal acquiescence. Others were too busy to help me with my study; they posted my letter among a myriad of other brochures and leaflets but did little to personally encourage women to participate. Several key contacts in local hospitals were very helpful and referred most of the 37 HIV-positive women who volunteered to be interviewed for my study.

Informed Consent

The protection of human subjects is of utmost importance when conducting any type of research. Thus although my study would require "only an interview" (as opposed to a clinical trial, e.g., where drugs and other treatments may be involved), obtaining *informed consent* from my participants was crucial. The IRB at Brown University provided the guidelines that helped me carry out my research in an ethical way, including providing potential participants with an understanding of the potential risks and benefits of participation. In my initial proposal, I was quick to note the benign nature of my study. I described the benefits of being able to share one's story with an interested listener and the potential impact of my results on social policy and medical care provision. I noted that I wanted to talk to women about their lives and about living with HIV/AIDS. Initially, I did not see any risks that could result from this. Comments from the IRB prompted a more careful reflection of the *potential* risks of participation in my study. Indeed, I realized my questions might elicit a negative emotional response; such uneasiness or distress

---

[2]This article was written for this text by Desirée Ciambrone and is published with permission. A complete description of her study and findings can be found in D. Ciambrone, *Women's Experiences with HIV/AIDS: Mending Fractured Selves*, New York: Haworth Press, 2003.

is, in fact, a risk. Thus, my informed consent was rewritten to include the possibility of emotional discomfort (see Appendix A).

Informed consent means that respondents demonstrate that they understand the study, what is expected of them, and how the information will be used. It also requires that participants *voluntarily* enter into a research project; that they haven't been coerced or duped into participation. Given previous research in which human subjects were misinformed and/or forced to participate in dangerous and life threatening experiments under the guise of science, the apprehensions of vulnerable populations are justified.

Women with HIV, while not unintelligent, are less likely to be highly educated and may not be knowledgeable about research; thus, it was my responsibility to explain the study in a clear, straightforward manner. I explained that I was a graduate student in medical sociology at Brown University who was conducting research in order to learn about the daily lives of women with HIV/AIDS. I told them that I wanted to interview them about their experiences in whatever setting was comfortable for them, their homes or a more neutral setting. I said I would record the interviews in order to have an accurate, complete record of their stories but that, in order to protect their privacy, I would not use their real names on the recording, on the transcripts, or in the reports and would keep the information confidential. I needed to describe what the study was and, perhaps more importantly, *what it was not*. I had to be clear that I was not a health care provider; I could not offer medical advice or prescribe medications. I was also not a social worker; I could not act as their case manager linking them to needed services. On the other hand, I felt I owed women answers to their questions. So I became educated about local services for persons with HIV/AIDS so that if a woman had a concern I might say, "Well, perhaps X organization can help. I know they deal with those types of issues." I was careful to make certain my respondents understood that while I was knowledgeable about many services, I did not know of *all* that was available nor could I guarantee the organization would be able to help. It was a challenge to be frank, honest, and share whatever I could *and* be sure that women did not perceive me as someone who would solve their problems; even with the best of intentions, that was not in my power.

An IRB follows research from its conception through its completion. Brown University's IRB requires annual reports in which researchers describe progress and explain any changes made to protocols or informed consent forms. This mechanism ensures that changes made are warranted and ethical. As an example, one of the things I reported in my annual progress report was my decision to offer respondents compensation for participation. I felt that this would attract those less accustomed to sharing their stories and offer all the participants some compensation for their time. To this end, I offered each participant a $10.00 gift certificate from a local supermarket chain, a decision approved by the IRB.

The IRB protocol outlines the appropriate storage of data. For example, I had to ensure that my respondents' personal identifying information was not attached to the data. Thus, instead of using the real names on the transcripts, pseudonyms were used. In accordance with federal regulations, transcripts, audiotapes, and other materials were stored under lock and key in a limited-access location.

(continues)

FOCAL RESEARCH (continued)

The Interview Experience

Women with HIV/AIDS, particularly those with substance abuse histories, face unique challenges including poverty, caring for young children and other adults, reproductive decisions, access to health care and social services, inadequate housing, negotiating condom use, sexual undesirability, and abandonment. Race and class inequities exacerbate these problems. I knew all of this from my review of the literature but had no *personal experience* with this population.

Being able to talk to the women in person was essential as I was able to listen to them speak, follow up on key issues, observe body language, and see where they lived. It was also important for me to learn that these women are knowledgeable, articulate people—experts on their experiences. I developed a newfound respect for these women's agency; in the face of poverty, drug use, and illness these women managed to live productive, meaningful lives. Several did public speaking and worked to help others with HIV/AIDS in their community.

Connecting with Women with HIV

In graduate school, I had learned about the importance of the researcher's role in the research process. Thus, at every stage of my research I made a conscious effort to empower the study participants and present them as experts on living with HIV. Prior to interviewing, I contemplated the worst-case interview scenarios, particularly access and rapport problems stemming from differential social positions. I did not encounter skepticism or scrutiny; none of my respondents questioned my interest in HIV/AIDS, nor did they appear uncomfortable sharing their stories with someone unlike themselves. This was probably due to the fact that many of these women were accustomed to talking to "outsiders" about their experiences via their participation in clinical trials of medical treatment at local health care centers. Women who had not participated in such projects may well have been skeptical of my intentions and less frank about their lives.

In addition to social distance and nondisclosure issues, I had also prepared myself for other possible problematic interview situations, such as lack of privacy and dire living circumstances. For most of the interviews, respondents and I had adequate privacy. Despite occasional commonplace interruptions (e.g., phone calls, demands of young children), the interviewees tried to create conducive interview situations.

The living conditions of some women were more troubling than relational issues. Several women lived in poor, traditionally high-crime neighborhoods or public housing developments, neighborhoods I was familiar with from a young age as "places not to go." In order to feel as safe as possible, I scheduled most of my interviews, especially those in "tough" neighborhoods, during the day. Entering questionable dwellings was difficult. By and large, the women's homes were adequately furnished and relatively well kept. It was distressing to witness the apparent poverty and the destitute living conditions of some of my respondents. For example, one woman's home was filthy—the floors appeared as if they had not been swept or washed for quite some time, there were dirty dishes and pans overflowing in the sink, and ashtrays were filled to the rim with cigarette butts. I felt somewhat uneasy in homes like this, and less accepting of women's hospitality (e.g., a cup of coffee). Not accepting food and beverages made me feel rude and

I never wanted to appear judgmental of women's homes. And although I had no intention of verbalizing my thoughts, I wondered how I could explain that it wasn't the HIV status that made me uncomfortable at times; it was the unsanitary home!

Eliciting Women's Stories and Confidentiality

Given the sensitive nature of my questions (e.g., about sexual relations and illicit drug use), I took great pains to make sure that the women knew the information provided was completely *confidential* and would not interfere with the receipt of medical/social services. I reiterated that they did not have to answer any questions that made them feel uncomfortable. While there were a couple of women who did not want to discuss certain topics, overall, the women I interviewed were incredibly open and often offered more intimate details than I had expected.

Many women were very interested in the purpose of the study. Often, women wanted to know how *their* responses compared to those of other women. Many also asked questions about my academic experience and personal life (e.g., What are you studying? Are you married? Do you have kids?). Recognizing the importance of establishing trust and minimizing power differentials between us, I answered the questions and told them about myself. When asked about the other interviewees, I was concerned that answering questions about the study would lead respondents to feel there was a normative or desired response to my questions that they should confirm. As I was willing to discuss my research, but didn't want women's narratives to be influenced by those of other respondents, I suggested that we talk about the study after the interview so as to be sure we had ample time to discuss their experiences and views. Evading interviewees' questions hinders efforts to establish trust and build rapport with respondents. My self-disclosure put my respondents at ease by reassuring them that the interview was a guided conversation and that it was their experiences and opinions that were important, not getting the "right" answer. The open-ended nature of my questions allowed women to highlight the issues that were important to them and explore areas that I may not have inquired about.

Interviewing women who were active drug users posed a challenge. I felt most uncomfortable around these women as their experiences were so far removed from mine. In addition to striving to appear calm and comfortable, I had to remind myself that my role was that of researcher, not social worker or care provider. Thus, I could not offer unsolicited advice about the dangers of drug use on their weakened immune systems and the potential ramifications of nonadherence to their medications or provide solutions for the apparent disrespect the women were shown by their partners or children. Furthermore, when women did ask me for my advice I was careful to make it clear to them that I was a sociologist, not a therapist or a counselor. While I answered their questions about HIV and services available to the best of my ability, I didn't feel it was appropriate to offer unsolicited advice even when I had strong feelings about their choices. For example, one woman was married to a devout Christian who had convinced her to leave her health in God's hands and forgo her treatment regimen. Sheila[3] felt torn between her husband's religious beliefs and her health care providers' recommendations. Despite a dropping T-cell count and increasing

(continues)

---

[3]Sheila is a pseudonym to protect the respondent's confidentiality.

FOCAL RESEARCH *(continued)*

viral load, Sheila stopped taking her medications at her husband's request. I found her situation troubling, and while I worried about Sheila's health, I felt it was not my role to encourage medication use or discredit her faith in religion.

Lessons Learned

In designing and conducting this study I learned a great deal about the *process* of conducting research, including the importance of following ethical practices, using informed consent procedures, protecting my respondents' privacy and confidentiality, and preserving the integrity of their narratives. My understanding of how women cope with HIV/AIDS was enhanced as a result.

Appendix A: Informed Consent
Informed Consent for Participation in Women with HIV Infection Study

You are being asked to participate in an important study to learn more about women's experiences with HIV infection. This study is being conducted by Desirée Ciambrone, a graduate student in the Sociology Department at Brown University.

During this interview, I will ask you about your illness, how HIV infection affects your daily life, and about the people who help you if, and/or when, you need assistance.

- This interview will be tape-recorded.
- In order to participate in this project I must be 18 years old or over.
- I may be contacted for further questions regarding this study.
- My participation in this project is completely voluntary; the decision to participate is up to me and no one else.
- I do not have to answer every question; I may choose not to answer certain questions.
- I can ask the interviewer to leave at any time. Even if I sign this form, I can change my mind later and not participate in the study.
- The information I provide will be kept confidential to the extent of the law.
- The only known risk to me is the possibility of some emotional discomfort.
- A direct benefit of participation in this study is the opportunity to talk with an interested listener and to think reflexively about my own experiences. I understand that the information obtained by this study may help other women who have had similar life experiences and may help professionals understand and treat women with HIV infection.

If I have any questions about this study, I may contact Desirée Ciambrone at (401) 863-1275 at any time. Or, I may contact Alice A. Tangredi-Hannon or Dorinda Williams at the Office of Research Administration at Brown University (401) 863-2777 with questions about my rights as a participant in this study.

I have read the consent form and fully understand it. All my questions have been answered. I agree to take part in this study, and I will receive a copy of this completed form.

Signature of Participant _____   Date _____

Signature of Interviewer _____   Date _____

PRINCIPLES FOR DOING ETHICAL RESEARCH

Ethical principles provide a moral compass to help researchers make decisions about their work as they are a set of moral and social standards that includes both prohibitions *against* and *prescriptions for* specific kinds of behavior in research. Although we can scientifically study people's values and their ethical standards, we cannot use scientific methods to determine how to behave ethically in our own work. Instead, social scientists must examine research behavior using other frames of reference. To act ethically, each researcher must consider many factors, including safeguarding the welfare of those who participate in their studies, protecting the rights of colleagues, and serving the larger society. As we consider specific ethical principles, we'll examine these issues.

RESEARCH IN THE NEWS

In 2012, the *New York Times* published a story about Sudhir Venkatesh, a well-known sociologist and sought-after speaker who through his research on gangs and prostitutes has made sociology seem hip (Kaminer, 2012). However, the *Times* notes that "[s]ome of his peers say that in search of a broader readership he takes liberties not appropriate for a scholar: sensationalizing his experiences, exaggerating the reliability of his memory and, in one case, physically assaulting someone... And at Columbia, where he briefly led the university's largest social science research center, he was the subject last year of a grueling investigation into a quarter-million dollars of spending that Columbia auditors said was insufficiently documented, misappropriated or outright fabricated" (Kaminer, 2012). Venkatesh (2012) responded that "The audit document discussed in the *Times* article was the beginning of an inquiry, not the end... The article also suggests that I work outside the boundaries of mainstream sociology. I plead guilty. My discipline is stuffy and losing relevance daily in the academic and public eye. But, I have never been anything other than scrupulous, honest and ethical in my research, and I have always safeguarded the risk of my research subjects at every moment."

Principles Regarding Participants in Research

Protect Participants from Harm

protecting study participants from harm, the principle that participants in studies not be harmed physically, psychologically, emotionally, legally, socially, or financially as a result of their participation in a study.

First and foremost among ethical standards is the obligation to **protect study participants from harm**, which means that the physical, emotional, social, financial, legal, and psychological well-being of those who participate in research must be protected, both during the time that the data are collected and after the conclusion of the study. Overall, there should not be negative consequences as a result of being in a study. The controversy over Venkatesh's behavior described above brings us to a consideration of *all* of the ethical issues in research and social science.

As in Venkatesh's work, social science data are frequently collected by asking study participants questions. In these kinds of studies, it's important to think about the impact of being asked and answering such questions.

Ciambrone, for example, knew that her participants were vulnerable so she felt it was appropriate for the IRB to require her to inform potential interviewees about possible emotional risk, even if the distress was short term.

One project that has been criticized as violating the "no harm" principle and other ethical standards is the Milgram research on obedience to authority figures. In a series of famous experiments, psychologist Stanley Milgram (1974) deceived his subjects by telling them that they were participating in a study on the effects of punishment on learning. The subjects, mostly men ages 20 to 50, were told that they would be "teachers," who, after watching a "learner" being strapped into place, were taken into the main experimental room and seated before an impressive shock generator with 30 switches ranging from 15 volts to 450 volts. The subjects were ordered by the experimenter to administer increasingly severe shocks to the "learner" whenever he gave an incorrect answer. They were pressed to do so even when the "learner" complained verbally, demanded release, or screamed in agony. After "receiving" 300 volts, the "learner" would pound on the wall and then no answer would appear on the panel. Subjects typically turned to the experimenter for guidance and were told to treat no response as a wrong answer and continue to shock according to the schedule of increasing volts. While the average subject showed a great deal of tension (including sweating, trembling, groaning, or nervous laughter), the majority continued to shock until 450 volts—two steps above the designation "Danger—Severe Shock." At debriefing, the subjects were told that the "learner," or victim, was part of the research team and had actually received no shocks at all. The intended purpose of the experiment was to see at what point, and under what conditions, ordinary people would refuse to obey an experimenter ordering them to "torture" others (Milgram, 1974).

The experiments results were unexpected as Milgram had asked groups of psychiatrists, college students, and middle-class adults for predictions and not one person predicted that any subject would give high-voltage shocks (Milgram, cited in Korn, 1997: 100). Milgram was taken aback to find that with "numbing regularity good people were seen to knuckle under to the demands of authority and perform actions that were callous and severe" (1974: 123).

It might surprise you that Milgram's work was consistent with the ethical principles for psychological research at the time. The APA's 1953 guidelines allowed for deception by withholding information from or giving misinformation to subjects when such deception was, in the researcher's judgment, required by the research problem (Herrera, 1997: 29). Not until the 1992 revision of the APA's statement on ethical principles were additional restrictions placed on the use of deception (Korn, 1997). When his study was published, Milgram found himself in the middle of a controversy over ethics that continued for decades. Detractors blamed him for deceiving his subjects about the purpose of the study and for causing his subjects a great deal of emotional stress both during and after the study (Baumrind, 1964), but supporters cited the importance of knowing the dark side of conforming to authority (Sokolow, 1999).

Milgram's study raises the issue of whether the ends can justify the means. Thinking about Milgram's work, you might feel that the study was acceptable because it made an important contribution to our understanding of the dangers of obedience. Or you might agree with those who have vilified Milgram's work because it violated the ethical principle of protecting subjects from harm.

STOP & THINK *An ethics review committee evaluated a research proposal that included interviews of romantic couples where the individuals in the couple sat together and answered questions about what they liked and disliked about their relationship. The committee decided that there was more than a minimum risk that couples might end their relationship as a result of this interview (Haggerty, 2004: 402). What do you think of their assessment?*

Whatever your perspective on Milgram's study, his research points out that the line between ethical and unethical methods is not always obvious. It can be difficult to design a study in which there is absolutely no risk of psychological or emotional distress to participants. For example, for at least a few people, being interviewed about their marriages or their relationships with siblings can dredge up unhappy memories. Similarly, for others, talking about experiences with sexual assault, divorce, or the death of a loved one can produce adverse effects or create difficulties for them in groups to which they belong. Sometimes methodological decisions can be based on the needs of subjects—that is, to protect them from harm that might occur. In a study of families with terminally ill children, for example, the researchers did not ask questions about the diagnosis, prognosis, or death and added a mentor to the research team to act as a resource in the event that the interviews became emotionally charged (Connolly and Reid, 2007: 1045).

It is impossible to anticipate each and every consequence of participation for every research participant. But even foreseeing possible negative reactions by subjects does *not* mean a study should be abandoned. Instead, a realistic and achievable ethical goal is for studies to create, *on balance*, more positive than negative results and to give participants the right to make informed decisions about that balance. Researchers should try to anticipate needs for help or support. In a study of children's experiences of violence, for example, the participants were informed of avenues of support, such as known and trusted adults, and given toll-free phone numbers to talk to counselors, which were, in fact, used by some children to talk about upsetting events (Carroll-Lind, Chapman, Gregory, and Maxwell, 2006: 985). In a study about partner violence, researchers learned that victims wanted referrals to and information about agencies that could provide help. The research staff networked with such agencies, visited them, and periodically checked the agency phone numbers to provide the women with accurate information about community resources (Logan, Walker, Shannon, and Cole, 2008: 1232).

Ciambrone told her participants about the possible emotional consequences of participating and occasionally offered referrals to agencies. She found that most seemed to enjoy the interview and liked the idea of telling their stories to help others with HIV. They said they appreciated her interest and in having the opportunity to share their views. A main theme was that the interview made them feel "normal," accepted, and valued rather than being defined exclusively by their HIV status.

STOP & THINK   *If you were asked to participate in a study, what would you like to know? One review board administrator, Robert L. Johnson of Appalachian State University, asks students about their concerns and says that the usual responses are "What are you doing? Will you use my name? Who's responsible if I get hurt? What's in it for me?" (Cohen, 2007). Do you agree with these students or is there other information you'd want before making a decision about participating?*

Voluntary Participation and Informed Consent

The ethical strengths of Ciambrone's study include the fact that all the interviewees volunteered to participate after being informed of the researcher's identity and affiliation and the study's purpose, risks, and topics. The ethical principles being followed were **voluntary participation** and **informed consent**. These principles require that, *before participating*, all potential participants be told that they are not obliged to participate; that they can discontinue their participation at any time; and that they be given accurate information about the study's purpose, the method of data collection, whether the information obtained will be collected anonymously or kept confidential, and the availability of results. For example, Gloria Gonzalez-Lopez (2011: 458) notes that in her study of sexual violence experienced by among Mexican women, she "always respected the assertiveness of informants who said they did not want to answer a question even when it challenged my curiosity as a researcher." Violations of ethical standards include not asking for consent, presenting inadequate information or misinformation, and using pressure or threats to obtain participation or continued participation.

One way to provide potential respondents with information about the study is through a written **informed consent form**, like the ones included in the Appendix of Ciambrone's article and in Chapter 10. Anonymous questionnaires that pose few risks for participants may be exempted from the requirement for informed consent, and some researchers rely on verbal rather than written permission as illustrated in Box 3.4. In recent years, it is more

voluntary participation, the principle that study participants choose to participate of their own free will.

informed consent, the principle that potential participants be given adequate and accurate information about a study before they are asked to agree to participate.

informed consent form, a statement that describes the study and the researcher and formally requests participation.

| BOX **3.4** | An Example of Verbal Parental Consent |

A research project by the National Alliance for Caregiving (2005: 45) was conducted to determine the impact of children's participation in caregiving within households. First an adult in the household was interviewed by phone and then asked for permission to speak directly to his/her child. A verbal consent statement was read that began: "Before we proceed, I would like to tell you about the interview so you feel comfortable having me talk to your child. I will be asking him/her about the kinds of things s/he typically does, including his/her responsibilities and free time activities. We will also talk about the kinds of things s/he does for the person she/he helps care for and how the responsibilities affect him/her. Please be assured that your responses and your child's will be kept strictly confidential. Individual results will not be released to anyone. Responses from many people will be grouped together so that no one person's answer can be identified. If you'd like, you may listen to the interview."

Source: http://www.caregiving.org/data/youngcaregivers.pdf

common to use written statements to provide potential interviewees with sufficient information to decide about participation. Informed consent should specify the purpose of the study, the benefits to participants and to others, the identity and affiliations of the researcher, the nature and likelihood of risks for participants, a statement of how results will be disseminated, an account of who will have access to the study's records, and a statement about the sponsor or study's funding source.

⬣ **STOP & THINK**   *Who should give informed consent in a study involving children under 18? The children? Their parents? Both?*

There is some debate about the issue of who can appropriately give informed consent. Of particular concern are children, adolescents, and individuals unable to act in an informed manner. Most agree that, for minors, it is necessary to obtain both the child's and a parent's permission. However, there is debate about the age at which people can decide to grant consent and who should be authorized to give consent for those judged unable to decide for themselves (Ringheim, 1995). Studying anorexic teenagers, Halse and Honey (2005: 2152) note that difficult family relationships or histories could make parental consent a "double-edged sword, protecting some girls and erasing other girls' potential for agency by increasing the opportunity for parental coercion." One Ph.D. candidate interested in interviewing gay teenagers he had contacted through clubs decided that his research was unfeasible when his university's ethics committee required parental consent before the interview (Crow, Wiles, Heath, and Charles, 2006: 89).

Some researchers are granted IRB approval to waive parental permission. Diviak, Curry, Emery, and Mermelstein (2004), noting that teen smokers might not be willing to admit their smoking habit to parents, surveyed 21 researchers engaged in smoking cessation research with adolescents and found that 13 had asked for parental permission to be waived. Four of the researchers' projects were authorized to collect teen permission only.

Methods of obtaining permission vary. In school settings, for example, active consent requires someone, such as a parent or guardian, to sign and return a form to give permission for their child to participate in a study, whereas **passive consent** requires them to sign and return a form if they do *not* want their child to participate. Getting consent forms returned can be quite difficult. In a study of high school students in Australia, de Meyrick (2005) decided on active consent and mailed 2,500 consent letters to parents, but only 200, or less than 8 percent, were returned. In comparison, researchers used passive consent at one U.S. high school; only 4 percent of parents notified by mail of a study on parent–child interaction contacted the school to refuse permission for their child's participation (Bogenschneider and Pallock, 2008). In addition to getting a good response rate, passive consent may protect children's rights. A study of child abuse among New Zealand children used passive consent procedures to obtain a participation rate of 93 percent as the researchers believed that parental rights to privacy regarding abuse were not paramount (Caroll-Lind, Chapman, and Raskauskas, 2011).

passive consent, when no response is considered an affirmative consent to participate in research; also called "opt out informed consent," this is sometimes used for parental consent for children's participation in school-based research.

STOP & THINK   *Have you ever been asked by a faculty member while you were in his or her class to participate in a study that he or she was conducting? Did you feel any pressure or did you feel you could decline easily? Do you think that students in this kind of situation who agree to participate are true volunteers?*

As a way of ascertaining whether participants are really *volunteers*, researchers should ask the following questions: Do potential participants have enough information to make a judgment about participating? Do they feel pressured by payments, gifts, or authority figures? Are they competent to give consent? Are they confused or mentally ill? Do they believe that they have the right to stop participating at any stage of the process? Those who are dependent on others, such as those in shelters or in nursing homes, might be especially vulnerable to pressure to cooperate. For example, in her discussion of research with homeless women, Emily Paradis (2000) notes the possibility of a dependent or coercive relationship when the institution in which the research takes place is the major source of subsistence, advocacy, or emotional support for the participants and when the participants are used to making accommodations to survive. Such relationships might make it harder for those approached to decline to participate.

Other studies raise different questions about consent. For a study on police interrogations, for example, researcher Richard Leo (1995) was required by his university's IRB to secure informed consent from the police officers, but not from the criminal suspects. In another research project, Julia O'Connell Davidson obtained informed consent from a prostitute and her receptionist but did not ask for permission from the prostitute's clients who did not know she was listening to their conversations in the hall or watching them leave the house and did not consent to having the prostitute share information about their sexual preferences or physical and psychological defects (Davidson and Layder, 1994). Davidson was not troubled by the uninvited intrusion into the clients' world, arguing that the prostitute had willingly provided knowledge that belonged to her.

New technologies can lead to ethical dilemmas. For example, what about using information from Facebook or similar sites? Some professional organizations have turned their attention to this issue (see Box 3.5). Although some consider Internet postings to be public and informed consent unnecessary, others disagree. Similar differences of opinion can be found about other sources of data such as field research, which is where the researcher observes daily life in groups, communities, or organizations.

 STOP & THINK   *What would you do if you wanted to study social interaction at parties? Would you do the observation without first asking your friends for their permission? How about observing people in public, such as watching crowd behavior at a sporting event? Would you or could you ask for permission? What would you do if you saw illegal behavior? How about if law enforcement personnel asked you for the notes you took while watching the crowd?*

New Media, Old Concerns

In 2000, international scholars from many disciplines came together to found the Association of Internet Research (AoIR) to promote research into the social, cultural, political, economic, and aesthetic aspects of the Internet. The organization's most recent set of recommendations for ethical decision making in Internet research was approved in 2012.

While some of the media that researchers study are new, the ethical concerns are not. Whether using media like blogs, Facebook, Google searches, personal e-mail or tweets as data, researchers must be aware of the obligation to protect the rights of subjects, including the right to privacy. The AoIR Ethics Working Committee (2012) used several guiding principles, among them "the greater the vulnerability of the community/author/participant, the greater the obligation of the researcher to protect the community/author/participant." The organization also reminds us that "[i]ndividual and cultural definitions and expectations of privacy are ambiguous, contested and changing. People may operate in public spaces but maintain strong perceptions of expectations of privacy."

Most researchers would argue that it's unnecessary, impractical, or impossible to get informed consent when observing in public places. Sometimes there is no simple distinction between "public" and "private" settings. In the study described in *My Freshman Year: What a Professor Learned by Becoming a Student*, 52-year-old anthropologist Rebekah Nathan obtained approval from her university's IRB and followed the ethical protocol of informed consent when conducting formal interviews with students (Hoover, 2005: 36). Some might argue that she acted unethically when in day-to-day interactions with students: Nathan let them assume she was one of them. Other research decisions are also debatable. For example, O'Halloran (2006) attended meetings of Alcoholics Anonymous without asking permission, taking field notes unobtrusively. Was that okay? The ASA Code of Ethics supports observing without permission in public places but suggests consultation with an IRB if there is any doubt about the need for informed consent.

Anonymity and Confidentiality

anonymity, when no one, including the researcher, knows the identities of research participants.

confidentiality, also called privacy, is when no third party knows the identities of the research participants.

Even if informed consent is not obtained, the ethical principles of anonymity and confidentiality do apply. **Anonymity** is when it is impossible for anyone, including the researcher, to connect specific data to any particular member of the sample. To collect data anonymously, data must be collected without names, personal identification numbers, or information that could identify subjects.

If it is not possible or practical to collect data anonymously, research ethics dictate keeping them confidential. **Confidentiality**, or privacy, is keeping the information disclosed by study participants, including their identities, from *all* other parties, including parents, teachers, and school administrators, and others who have given permission or aided participation in the research. To keep material confidential, the data collected from all respondents must be kept secure. Identifying respondents by code numbers is useful. When study results are made public, confidentiality can be achieved by using pseudonyms or by grouping all the data together and reporting summary statistics for the whole sample. In addition, researchers can change identifiers like specific occupations or industries and names of cities, counties, states, and organizations when presenting study results.

Anonymity and confidentiality are at the very heart of ethical research. One way to ensure the security of participant data is to destroy all identifying information as soon as possible. Such a step, however, eliminates the possibility of doing future research about those participants because collecting additional data from the same participants means being able to identify them. If a researcher wants to keep in touch with study participants over a number of years, then records of names and ways to contact them must be kept. However, keeping identifying information means that confidentiality could be at risk because it is possible through subpoena to link a given data record to a specific person (Laumann, Gagnon, Michael, and Michaels, 1994: 72).

One very controversial study, *Tearoom Trade*, by Laud Humphreys (1975), illustrates several ethical principles in doing research with human subjects—perhaps more by violation of than by compliance with the principles. This well-known study gives an account of brief, impersonal, homosexual acts in public restrooms (known as "tearooms"). Humphreys was convinced that the only way to understand what he called "highly discreditable behavior" without distortion was to observe while pretending to "be in the same boat as those engaging in it" (Humphreys, 1975: 25). For that reason, he took the role of a lookout (called the "watchqueen"), a man who watches the interaction and is situated so as to warn participants of the approach of anyone else.

Humphreys observed in restrooms for more than 120 hours. Once, when picked up by the police at a restroom, he allowed himself to be jailed rather than alert anyone to the nature of his research. He also interviewed 12 men to whom he disclosed the nature of his research and then decided to obtain information about a larger, more representative sample. For this purpose, he followed men from the restrooms and recorded the license plate numbers of the cars of more than 100 men who had been involved in homosexual encounters. Humphreys obtained the names and addresses of the men from state officials who were willing to give him access to license registers when he told them he was doing "market research." Waiting a year, and using a social health survey interview that he had developed for another study, Humphreys and a graduate student approached and interviewed 50 men in their homes as if they were part of a regular survey. Using this deception, Humphreys collected information about the men and their backgrounds.

Humphreys viewed his primary ethical concern as the safeguarding of respondents from public exposure. For this reason, he did not identify his subjects by name and presented his published data in two ways: as aggregated, statistical analyses and as vignettes serving as case studies. The publication of Humphreys's study received front-page publicity, and because homosexual acts were against state law at the time, he worried that a grand jury investigation would result. Humphreys (1975: 229) spent weeks burning tapes, deleting passages from transcripts, and feeding material into a shredder and resolved to go to prison rather than betray his research subjects. Much to his relief, no criminal investigation took place.

STOP & THINK
Review the principles of ethical research that have been discussed in this chapter. Before we tell you what we think, decide which of the principles you believe Humphreys violated. Make some guesses about why he behaved the way he did.

Two principles that Humphreys violated are voluntary participation and informed consent. By watching the men covertly, tracking them through their license plates, asking them to participate in a survey without telling them its true purpose, and lying to the men about how they were selected for the study, Humphreys violated these principles several times over. After the study's publication, critics took him to task for these actions.

Humphreys's work provided insights about sexual behavior that had received little scholarly attention and told about groups that were often targets of hostile acts. Because of the usefulness of the work and his observational setting, Humphreys maintained that he suffered minimal doubt or hesitation. He argued that these were *public* restrooms and that his role, a natural one in the setting, actually provided extra protection for the participants (Humphreys, 1975: 227). But, reevaluating his work in the years after the study's publication, Humphreys came to agree with critics about the use of license plate numbers to locate respondents and approach them for interviews at their homes. He decided he had put the men in danger of public exposure and legal repercussions. We will never know how many sleepless nights and worry-filled days the uninformed and unwilling research subjects had after the publication of and publicity surrounding Humphreys's book. However, no subpoena was ever issued; the data remained confidential, and no legal action was taken against the subjects.

Decades later, the Humphreys study continues to be a source of debate about where to draw the line between ethical and unethical behavior. Although many of its substantive findings have been forgotten, Humphreys's work has become a classic because of the intense debate about ethics that it generated. Discussing the choices that researchers like Humphreys made can help others to understand more clearly the legal and ethical issues implicit in doing social research. Debate about the costs and benefits of deception continues. For a less common point of view, see Box 3.6.

BOX **3.6**   Another Point of View

In "The Ethics of Deception in Social Research: A Case Study," Erich Goode argues a different point of view:

[P]rotecting our respondents from harm is the researcher's *primary* interest ... [but] I intend to argue that it *is* ethical to engage in *certain kinds* of deception ... the ethics of disguised observation [should be] evaluated on a situational, case-by-case basis.... In *specific* social settings, some kinds of deception should be seen as entirely consistent with good ethics....

I strongly believe every society *deserves* a skeptical, inquiring, tough-minded, challenging sociological community. It is one of the *obligations* of such a society to nurture such a community and to tolerate occasional intrusions by its members into the private lives of citizens.... I have no problem with the *ethics* of the occasional deceptive prying and intruding that some sociologists do. I approve of it as long as no one's safety is threatened. (1996: 14 and 32)

Ethical Issues Concerning Colleagues, the Community, and the General Public

STOP & THINK *Assume that you have designed a study and you feel confident that the participants will face minimal risks. Can you think of any other ethical responsibilities that should concern you?*

In addition to treating research participants ethically, researchers have responsibilities to colleagues, the community, and the public at large. Soliciting input from groups involved in the research topic, such as local advocacy groups or community leaders, is an important part of ethical behavior (Dickert and Sugarman, 2005). At the end of a study the researcher can partner with community groups to use the findings to promote better conditions. For example, after studying mental health services in Canadian immigrant communities, researchers presented the concerns of the participating communities about mental health services to community service providers, policy makers, and other decision makers (Malter, Simich, Jacobson, and Wise, 2009: 320). Similarly, in Chapter 14, you'll read Harrison and Norton-Hawk's assessment of a prison in Ecuador. After completing their research, they met with prison officials, quasi-governmental organizations, and medical personnel to help improve the lives of the inmates.

honest reporting, the ethical responsibility to produce and report accurate data.

A central ethical principle is **honest reporting**, which is the responsibility to produce accurate data, report honestly, and disseminate it in both professional and community forums. Recent scandals in the biological sciences and a survey of scientists reporting that a significant number anonymously admitted to committing some form of scientific misbehavior (Singer, 2005) indicate that dishonest reporting might be more widespread than previously thought. Important responsibilities to colleagues, the community, and the general public include presenting data without plagiarism, fraud, falsification, distortion, or omission, disclosing the research's limitations, and identifying hypotheses not supported by the data.

Actions of an unethical researcher can affect the reputations of other social scientists and the willingness of potential research participants in the future, so acting ethically is part of our responsibility to colleagues. Research funding can be a source of concern as funding agencies and sponsors might influence research agendas and methodologies by making awards on the basis of certain criteria. In addition, they can limit the freedom of researchers by imposing controls on the release of research reports and limit public disclosure of findings. A researcher hired and paid by an organization can face expectations to publish findings that support its goals. Criminologist Charles Thomas, for example, studied the impact of privatizing prisons while employed by the Florida Correctional Privatization Commission. After reporting results exaggerating the positive impact and reduced recidivism rates of juveniles leaving private versus public facilities, he was criticized by other criminologists for acting unethically and producing tainted results because of his funding source (Fogel, 2007: 114).

Sometimes a conflict of interest exists between the aims of the study's sponsor and those of the larger community. For example, in one city, a team of researchers representing a city planning commission, a community welfare council, a college, and several cooperating organizations, including local Boy Scout officials, studied the metropolitan area's social problem rate. At the conclusion of the study, the planning commission, the welfare council, and a local council of churches attempted to suppress the research findings that did not fit their groups' expectations or images. The researchers had to press for full dissemination of the findings to the public at large (Gorden, 1975: 143–144).

ETHICAL DILEMMAS DURING AND AFTER DATA COLLECTION

> Who owns the data? ... Whose interpretation counts? Who has veto power? ... What are the researcher's obligations after the data are collected? Can the data be used against the participants? Will the data be used on their behalf? Do researchers have an obligation to protect the communities and the social groups they study or just to guard the rights of individuals? (Marecek, J., M. Fine, and L. Kidder. 1997. Working between worlds: Qualitative methods and social psychology. *Journal of Social Issues* 53: 631–645)

Most examples of unethical behavior in the social and behavioral sciences research literature are rooted in ethical dilemmas rather than in the machinations of unethical people. Many of the dilemmas involve *conflicts*—conflict between ethical principles, between ethical principles and legal pressures, or between ethical principles and research interests.

Conflict Between Ethical Principles

In some research situations, researchers encounter behavior or attitudes that are problematic or morally repugnant. Although ethical responsibilities to subjects might prohibit us from reporting or negatively evaluating such behavior, other ethical principles hold us responsible for the welfare of others and the larger society. James Ptacek encountered an example of this ethical dilemma in his study of wife batterers. He recruited his sample from a counseling organization devoted to helping such men, paid each participant a nominal fee, and told the men that participation would help the organization and that the interview was not a formal counseling session (Ptacek, 1988: 135). He used a typical sociological interviewing style—staying dispassionately composed and nonjudgmental even when the men described bloody assaults on women. Even though Ptacek's goal was to facilitate their talking rather than to continually challenge the men, he did worry about the moral dimension of his impartial approach. In the three interviews where men reported ongoing if sporadic violence, Ptacek switched from an interviewer role to a confrontational counselor role after the formal questions had been completed. He noted, "At the very least, such confrontation ensures that among these batterers' contacts with various professionals, there is at least one place where the violence, in and of itself, is made a serious matter" (Ptacek, 1988: 138). Facing a conflict between ethical principles, Ptacek gave a higher priority to the elimination of abuse than to the psychological comfort of study participants but others may have made different choices. In contrast, Yassour-Borochowitz (2004) interviewed a group of men who had abused their partners but worried about the ethics of the pain and hurt that she caused her subjects by talking to them about their behavior.

STOP & THINK   *What would you do if someone you were interviewing told you about illegal behavior they'd engaged in? What about if they were planning self-destructive acts or to hurt others? Would you keep such information confidential or report it to someone?*

Conflict Between Ethical Principles and Legal Concerns

Sometimes during the course of research, observing others or interviewing study participants, the researcher finds out about illegal behavior. In his study of the nighttime economy in Manchester, England, David Calvey (2008) took the role of a "bouncer" in a club and hid his role as a researcher. He observed drug taking, violence, and the withholding of information from the police without comment. He was able to keep the data confidential but knew that his situation might be problematic and that the encounters might lead to a difficult legal position. He felt fortunate because he felt he was doing "fingers crossed" research where his luck could have run out (Calvey, 2008: 913).

Counting on luck isn't wise as researchers can face pressure or feel compelled to violate confidentiality. Although ethical responsibilities to subjects prohibit the reporting of what study participants tell us, other ethical principles hold us responsible for the welfare of others and to the larger society. If told about someone planning to hurt themselves or others, we would need to consider our ethical and legal responsibilities. For example, in 18 states, studying a topic such as child abuse or neglect means a researcher will face the legal requirement that anyone having reason to suspect maltreatment must report it to the authorities (Child Information Gateway, 2011). This has significant implications for the informed consent process, as there are consequences of participating, including legal risks for subjects and legal liabilities for investigators. Some researchers have tried to avoid a legal obligation to report suspected child maltreatment by collecting data anonymously, but others would argue that ethical obligations remain. The ASA's Committee on Professional Ethics has considered the choice that must be made if, for example, a researcher who gets permission from a family to observe their interaction discovers child abuse. Should the child abuse be reported and the confidentiality that might have been promised to the parent be breached, or should the confidentiality be maintained and risk future harm to the child? The ASA (1999: 12) and other organizations conclude that in these cases, sociologists must balance guarantees of confidentiality with other principles in the Code of Ethics, standards of conduct, and applicable law.

Even if the researcher decides to keep data confidentiality, there may be legal pressures. Russell Ogden, a graduate student studying assisted suicide, was asked by a Vancouver coroner to turn over research materials.

Ogden used a common-law claim of privilege and became the only researcher in Canada to win privilege and avoid contempt of court (Palys and Lowman, 2006). In the United States, social scientists do not have the right to refuse to turn over materials by claiming privileged relationships with participants. For example, in 2012, a U.S. federal appeals court upheld a lower court's ruling ordering Boston College to turn over materials, including recordings and transcripts, to British authorities from its Belfast Oral History Project on Northern Ireland even though the researchers had promised former IRA members confidentiality (Anderson, 2012).

Not all researchers comply with such subpoenas. Sociologist Rik Scarce (1995) went to jail to protect the confidentiality of participants in his research. After completing a study of radical environmental activists, Scarce remained in contact with some of them. At one point, one activist, Rod Coronado, stayed at Scarce's house for several weeks while Scarce was away. The time in question included the evening that the Animal Liberation Front took an action resulting in the release of 23 animals and $150,000 in damages at a local research facility. Scarce was subpoenaed to testify about Coronado before a grand jury. Even after being granted personal immunity, Scarce (2005: 16) refused to answer some questions because they called for research information he had by virtue of confidential disclosure. The district court judge held Scarce in contempt of court, ruling that if the government has a legitimate reason for demanding cooperation with a grand jury, a social scientist cannot claim academic privilege; the appeals court upheld the ruling. After Scarce had spent 159 days in jail, the judge decided that further incarceration was not likely to pressure him into giving grand jury testimony and released him (Scarce, 2005: 207).

Situations involving subpoenas are rare but it is important to be aware of their potential so that researchers can specify the limits of confidentiality. For example, in a study of parents and children in 70 households, the researchers indicated the legal limits of confidentiality in the informed consent form and in discussion with participants. During the course of the study, because of serious concerns about the welfare of the children in two households, the researchers made two referrals to Children's Services (Gorin, Hooper, Dyson, and Cabral, 2008).

At the request of the IRB at Brown University, Ciambrone included in her informed consent form the statement that all information provided by interviewees would be kept confidential to the extent permitted by law. She tells us that she was fortunate that no interviewee told her anything of concern, so that she did not have to consider breaking confidentiality.

Conflict between Ethical Principles and Research Interests

STOP & THINK   *You're thinking of a research project that involves studying people's behavior by observing them. You wonder what the effect of asking for permission will be. Could there be situations where it might be better for the outcome of the study if researchers don't ask for informed consent?*

What if you want to study powerful organizations? And what if you want to study something that organizations do not want you to study? Would you be tempted to be deceptive to open closed doors? Robert Jackall (1988) was turned down by 36 corporations when he asked for permission to study how bureaucracy shapes moral consciousness and managerial decisions. His analysis of the occupational ethics of corporate managers was made possible only when he used his personal contacts to gain access and recast his proposal using "bland, euphemistic language" (Jackall, 1988: 14). He was able to do field work and interviews in a chemical company, a defense contractor, and a public relations firm because the companies thought his interest was in issues of public relations. Jackall completed an analysis of the "moral mazes" navigated by executives only after he made a few twists and turns of his own. His work answers some questions and raises others: Can the knowledge gained by a study outweigh the ethical costs? What about choosing between using ethically compromised methods and not doing the research? We must ask at what point the researcher should refrain from doing a study because of the ethical problems. Did Milgram's work contribute so much to our understanding of human nature that the violations of ethics are worthwhile? How important were Humphreys's contributions? At what point should the Tuskegee study have been stopped?

Blee and Currier (2011: 401) ask us to consider "the ethical stakes of what is *not* studied, the questions that are *not* asked, the social groups and communities that are *not* the subject of research." Sociologist Howard Becker concurs. Thinking back over his long career, he notes that had such approval been required, many of his projects would not have been approved as "I could not have studied musicians or medical students or college students—because all that work involved me and my colleagues just hanging out with a lot of people to whom we never could have introduced ourselves and recited the proper warnings" (Becker, 2004: 416).

In future chapters we'll discuss issues of study design and methods of data collection and see how even the most benign of studies can run into ethical challenges. At this point, we'll present a few examples of potential conflicts between methodological and ethical considerations.

Dilemmas can occur when recruiting a sample of participants for a study if disclosing the specific focus of a project will lead to many more refusals to participate than would a more general description of a study. Ruth Frankenberg (1993), for example, originally told people, accurately, that she was "doing research on white women and race" and was greeted with suspicion and hostility. When she reformulated her approach and told white women that she was interested in their interactions with people of different racial and cultural groups, she was more successful in obtaining interviews.

In observing groups or organizations meeting in private settings, it is ethical to ask for permission of those you'll be observing. If, however, those being observed are likely to change their behavior if they know about the observation, then you will need to make a choice. Judith Rollins (1985) faced this issue and decided not to disclose her identity as a graduate student researcher when she also worked as a domestic worker and collected data about her interactions with her employers. Rollins acknowledged feeling deceitful and guilty, especially when her employers were nice. However, she feels that her behavior was acceptable because there was no risk of harm or trauma to the subjects.

Matthew Lauder also feels that deception can be ethically defensible. In a study of a Canadian neo-National Socialist organization that had a history of racism and using violence, Lauder (2003) first tried collecting information by interviewing people using informed consent. But many group members refused to sign informed consent forms and his formal interviews produced little useful information about the anti-Semitic and religious character of the organization's ideology. Rather than abandoning the project, Lauder decided to change tactics. Using deception, he moved from an overt to a covert role by pretending to have converted to the group's worldview (Lauder, 2003: 190–191) and then by talking to members informally. He collected a great deal of information covertly and protected confidentiality when using the data. Lauder feels that he balanced the needs of society against the needs of individuals and concluded that he was morally obligated to conduct the study.

Gerontologists in a Harvard-led research project also put research interests before ethics in a study of preventing hip fractures among nursing home residents, resulting in an order from Federal regulators to develop corrective action. Studying whether hip protectors (underwear that is padded around the hips) decreased hip fractures, the researchers used protectors that were padded on only one side. Their research design called for recording differences in fractures on the protected versus unprotected side. While the participants had given initial consent to be in the study, as new participants were enrolled, data increasingly indicated that because it was padded on only one side, the garment itself was causing instability and increasing the likelihood of falls. However, the research continued and participants were not warned of this hazard (Lazar, 2012).

Others come down on the side of ethics over research considerations. Ward and Henderson (2003) conducted a long-term study of young people leaving the care of state service agencies. Noting that study participants had faced painful childhood experiences and emotional upset, they felt some were too vulnerable to participate in the follow-up interview; they gave up research data because of their responsibility to not harm participants by asking questions about sensitive issues (Ward and Henderson, 2003: 258).

ETHICS AND PUBLISHING RESULTS

When a study is completed and the results are made public, if there is a possibility that the participants can be identified, the researcher faces another dilemma. Published material can have an impact on the study's participants. Clark and Sharf (2007), for example, cite several researchers who made decisions to omit material from their publications because they felt they might be harmful to their research participants. Other researchers make different decisions. When Carolyn Ellis studied isolated fishing communities, she didn't worry about how the people would respond. She thought, "They're illiterate; they'll never read this

BOX **3.7**	Interviewing and Confidentiality by Sandra Enos*

In 1979, Rhode Island closed the state orphanage that had provided institutional care for thousands of dependent and neglected children since 1885. In 2002, an interdisciplinary team began a study of the state's experiment to provide care at the State Home and School for children who otherwise would have been placed in local and state almshouses or in private orphanages.

Typically in sociological research, data are aggregated and presented statistically or pseudonyms are used. Because we wanted to use historical documents and excerpts from some of the audio- and video-taped interviews in publications and public programming, we faced particular challenges concerning confidentiality. Our solution, approved by a local IRB, was to create a declaration of participants' rights and a disposition form with options. We allowed the former residents in our sample to review and edit transcripts, to decide whether to have some or all of their

information closed to public view and to have it released at a later date, such as after their deaths.

We were able to balance our research goals with participant rights. Participation in our study was voluntary and all members of the sample controlled the privacy of their data. As we expected, the process of being interviewed about this time in their lives was hard for some participants. Some had never before told the story of their years at the State Home. We found that residents who came to the State Home in the 1920s through the 1940s thought it provided them a good and safe place to live; those who were residents in the 1960s and later were critical of the care they received. The result of our research is the audiobook *Let Us Build a Home for Such Children: Stories from the State Home and School.*

*Sandra Enos, Ph.D., is an Associate Professor of Sociology at Bryant University.

work," but found out later that it was painful for many members of the community when they learned what she had written (Ellis et al., 2008: 271).

It is better for researcher to anticipate ethical problems than to face difficulties at publication. Sandra Enos and a group of researchers studying an orphanage thought ahead about the way they wanted to present their results and thereby saved themselves from breaking confidentiality (see Box 3.7).

MAKING DECISIONS TO MAXIMIZE BENEFIT AND MINIMIZE RISK

Although social scientific research does not usually place subjects in situations that jeopardize their health and well-being, most social research involves some risk. Discomfort, anxiety, reduced self-esteem, and revelation of intimate secrets are all possible costs to subjects who become involved in a research project. No investigator should think it possible to design a risk-free study, nor is this expected. Rather, the ethics of human subject research require that investigators calculate a risk–benefit equation by seeking to balance the risks of subjects' involvement against the possible benefits of a study for the individual and for the larger society.

Many colleges and universities have research review boards and most professional organizations have codes of ethics. The guidelines they provide can make us aware of the ethical issues involved in conducting a study and offer advice for appropriate behavior. Ultimately, however, the responsibility for ethical research lies with each individual researcher.

To make research decisions, it's wise to use separate but interconnected criteria focusing on the ethics, practicality, and methodological appropriateness of the proposed research. In selecting a research strategy, we should ask these questions: Is this strategy practical? Is it methodologically appropriate? Is it ethically acceptable? When we answer "yes" to all three questions, there is no dilemma. But if we feel that it will be necessary to sacrifice one goal to meet another, we must determine our priorities. As Bruce Macfarlane (2009: 32) points out in his book on researching with integrity, codes of ethics operate through the researcher's value system where practice might be "the 'least bad' course of action rather than any ideal."

Think about the consequences of doing a given study as opposed to *not* doing a study and consider *all* options and methods to find research strategies that balance ethical and practical concerns with the likelihood of obtaining good quality data. Each researcher must balance the rights of study participants against a desire for research conclusions that can be published with confidence.

SUMMARY

Even though researchers often face unanticipated situations, most aspire to conduct their studies ethically and to apply the values and standards of ethical behavior to all stages of the research process. Recognizing the need to use feasible and methodologically acceptable strategies, researchers strive for appropriate and acceptable conduct during the planning, data collection, data analysis, and publication stages of research, as summarized in Box 3.8.

Being ethical in the treatment of study participants means collecting data anonymously or keeping identities confidential, providing adequate information about the study to potential participants, and, most important, working toward keeping negative consequences of participation to a minimum. Many ethical dilemmas can be anticipated and considered when planning a research strategy. IRBs and the codes of ethics of professional associations provide guidelines for individual researchers. When researchers face unusual ethical and legal dilemmas, they need to take special care in planning and conducting their studies.

BOX **3.8** Summing Up Ethical Principles

- Protect research participants from harm.
- Get informed consent.
- Be sure that study participants have not been pressured into volunteering.
- Collect data anonymously or keep data confidential.
- Submit the research proposal to a review board.

- Provide accurate research findings.
- Consider responsibilities to research participants, colleagues, and the general public at all stages of research including after project's completion and publication.
- Maximize benefits and minimize risks.

In addition to the treatment of participants, ethical principles require researchers to conduct themselves in ways that demonstrate that they are responsible to other scholars and to society as a whole. Scholars contribute an important service to their fields and the general public by providing and disseminating accurate information and honest analyses. Researchers will confront difficult situations if ethical principles conflict with each other, when legal responsibilities and ethical standards clash, or when collecting high-quality data means violating ethical principles. These situations must involve careful decision making. The best way to prepare for ethical dilemmas is to try to anticipate possible ethical difficulties and to engage in open debate of the issues.

EXERCISE 3.1

Knowing Your Responsibilities

All institutions that receive federal funds and most that conduct research have a committee or official group to review the ethical issues involved in studies done on campus or by members of the campus community. Called an IRB, a Human Subjects Committee, or something similar, the group must approve research involving human subjects before such research begins. Your college or university almost certainly has such a committee.

Find out if there is a committee on your campus that evaluates and approves proposed research. If there is no such group, check with the chairperson of the department that is offering this course and find out if there are any plans to develop such a committee; if they exist, describe those plans. If there is an IRB, contact the person chairing this committee and get a copy of the policies for research with human subjects. Answer the following questions based on your school's policies. Attach a copy of the policies and guidelines to your exercise.

1. What is the name of the review committee? Who is the committee's chairperson?
2. Based on either written material or information obtained from a member of the committee, specify the rules that apply to research conducted by *students* as part of course requirements. For example, does each project need to be reviewed in advance by the committee or is it sufficient for student research for courses to be reviewed by the faculty member teaching the course? If the committee needs to review all student projects, how much time would you need in advance of the project's start date to allot to the review process?
3. What are some advantages and disadvantages, in terms of your ability to collect data, of your institution's procedures for reviewing the ethics of student research?
4. Based on the information you've received from the committee, describe at least three of the ethical principles that are included in the standards at your school.

EXERCISE 3.2

Ethical Concerns: What Would You Do?

In the following examples, the researcher has made or will need to make a choice about ethical principles. For each situation, identify the ethical principle(s), speculate about why the researchers made the decisions they did, and then state whether you agree with their decisions.

A. Professor Ludwig's research question is whether there is a connection between experiencing family violence and academic achievement among students. He contacts the principal of a local high school who agrees to the school's participation. At a school assembly, the principal hands out Professor Ludwig's questionnaire, which includes questions about family behavior including violence between family members. The principal requests that the students fill out the survey and put their names on it. The following day, Professor Ludwig uses school records to obtain information about each student's level of academic achievement so that he can answer his research question. He keeps all information that he obtains confidential and presents only a statistical analysis of the data.

1. What ethical principle(s) do you think was (were) violated in this research?
2. Describe the probable reason(s) Professor Ludwig violated the principle(s) you noted in your answer to Question 1.
3. Describe your reactions to the way this study was done.

B. Professor Hawkins receives funding from both a state agency and a national feminist organization to study the connection among marital power, spouses' financial resources, and marital violence. Deciding to interview both spouses from at least 100 married couples, she contacts potential participants by mail. Because she is a university professor, she writes to each couple on university letterhead stationery and asks each couple to participate in her study. She specifies that she will be interviewing them about their marriage but does not state the specific topics that the interview will cover or the source of her funding. She specifies that everything the couple tells her will be kept confidential.

1. What ethical principle(s) do you think was (were) violated in this research?
2. Describe the probable reason(s) Professor Hawkins violated the principle(s) you noted in your answer to Question 1.
3. Describe your reactions to the way this study was done.

C. To study a radical "right-to-life" group, Professor Jenkins decides to join the group without telling the members that he is secretly studying it. After each meeting, he returns home and takes field notes about the members and their discussion. After several months of attending meetings, he is invited to become a member of the group's board of directors and he accepts. One evening at the group's general membership meeting, one member of the board becomes enraged during the discussion about abortion information and procedures available at a local women's health clinic. He shouts that the clinic staff deserves to die. Later that week, someone fires a gun through several windows of the clinic. Although no one is killed, several clinic staff and one patient are injured. The police find out that Professor Jenkins has been studying the local group and subpoena his field notes. Professor Jenkins complies with the subpoena, and the board member is arrested.

1. What ethical principle(s) do you think was (were) violated in this research?
2. Describe the probable reason(s) Professor Jenkins violated the principles you noted in your answer to Question 1.
3. Describe your reactions to the way this study was done and the researcher's response to the subpoena.

Planning a Research Project

CHAPTER 4

CHAPTER OBJECTIVES

This chapter is designed to help students

- **Develop the ability to create research questions**
- **Identify sources of research questions**
- **Consider the relationship between values and science**
- **Understand sources and impacts of research funding**
- **Learn to do a literature review**
- **Distinguish the parts of a research article**
- **Estimate feasibility (access, time, and costs) in planning a study**

CHAPTER OUTLINE

SOURCES AND EXAMPLES OF RESEARCH QUESTIONS

research questions, questions about one or more topics or concepts that can be answered through research.

In Chapter 1 we introduced the topic of **research questions,** questions about topics that can be answered through research. So where do these topics or questions come from? **Research questions** can come from our own lives, from social patterns we observe, or from significant world events. Developing research topics is not a new endeavor. In 1886, for example, after widespread rioting in London, sociologist Charles Booth published a paper reporting the results of his survey of poverty in London's East End (Bales, 1999). The press at the time was warning of revolution in the streets, but Booth argued that the working class was not well organized and did not pose a significant threat to the social order. His arguments for social reform to address the poverty he documented were widely reported in newspapers around the world. In Booth's work, we find an early example of how events can shape research and, conversely, how research can shape social and political responses to events and underlying social problems. Booth ultimately asked and answered three questions: How many people were living in poverty? What brought them to poverty and kept them impoverished? What could be done to alleviate the poverty? (Bales, 1999).

One hundred and thirty years later, other research questions illustrate the connection between social problems and social research. In 2005, Hurricane Katrina hit the United States. Nearly 5 million people lived directly in the hurricane's path and an additional 1.3 million lived in the New Orleans metropolitan area at the time (Hurricane Katrina Community Advisory Group, 2013). The suffering caused by this disaster motivated sociologist Ronald Kessler and others to find out about the survivors—the adequacy of their medical care, the long-term mental health consequences for them, and the extent to which survivors would rebuild their cities and their lives (Allen, 2006). Kessler and his colleagues asked a variety of research questions about topics such as the rates of hurricane-related mental illnesses including posttraumatic stress disorder, the extent to which PTSD onset was delayed, and the recovery time for illnesses (McLaughlin et al., 2011).

Research questions can be about local or global phenomena; can be about individuals, organizations, or entire societies; can be very specific or more general; can focus on the past, the present, or the future; and can ask basic questions about social reality or seek solutions to social problems. Research questions are similar to hypotheses, except that a **hypothesis** presents an *expectation* about the way two or more variables are related, but a research question does not. Both research questions and hypotheses can be "cutting edge" and explore new areas of study, can seek to fill gaps in existing knowledge, or can involve rechecking things that we already have evidence of. Whereas research projects that have explanatory or evaluation purposes typically begin with one or more hypotheses, most exploratory and some descriptive projects start with research questions.

hypothesis, a testable statement about how two or more variables are expected to be related to one another.

The American Sociological Association (ASA), founded in 1905, is a nonprofit organization dedicated to advancing sociology as a scientific discipline

BOX **4.1**   **Examples of Research Questions Asked in Papers Presented at Recent ASA Meetings**

Charles Gallagher (2012) asked about the perceptions in the U.S. and England about the recent rise in the number of immigrants. Do whites in both countries feel that immigration poses a serious threat to maintaining a shared common culture, compromises national security, or undermines employment opportunities? What social factors underlie hostility towards immigrants?

Eszter Hargittai and Eden Litt (2012) asked questions about Twitter. What are the predictors of using Twitter? How does its use relate to other online activities?

Barbara Risman and Rachel Allison (2012) asked about sexual double standards. Noting that the stigma attached to premarital sex is gone, does a double standard about casual sex or "hooking up"

exist on college campuses? Are gender and Greek affiliation associated with college students' sexual attitudes?

Grant Tietjen (2012) asked about the obstacles that ex-convicts face when trying to acquire and/or use higher educational credentials. Is there a bias against ex-con status during their academic careers, does it tend to decrease over time, and does type of crime and the class and race of the criminal matter?

Lisa Williams and Claudia Buchmann (2012) asked about school bullying. What is the impact of the school's bureaucratic structure? Does the racial composition of the student population and other populations such as the teaching staff affect who is bullied?

and serving the public good (ASA, 2013). With over 14,000 members, 20 percent of whom work in government, business, or nonprofit organizations, the ASA encourages professional interaction among members on a wide range of topics, including aging and the life course, animals and society, education, family, gender, and health. A few of the research questions covered in papers presented at a recent ASA conference are found in Box 4.1.

⬣ STOP & THINK   *We're willing to bet that by this point in your college career you've read dozens of articles and most likely some books that report the results of research. You can probably identify the research questions or hypotheses that the studies focus on. But try to recall if there was any information about the actual genesis of each project in the articles. That is, do you know why those particular social or behavioral scientists did those particular research projects at that point in time?*

Most research articles or papers, including the ones described in Box 4.1, usually include a review of the literature on the research topic and make connections between the research question(s) and at least some of the previous work in the field. However, research questions and hypotheses are usually presented as if they've emerged dispassionately simply as the result of a careful and thorough review of the prior work on a given topic.

As a result of the formal and impersonal writing style that is often used in academic writing, we often don't learn much about the beginnings and the middles of many projects. Although real research is "messy"—full of false starts, dead ends, and circuitous routes—we frequently get sanitized versions of the process. There are some wonderful exceptions,[1] but the "inside story" of the research process is rarely told publicly. More typically, written accounts of research projects are presented in formalized ways. Articles often include clearly articulated hypotheses, summaries of the work in the field that the study builds on, descriptions of carefully planned and flawlessly executed methods of collecting data, and analyses that support at least some of the hypotheses. Researchers will frequently comment on the limitations of their work, but only sometimes on problems encountered, mistakes made, and pits fallen into. Most articles in scholarly journals present the research process as a seamless whole. There is little sense that, to paraphrase John Lennon, real research is what happens to you when you're busy making other plans. Nor is there a sense that researchers are multidimensional people with passions, concerns, and lives beyond that of "researcher."

Because our primary goal is to explain research techniques and principles, we've tried to do more than reproduce conventional social science writing in the excerpts of research that we've included. To give you a realistic sense of the research process, we've mostly focused on small, manageable research projects of the kind being done by many social scientists. In each case, we wanted to get "the story behind the story." We have been most fortunate to have found a wonderful group of researchers, each of whom was willing to give us a candid account. From them, we learned about the interests, insights, or experiences that were the seeds from which their projects grew, the kinds of practical and ethical concerns they had at the start, and how "real life" shaped their data collections and analyses. As a result, in each chapter, we are able to share at least a little of "the inside scoop."

In the focal research that follows Michele Hoffnung tells about her study of women's lives with some details from the beginning, the middle, and the end.

FOCAL RESEARCH

Career and Family: From College Expectations to Late-Thirties Outcomes

By Michele Hoffnung[2]

The Research Questions

In 1992, I began a long-term study of women's employment and family choices. Several factors led me to undertake this research project. First and foremost, my personal experience as a woman committed to both career and motherhood gave me an understanding of what such a commitment entails. Second, my

---

[1]See, for example, some of the classics in social research including *Street Corner Society* (Whyte, 1955), *Sociologists at Work* (Hammond, 1964), and *Tell Them Who I Am* (Liebow, 1993) as well as the more recent *Moving Up and Out: Poverty, Education and the Single Parent Family* (Holyfield, 2002).

[2]This article was written for this text by Michele Hoffnung, PhD, Quinnipiac University, and is published with permission.

concern about the unequal status of women in the labor force led me to focus on the costs of the conflicting demands of employment and family for women. Third, the scarcity of theories and longitudinal research about women's adult development presented a challenge to me as a feminist researcher.

Even as a girl, I always knew I wanted children. My mother was one of eleven children; my father one of eight. While I was one of three children, I wanted to have *at least* four. I also expected to go to college. During my college years, I became very interested in psychology. My academic success led my professors to suggest graduate school. I went and loved it. At the University of Michigan I committed myself to an academic career and I also got married. My desire for motherhood remained unchanged; with my course work completed, I had my first child at the end of my third year of graduate school.

The first year of motherhood, I worked on my dissertation. Since my son was a good sleeper, my husband was willing and able to share infant care, and we were able to hire a student to help us for a couple of hours a day, I managed. But, the next year was far harder. Starting as a new assistant professor was more taxing and less self-scheduled than researching and writing a dissertation. I found balancing the new career and caring for the baby extremely difficult. I spent three days at school and four days at home with my son, writing lectures during nap time and late into the night. By mid-year I was exhausted. Luckily, I had an intersession in which to partially recover, so I survived.

What, I wondered, happened to women with less flexible jobs, husbands with less time or willingness to share family responsibilities, less money for outside help, and no intersessions? The social science literature focused on what women should be doing or feeling rather than describing what women *actually* did and felt. When my second child was out of infancy, I found the time to study mothers' lives. I conducted in-depth interviews with forty mothers of at least one preschool child and discovered that the mothers who were pursuing careers were more likely to plan family life with rigorous use of birth control, systematic preparation for childbirth, and careful selection of child care. They formed marital relationships with a less gender-stereotyped division of labor than the non-career mothers; their husbands shared parenting. I also found that all of the women were grappling with the issue of employment; even at-home mothers were concerned with when and whether they would be able to work again (Hoffnung, 1992).

The women I interviewed were products of the 1950s and 1960s when women with career aspirations were not common. What, I wondered, were the realities for women growing up in more recent decades? Studies done during the 1980s and 1990s indicated that a vast majority of college women wanted it all: marriage, motherhood, and career. While most college women say that they plan to have a career as well as a family, we have little data on how they will balance career and family. When Kristine Baber and Patricia Monaghan (1988) questioned 250 college women, they found that all planned careers, 99 percent planned to marry as well, and 98 percent also planned to be mothers. When asked how they would accomplish their goals, 71 percent said they planned to establish their careers before having their first child. The more career-oriented the women, the later they projected their first birth. The most common plan was to return to work part time after the child was at least a year old. We know that "combining" in this way typically has negative impact on women's career development. We also know that, regardless of their intentions, most women are employed these

(continues)

FOCAL RESEARCH *(continued)*

days, even mothers of preschool children. Many women have jobs rather than careers; others are underemployed given their talents and training, often because of their family commitments.

My experiences, previous research, and current interests led me to plan this long-term study of a group of young women. I thought that starting when they had plans to "do it all," and following them as they made life choices, would reveal patterns in their lives and the efficacy of various choices. For example, does the level of career commitment affect family choices? Does it affect work satisfaction? My research goal was to see how women's ideas and expectations change as they become more educated, face the world of graduate and professional school and employment, and make the choices of whether and when to marry, whether and when to have children, and whether and how to combine career and motherhood. These are important questions for our time, as most mothers work outside the home, even though the family continues to have primary responsibility for children and their care.

The Study Design

My earlier study was retrospective, with adult respondents recalling their childhood hopes and dreams; this was prospective, first asking young women still in the relatively sheltered environment of college to look toward their futures and then surveying them over time. My strategy was designed to avoid the problem of having respondents' memories altered by the choices they've actually made, but it required a long-term commitment with methodological challenges.

In 1992–1993, I conducted in-person interviews with 200 seniors at five New England colleges about their expectations and plans for the future regarding employment, marriage, and motherhood. Then I followed their progress for sixteen years, most years with a brief mailed questionnaire. During the seventh and fourteenth years of the study, I interviewed the participants by telephone. This regular contact helped me maintain a high response rate over the years. In 2009, I ended the study with the exceptionally high retention rate of 155 women, 77.5% of the original sample.

Collecting Data

My study required that respondents participate over many years, so I had to interest the women enough in my project to commit to responding annually. Because American young adults move frequently, I devised ways to contact them as they travelled and relocated. In this process, I saw my relationships with the women as more ongoing, personal, and collaborative than is typical of more traditional and impersonal survey procedures. At the same time, I worked to maintain a value-free, open-ended environment in which they could honestly tell about their lives.

A major methodological concern was to select a sample that included white women and racial/ethnic women from a wide variety of backgrounds. This was important because until recently, most social science research, even research on women, has focused on members of the white middle class. I wanted a sample that would be varied and accessible to me. I decided to select a stratified random sample of female college seniors from five different colleges. While my sample excludes the lower ranges of the socioeconomic spectrum, by including a small private college of mostly first-generation college students, a large state university, two women's colleges, and an elite private university, I reached women with a wide range of family, educational, and economic backgrounds. I obtained approval from my university's IRB and followed procedures required by the other colleges to gain access to the senior class lists, from which I picked random samples of both white and racial/ethnic women. Institutional approval gave me credibility when I called to ask the women to participate and allowed alumni associations to help me locate students in the years after graduation.

Since for the initial in-person interview I could expect students to come to me only if I made it convenient, I needed an office to do my interviews in private and a phone for people to call if they couldn't come. One college was my home institution, where I already had an office. I received a fellowship and office space at a research institute housed on another's campus. For the remaining campuses, I used contacts I had to find space. Once I solved the office problem, I called students, recruited, scheduled, reminded, and, if they did not show, rescheduled. Each interview took about an hour; many more hours went into phone calls and waiting for those who forgot or skipped their appointments.

I quickly learned to make reminder phone calls, and I always called and rescheduled interviews for those who missed an appointment. As a result, most of the random sample I contacted completed the interview. I suspect a couple of women finally showed up to stop my phone calls!

In the initial interview, I used a set of questions that was consistent from respondent to respondent, yet was flexible enough to allow respondents to explain themselves. I asked questions about their families of origin, attitudes about work, marriage, motherhood, and short- and long-term plans for the future. Because the women married and became mothers at their own pace, subsequent interviews took different forms with different participants. In the mailed surveys, I asked about relationships, life satisfaction, work, school, and some open-ended questions about aspirations, feelings, and concerns.

Findings

What have I found? (for fuller discussion, see Hoffnung & Williams, in press). As college seniors, these women said they wanted it all – career, marriage, and motherhood. By 2009 (average age 38) nearly everyone had married (91%). More are likely to marry, as single college-educated women have a 20% rate of marriage after age 40 (Isen & Stevenson, 2011). About 18% of the women had separated or divorced, but most of them had new partners. Almost two thirds (63%) were employed full time. About three quarters were mothers. Clearly career, marriage and motherhood continued to be important to these women college graduates.

About 57% of the women were combining full-time employment and motherhood. Still, that breakdown does not tell the full story, because almost all of them continued to want it all. Some of the women without children said they may have or adopt them. Lack of a suitable partner, reproductive difficulties, and delaying were the reasons most frequently given for not yet having children. Likewise, most of the mothers who were not full-time employed were still concerned about their careers. Most were employed part-time and had every intention of resuming or jump starting their career full-time, as soon as their children were older.

(continues)

FOCAL RESEARCH (continued)

Why did some mothers continue full-time employment and others not? Attitudes about being a good mother seemed to be a predictive factor. Full-time employed mothers had less anxiety about being separated from their children in order to be employed. But other forces seem to be at work as well. Employed mothers were also more likely to have spouses with less education than themselves. Either they had more often chosen partners who were more family than career-oriented, or life circumstance had led them to partners with less education than themselves which promoted their continuing in full-time employment. It is likely that the combination of personal preferences and situational factors varied among these full-time employed mothers.

Most of the women were overwhelmingly successful at doing what they had wanted to do. Sixteen years after college graduation, more than half were combining full-time career and motherhood, and another 21 percent were combining part-time employment and motherhood. One finding surprised me. I had expected that, despite the challenges, combining a full-time career and being a mother would lead to the highest life satisfaction because multiple roles provide multiple benefits. Instead I found that motherhood was the variable that made the biggest difference, as mothers scored significantly higher in life satisfaction than nonmothers, irrespective of employment status. This provides evidence that some roles are more important than others and that people look to their private lives to find happiness and buffer them from impersonal work roles. Career was highly valued, but it did not outweigh the benefits of family. Although motherhood is not the only family role, it is a compelling one. Not a single person regretted having children, although several were surprised at how much they enjoyed being mothers.

References

Baber, K. M., and Monaghan, P. 1988. College women's career and motherhood expectations: New options, old dilemmas. *Sex Roles* 19: 189–203.

Hoffnung, M. 1992. What's *a mother to do? Conversations on work and family.* Pasadena, CA: Trilogy.

Hoffnung, M., & Williams, M.A. (in press). Balancing act: Career and family during college-educated women's 30s. *Sex Roles.*

Isen, A., & Stevenson, B. 2011. Women's education and family behavior: Trends in marriage, divorce, and fertility. In J. Shoven (Ed.), *Demography and the economy* (pp. 107–140). IL: University of Chicago.

THINKING ABOUT ETHICS

Before considering other issues, let's see how Hoffnung did in making sure that her research was ethical. She submitted her proposal to her university's Institutional Review Board (IRB) where it was approved; study participation was voluntary for 17 years; she believes that sharing information about their lives has not harmed the participants; she assured her participants of confidentiality and has kept all materials confidential; when writing about her study, she uses aggregated data, such as percentages, and when talking about individual study participants uses pseudonyms rather than their real names.

In the Focal Research box, Hoffnung tells us the genesis of her study: her personal experiences and values, her choice of a professional specialization, and the lack of research and theory on women's lives when she began the study. The selection of a research question is often the result of many factors, including personal interests, experiences, values, and passions; the desire to satisfy scientific curiosity; previous work—or the lack of it—on a topic; the current political, economic, and social climates; being able to get access to data; and having a way to fund a study.

STOP & THINK   *Think about your life, the issues you're facing, the world today, and the concerns you have. Are there any questions you'd like to have answers to? This is one way to create a research question.*

THE IMPACT OF VALUES ON SCIENCE

During the nineteenth century and for a good part of the twentieth, it was commonly thought that all science was "value free," but today it is more widely believed that values—social, moral, and personal—are part of all human endeavors, including science. As science is located within its social and cultural context, group interests and values can subtly influence scientific investigations. These influences are especially important during the creation and evaluation of research questions and hypotheses.

Social and personal values do not necessarily harm science but researchers should be aware of their values and how they influence their work. Hoffnung, for example, knows that her values include a positive view of combining motherhood and career. Rose Weitz (1991) tells of beginning a study of people living with HIV or AIDS in the 1980s when she learned about a governmental policy she opposed—the mandatory reporting of people with these diseases.

The desire to produce knowledge is a social value. So is the desire to do accurate work and the belief that such work can ultimately benefit rather than harm humankind. We know that values like these contribute to the motivations and conceptual outlook of scientists. The danger comes when scientists allow their values to introduce biases into their work that distort the results of scientific investigation (National Academy of Sciences, 1993: 343). Articulating values can help others evaluate a researcher's work, including the choice of research questions and methodological decisions, and may make the researcher more aware and perhaps more objective as a result.

PERSONAL EXPERIENCES

STOP & THINK   *Compare your life today to your parents' lives when they were your age. How did they keep in touch with their friends? What are some of the differences in the technologies that you use that they didn't (such as cell phones, e-mail, Twitter, texting, blogging, and Facebook)? Think about the social and psychological impacts of the widespread use of the new technologies, and create a research question about one or more of these technologies.*

Researchers often undertake studies to satisfy intellectual curiosity and to contribute to society and to social science. However, personal interests and experiences can influence the selection of a specific research topic or question. It shouldn't surprise us that researchers often study topics with special significance for them. After all, research projects can take months or years to complete; having a strong personal interest can lead to the willingness to make the necessary investment of time and energy.

Hoffnung tells us how her personal experiences in juggling career and motherhood influenced her work. Jane Bock writes of conducting a study of single mothers by choice as a "midlife single woman who had long considered single motherhood but had not yet taken any action" (2000: 66). One important study came about through very difficult personal circumstances. In 1984, anthropologist Elliot Liebow learned that he had cancer and a limited life expectancy. Not wanting to spend what he thought would be the last months of his life working at his government job, Liebow (1993: vii) retired on disability but found himself with a great deal of time on his hands. He became a volunteer at a soup kitchen and later at an emergency shelter for homeless women and found he enjoyed meeting the residents. Liebow (1993: viii) was struck by the enormous efforts that most of the women made to secure even elementary necessities and he appreciated their humor and lack of self-pity. Liebow

BOX **4.2**   Reflections on "Studying Up"
By Jessica Holden Sherwood*

I wasn't one of those people who had a plan right from the beginning. Approaching my doctoral dissertation, I was like any student in need of a paper topic. At my university, graduate students must choose specialty areas within sociology. I picked social inequality, as in race, class, and gender. Often, studying inequality means studying minorities, the poor, and women. Missing from this picture is the other side of each equation: the white, the rich, and men. Sometimes I would read or hear a social scientist point out the importance of "studying up" as well as "down" regarding inequality. While these groups are not what we usually think of when we think of race, class, and gender, that's part of the point. If they *don't* get studied, we'll never have a complete understanding of inequalities—who maintains them, why, and how.

The call to "study up" or study the "unmarked" (the categories that seem standard and go unnoticed) resonated with me. My graduate experience at a large state university has been eye-opening in terms of my own position on the socioeconomic scale. Coming from an

upper-middle-class background, my socioeconomic position had been unmarked for me. Now I recognize its significance: first, that as a member of the unmarked group (like "white"), one of my privileges has been never to think about that group membership**; and second, that my life was organized in such a way (such as attending an expensive private college) as to preserve that obliviousness well into my twenties.

One reason that "studying up" is rare is that it's considered difficult. Especially when it comes to class, advantaged group members are thought of as elusive subjects of study. I thought that my own social location—being white, knowing some people who are rich and/or powerful, and having the cultural capital*** to interact with them effectively and comfortably— might enable me to overcome that elusiveness.

Another facet of my biography is that I have one Jewish parent and one WASPy or white Anglo-Saxon Protestant parent (WASP is a common label in the Northeast). So I have a tenuous relationship with the mythical "old-WASPy-Yankee" peoples. It is not volunteered for two years, but then, out of either habit or training, during the last three years of his work at the shelter, he collected data as a participant observer—talking to, observing, and taking notes about the women and their lives. Before his death, Liebow completed an important contribution to helping others, the thoughtful and sympathetic analysis *Tell Them Who I Am: The Lives of Homeless Women* (1993).

Other less dramatic examples of personal interests and experiences can be found in many research projects. Jonah Berger, co-author of a study on when losing leads to winning, said the idea for the project came to him when he was coaching youth soccer and "always felt like the kids worked harder when we were slightly behind at halftime" (Cammorata, 2009: C2). In Chapter 11, you'll read about Marci Cottingham's research about fans of the Pittsburgh Steelers football team. In graduate school in the Pittsburgh area, she was struck by the importance of the Steelers to so many people who lived there and decided to do research on the fans (Cottingham, personal communication). Emily's first major study was a study of the division of labor and the division of power in marriage. She began the study in the first few years of her own marriage, as half of a two-career couple. And, perhaps because both of us, Emily and Roger, are parents who have read hundreds of books to children over the years, we've produced, separately and together, more than a dozen studies on children's literature. In Box 4.2, Jessica Holden Sherwood recounts

hopelessly foreign and distant, but it's also not a circle I wholly belong to. So I am faced with the option of making efforts to belong to that circle, and also asking myself why I even want to. This tenuous relationship is probably one reason that I care about, and have chosen to study, this particular slice of social life.

Next, I needed a place to find these people, a context in which to study them. I attended an elite boarding school, and I considered doing research at that school or a similar one. But, I was more interested in the doings of adults than teenagers. I'm familiar with a small island nearby that's a destination for wealthy vacationers, and I considered studying its social life. But anybody can visit that island, and the wealthy vacationers don't all know each other, so it's not exactly a coherent community. I've settled on something that's convenient and concrete: a study of country clubs and other exclusive social clubs in one section of the Northeast. These private social clubs provide a specific context for the social lives of people who are "unmarked" and advantaged rather than disadvantaged.

I'm finding a unique blend of pros and cons as I do this research. On the one hand, I'm having a

perfectly smooth time securing interviews with the people who are members of the clubs. On the other hand, though I expect my dissertation to be critical, I'll hesitate to condemn the people I've studied, because some of them are neighbors and friends. But in any case, I'm pleased with the research topic I've settled on and feel it's enriched by my consideration of its connection to my biography.

*Jessica Holden Sherwood wrote this while she was a graduate student in the Department of Sociology, North Carolina State University. Her dissertation, *Talk about Country Clubs: Ideology and the Reproduction of Privilege*, can be found at http://www.lib.ncsu.edu/theses/available/etd-04062004-083555. This article is published with permission.

**Peggy McIntosh, 1992, "White Privilege and Male Privilege: A Personal Account of Coming to See Correspondences through Work in Women's Studies," in *Race, Class and Gender: An Anthology*, ed. Margaret Anderson and Patricia Hill Collins. Belmont, CA: Wadsworth.

***Cultural capital is an acquired cluster of habits, tastes, manners, and skills, whose social display often sends a message about one's class standing. From Pierre Bourdieu, 1977, "Cultural Reproduction and Social Reproduction," in *Power and Ideology in Education*, eds. Jerome Karabel and A. H. Halsey. New York: Oxford University Press.

her search for a research topic and how she came to study members of exclusive country clubs, a social world to which she had some connection.

> **STOP & THINK**   *There are frequent news stories about social phenomena such as an article about families dealing with parental unemployment, a graph showing changes in the crime rate, a story about using Facebook to encourage voting, or an analysis of the job markets for college graduates majoring in different fields. Do you have a personal interest in any of these topics? If you needed to design a research project for a class assignment, which of these or other topics might you select to study?*

RESEARCH AND THE SOCIAL, POLITICAL, AND ECONOMIC WORLD

Personal interests alone rarely account for the selection of a research topic. As knowledge is socially constructed and situated, social relations produce individual research projects and the enterprise of research itself. In fact, all knowledge claims "bear the fingerprints of the communities that produce them" (Harding, 1993: 57). As social, political, and economic climates change, the kinds of research questions posed and the ones that are of interest to the research community and the larger society also change.

Some topics have been largely invisible to the general public and generally ignored by social scientists until recently. In Chapter 12, we'll read about an interesting study that compares homicide rates of different groups. While in graduate school, one of the study's authors, Ramiro Martinez, was told that homicide was the leading cause of death among young African American males. But when he asked about the Latino-specific rate, he was told that this information was not available. This omission became one of the reasons he began his work (Martinez, personal communication).

Before the 1970s, few studies focused on women and, when considered at all, data about women were analyzed using male frames of reference (Glazer, 1977; Westkott, 1979).[3] However, as a result of the women's movement, which first developed outside of the university in the late 1960s and early 1970s, scholars began to study women and their lives. Hoffnung, for example, noticed that few of the long-term studies in her specialty, adult development, included women as subjects. She found herself perplexed that the theories that she had studied as a student focused on only men and their lives. In her professional work, she sought to redress the imbalance and began using new theories being developed by feminist thinkers such as Nancy Chodorow, Carol Gilligan, and Jean Baker Miller (Hoffnung, personal communication).

Critiques of gender imbalance are found frequently in social science. For example, in the mid-1990s, when three out of five American women were

---

[3]The natural, no less than the social, sciences offer many interesting examples of such male bias. In one federally funded study examining the effects of diet on breast cancer, only men were used as sample subjects (Tavris, 1996).

employed, Weaver (1994) noted that there were few studies focusing on women's retirement. Studies of the homeless have followed a similar pattern, with early efforts ignoring women's experiences (Liebow, 1993; Lindsey, 1997). Given the recent changes in the social climate, current research is much more likely to focus on both genders.

Research on other topics has also been affected by social changes. In 1995, Allen and Demo argued that sexual orientation was understudied, especially in research on families. Analyzing nine journals that published more than 8,000 articles on family research between 1980 and 1993, they found that only 27 articles focused explicitly on gay men, lesbian women, or their families (Allen and Demo, 1995). The data led them to conclude that sexist and heterosexist assumptions continued to underlie most of the research on families, with the research focusing mainly on heterosexual partnerships and parenthood. Allen and Demo (1995: 112) felt that studying lesbians and gay men as individuals, but not as family members, reflected a society-wide belief that "gayness" and family are mutually exclusive concepts. To see if this criticism was still valid, Emily did a search of *SocINDEX*, a database of scholarly sources that we'll talk about later in this chapter. Her search found citations for 650 articles about gay men and lesbians as family members published between 1993 and 2002 but over 7,000 published between 2003 and 2013. If this particular omission has been rectified, there will always be topics that need more research. Chito Childs (2005), for example, noticing that black women's voices were overlooked in research on interracial relationships, designed a study focusing on them; Hagestad and Uhlenberg (2005) point out that while ageism has been studied extensively, age segregation is a neglected topic; Fikkun and Rothblum (2011) argue that while research on weight-based stigma is readily available, much less attention has been paid to the disproportionate bias experienced by fat women.

Topics and issues typically become the focus of research when they become visible, more common, and/or a source of concern to the public or the scholarly community. We've seen, for example, more studies on the elderly as the proportion of the population older than 70 has increased. Similarly, as the number of children living in single-parent and in two-worker households rose, the impacts of family structure and workplace demands on children received more scholarly attention. Finally, consider the fact that in the U.S., state and local spending on criminal justice, including law enforcement, prosecution, courts and corrections, was $186.2 billion in 2006 (U.S. Department of Justice, 2012: 2)—and then note the multitude of articles on crime and criminal behavior that are published annually.

Social changes and current events can lead to specific research questions. In the next chapter, you'll read about research on cell phones done by Scott Keeter and his colleagues at the Pew Research Center for the People & the Press. They started this work because reporters and political observers were constantly asking about the impact of cell phone use on public opinion polling (Keeter, personal communication). Similarly, the dramatic increase in the use of social networking sites like Facebook has led to a variety of research projects including one covered by NBC News.

RESEARCH IN THE NEWS

In September 2012, NBC reported on an experiment using Facebook. On U.S. Election Day 2010, 61 million Facebook users saw a "get out the vote" message with pictures of six of their friends while others saw an identical message without the pictures. Both messages included links to local polling places and a clickable "I voted" button. The researchers conducting the experiment concluded that the message directly influenced 60,000 people to vote and indirectly influenced 280,000 more. James Fowler, lead researcher for the project, was quoted saying "A single message directly mobilized millions of acts of online political communication ... [and] people who saw the message were more likely to show up at the polls."

Source: http://www.msnbc.msn.com/id/49005643/ns/technology_and_science-science/t/
scientists-show-how-facebook-friends-can-get-out-vote/#.UQAwJvImSAo

Although the social and psychological aspects of natural disasters have been studied for decades, recent events are probably why research on the social dimensions of disasters has greatly increased in past few decades. A quick search of articles in *socINDEX* finds the number presenting research on the social impact of hurricanes, earthquakes, and tsunamis to have exploded in recent years: 20 articles between 1994 and 2003, but 408 articles between 2004 and 2013. As social problems come to the fore, social scientists are more likely to select them as research topics and to consider possible solutions.

SOURCES AND IMPACTS OF RESEARCH FUNDING

As one group of researchers has noted, "It is always difficult to raise money for social science research; after all, it neither directly saves thousands of lives nor enables one to kill thousands of people" (Fischman, Solomon, Greenspan, and Gardner, 2004: x). Even a small study can be expensive and the cost of doing research can run into hundreds, thousands, or millions of dollars, so funding is an important consideration. Many research projects, including those in the social sciences, are funded by federal agencies, private foundations, local and state institutions, or corporate sponsors.

National funding for behavioral and social science research has been significant since the middle of the twentieth century. In the United Kingdom, the government-supported Economic and Social Research Council, the country's leading social and economic research and training agency, had a total budget for 2012/13 of £205 million, which supported over 4,000 researchers and postgraduate students in academic institutions and independent research institutes at any one time (ESRC, 2013).

In the United States, social and behavioral research is funded by the National Institutes of Health (NIH), the National Science Foundation (NSF), the National Institute of Justice, the Department of Education, the National Institute on Aging, and other agencies. The NSF supports 57 percent of the academic basic research in the social sciences and for the past few years has provided over $250 million in funding annually to these fields (Silver et al., 2013: 206). The National Institutes of Health supports research on the roles that behavioral and social factors play in illness and health including projects such as studies on the mental health effects of Hurricane Katrina on the survivors (Allen, 2006) and research to understand and reduce health disparities including poorer health among racial/ethnic populations, lower socioeconomic classes, and rural residents (OBSSR, 2013).

Some private foundations offer support for the social sciences. The Foundation Center (2012) notes that in 2010, there were 1,194 grants of $10,000 or more given to social scientists for a total of $197,864.000. However, this amount was only 1 percent of the grant monies awards for all kinds of research. Some foundations support specific kinds of social research. For example, in 2005, the Robert Wood Johnson Foundation gave over $100 million in grant awards to support 329 projects involving research and evaluation in the areas of health and heath care with a focus on finding answers relevant to policy decisions (Knickman, 2005).

Funding research expresses a value choice—the judgment not just that knowledge is preferable to ignorance, but that it is worth allocating resources to research as opposed to other priorities (Hammersley, 1995: 110). The particular values associated with specific projects affect funding as the appropriations process for research is part of a larger political process. Each year, for example, after the president submits the budget, constituent institutes appear before appropriations committees to defend it. Interest groups, such as the Consortium of Social Science Associations, the umbrella organization of the professional societies in the behavioral and social sciences, testify as well. Research that is perceived to be threatening to those making decisions might have a more difficult time getting financial support (Auerbach, 2000).

Funding can be in the form of grants or in the form of a contract for an agency or sponsor. Grants typically allow the researcher more control of the topic, methods of data collection, and analysis techniques, whereas contracts usually commit the researcher to a specific topic, timetable, and methods. Although some sources are more restrictive than others, the availability of funding and economic support can influence a study, specifically the questions asked, the amount and kinds of data collected, and the availability of the resulting research report.

Research funding can be affected by political decisions and lobbying. For example, in the past, the Center for Disease Control (CDC) funded a great deal of research on gun violence. The funding peaked at about $2.6 million in 1996 and resulted in important studies, including ones that concluded that homicides were significantly more likely to occur in households with guns (Marshall, 2013). Giving in to pressure from the gun lobby, Congress cut funding for the agency and put legal restrictions on agency actions by specifying that "none of the funds … may be used to advocate or promote gun control"— effectively ending gun-related social research (Marshall, 2013). In 2013, President Obama reversed the policy by giving Kathleen Sebelius, secretary of the Department of Health and Human Services, the authority to direct CDC to "conduct or sponsor research into the causes of gun violence and the ways to prevent it" (Marshall, 2013).

Another example of the political nature of research can be found in the 2003 action of U.S. Representative Patrick Toomey, who sponsored an amendment to a bill that, had it succeeded, would have rescinded the National Institute of Health funding for five already approved grants on sexual behavior and health with goals such as preventing the spread of AIDS (Silver, 2006: 5). A survey of NIH researchers working on the grants found the political controversy led to a "chilling effect" with many researchers engaging in self-censorship practices, such as eliminating controversial language from grant applications, reframing studies, and removing research topics from their agendas (Kempner, 2008).

PRACTICAL MATTERS

Regardless of the source of an idea for a project—scientific curiosity, personal experiences, social or political climate, the priorities of funding agencies, or even chance factors—once a researcher has a topic, there's still work ahead.

STOP & THINK *Let's say you were interested in the topic mentioned earlier and created a research question about the social and psychological impacts of the use of technologies such as cell phones or Facebook. Where would you go from here?*

researchable question, a question that is feasible to answer through research.

Several steps are needed to turn a research question into a **researchable question,** a question that is feasible to answer through research. The first step is to narrow down the broad area of interest into something that's manageable. You can't study *everything* connected to cell phones, for example, but you could study the effect of these phones on family relationships. You can't study *all* age groups, but you could study a few. You might not be able to study people in *many* communities, but you could study one or two. You might not be able to study *dozens* of behaviors and attitudes and how they change over time, but you could study some current attitudes and behaviors. While there are many research questions that *could* be asked, one possible researchable question is: In the community in which I live, how does cell phone use affect parent–child relationships; more specifically, how does the use of cell phones affect parents' and adolescents' attempts to maintain and resist parental authority?

Reviewing the Literature

STOP & THINK *Let's assume that we've selected the impact of cell phones on family relationships as our research topic, and we want to begin a review of the literature. Where should we start?*

The accumulation of scientific wisdom is a slow, gradual process. Each researcher builds on the work of others and offers his or her findings as a starting point for new research. Reviewing previous work helps us to figure out what has already been studied, the conclusions that were reached, and the remaining unanswered questions. Although a thorough review is time consuming, it can generate a more cumulative social science, with each research project becoming a building block in the construction of an overall understanding of social reality. Reading, summarizing, and synthesizing at least a significant portion of the work that has preceded our own is an important step in every project. We can look for warnings of possible pitfalls and reap the benefits of others' insights.

literature review, the process of searching for, reading, summarizing, and synthesizing existing work on a topic or the resulting written summary of a search.

Because both the *process* of reviewing previous work and the resulting *written summary* of the work that has been reviewed are called a **literature review,** we'll discuss both uses of that term in the sections that follow.

Academic Sources

To start the *process* of reviewing the literature means figuring out what literature or sources you want to search. Books, articles, and government documents are among the most commonly used sources of research materials. Popular literature, including newspapers and magazines, might be good sources of research ideas, but academic journals will be more useful for your literature review.

A good place to start looking for books and published government documents is in your university's library. Many libraries allow others in addition to their patrons to search their catalogs online. If you find a book, article, or government publication that you want to use at a library other than your own, you can often request it through an interlibrary loan system. In addition, most government agencies have a great deal of material that can be downloaded from their websites. In addition to the sites of specific agencies, www.fedstats.gov provides links to information from more than 100 government agencies.

Hundreds of print and online journals publish research in the social sciences and related fields. Many of them are peer-reviewed journals, meaning that the articles have been evaluated by several experts in the field before publication. Some well-known sociology journals are *American Journal of Sociology, American Sociological Review, Crime and Delinquency, Gender & Society, Gerontology, Journal of Health and Social Behavior, Marriage and Family, Social Psychology Quarterly,* and *Sociology of Education.*

Almost every university library has electronic databases that allow you to search many journals or books simultaneously. Some databases list citations and abstracts and others provide full text versions. *SocINDEX* and *Sociological Abstracts* are the two major databases that index the international literature in sociology and related disciplines in the social and behavioral sciences. Other useful databases include *Abstracts in Anthropology, Academic Search, Criminal Justice Abstracts, eBook Collection, ebrary, Education Resources Information Center (ERIC), JSTOR, LexisNexis Academic, Criminal Justice Abstracts, PsycArticles,* and *PsycInfo.* You might also want to check out Google Scholar at http://scholar.google.com/, a free search engine that covers peer-reviewed papers, books, and articles from academic publishers, professional societies, and universities.

keywords, the terms used to search for sources in a literature review.

Although the databases vary, most allow you to search titles and abstracts only or to search the entire article using one or more **keywords,** which are the terms and concepts of interest. You'll generate numerous citations (title of the article, name of the author, name of the journal, volume, date of publication, and pages) with commonly used keywords. With common keywords, narrow the focus by using multiple keywords—such as by including "and" between the terms. For more unusual topics, think of broader terms or synonyms that could be used as keywords in additional searches. If you want articles that focus on research findings, use the word "research" and/or "study" as a key term in the search.

STOP & THINK   *If you were going to use a database to look for scholarly articles on the impact of the cell phone on family relationships, what keywords might you use?*

Using *SocINDEX* as our database, we searched for articles published between 2003 and 2013 by using the terms "mobile phone" or "cell phone" and "research" or "study," and searching the entire article record. We obtained 408 articles as results, but when we added other search words, such as "parent," "family," or "adolescent," the number of results dropped significantly. Once a list of citations is generated, the titles and abstracts can help you decide if the book or article is relevant to your work. In our search, we found some interesting results including a citation for "Communication Technology and Postdivorce Coparenting"—an article published in the July 2012 issue of *Family Relations*.

Most university libraries have many thousands of books, documents, and academic journals (in print, in microforms, or online). If a library doesn't subscribe to a specific journal or own the book, you can usually request a copy through interlibrary loan. When reading a journal article, you'll find that it will usually include a literature review with references to additional sources that you might want to check out. Box 4.3 describes the sections in a typical journal article, including the review of the literature.

A thorough literature search is a process of discovery. In addition to providing useful information, each article, book, or other item can direct you to new ideas and sources. A literature search is like taking a walk in the woods with only a general idea of the direction that you want to travel. For both you need a good compass and some common sense to find your way.

Using the Literature in a Study

When you read our focal research sections, you'll find that most of the researchers have used the existing literature to develop their own research questions or hypotheses. Some researchers, however, postpone a complete literature review until after data collection so that they can develop new concepts and theories without being influenced by prior work in the field. In both approaches, prior work in the field is used at *some point* in the research process.

Existing research helps put new work into context. What comes before determines whether a study is breaking new ground, using an existing concept in a new way, expanding a theory, or replicating a study in a new setting or with a new sample. When writing your own literature review, you will want to be up to date and thorough, but you will have to choose which sources to use and *give credit to*, using an appropriate method of citing sources. In writing a literature review for a research proposal or project, you should pick the

The title tells what the research is about. It usually includes the names of at least a few of the major concepts.

The **abstract** is a short account of what the research is about including a statement of the paper's major argument and its methods.

The **introduction** will introduce the topic, discuss previous work and, sometimes, present relevant theories. Part of the introduction is a **literature review** as the author discusses how this research fits with existing research and gives credit to those who have contributed ideas, theories, conceptual definitions, or measurement strategies.

Data and Methods sections describe and explain the research strategies and usually cover the population, sample, sampling technique, study design, method of data collection and measurement techniques that were used, and any practical and ethical considerations that influenced the choices.

Results or Findings are usually the point of the paper. In findings sections, researchers incorporate data or summaries of data.

The **discussion** section highlights the major findings and typically connects the findings to the literature review.

Conclusions usually sum up the current study, point out the directions for future work, and sometimes give cautions about the study's limitations.

References list the books, articles, and other sources used in the article.

materials that focus on the most relevant concepts, theories, and findings to provide a context for your own work.

You can use the literature review sections of published articles and research monographs as guides for writing your own literature reviews. A good literature review provides a perspective for the study you are proposing, in much the same way that a frame helps to focus attention on a painting. They can serve many purposes, among them a discussion of the significance of a research question, providing an overview of the thinking about a research question as it has developed over time, and presenting one or more theoretical perspectives on the topic. One endpoint of a literature review could be one or more research questions to be answered with data. In addition, a literature review can include definitions of useful concepts and variables, identify testable hypotheses deduced from prior research, and suggest some useful strategies and methodologies. Desirée Ciambrone, in her study of women with HIV/AIDS described in Chapter 3, found that doing a thorough literature review was one of the most important steps in the research process, as it honed her thinking on the topics she selected for her own interviews. For a discussion of writing a research report including some tips on writing the literature review, see Appendix C of this text.

STOP & THINK   *Let's assume that we've picked a topic and developed a research question about the effect of technologies, like cell phones, on family relationships. Let's also assume that as a result of the literature review, we have some ideas for a research strategy. Now, we need to see if our study is "doable."*

Even with a general idea of a research strategy, it's important to make sure that the strategy is feasible before beginning the actual data collection. No matter how important or interesting a question is, we don't have a researchable question unless it is feasible or practical to answer the question through research. **Feasibility** means considering three separate but related concerns: access, time, and money.

feasibility, whether it is practical to complete a study in terms of access, time, and money.

Access

Access is the ability to obtain the information needed to answer a research question. Although it's very easy to get some data, other kinds are much more difficult to secure. The interviewers in Robert Sampson's study on human development surveyed almost 12,000 Chicago residents, an undertaking that ran into some problems when a few interviewers were robbed at gunpoint, sexually harassed, or had to bribe doormen for access to residents closed off in high-rises (Parry, 2012). In her study of college seniors, Hoffnung had relatively easy access. Once her research proposal was approved by an IRB, Hoffnung was permitted to obtain a random sample of names and phone numbers of seniors. When contacting students on her list to request participation, Hoffnung could mention the university's "stamp of approval." Relatively easy access, IRB approval, and Hoffnung's personal and professional skills all contributed to the success of the project, as most of the women selected agreed to participate in the initial and later phases of the study. In a study of football fans that is the focal research in Chapter 11, Marci Cottingham found she could watch behavior at games, while tailgating, and in sports bars because of the public nature of the settings. For the "Research in the News" study about promoting voting, James Fowler used his introduction to Cameron Marlow, the director of a new "data science team" at Facebook, to create a project. As Facebook was already planning on a nonpartisan "Get out the Vote" campaign, Fowler's suggestion to modify it to become a massive controlled social science experiment was accepted (Bohannan, 2012).

access, the ability to obtain the information needed to answer a research question.

STOP & THINK   *College students are the subjects of many studies in the social and behavioral sciences, partly because of convenient access. Are there other groups or situations that you think are also easily accessible? Any that you think are particularly difficult to study?*

Most public information is easy to obtain and therefore practical to analyze. Typically, materials produced by government or public agencies and most media, such as newspapers, children's books, and television commercials, are quite easy to access. For example, if you were interested in comparing states by the percentage of residents of Hispanic or Latino origin, you would find this information very easy to obtain (see, for example, http://quickfacts.census.gov/qfd/index.html).

People and information that are connected in some way to organizations and institutions may be fairly accessible, although the researcher will often have to go through a **gatekeeper** to gain access or get permission to do a study. Some gatekeepers will deny access because they see the research as disruptive of

gatekeeper, someone who can get a researcher into a setting or can facilitate access to participants.

organizational or institutional activities or feel that participation would be risky for some reason. Examples of individuals who may be easily accessed through organizations once the gatekeeper agrees include employees of businesses, members of professional unions or associations, patients in a hospital, residents of a nursing home, college students, pet owners whose pets get care from veterinarians, and families whose children attend a private school. In each case, there is a place that allows us to easily meet or contact people to ask for their participation and an organizational structure that could facilitate the request. For Desirée Ciambrone, whose research was highlighted in the previous chapter, finding women with HIV/AIDS to interview was the most difficult part of her study. She found enough study participants only when the gatekeepers, her contacts at a local hospital, started referring women who were participating in an ongoing clinical trial of a medical treatment.

Sometimes gaining access to data can be quite challenging. (See Box 4.4 for some examples of difficult access groups and records.) In a study of institutional emergency plans and procedures, Rachel Filinson had easy access to a list of long-term care facilities but had great difficulty in getting the information she wanted from the facility directors (see Box 4.5). In a study of commercial fishing families, researchers mailed out 2,000 surveys but had only 43 returned. Phone calls inquiring about reasons for refusal prompted fishermen to reply in anger that the study was a waste of taxpayers' money (McGraw, Zvonkovic, and Walker, 2000: 69). In a study of Australian medical practitioners, researchers

BOX 4.4 Groups or Records with Difficult Access

Group or Record	Examples
People without an institutional connection	Homeless people
	Those not registered to vote
	Atheists
People who are not easily identifiable	Undocumented immigrants
	Italian Americans
	HIV-positive individuals
People who don't want attention from outsiders	Members of gangs
	Anonymous philanthropists
	The Amish
	Child abusers
People who are very busy or who guard their privacy	U.S. Senators
	Celebrities
	Chief executive officers
Records that are private or closed	Personal diaries
	Tax records
	Recent Census data for individual households

BOX **4.5**   **Access Denied**

By Rachel Filinson*

The impact of Hurricane Katrina demonstrated the lack of effective emergency preparedness by governments and institutions. The negative consequences were particularly pronounced for vulnerable populations such as the poor, minorities, disabled, and older adults. Because of the hurricane and its aftermath, students enrolled in a course entitled "Community" at Rhode Island College decided to study the level of disaster preparedness in nursing homes in RI.

Federal law requires Medicare- and Medicaid-certified facilities (which are virtually all long-term care facilities) to have written plans and procedures to meet all potential emergencies and to provide emergency training for employees. The student researchers wanted to compare the content of the emergency plans with the model recommended by the Office of the Inspector General after the devastating Gulf coast hurricanes. They obtained the list of the 86 facilities in Rhode Island from the Center for Medicare and Medicaid Services Nursing Home Compare website that provides a complete listing of Medicare/Medicaid certified facilities for each state. The students sent a letter to each facility director

explaining the study's purpose, assuring confidentiality, and offering to send a copy of the study's aggregated results. The letter further indicated that a student would be in contact with the facility to arrange retrieval of its emergency plan. Despite repeated efforts to procure the plans—and many promises by the facilities to provide them—only 26 of 86 facilities ultimately did so.

Perhaps the difficulty in obtaining the data was because the law does not require the plan to be publicly accessible even though it (the plan) is required by law. It may be that administrators had more pressing responsibilities or they worried about disclosing deficient plans. While our response rate was disappointing, we were able to compare the respondents with the non-respondents on a variety of characteristics documented on the Nursing Home Compare website. On these variables, we found no significant differences between the nursing homes that shared their disaster plans and those that didn't.

*Rachel Filinson, Ph.D., is the director of the Gerontology Program at Rhode Island College. This article is published with permission.

sent an online survey followed by two mailed questionnaires on injecting drug users to 609 registered pharmacotherapy prescribers; only 62 were returned—perhaps because the topic constitutes only a small portion of the physicians' busy workloads (Aitken, Power and Dwyer, 2008).

In some situations, gaining access is an ongoing process as people "check out" researchers, find out whether they are to be trusted, and then gradually let them into various parts of their lives and organizations. Jessica Holden Sherwood, in the research described earlier in this chapter, was interested in an elite group: members of exclusive country clubs. Beginning with a few personal acquaintances who were club members, she asked them to identify others and then had no trouble securing interviews, most likely because she always introduced herself using the names of others who were participating (Sherwood, 2004). In Chapter 12, you'll learn about a study of homicide by Ramiro Martinez and Matthew Lee. Martinez was introduced to the chief of police for Miami through a colleague of a colleague and eventually received permission to do the study (Martinez, personal communication). By spending countless hours accompanying detectives on calls and demonstrating that he was trustworthy, Martinez was able to gain their confidence and that of the administrators and, as a result, access to the data (Lee, personal communication).

Time and Money

Research can be time consuming and expensive. As discussed in the previous chapter, research with human subjects requires IRB approval, as does planning and conducting all parts of a project. For example, in a study to examine the circumstances of hospitalized children with complex health needs (Stalker, Carpenter, Connors, and Phillips, 2004), the researchers first obtained funding from a foundation and then ran into access problems. With a variety of health authorities involved, the complex hierarchy of gatekeepers made getting permissions take much longer than the researchers had anticipated.

Costs will vary depending on the method of data collection used. Studies that interview participants must budget for interviewers. Sometimes interviewees are paid or given gifts as compensation for their time and encouragement for participation. In the last chapter, we read of Ciambrone giving $10 gift certificates to each woman in her study. Compensation was more expensive in a study of academics and professionals in which participants were given $1,000 apiece to participate in several focus groups (Stenmark et al., 2010). If phone surveys are done, the researchers need to pay for interviewers and phone lines. As Scott Keeter and his colleagues at the Pew Research Center for the People & the Press found, calling cell phones involves additional cost, including the $10 that they gave to each respondent to defray the costs of receiving their calls. In a typical poll, the Pew Center contacts about 1,500 Americans, doing around 1,125 landline interviews and 375 cell phone interviews for a cost of between $75,000 and $80,000 (Keeter, personal communication).

If mailed questionnaires are planned, costs include paper, printing, and postage. If historical or archival documents are used, there could be copying costs or time expenditures. In the study using homicide records that you'll read in Chapter 12, the researchers needed to fly to where the records were, rent a car, and stay in hotels in that city. They even ended up buying their own copy machine because they found the one in the police department frequently out of order or unavailable (Martinez, personal communication).

Some projects can be completed quickly and very inexpensively. For example, the data collection time and costs might be negligible when using easily obtained public documents or books that can be borrowed from libraries. In the study of long-term facilities and their emergency plans, neither Filinson nor her students received financial compensation for their time, so the study's data costs were limited to funds for postage for the letters to facilities and phone charges for follow-up calls. Doing surveys where data is collected using self-administered questionnaires, especially if they can be handed out in groups, can be relatively inexpensive. Even less expensive than doing a survey is using survey data that were collected by others and are archived on a database such as the one maintained by the Inter-university Consortium for Political and Social Research (ICPSR).

Typically, most web-based information is easily accessible and inexpensive. For example, the Statue of Liberty-Ellis Island Foundation offers a great deal of information, including ship manifests and personal stories, about many of the millions of immigrants and passengers that came through Ellis Island from

1892 to 1924 through their site http://www.ellisisland.org/. Although there is no charge for the data, it can take many hours to find information about specific individuals.

Projects that collect data over time can be much more costly than those with only one data collection. Similarly, projects that study large numbers of people or groups are more expensive than those that study smaller samples. Researchers can sometimes rely on donated materials and volunteer workers to make a study feasible. Other times, unanticipated costs and events can lead to a scaled-down project.

research costs, all monetary expenditures needed for planning, executing, and reporting research.

Although each method of data collection has unique costs and time expenditures, we can identify some general categories of expense. **Research costs** include the salaries needed for planning the study; any costs involved in pretesting the data collection instruments; the actual costs of data collection and data analysis such as salaries of staff, payments to subjects for their time, rent for office space or other facilities; the cost of equipment (such as recording devices and computers) and supplies (such as paper or postage); and other operating expenses (such as paying for cell phone calls). **Time expenditures** include the time to plan the study, obtain approval, complete the data collection process, and organize and analyze the data.

time expenditures, the time it takes to complete all activities of a research project from the planning stage to the final report.

The focal research in this chapter is an example of an ambitious and a "doable" study. Hoffnung invested a great deal of her own time in the project over many years. By doing the interviewing herself, she was able to control the quality of the interviews and avoid the expense of hiring interviewers. She obtained small research grants to cover costs of postage, recording costs, and the largest expense, interview transcription. She kept costs to a minimum by employing work-study students as assistants for the follow-up questionnaires, by using her existing and donated office space, and by working on her personal computer.

SUMMARY

Research questions can vary in scope and purpose. Some projects are designed to address very broad and basic questions whose answers will help us describe and understand our social world more completely. Other projects work toward solving social problems or creating information that can be used directly by members of communities and participants in programs or organizations. Questions can focus on one or more concepts, can be narrowly or broadly defined, and concentrate on the present or on changes over time. Questions can be generated from a number of sources, such as values, interests, and experiences; the social, political, economic, and intellectual climates; previous research; or the ability to get funding. The availability of funding can determine not only the size and scope of a project, but also the specific question, the timetable, and the methodology.

A review of the literature is essential at some point in the research process. By reviewing the work that other researchers have done, each new study builds upon that work and provides a foundation for the next wave of data collection and analyses.

In order to plan a feasible study considering the practical matters of time, money, and access to data is critical. No matter how important or interesting a research question is, unless it is practical, a study will not be completed. Hoffnung gained access to a sample of women to interview and she kept in contact with most of them for 16 years. She had sufficient resources of time and funding to be able to continue to collect and analyze data so that she could answer the research questions she set for herself many years ago.

EXERCISE 4.1

Selecting a Topic

1. Get a copy of a local or regional newspaper or go to a newspaper's website. Find three articles that deal with social behavior or characteristics. List the name of the paper, the date of publication, and the headline for each of these articles.
2. Select one of the stories and make a copy to turn in with this exercise.
3. Create one research question on the topic of the article that could be answered by social research and state it.

4. Assuming you could get enough funding and had the time to do research to answer the question you stated, would you do the research? Describe why you would or would not do the research by focusing on your personal interests, today's social and political climate, and the contribution you think the research would make to social science or society.

EXERCISE 4.2

Starting a Literature Review

Select a topic you are interested in or one of the topics from Hoffnung's study, such as career aspirations, self-concept, two-career family, family planning attitudes, or role conflicts. Using your library's resources, find five citations of original social research from social science journals on this topic that are no more than five years old. (Be careful not to get an article that summarizes other research or is solely a discussion of theory.) Complete the following exercise:

1. Write down the topic you selected and the database(s) you used for your search.
2. List the five citations, giving full bibliographic information for each including name(s) of the author(s), title of the article, name of the journal, date of publication, and pages of the article.

3. Select one of the articles from your five citations that is available to you either online or in print and read it. Put an asterisk (*) next to the name of the article that you've read.
4. Describe the review of the literature that the researcher(s) presented in the article. For example, is the literature review extensive? To what extent does it focus on theories and theoretical perspectives? To what extent does it present the research findings from previous research? How adequate is the literature review in covering previous work on the topic?
5. Does (do) the researcher(s) start with a research question or hypothesis? If so, list one research question or hypothesis. Otherwise, list the topic of the study. Note the page number of the article where you found the research question, hypothesis, or the topic.

EXERCISE 4.3

Using the Internet and the World Wide Web

1. Select one of the following topics: two-career family, death penalty legislation, or unemployment. Write down the name of your topic.
2. Using a search engine on the web, such as Google or Yahoo, do a search for research on this topic by searching for an exact phrase that's of interest (such as "death penalty legislation") and the word *research*.

3. How many results did you obtain in your search? If you ended up with more than a few hundred items, narrow down your topic in some way. Describe what you did to narrow the search. Describe the number and quality of the results you obtained.
4. Check out at least three of the sites that you found in your search. Identify the website, describe the information on the site, and evaluate the usefulness of this information for you and/or other researchers.

EXERCISE 4.4

Summing Up Practical Matters

Think about the following research questions and the suggested ways that a researcher might collect data. For each, make an estimate of the cost and time and your evaluation of whether access will be easy or hard.

Research question	Will access be easy or hard?	Time consuming or not?	What are some expenses?
1. Do teachers' characteristics and teaching styles have an impact on the amount of student participation in class? (Suggested method: Observe 50 classes for several hours each.)			
2. What kinds of countries have the highest infant mortality rates? (Suggested method: Use government documents for a large number of countries.)			
3. Is the psychological impact of unemployment greater for people in their twenties or fifties? (Suggested method: Do a survey using a questionnaire e-mailed to 300 people who have contacted the unemployment office in one state.)			
4. Which regions of the United States have experienced the greatest population growth over the past 20 years? (Suggested method: Use data from the U.S. Bureau of the Census.)			
5. Are people who are more religiously observant more satisfied with their lives than those who are less observant? (Suggested method: Interview 100 people for approximately one hour each.)			

CHAPTER **5**

Sampling

CHAPTER OBJECTIVES

This chapter is designed to help students

- **Appreciate the reasons for sampling**
- **Understand the kinds of error associated with sampling**
- **Get a sense for the advantages and disadvantages of probability and nonprobability sampling**
- **Learn how to do five kinds of probability sampling**
- **Learn how to select four kinds of nonprobability samples**

CHAPTER OUTLINE

Have you ever asked yourself, "How do pollsters figure out what Americans feel about things like the economy, gay marriage, presidential candidates, and all those other things they tell us about on the news? They surely don't ask *all* Americans about their feelings, do they?" An important part of the answer to this question involves the topic of this chapter: sampling.

sampling, the process of drawing a number of individual cases from a larger population.

Sampling is the process of drawing a number of individual cases from a larger population. Researchers sample in the hopes that they can gain insight into a larger population without studying each member of the population. Presidential polls are based upon samples of the population that might vote in an election. We'd like to begin this chapter on sampling with three stories about presidential election polls (see Box 5.1): two that are examples of how misleading sampling can be and the other of how informative it can be.

STOP & THINK   *How much better were the 2012 presidential polls at predicting the winner of the presidential election than the 1936 and 1948 polls?*

Why did the *Literary Digest* poll, with two million respondents, do so poorly in 1936? Why did Gallup, with his apparently superior technique, miss

---

BOX 5.1   **Three Different Sets of Pre-Presidential Election Polling Results**

1936 *Literary Digest* Poll Predicts That Alf Landon Will Beat Franklin Delano Roosevelt, 57 Percent to 43 Percent

Based on a survey of over two million people, a survey that had predicted with "uncanny accuracy" (its own words), the outcome of the 1924, 1928, and 1932 presidential elections, the *Literary Digest* predicted that Alf Landon would win the 1936 election by a margin of 14 percentage points over then President Franklin Delano Roosevelt: 57 percent to 43 percent. Imagine the *Digest's* surprise when Roosevelt actually won—and won big: 61 percent to 39 percent.

1948 Gallup Poll Predicts That Thomas Dewey Will Beat Harry Truman

Using a technique called quota sampling, George Gallup correctly predicted that Roosevelt would win the 1936 election, even if his prediction that Roosevelt would receive 54 percent of the vote was somewhat wide of the 61 percent Roosevelt actually received. Gallup got it completely wrong in 1948, however, when he predicted that Thomas Dewey would beat Harry Truman in that year's presidential

election. Truman actually won. It was a very good day for Truman, but a very bad day for Gallup.

2012 Pre-Election Polls Say Obama Will Win Election

Based on an average of the nine national surveys taken between November 3 and November 5, 2012, the RealClearPolitics (2012) webpage predicted, the day before the election, that Barack Obama would win the 2012 Presidential election over Mitt Romney by a 48.8 percent to 48.1 percent margin. The predicted spread, of 0.7 percent, missed the real mark by 3.2 percent, as Obama actually won the general election on November 6 by 51.1 percent to 47.2 percent, a 3.9 percent spread. The RealClearPolitics result wasn't as close as it had been for the 2008 election, between Obama and John McCain, when it been off the final difference by only 0.3 percent. But it did predict the actual winner of the close 2012 Presidential election. Why was RealClearPolitics prediction so much better than the *Literary Digest* prediction in 1936 and the Gallup Poll prediction in 1948? The answer is improvements in the science of sampling.

the mark in 1948? And how was it possible that the RealClearPolitics average of the nine pre-election polls could come with 3.2 percent of the actual spread in the race between Obama and Romney in 2012? The main reason has to do with the topic of this chapter: sampling. The science of sampling has progressed dramatically since the 1930s, to the point where even a basic introduction must touch on some relatively technical matters. But the logic of sampling can be easily grasped, and we think you'll actually enjoy learning something about it. Let's start at the beginning by addressing a question that has probably already occurred to you: why sample?

WHY SAMPLE?

Sampling is a means to an end: to learn something about a large group without having to study every member of that group. The *Literary Digest*, George Gallup, and the authors of the 2012 polls mentioned in Box 5.1 wanted to know how Americans were going to vote for president. Why didn't the *Literary Digest*, Gallup, and the 2012 poll takers simply survey the whole of the American voting-age population? Put this way, one answer to the question, "Why sample?" is obvious: We frequently sample because studying every single instance is impractical or too expensive. The *Literary Digest*, Gallup, and 2012 poll takers couldn't afford to contact every American of voting age.

We sample because studying every single instance is beyond our means. Suppose, for example, you wanted to know how all Americans felt about something? Why not ask them all? The U.S. Census Bureau is, in fact, obliged by law to learn something about every American every 10 years. But it's very expensive. The 2010 census, during which 308,745,538 Americans were enumerated (U.S. Census Bureau, 2011), cost around $13.1 billion (Roberts, 2010), more than half of which was spent to employ the 635,000 temporary workers used to follow up on the (28 percent of) households that failed to return their forms by mail. Such costs, in terms of money and people, are beyond the means of even the wealthiest research organizations.

So we sample to reduce costs. But we also sample because sampling can improve data quality. Sandra Enos, whose study of how imprisoned women handle mothering we present in Chapter 10, wanted to understand women's approaches to mothering in an in-depth fashion, with a real appreciation of each individual woman's social context. Enos intensively interviewed 25 incarcerated women, not only because studying the whole population of such women would have been expensive, but also because she wanted to spend her time gathering in-depth information about each of her subjects, rather than more superficial data on all of the women in that prison.

In short, we sample because we want to minimize the number (or quantity) of things we examine or maximize the quality of our examination of those things we do examine.

STOP & THINK   *Using the logic of this section, when do you think sampling is unnecessary?*

We don't need to sample, and probably shouldn't, when the number of things we want to examine is small, when data are easily accessible, and when data quality is unaffected by the number of things we look at. For instance, suppose you are interested in the relationship between team batting average and winning percentage of major league baseball teams. Because there are only 30 major league teams, and because data on team batting averages and winning percentages are readily available in almanacs, you'd simply be jeopardizing the credibility of your study by using, say, 10 of the teams. So we would examine all 30 of them or, in other words, use the whole population.

We need to remind ourselves of some terms, and define a few others, before proceeding. First, we need a term to describe the kind of thing about which we want information—for example, a major league baseball team, an adult American of voting age, all Americans. As you learned in Chapter 1, social scientists usually refer to such things as units of analysis, or the things about which information is collected. Sometimes, however, they call these things **elements**, especially at the stage of sampling. Social scientists normally use the term *element* when they're referring to members of a **population** (the entire group of interest). In the case of presidential elections in the United States, the entire group of interest, or the population, is the voting population. Elements of this population are voters. Lastly, one can define a **sample** as a subset of a population. As you'll see later, there are a number of ways of picking a sample from a population. When selecting members of a population to include in the sample, those selected members are referred to as elements of the population. Once one collects information about those members, and analyzes it, they become units of analysis.

element, a kind of thing a researcher wants to sample.

population, the group of elements from which a researcher samples and to which she or he might like to generalize.

sample, a number of individual cases drawn from a larger population.

 STOP & THINK   *Suppose you're interested in describing the nationalities of Nobel prize-winning scientists. What would an element in your study be? What would the population be?*

SAMPLING FRAMES, PROBABILITY VERSUS NONPROBABILITY SAMPLES

We can now begin to address the question of why the *Literary Digest* poll did so poorly in 1936, why Gallup did badly in 1948, and why the 2012 final pre-election polls did relatively well at predicting the outcome of the 2012 election. All of these polls sought information about the same population: people who would vote on Election Day. But the concept of "people who would vote on Election Day" is elusive, just as many target populations are. A **target population** is the population of theoretical interest. In the case of "people who would vote on Election Day," the target is notoriously difficult to hit, especially *before* an election, when even the most well-intentioned citizens can't be sure what might keep them from voting on Election Day (e.g., weather, natural or social disaster, death). One thing most researchers must do, then, is come up with a

target population, the population of theoretical interest.

sampling frame or **study population,** the group of elements from which a sample is actually selected.

list or description of elements that approximates the elements of the population of interest. This list is referred to as a **sampling frame** or **study population.** A large part of the reason why today's pre-election polls are so much better than the ones of the past has to do with refinements in sampling frames. The list or sampling frame that the *Literary Digest* used for its 1936 poll consisted of names from the list of *Literary Digest* readers, automobile registrations, and telephone directories. But, in 1936, at the height of the Great Depression, substantial numbers of voters weren't readers of the *Literary Digest* and owned neither a car nor a telephone. So, even though the *Literary Digest* received two million responses to the (10 million) ballots it sent out, the best it could hope to do was get responses from people either who were readers (of the *Digest*) or who owned cars or telephones—people who, because of their wealth, were more likely to vote for the Republican Landon than the Democrat Roosevelt. All of the 2012 polls were of people who had telephones but, because the percentage of U.S. households with telephones was 97.8 percent in 2012 (Blumberg and Luke, 2012), this sampling frame provided a much greater chance of reaching the desired population than the one used by the *Literary Digest* in 1936 had. Modern surveys conducted over the Internet, when intended to provide insights about the total American population, are a more recent example of surveys that are subject to (admittedly decreasing) inaccuracy because of a discrepancy between the target population and the sampling frame. As of 2011, only about 78 percent of the U.S. population used the Internet (Zickuhr and Smith, 2012).

So, the main reason the *Literary Digest* poll did so poorly is that it employed a relatively poor sampling frame. A major problem with Gallup's poll in 1948 was that he, like the *Literary Digest* pollsters, employed a nonprobability, as opposed to a probability, sample. **Nonprobability samples** are those in which members of the population have an unknown (or perhaps no) chance of being included. But Gallup used a fairly sophisticated type of nonprobability sampling called quota sampling. Quota sampling begins with a description of the population: its proportion of males and females, of people in different age groups, of people in different income levels, of people in different racial groups, and so on. Based on this description, which in Gallup's case was from the U.S. Census, the researcher draws a sample that proportionately represents the sampling frame in all ways that are deemed important (e.g., gender, age, income levels, race). The main problem in 1948 was that the description of the American population that Gallup used was one based upon the 1940 census, a census that, because of all that had gone on between 1940 and 1948 (World War II, etc.), didn't provide a particularly good profile of the target population—adult Americans who would vote.

nonprobability samples, samples that have been drawn in a way that doesn't give every member of the population a known chance of being selected.

A better approach for predicting elections is to use a probability rather than nonprobability sample of Americans. **Probability samples** are selected in a way that gives every element of a population a known (and nonzero) chance, or probability, of being included. The most recent presidential election polls used probability samples that were designed to give every likely voter of the United States a (known) chance of being included in the samples.

probability samples, samples drawn in a way to give every member of the population a known (nonzero) chance of inclusion.

The advantage of probability over nonprobability samples is not, as you might hope, that they ensure that a sample is representative or typical of the population from which it is drawn. Still, probability samples are *usually* more representative than nonprobability samples of the populations from which they are drawn. To use the jargon of statisticians, probability samples are less likely to be **biased samples** than are nonprobability samples. Biased means that the cases that are chosen are unrepresentative of the population from which they've been drawn. And because probability samples increase the chances that samples are representative of the populations from which they are drawn, they tend to maximize the **generalizability** of results. Generalizability here refers to the ability to apply the results of a study to groups or situations beyond those actually studied.

biased samples, samples that are unrepresentative of the population from which they've been drawn.

generalizability, the ability to apply the results of a study to groups or situations beyond those actually studied.

But the greatest problems afflicting the *Literary Digest* and Gallup polls were likely that their sampling frames didn't match their target populations very well. Finding the right sampling frame is a problem that continues to plague polling organizations. This chapter's focal research piece, by Scott Keeter, Michael Dimock, and Leah Christian, describes how the Pew Research Center, like other polling organizations, dealt with another sampling frame issue: that associated with the growing percentage of the American population that uses cell phones and doesn't own landline phones.

STOP & THINK *Can you think why researchers haven't used cell phone numbers in polling until recently? What problem may result from using only landline numbers?*

FOCAL RESEARCH

Calling Cell Phones in '08 Pre-Election Polls

By Scott Keeter, Michael Dimock, and Leah Christian*

Public opinion polling faced many challenges during the 2008 presidential election. None was more daunting than the rising number of "cell phone only" voters who could not be reached over landline telephones. In previous elections, polling was generally limited to people with landline telephones, partly because cell phone owners in the U.S. are charged for calls and partly because relatively few adults with cell phones didn't have landlines. Cell phone only adults constituted only two percent of those with telephones in 2001 (Weisberg, 2005: 210). The latest estimates from the National Health Interview Survey (Blumberg and Luke, 2008)—the most comprehensive measure available—suggest that in the fall of 2008 nearly 18% of households were wireless only; in the National Exit Poll (Mokrzycki, 2008) conducted on Nov. 4, 20% of Election Day voters identified themselves as "cell-only" voters. What made this growth so important in the

* This is an abridged and revised version of a report that appeared as a release from the Pew Research Center for the People & the Press on December 18, 2008. It is reprinted with permission from the Pew Research Center.

recent election is that many news commentators (e.g., Wihbey, 2008) suggested that Barack Obama would prove more popular among cell phone users, because of their youth and education, than his opponent in the presidential election, John McCain. This paper, in fact, examines the hypothesis than Barack Obama fared better in probability samples including landline- and cell phone-users than in samples including landline users alone.

Method

Results for this report are based on combined data from six surveys conducted from September 9 through November 1, 2008. The data are from telephone interviews conducted under the direction of Princeton Survey Research Associates International and Abt SRBI, Inc., using a nationwide probability sample of adults, 18 years of age or older. In total, 10,430 adults were interviewed on a landline telephone and 3,460 on a cell phone, including 1,160 who had no landline telephone. This sample composition is based on a ratio of approximately three landline interviews to each cell phone interview. This ratio is based on an analysis that attempted to balance cost and fieldwork considerations in addition to resulting demographic composition of the sample and the need for a minimum number of cell-only respondents in each survey.

Pew's cell phone and landline contact and cooperation rates for election surveys conducted between September and November were essentially the same. (Contact rates for landline respondents were 77%, while for cell phone respondents, they were 76%; cooperation rates for landlines were 31% and 30% for cell phones. Thus, response rates for landlines were 22%, while for cell phones, they were 21%.) The response rate in Pew's cell phone surveys may benefit from the decision to offer cell respondents a modest $10 reimbursement for potential phone charges they might incur. In contrast, landline respondents were not offered any reimbursement.

Results

An analysis of six Pew surveys conducted from September through the weekend before the election shows that estimates based only on landline interviews were likely to have a pro-McCain tilt compared with estimates that included cell phone interviews (see Table 5.1). But the difference, while statistically significant, was small in absolute terms. Obama's average lead across the six surveys was 9.9 points among registered voters when cell phone and landline interviews were combined. If estimates had been based only on the landline samples, Obama's average lead would have been 7.6 points, indicating an average bias of 2.3 percentage points. Limiting the analysis to likely voters rather than all voters produced similar results. Obama's average lead among likely voters was 8.2 points across all six surveys versus 5.8 points in the landline sample.

Obama enjoyed a large advantage over McCain among younger people, but this advantage would have been underestimated by polls using landline samples only. Obama led McCain by 33 points (63% – 30%) in the full-frame sample of registered voters, compared with his 29-point advantage in the landline sample. The sample difference among likely voters under 30 was even larger, where the samples involving both cell phone- and landline-owning respondents set the difference at 33 percentage points, but the landline-only sample set it at 26 points.

(continues)

FOCAL RESEARCH *(continued)*

According to the national exit poll, Obama won this age group by 34 points, 66%–32%.

TABLE **5.1**

Samples with Cell Phones Showed a Larger Obama Advantage

	Landline/Cell Phone Sample	Landline Sample
Registered Voters	%	%
Obama	49.9	48.5
McCain	40.0	40.9
Other/DK	10.1	10.6
Obama Adv.	+ 9.9	+ 7.6
Sample size	(11,964)	(9,228)
Likely Voters		
Obama	49.8	48.5
McCain	41.6	42.7
Other/DK	8.6	8.8
Obama Adv.	8.2	5.8
Sample size	(10,819)	(8,143)

Discussion

The wireless-only population has been growing steadily as many young people enter adulthood without ever getting a landline and others drop their landline telephone service. Our projection for the fall of 2008, based on previous data from the National Health Interview Survey, is that 18% of adults are wireless only, with a rate of increase of roughly 3% annually.

Although there is little evidence of sizable non-coverage bias in landline surveys of the general public, the potential for bias exists and has been documented in studies of certain sub-populations including young and low-income people (Blumberg and Luke, 2007). Pew's research found that differences in presidential preferences expressed by samples of landline respondents alone and those involving both Land-line and cell phone respondents were small but could have been important in the 2008 election cycle had the election been closer. Still, it is not possible at this stage to offer a blanket recommendation to researchers regarding the inclusion of cell phones. There are considerable extra costs involved in cell phone interviewing; our cell phone interviews cost approximately two to two-and-one-half times as much as a landline interview. If present trends in the growth of the cell-only population continue, it is likely that the bias involved in contacting only landline users in the future will make these costs a necessary part of doing polling.

References

Blumberg, Stephen J. and Julian V. Luke. 2007. Coverage Bias in Traditional Telephone Surveys of Low-Income and Young Adults. *Public Opinion Quarterly* 71 734–749. http://poq.oxfordjournals .org/cgi/reprint/71/5/734

Blumberg, Stephen J. and Julian V. Luke. 2008. Wireless Substitution: Early Release of Estimates from the National Health Interview Survey, January–June 2008. *National Center for Health Statistics*. Retrieved February 6, 2009. http://www.cdc.gov/nchs/data/nhis /earlyrelease/wireless200812.htm

Mokrzycki, Mike. 2008. Exit Poll: 1 in 5 Voters Only Have Cell Phones. *News-day*. November 7. http://www.newsday.com/news/local/politics/ny-bc-polls-cellphones1107nov07,0,7248380.story

Weisberg, H. F. 2005. *The Total Survey Error Approach*. Chicago: The University of Chicago Press.

Wihbey, John. 2008. The Cell Phone Polling Gap: An Artificially "Close" Race? *The Huffington Post*. October 1. http://www.huffingtonpost.com/john-wihbey/the-cell-phone-polling-ga_b_131021 .html

THINKING ABOUT ETHICS *Because of the sampling technique employed, the Pew pollsters never knew the identity of their respondents, so respondent anonymity was never in danger. Moreover, participation in the survey was voluntary.*

SOURCES OF ERROR ASSOCIATED WITH SAMPLING

Keeter, Dimock, and Christian report that Obama's advantage over McCain was slightly, and significantly, greater in samples that included both landline and cell phones than in samples that included landline phones alone. And then they go on to suggest that part of the reason for this might be that young people are more likely to be cell phone users only and that young people were more likely to be Obama supporters.

Keeter, Dimock, and Christian's larger point, however, is this: that one does not get exactly the same results in polls from samples using owners of landline phones as a sampling frame and samples using owners of landline phones and cell phones as a sampling frame.

Increasing percentages of Americans are using cell phones only. In fact, the number grew from the 18 percent reported by Keeter et al. in 2008 to 34 percent of all adults in 2012 (Blumberg and Luke, 2012). Consequently, it makes sense that using sampling frames that exclude cell phones users will increasingly "miss the mark." But this raises the larger question of why there might be errors involved in surveys using samples, and it is to this question that we'd now like to turn.

The overarching reason that samples fail to represent accurately the populations we're interested in is that populations themselves aren't uniform or *homogeneous* in their characteristics. If, for instance, all human beings were identical in every way, any sample of them, however small (say, even one person), would give us an adequate picture of the whole. The problem is that real populations tend to be heterogeneous or varied in their composition—so any given sample might not be representative of the whole. Obviously, not all

Coverage errors	Nonresponse errors	Sampling error

FIGURE **5.1** Sources of Error Associated with Sampling

Source: Adapted from Weisberg (2005).

---

coverage error, an error that results from differences between the sampling frame and the target population.

nonresponse error, an error that results from differences between non-responders and responders in a survey.

sampling error, any difference between sample characteristics and the equivalent characteristics in the sampling frame, when this difference is not due to nonresponse error.

parameter, a summary of a variable characteristic in a population.

statistic, a summary of a variable in a sample.

eligible voters in the United States vote for the same candidate in every election. If they did, and did so reliably, any sample, however selected, would perfectly predict the election's outcome. Weisberg (2005) lists three sources of survey error that are associated with sampling for populations that are not homogeneous, however: **coverage error, nonresponse error,** and **sampling error** (see Figure 5.1).

Coverage Errors

Coverage errors are attributable to the difference between a sampling frame and a target population. It's useful to introduce a little more terminology here. A **parameter** is a summary of a variable *in a population*. In the 1936 election, about 61 percent of the voting public favored Roosevelt over Landon for president. This percentage is a parameter of the population that voted for president in 1936. A **statistic**, on the other hand, is a summary of a variable *in a sample*. The best the *Literary Digest* could do was come up with a statistic: that Roosevelt had received 43 percent of the ballots cast in their poll. But the largest problem was that the sample with which that statistic was calculated was drawn from a sampling frame, lists of *Digest* readers, telephone owners, and car owners, which bore only a remote resemblance to the target population. Hence the coverage error, defined precisely as the difference between the parameter for the sampling frame, or study population, and the parameter for the target population, was necessarily large.

STOP & THINK   *Can you imagine any sources of coverage error in the 2012 election-eve polls?*

The 2012 polls also involved some coverage error. After all, they were based upon sampling frames that included only people with telephones and then only people who said they were "likely voters" in those households. When we recall that the percentage of U.S. households with phones is just under 98 percent, not 100 percent, and that even people who thought they were "likely" voters couldn't be sure they'd vote (and those who thought they weren't likely voters couldn't be sure they wouldn't), we can see that the sampling frames for these polls were imperfect, if pretty good, matches for their target populations (people who did vote in the election).

STOP & THINK   *What do you think survey organizations do about the problem of unlisted telephone numbers?*

random-digit dialing, a method for selecting participants in a telephone survey that involves randomly generating telephone numbers.

One of the great potential sources of coverage errors is that many telephone numbers are not listed in telephone books. Survey organizations have gotten around this problem through a process called **random-digit dialing.** Incidentally, the Pew Research Center drew probability samples using random-digit dialing for both landline and cell phones. In doing so, they used random telephone numbers to select the people they talked to. We'll be saying more about random-digit dialing a little later. For now, notice that one of the great advantages of random-digitdialing is that it avoids the problem of unlisted telephone numbers, even if it brings into play numbers, like those of businesses, that are best left out of surveys of households.

Most of the time, when we think of coverage errors, we think of the eligible elements of a target population that are excluded in the sampling frame. The eligibles who were excluded by the *Literary Digest* poll, for instance, were voters who weren't *Digest* readers or didn't own a car or didn't own a telephone. But coverage errors also might occur when elements that are ineligible for the target population are included in the sampling frame. Thus, a voting survey of households with telephones could contact noncitizens—ineligibles in U.S. elections. Similarly, a household survey, especially one based, as many large-scale surveys are today, on random-digit dialing, could contact some people at non-residential (perhaps business) numbers. The problem of ineligibles, however, can, in most cases, be dealt with by the use of screening questions ("Are you an eligible voter?" "Is this your personal phone?"). The problem of excluded eligibles, such as those without a personal phone, is more challenging, but not always insurmountable.

The issue of excluded eligibles had become particularly problematic with the increased popularity of cell phones. Increasingly, as pointed out by Keeter, Dimock, and Christian, cell phone–only users possess only a cell phone, with no landline. The figure was only 2 percent in 2001, but had reached nearly 18 percent by 2008, and may have been even higher among those who actually voted, according to exit polls on Election Day. By 2012, as we've mentioned, 34 percent of American adults had cell phones but not landlines. Because many cell phone plans charge users for calls they receive, survey organizations used to avoid calling numbers assigned to cell phones. But several polling organizations, including the Pew Research Center, offered cell phone users compensation for receiving their calls. The fact that the RealClearPolitics prediction that the spread between Obama and McCain (7.6 percent—a statistic) was so close to the actual spread in the election (7.3 percent—a parameter) indicates, among other things, that coverage error for this election was quite small.

The list of technological innovations that currently frustrate survey researchers does not stop at cell phones alone, however. Answering machines and caller ID, their functional equivalents on cell phones, and other devices used to screen calls are increasingly used by "eligibles" to shield themselves from participation in surveys. And one can imagine that the increasing use of Internet technology as a substitute for traditional landline telephone connections will, in future, create a new coverage problem for telephone surveys. But now our discussion is bleeding into the next source of survey error: nonresponse error.

RESEARCH IN THE NEWS

In November, 2012, *The New Republic* broke a story that internal pre-election polls by the Romney organization had shown Romney winning the election fairly handily and that the candidate himself genuinely believed he would win, despite publicly available polls showing that he would narrowly lose. The story's author, Noah Scheiber, suggested several reasons for the problems of the internal polls, but they all boiled down to a belief that Romney's internal polls entailed more coverage error than the publicly available polls. Even the main researcher for Romney suggested the problem "may [have been] a function of Sunday polling," Saturday and Sunday being days when younger and more active people (people who tended to vote for Obama) were less likely to take telephone calls.

For the complete article see: http://www.newrepublic.com/blog/plank/110597/exclusive-the-polls-made-mitt-romney-think-hed-win#

Nonresponse Errors

Another source of error in both nonprobability and probability samples is nonresponse error. Nonresponse refers to the observations that cannot be made because some potential respondents refuse to answer, weren't "there" when contacted, and so forth. Nonresponse error is the error that results from differences between those who participate in a survey and those who don't. Nonresponse can involve large amounts of bias, as it was suspected to do in the *Literary Digest*'s 1936 presidential poll, when about 8 million of the *Digest*'s 10 million contacts simply didn't return their ballots. In this case, responding took a good deal of work (filling out forms, affixing stamps, which cost money, and mailing) and probably reflected unusual degrees of commitment to the candidates that respondents favored. The response rates for the Pew Research polls on which Keeter, Dimock, and Christian report (22 percent for landline phones and 21 percent for cell phones) weren't much higher than that of the *Literary Digest* and seem to be in line with long-term trends toward nonresponse. Curtin, Presser, and Singer (2005) report, for instance, that response rates to the University of Michigan's telephone Survey of Consumer Attitudes dropped from about 72 percent in 1979 to about 48 percent in 2003. But recent pre-election polls have been done on the (evidently plausible) assumption that there really isn't much difference between those who respond and those who don't. In fact, several scholars, including Keeter, Miller, Kohut, Groves, and Presser (2000), have found that, with particular topics, there aren't necessarily very large differences in the findings of probability samples with substantially different levels of response. Moreover, Groves (2006) and Groves and Peytcheva (2008) demonstrated that nonresponse errors are only indirectly related to nonresponse rates and are particularly likely to exist when the survey's variable of interest is strongly associated with the likelihood of responding. Thus, if people are simply more averse than they once were to *surveys of all kinds,* lower response rates, generally, don't necessarily mean greater levels of nonresponse bias or error. As Keeter says, with respect to political polls:

> Basically we think that non-response is largely random and not correlated with the things we want to study—e.g., non-respondents are not disproportionately conservatives who hate the media, or more affluent people who are too busy to talk on the phone, etc. Instead, people have varying propensities to participate, and this propensity probably varies over time—you might be a willing respondent one day but too busy the next. (personal communication)

But when one has reason to believe that the likelihood of responding is affected by the variable of interest, all bets are off. Thus, Travis Hirschi (1969) aptly argued that nonresponse bias is enormous in studies of delinquency among school-aged adolescents when the questionnaire is given in school because delinquents are much more likely than their nondelinquent counterparts to have dropped out of school or to skip sessions when questionnaires are administered.

⬤ STOP & THINK   *What kinds of people might not be home to pick up the phone in the early evening when most survey organizations make their calls? What kinds of people might refuse to respond to telephone polls, even if they were contacted?*

If sampling is done from lists, as is frequently done in probability sampling, it is sometimes possible to study the nature of nonresponse bias, especially if the lists include some information about potential sampling units. Thus, in a study of married people, Benjamin Karney and his colleagues (1995) were able to determine that people who did not respond to their questionnaire survey generally had less education and were employed in lower-status jobs than those who did respond. Karney and his colleagues used information from the marriage certificates of those who constituted their sampling frame. Weisberg (2005) points out that nonresponse can come at the unit level (when there is no response at all from the respondent) or the item level (when a respondent fails to answer a particular question) and that the former can be due to an inability to contact the respondent, an incapacity on the part of the respondent to be interviewed, or noncooperation on the part of the respondent. The latter, noncooperation, is most likely to be associated with error, especially insofar as respondents' lack of cooperation is systematically related to their attitudes toward the subject of the survey, as, for instance, when people with conservative social values refuse to answer a survey on sexual behavior (Weisberg, 2005: 160). Keeter, Dimock, and Christian point out that some nonresponse in the Pew 2008 pre-election surveys was due to noncontact (the contact rates were 77 percent for landline respondents and 76 percent for cell phones) but that cooperation rates for those contacted (31 percent for landlines; 30 percent for cell phones) were lower. A key issue is whether cooperation rates had more to do with people's attitudes toward presidential candidates or with a general unwillingness to be bothered. The evident accuracy of the Pew poll in 2008, and many pre-election polls in 2012, suggests it had more to do with the latter than the former.

Sampling Errors

Sampling error is the error that arises when sample characteristics are different from the characteristics of the sampling frame, assuming this difference is not due to nonresponse error. Here the question isn't whether there's a perfect match between the sampling frame and the target population or whether non-responders are different from responders. Theoretically, if not practically, one can imagine no coverage errors (due to differences between the sampling frame and the target population) or nonresponse errors (due to differences between responders and non-responders) and still imagine that a sample does not perfectly represent the population from which it is drawn. In fact, because the nine 2012 election-eve polls of likely voters conducted by various nonpartisan polling agencies and used by RealClearPolitics to compose its final average showed "spreads" between candidates Obama and Romney of between 1 percent in Romney's favor and 3 percent in Obama's favor (see Box 5.2), it is clear that at least some of the samples on which the polls were based did not give a perfectly accurate picture of the spread among all likely voters (the sampling frame for these polls). If they all did, their results would all be the same.

Some of the differences in the election-eve polls were probably due to differences in the sampling frames (e.g., some of the polls, like Pew, polled both landline and cell phone users; some, just landline users; one sampled just "battleground states," while the rest sampled the whole country). And some were probably due to different nonresponse errors. But much of the difference was probably due to another source of error: **sampling variability**, or the variability in sample statistics that can occur when different samples are drawn from the same study population or sampling frame.

sampling variability, the variability in sample statistics that can occur when different samples are drawn from the same population.

BOX **5.2**	The Final Presidential Poll Results by Non-Partisan Polling Agencies, 2012 (RealClearPolitics, 2012)

Poll	Obama	Romney	Spread
Politico/GWU/Battleground	47	47	Tie
Rasmussen	48	49	Romney +1
IBD/TIPP	50	49	Obama +1
CNN/Opinion Research	49	49	Tie
Gallup	49	50	Romney +1
ABC News/Washington Post	50	47	Obama +3
Monmouth/Survey Research/Braun	48	48	Tie
NBC News/Wall St. Jrnl	48	47	Obama +1
Pew Research	50	47	Obama +3
RealClearPolitics Ave.	48.8	48.1	Obama +0.7
Actual Election Results	51.1	47.2	Obama +3.9

Even probability samples entail sampling variability and, therefore, frequently involve sampling error. To demonstrate, we'd like you to imagine drawing random samples of two elements from a population of four unicorns: a population with four members named, let us say, Leopold, Frances, Germaine, and Quiggers. Let us also say that these unicorns were one, two, three, and four years of age, respectively.

● STOP & THINK   *How could you draw a sample of two unicorns, giving each unicorn an equal chance of appearing in the sample?*

---

simple random sample, a probability sample in which every member of a study population has been given an equal chance of selection.

A sample in which every member of the population has an equal chance of being selected is called a **simple random sample**. Simple random samples, because they give every member of the population an equal chance of being selected, also give every member a known chance and are therefore probability samples. Perhaps we should emphasize that the word *random* in the phrase *simple random sample* has a special meaning. It does not mean, as it can in other contexts, haphazard, erratic, arbitrary, funny, or crazy; here it connotes a "blind" choice, almost lottery-like, of equally likely elements from a larger population. One way of drawing (selecting) a simple random sample of two of our unicorns would be to write the four unicorns' names on a piece of paper, cut the paper into equal-sized slips, each with one unicorn's name on it, drop the slips into a hat, shake well, and, being careful not to peek, pick two slips out of the hat. Let's say we have just done so and drawn the names of Frances and Quiggers (ages two and four). We thereby achieved our goal of a sample of unicorns whose average age ([2 years + 4 years]/2 = 3 years) was *not* the same as that of the population (2.5), hence proving that it is possible to draw a probability sample that misrepresents the larger population and therefore entails sampling error.

Of course, if we had selected a sample of Frances and Germaine, the average age of sample members ([2 years + 3 years]/2 = 2.5 years) would have been the same as the average age for the whole population of unicorns. But we would also have proven that it is possible not only, as we did in our first attempt, to draw a sample that misrepresents the larger population, but also to draw probability samples from the same population that have different statistics (and so show sampling variability).

● STOP & THINK   *How many possible unique samples of two unicorns could be drawn from our population?*

You could randomly select six possible unique samples of two unicorns from our population. These would be: Leopold and Frances (average age = 1.5), Leopold and Germaine (2.0), Leopold and Quiggers (2.5), Frances and Germaine (2.5), Frances and Quiggers (3.0), and Germaine and Quiggers (3.5).

sampling distribution, the distribution of a sample statistic (such as the average) computed from many samples.

We could make a visual display of these samples, called a **sampling distribution,** by placing a dot, representing each of the sample averages, on a graph big enough to accommodate all possible sample averages. A sampling distribution is a distribution of a sample statistic (like the average) computed

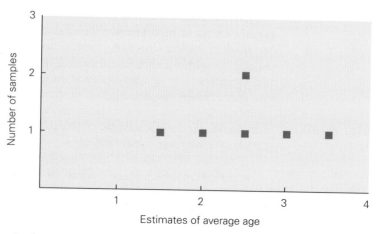

FIGURE **5.2** The Sampling Distribution of Unique Samples of Two Unicorns

from more than one sample. The sampling distribution of any statistic, according to probability theory, tends to hover around the population parameter, especially as the number of samples gets very large. Figure 5.2, a sampling distribution for the possible unique samples of two unicorns, suggests this property by showing that the sampling distribution of the average age for each possible pair of unicorns is centered around the average age for unicorns in the population (2.5 years).

STOP & THINK   *You're obviously pretty imaginative if you've followed us up to here. But can you imagine the sampling distributions of all possible unique samples of one unicorn? Of four unicorns? What, if anything, do you notice about the possible error in your samples as the number of cases in the samples grows?*

The sampling distribution of the average ages of samples of one unicorn would have four distinct points, representing one, two, three, and four years, all entailing sampling error. The sampling distribution of the average age for a sample of four unicorns would have one point, representing 2.5 years, involving no sampling error at all. Figure 5.2 suggests that the sampling distribution of samples with two cases generally has less average sampling error than that for samples with one case, but generally more average sampling error than the sampling distribution with four cases. This illustrates a very important point about probability sampling: Increases in sample size, referred to as n, improve the likelihood that sampling statistics will accurately estimate population parameters.

A major point of this section has been to demonstrate that even when one employs probability sampling, sampling error can and frequently does occur. This means, unfortunately, that no sample statistic can be relied on to give a perfect estimate of a population parameter. Another point of this section has been to suggest something about the nature of sampling distributions. Notably, they tend to be centered around the population parameter when the number of samples is large. Mathematicians have given extensive thought to the nature of sampling distributions. A practical offshoot of this consideration is a branch of mathematics known as probability theory.

A tremendous advantage of probability samples is that they enable their users to use probability theory to estimate population parameters with a certain degree of confidence. All of the final polls reported by RealClearPolitics (2012), for instance, were associated with an estimate of **margin of error**, a suggestion of how far away the *actual* population parameter (say, the spread among all likely voters) was likely to be from their statistic (of the spread within its sample). Thus, when the Pew Research poll told us on the eve of the 2012 election that Obama would win 50 percent of the vote, with a margin of error of 2.2 percent, it meant that there was a 95 percent chance that the actual percentage was somewhere between 47.8 and 52.2 percent of the votes. (And, in fact, Obama won 51.1 percent of the votes.)

We haven't shown you how to calculate margins of error here, but they are not difficult to calculate and perhaps you'll learn to do that in a statistics class. As our unicorn example suggests, though, the accuracy with which sample statistics estimate population parameters increases with the size of the sample. Probability theory tells us that if we want to achieve a margin of error of 10, we need a sample of about 100—and it doesn't matter what the size of the population is. If you'd like to reduce the margin of error from 10 percent to 5 percent, you'd need a sample of 400. To achieve its 2.2 percent margin of error, Pew Research interviewed 2,709 likely voters. In general, *one needs to increase the sample size by about four times to achieve a halving of the margin of error.* Because of the costs involved in sampling more people, Weisberg (2005: 229) sees little reason for trying to achieve, say, a reduction in the margin of error from 3 percent to 1.5 percent by surveying 4,800 respondents rather than 1,200. It's far better, once sampling error has been reduced to, say, 3 percent or even 5 percent, to spend time and energy trying to reduce other sources of error, such as coverage, nonresponse, and measurement error, a kind of error we'll be focusing on in Chapters 6 (on measurement) and 9 and 10 (on questionnaires and interviews). These kinds of error are normally far greater contributors to what Weisberg calls "total survey error" than can possibly be offset by increases in sample size beyond, say, 1,200.

margin of error, a suggestion of how far away the *actual* population parameter is likely to be from the statistic.

TYPES OF PROBABILITY SAMPLING

In the previous section, we introduced you to simple random sampling, a type of probability sampling that is appropriate for many statistical procedures used by social scientists. Here, we'd like to introduce you to common ways of doing simple random sampling and to other kinds of probability sampling and their relative advantages.

Simple Random Sampling

Simple random samples are selected so that each member of a study population has an equal probability of selection. To select a simple random sample, you need a list of all members of your population. Then, you randomly select elements until you have the number you want—perhaps using the blind drawing technique we mentioned earlier.

To show you another way of drawing a simple random sample, we'd like you to consider the list (or sampling frame) of Pulitzer Prize winners in Journalism for Commentary until 1993. The list of such award winners, given only since 1970, is relatively short and so is useful for illustrative purposes (though if, in fact, your study population included so few elements, you would probably want to use the entire population). From 1970 to 1993, there were 24 members of this population:

01. Marquis Childs	09. William Safire	17. Jimmy Breslin
02. William Caldwell	10. Russell Baker	18. Charles Krauthammer
03. Mike Royko	11. **Ellen Goodman**	19. Dave Barry
04. David Broder	12. Dave Anderson	20. Clarence Page
05. Edwin Roberts	13. Art Buchwald	21. Jim Murray
06. **Mary McGrory**	14. Claude Sitton	22. Jim Hoagland
07. Red Smith	15. Vermont Royster	23. **Anna Quindlen**
08. George Will	16. Murray Kempton	24. **Liz Baalmaseda**

STOP & THINK   *We have shown the female members of the population in boldface print. Because we've presented the population in this way, it is possible to calculate one of its parameters: the fraction of award winners who have been female. What is that fraction?*

Let us draw a simple random sample of six members from the list of award winners (i.e., we want an *n* of 6). If we don't want to use the "blind" drawing technique, we could use a random number table. To do so, we would, first, need to locate a random number table, such as the one we've provided in Figure 5.3, a larger version of which exists in Appendix E of this book. (Such tables, incidentally, are generally produced by computers in a way that gives every number an equal chance of appearing in a given spot in the table.) Second, we'd need to assign an identification number to each member of our population, as we have already done by assigning the numbers 1 through 24 to the list of award winners. Now we need to randomly select a starting number whose number of digits is equal to the digits in your highest identification number. The highest identification number in our list was 24 (with two digits), so we need to select (randomly) a number with two digits in it. Let's say you did this by closing your eyes, pointing toward the page, and (after opening your eyes again) underlining the closest two-digit number—perhaps, the number 96 (see the "96") underlined in Figure 5.3.

5334	5795	2896	3019	7747	0140	7607	8145	7090	0454	4140
8626	7905	3735	9620	8714	0562	9496	3640	5249	7671	0535
5925	4687	2982	6227	6478	2638	2793	8298	8246	5892	9861
9110	2269	3789	2897	9194	6317	6276	4285	0980	5610	6945
9137	8348	0226	5434	9162	4303	6779	5025	5137	4630	3535
4048	2697	0556	2438	9791	0609	3903	3650	4899	1557	4745
2573	6288	5421	1563	9385	6545	5061	3905	1074	7840	4596
7537	5961	8327	0188	2104	0740	1055	3317	1282	0002	5368
6571	5440	8274	0819	1919	6789	4542	3570	1500	7044	9288
5302	0896	7577	4018	4619	4922	3297	0954	5898	1699	9276

FIGURE **5.3** Random Number Table Excerpt

Then, proceeding in any direction in the table, you could select numbers, discarding any random number that does not have a corresponding number in the population, until you'd chosen the number of random numbers you desire. Thus, for instance, you could start with 96 and continue across its row, picking off two-digit numbers, then down to the next row, and so forth, until you'd located six of them (remember our desired *n* numbers).

STOP & THINK *What numbers would you select if you followed the procedure described in the previous sentence?*

The numbers you'd select, in this instance, would be the following:

96, 30, **19**, 77, 47, **01**, 40, 76, **07**, 81, 45, 70, 90, **04**, 54, 41, 40, 86, 26, 79, **05**, 37, 35, 96, **20**

Of course, you'd keep only six of these: those in bold print in the previous array—19, 01, 07, 04, 05, 20. Incidentally, you'd be working with a definition of simple random selection that assumes that once an element (corresponding to, say, the number 19) is selected, it is removed from the pool eligible for future selection. So if, say, during your selection number "07" had appeared more than once, you would have discarded the second "07." We'll also be using this definition later on in our discussions.

Your resulting sample of six journalists would be: Marquis Childs (case 01), David Broder (04), Edwin Roberts (05), Red Smith (07), Dave Barry (19), and Clarence Page (20).

STOP & THINK *What fraction of this sample is female? Does this sample contain sampling error relative to its gender makeup?*

STOP & THINK AGAIN

Pick a different simple random sample of six journalists. (Hint: Select another starting number in the random number table and go from there.) Who appears in your new sample? Does this sample contain sampling error relative to gender makeup?

A quicker way of drawing a random sample is to use Urbaniak and Plous's (2013) website, the Research Randomizer (http://randomizer.org). To random sample the same Pulitzer Prize winners, you'd simply tell the *Research Randomizer* you wanted one set of six numbers from the numbers 1 through 24 and, presto, you'd have your set of random numbers.

Systematic Sampling

systematic sampling, a probability sampling procedure that involves selecting every *k*th element from a list of population elements, after the first element has been randomly selected.

Even with a list of population elements, simple random sampling, done manually, can be tedious. A much less tedious probability sampling technique is **systematic sampling.** In systematic sampling, every *k*th element on a list is (systematically) selected for inclusion. If, as in our previous example, the population list has 24 elements in it, and we wanted six elements in our sample, we'd select every fourth element (because 24/6 = 4) on the list. (If the division doesn't yield a whole number, you should round to the nearest whole number.) To make sure that human bias wasn't involved in the selection and that every element actually has a chance of appearing, we'd select the first element randomly from elements 1 through *k*. Thus, in the example of the Pulitzer Prize winners, you'd randomly select a number between one and four, then pick the case associated with that number and with every fourth case after that one on the list.

STOP & THINK

Try to draw a sample of six award winners using systematic sampling. What members appear in your sample?

In general, we call the distance between elements selected in a sample the selection interval (4 in the award-winner example but this is generally referred to by the letter "k"), which is calculated using the formula:

$$\text{selection interval (k)} = \frac{\text{population size}}{\text{sample size}}$$

We've mentioned the relative ease of selecting a systematic sample as one of the procedure's advantages. There is at least one other advantage worth mentioning: If your population is physically present (say, file folders in an organization's filing cabinets), you don't need to make a list. You need only determine the population size (i.e., the number of folders) and the desired sample size. Then, you'd just need to look at every *k*th element (or folder) after you'd randomly selected the first folder. Ian Rockett, a colleague in epidemiology, tells us that he has used a modified form of systematic sampling in a hospital emergency use study. A potential interviewee was the first eligible patient entering the emergency room after the hour. Eligibility criteria included being 18 years of age or older and able to give informed consent to participate in the study. In each hospital, interviewing lasted three weeks and covered each of three daily eight-hour shifts twice during the three-week period.

The only real disadvantage of systematic sampling over simple random sampling, in fact, is that systematic sampling can introduce an error if the sampling frame is cyclical in nature. If, for instance, you wanted to sample months of several years (say, 1900 through 2006) and your months were listed, as they frequently are, January, February, March, and so on for each year, you'd get a very flawed sample if your selection interval turned out to be 12 and all the months in your sample turned out to be Januarys. Imagine the error in your estimates of average annual snowfalls in North Dakota if you collected only data about Januarys and multiplied by 12! A word to the wise: When you're tempted to use systematic sampling, check your population lists for cyclicality.

Stratified Sampling

One disadvantage of simple random sampling is that it can yield samples that are flawed in ways you want to avoid and can do something about. Take the earlier example of the Pulitzer Prize winners. Suppose you knew that approximately one sixth of that population was female and you wanted a sample that was one sixth female (unlike the one that we drew in the example). We could use a procedure called **stratified random sampling**. To draw a stratified random sample, we group the study population into strata (groups that share a given characteristic) and randomly sample within each stratum. Thus, to draw such a sample of the prizewinners, stratified by gender, we'd need to separate out the four female winners and the 20 male winners, enumerate each group separately, and random sample, say, a quarter of the members from each group.

stratified random sampling, a probability sampling procedure that involves dividing the population in groups or strata defined by the presence of certain characteristics and then random sampling from each stratum.

 STOP & THINK *See if you can draw a stratified random sample of the prizewinners, such that one female and five males appear in the sample.*

More complexity can be added to the basic stratifying technique. If, for instance, you were drawing a sample of students at your college, using a roster provided to you by the registrar, and the roster provided information not only about each student's gender but also about his or her year in school, you could stratify by both gender and year in school to ensure that your sample wasn't biased by gender or year in school.

Stratified sampling can be done proportionately and disproportionately. The Pulitzer Prize sample you drew for the "Stop & Think" exercise was a proportionately stratified random sample because its proportion of females and males is the same as the proportion of females and males in the population. However, if you wanted to have more female prizewinners to analyze (maybe all of them), you could divide your population by gender and make sure your sample included a number of females that was disproportionate to their number in the population.

STOP & THINK *How would you draw a disproportionate stratified random sample of eight prizewinners, with four females and four males in it?*

Disproportionate stratified sampling makes sense when the given strata, with relatively small representation in the population, are likely to display characteristics that are crucial to a study and when proportionate sampling might yield uncharacteristic elements from those strata because too few elements were selected. Suppose, for instance, you had a population of 100, 10 of whom were women. You probably wouldn't want to take a 10 percent proportionate sample from men and women because you'd then be expecting one woman to "represent" all 10 in the sample. You could oversample the women and then make an adjustment for their over-representation in the sample during your analysis. This is what Hoffnung did in her study of student views of their futures focused on in Chapter 4: she oversampled African American, Hispanic, and Asian American women to have large numbers of women of color in the study (personal communication).

It's also what Bostock and Daley (2007) did when they wanted to estimate lifetime and current sexual assault and harassment victimization rates of active-duty women in the Air Force. They drew a disproportionate random sample of women with senior ranks to make up for their relatively small numbers. Among their findings was that the prevalence of lifetime rape among Air Force women (28 percent) was more than twice as high as it is in national samples of women (13 percent).

Cluster Sampling

cluster sampling, a probability sampling procedure that involves randomly selecting clusters of elements from a population and subsequently selecting every element in each selected cluster for inclusion in the sample.

Simple random, systematic, and stratified sampling all assume that you have a list of all population elements and that the potential distances between potential respondents aren't likely to result in unreasonable expense. If, however, data collection involves visits to sites that are at some distance from one another, you might consider cluster sampling. **Cluster sampling** involves the random selection of groupings, known as clusters, from which all members are chosen for study. If, for instance, your study population is high school students in a state, you might start with a list of high schools, random-sample schools from the list, ask for student rosters from each of the selected schools, and then contact each of those students.

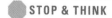 **STOP & THINK** *How might you cluster sample students who are involved in student organizations that are supported by student activity fees at your college?*

multistage sampling, a probability sampling procedure that involves several stages, such as randomly selecting clusters from a population, then randomly selecting elements from each of the clusters.

Multistage Sampling

Cluster sampling is used much less frequently than **multistage sampling,** which frequently involves a cluster sampling stage. Two-stage sampling, a simple version of the multistage design, usually involves randomly sampling clusters and then randomly sampling members of the selected clusters to produce a final sample. Ozbay and Ozcan (2008), for instance, took a random sample of high schools in Ankara, Turkey, and then a random sample of students from each of those high schools. With their sample of 1,710 students, they found that Travis Hirschi's social bonding theory worked better for explaining the delinquent behavior of boys than of girls.

Random-digit dialing is generally a form of multistage sampling, especially when nationwide surveys are being conducted. The Pew Research Center for People & the Press (2013) uses slightly different approaches to random-digit dialing of landlines and cell phones, but both involve multistage sampling. When it comes to landlines, for instance, the Pew procedure is to random-sample telephone numbers within a given county in proportion to the county's share of telephone numbers in the United States. (This process is imaginative. You might have a look at http://people-press.org/methodology/sampling/ for more details.) Then, once a number is contacted, the interviewer asks to speak with "the youngest male, 18 years or older, who is now at home." If there is no eligible male at home, the interviewer asks to speak with "the youngest female, 18 years or older, who is now at home." Pew has found that this procedure, once a number is contacted, is a good way of assuring that all adult members of a household have something like an equal chance of appearing in the sample.

STOP & THINK   *What coverage error might result from questioning the first adult to answer the phone?*

Some people who are more likely than others to answer phones are the able-bodied, those who have friends they like to chat with on the phone, those who aren't afraid that creditors might be calling, those without caller ID, and so forth. Any number of biases might be engendered if you queried only those who answered the phone. But if you, say, asked to speak to the adult who had the most recent birthday (as Bates and Harmon, 1993, did), you'd be much more likely to get a representative sample of the adults living in the residences you contacted. Keeter admits that, theoretically, the Pew procedure of asking first for the youngest adult male and then for the youngest adult female has no clear basis for yielding probability samples at the household level; however, he argues that, in fact, it's better at doing so than asking for the person with the most recent birthday.

> We get a lot of criticism for this method (though it is used by several other national polling organizations). Critics rightly charge that there is no statistical theory on which to base the selection, and thus we throw a wrench into the probability nature of the sample. We agree at some level, but the household enumeration that is required to do a true probability selection is very onerous and likely to result in lower response rates. "Last birthday" methods that have at least some claim to randomness don't work very well in practice—you get too many women, indicating that many respondents are not complying with the request. (personal communication)

The Pew Research Center multistage design, like all probability sampling designs, is much better at ensuring representativeness (and generalizability) than any nonprobability sampling design of a comparable population is. There are times, however, when nonprobability sampling is necessary or desirable, so we now turn our attention to some nonprobability sampling designs.

TYPES OF NONPROBABILITY SAMPLING

We've already suggested that although nonprobability samples are less likely than probability samples to represent the populations they're meant to represent, they are useful when the researcher has limited resources or an inability to identify members of the population (say, underworld or upperworld criminals) or when one is doing exploratory research (say, about whether a problem exists or about the nature of a problem, assuming it exists). Let us describe each of these advantages as we describe some ways of doing nonprobability sampling.

Purposive Sampling

For many exploratory studies and much qualitative research, **purposive sampling**, in any of a number of forms, is desirable. In purposive sampling, the researcher selects elements based on his or her judgment of what elements will facilitate an investigation. Ghose, Swendeman, Sheba, and Chowdhury (2008), for instance, drew upon sex workers who had been mobilized by an organization for sex workers in Kolkata, India, to see why there was such a high rate of condom use among sex workers in that city. Because it can be used to study elements of particular interest to the researcher, even when comprehensive lists of or reliable access to such elements is nonexistent, purposive sampling is used with great frequency by social scientists.

purposive sampling, a nonprobability sampling procedure that involves selecting elements based on the researcher's judgment about which elements will facilitate his or her investigation.

Quota Sampling

Quota sampling was the type of sampling used by Gallup and others so well in presidential preference polls until 1948. (You might remember that Gallup's projection, picking Roosevelt over Landon, was better for the 1936 election than was the *Literary Digest*'s, which depended on surveying owners of cars and telephones.) Quota sampling begins with a description of the target population: its proportion of males and females, of people in different age groups, of people in different income levels, of people in different racial groups, and so on.

quota sampling, a nonprobability sampling procedure that involves describing the target population in terms of what are thought to be relevant criteria and then selecting sample elements to represent the "relevant" subgroups in proportion to their presence in the target population.

STOP & THINK   *Where do you think Gallup and other pollsters might have gotten such descriptions of the American public?*

Until 1948, Gallup relied on the decennial censuses of the American population for his descriptions of the American voting public. Based on these descriptions, a researcher who used quota sampling drew a sample that "represented" the target population in all the relevant descriptors (e.g., gender, age, income levels, race).

Quota sampling at least attempts to deal with the issue of representativeness (in a way that neither convenience nor snowball sampling, discussed later, do). In addition, quota samples are frequently cheaper than probability samples are. Finally, when it yields a sample that consists of groups whose members are distinguished from one another in a theoretically significant way, it can be used as a way of making useful tests. Thus, Moller, Theuns, Erstad, and Bernheim (2008) took a quota sample of two groups in South Africa—regular community residents and patients at a tuberculosis (TB) hospital—to examine the usefulness of a new measure of subjective wellbeing. They found that the new measure showed that patients in the TB hospital had, as one might expect theoretically, lower subjective wellbeing than community residents, and that it did so more sensitively than better-established measures of life satisfaction and happiness. Quota sampling is, however, subject to some major drawbacks. First is the difficulty of guaranteeing that the description of the target population is an accurate one. One explanation given for the inability of Gallup and other pollsters to accurately predict Truman's victory in the 1948 presidential election is that they were using a population description that was eight years old (the decennial census having been taken in 1940), and much had occurred during the forties to invalidate it as a description of the American population in 1948. Another drawback of quota sampling is how the case selection is made. Thus, the researcher might know that she or he needs, say, 18 white males, aged 20 to 30 years, and select these final 18 by how pleasant they look (in the process, avoiding the less pleasant-looking ones). The result is that pleasantness of appearance is likely to be one source of bias in the final sample.

Snowball Sampling

snowball sampling, a nonprobability sampling procedure that involves using members of the group of interest to identify other members of the group.

When population listings are unavailable (as in the case of "community elites," gay couples, upperworld or underworld criminals, and so forth), **snowball sampling** can be very useful. Snowball sampling involves using some members of the group of interest to identify other members. One of the earliest and most influential field studies in American sociology, William Foote Whyte's (1955) *Street Corner Society: The Social Structure of an Italian Slum*, was based on a snowball sample: Whyte talked with a social worker, who introduced him to Doc, his main informant, who then introduced him to other "street corner boys," and so on. More recently, Jacinto, Duterte, Sales, and Murphy (2008) used a snowball sample of 80 women and men who sold Ecstasy in private settings to demonstrate how such sellers denied the identity of being "dealers" on three grounds: Ecstacy's benign reputation, the fact that they sold it in private places, and the fact that they sold it mainly to friends. With snowball sampling, the disadvantage of not being able to generalize can be more than compensated for by the capacity to provide an understanding of a previously unknown phenomenon.

Convenience Sampling

convenience sample, a group of elements that are readily accessible to the researcher.

A **convenience sample** (sometimes called an available-subjects sample) is a group of elements (often people) that are readily accessible to, and therefore convenient for, the researcher. Psychology and sociology professors sometimes use their students as participants in their studies. Tang, Carr Copeland, and Wexler (2012) used a convenience sample of 180 black and white elderly people in Pittsburgh to find out differences in the ways they volunteered in their communities. A higher percentage of whites claimed to do such volunteer work, but blacks who volunteered committed more time and received more psychosocial benefits from volunteering.

Convenience samples are relatively inexpensive and they can yield results that are wonderfully provocative and plausible. Nevertheless, the temptation to generalize beyond the samples involved is dangerous and should be tempered with caution.

STOP & THINK   *A student who hands out 100 questionnaires in a campus café might be tempted to say he or she collected a "random" sample. Why is his or her really not a random sample?*

CHOOSING A SAMPLING TECHNIQUE

Choosing an appropriate sampling technique, like making any number of other decisions in the research process, involves a balancing act, often among methodological, practical, theoretical, and ethical considerations. First, one needs to consider whether it's desirable to sample at all, or if the whole population can be used. Then, if it is methodologically important to be able to generalize to some larger population, as in political preference polls, choosing some sort of probability sample is of overwhelming importance. The methodological importance of being able to generalize from such a poll can clash with the practical difficulty of finding no available list of the total population and so dictate the use of some sort of multistage cluster sampling, rather than some other form of probability sampling (e.g., simple random, systematic, or stratified) that depends on the presence of such a list. Or, however desirable it might be to find a probability sample of corporate embezzlers (a theoretical concern), your lack of access to the information you'd need to perform probability sampling (a practical concern) could force you to use a nonprobability sampling technique—and perhaps finally settle for those you can identify as corporate embezzlers and who volunteer to come clean (ethical).

You might find that your study of international espionage would be enhanced by a probability sample of the list of Central Intelligence Agency (CIA) agents your roommate has downloaded after illegally tapping into a government computer (methodological and practical considerations), but that your concern about going to prison (a practical consideration) and for jeopardizing the cover of these agents (practical and ethical) leads you to sample passages in the autobiographies of ex-spies (see Chapter 13 on content analysis) rather than to survey the spies themselves. In another study, your exploratory (theoretical) interest in the psychological consequences of adoption for non-adopted adult siblings could dovetail perfectly with your recent introduction to three such siblings (practical) and their obvious desire to talk to you about their experiences (practical and ethical).

Method, theory, practicality, and ethics—the touchstones of any research project—can become important, then, as you begin to think about an appropriate sampling strategy.

SUMMARY

Sampling is a means to an end. We sample because studying every element in our population is frequently beyond our means or would jeopardize the quality of our data—by, for instance, preventing our close scrutiny of individual cases. On the other hand, we don't need to sample when studying every member of our population is feasible.

Probability samples give every member of a population a chance of being selected; nonprobability samples do not. Probability sampling is generally believed to yield more representative samples than nonprobability sampling. Both probability and nonprobability sampling can be associated with coverage errors, nonresponse errors, and sampling error itself. Coverage errors are caused by differences between the sampling frame, or study population, and the target population. Nonresponse errors are caused by differences between non-responders and responders to a survey. Sampling error is caused by differences between the sample and the study population, or sampling frame, from which it is drawn. A great advantage of probability samples, however, is that they allow the estimation of sampling error and thereby generate estimates of things like margins of error. Keeter, Dimock, and Christian's article reminds us that new sources of coverage errors may be entailed by technological change, as it demonstrates that the increased reliance on cell phones by the American public means that sampling frames that ignore cell phone–only respondents may become increasingly inadequate ways of estimating public attitudes.

We looked at five strategies for selecting probability samples: simple random sampling, systematic random sampling, stratified random sampling, cluster sampling, and multistage sampling. We then looked at four common strategies for selecting nonprobability samples: purposive sampling, quota sampling, snowball sampling, and convenience sampling. We emphasized that the appropriate choice of a sampling strategy always involves balancing theoretical, methodological, practical, and ethical concerns.

EXERCISE 5.1

Practice Sampling Problems

This exercise gives you an opportunity to practice some of the sampling techniques described in the chapter.

The following is a list of the states in the United States (abbreviated) and their populations in 2005 to the nearest tenth of a million.

State	Population	State	Population
AL	4.5	CO	4.5
AK	0.6	CT	3.5
AZ	5.5	DE	0.8
AR	2.7	FL	17.0
CA	35.5	GA	8.6

State	Population	State	Population
HI	1.3	NM	1.9
ID	1.4	NY	19.2
IL	12.7	NC	8.4
IN	6.2	ND	0.6
IA	2.9	OH	11.4
KS	2.7	OK	3.5
KY	4.1	OR	3.6
LA	4.5	PA	12.4
ME	1.3	RI	1.1
MD	5.5	SC	4.1
MA	6.4	SD	0.8
MI	10.0	TN	5.8
MN	5.1	TX	22.1
MS	2.9	UT	2.4
MO	5.7	VT	0.6
MT	0.9	VA	7.4
NE	1.7	WA	6.1
NV	2.2	WV	1.8
NH	1.3	WI	5.4
NJ	8.6	WY	0.5

1. Number the states from 01 to 50, entering the numbers next to the abbreviated name on the list.
2. Use the random number table in Appendix E and select enough two-digit numbers to provide a sample of 12 states. Write all the numbers and cross out the ones you don't use.
3. List the 12 states that make it into your random sample.
4. Now, if you have easy access to the Internet, locate the *Research Randomizer* (http://randomizer.org/) and draw one set of 12 numbers from 01 to 50 from it. Then list the 12 states that would make it into your random sample this time.
5. This time, take a stratified random sample of 10 states, one of which has a population of 10 million or more and nine of which have populations of less than 10 million. List the states you chose.
6. How might you draw a quota sample of 10 states, one of which has a population of 10 million or more and nine of which have populations of less than 10 million?
   a. Describe one way of doing this.
   b. Describe, in your own words, the most important differences between the sampling procedures used in Questions 5 and 6a.

EXERCISE 5.2

Examining Sampling in a Research Article

Using your campus library (or some other source), pick a research article from any professional journal that presents the results of a research project that involves some sort of sampling. Answer the following questions.

1. Name of author(s):
   Title of article:
   Journal name:
   Date of publication:
   Pages of article:
2. What kind of sampling unit or element was used?
3. Was the population from which these elements were selected described? If so, describe the population.
4. How was the sample drawn?
5. What do you think of this sampling strategy as a way of creating a representative sample?
6. Do(es) the author(s) attempt to generalize from the sample to a population?
7. Did the author(s) have any other purposes for sampling besides trying to enable generalization to a population? If so, describe this (these) purpose(s).
8. What other comments do you have about the use of sampling in this article?

CHAPTER **6**

Measurement

CHAPTER OBJECTIVES

This chapter is designed to help students

- **Create measures for quantitative research**
- **Conceptualize variables**
- **Operationalize variables**
- **Generate simple and composite measures**
- **Evaluate quantitative measures**
- **Appreciate the importance of exhaustive and mutually exclusive categories**
- **Assess the reliability and validity of variables**
- **Identify nominal, ordinal, interval and ratio levels of measurement**
- **Plan a measurement strategy for visual analysis**
- **Comprehend basic principles of measurement in qualitative research**

MEASURING VARIABLES

measurement, the process of devising strategies for classifying subjects by categories to represent variable concepts.

Measurement generally means classifying units of analysis (say, individuals) into categories (say, satisfied with life or not satisfied with life) to represent variable concepts (say, life satisfaction). Measurement in quantitative research and qualitative research is often different, relating to differences in the purposes of the two enterprises. While distinctions between quantitative and qualitative research can be overdrawn, one thing is quite clear: a defining characteristic of **quantitative research** is its focus on variables. The Pew polling of prospective voters before the 2008 elections (see Chapter 5) is an example of quantitative research inasmuch as its primary function was to determine who was going to vote for Barack Obama and who was not—a variable characteristic of the American adult population. Consequently, measurement, with its focus on sorting cases into categories of variables, is central to quantitative research. **Qualitative research,** on the other hand, tends to be less focused on variables per se and more concerned with interpreting action and sorting out meaning—and is typically more focused on individual cases than on variable characteristics of many cases. Ciambrone was primarily interested in eliciting the stories of individual women with HIV/AIDS (see Chapter 3) to learn what it was like to deal with the disease. Eventually, however, Ciambrone did become interested in variable characteristics that distinguished the women—and it was at that point that she became interested in measurement.

quantitative research, research focused on variables, including their description and relationships.

qualitative research, research focused on the interpretation of the action of, or representation of meaning created by, individual cases.

Often, in quantitative research, the researcher begins the measurement process by identifying a variable of concern, works out categories of that variable that make sense, and then classifies, through observation, various units of analysis. In contrast, in qualitative research, the process typically begins with close observation of individual units. The researcher may move to composing categories that help make sense of the units and sometimes goes on to making decisions about variables that might link the categories. We devote more of the current chapter to quantitative measurement because measurement has received more attention in quantitative than in qualitative research. But by the end, you'll know quite a bit about both quantitative and qualitative measurement.

measure, a specific way of sorting units of analysis into categories.

A **measure** of a variable is a specific way of sorting units of analysis into categories. For example, we could decide that our measure of gender would be people's answer to the question, "What is your gender?" on a questionnaire. But measuring a variable can, and sometimes should, be considerably more complicated. And since measurement can be an important source of both insight and error in research, it is a process that requires considerable care during the preparation of research, and attention when we read a research report. The building blocks of research are concepts or ideas (like "happiness," "suicide," "satisfaction with life," "fatherhood," and "power") about which we've developed words or symbols. Such concepts are almost always abstractions that we make from experience. Concepts cannot often be observed directly. What quantitative researchers do during the measurement process is to define the concept, for the purposes of the research, in progressively precise ways so that they can take indirect measurements of it.

Suppose we want to determine (or measure) whether a death was a suicide. We might, for instance, begin with a dictionary definition or something we will call a theoretical definition of suicide as "the act of killing oneself intentionally." Then, we might decide to take any clear evidence of a deceased person's intention to take his or her life (e.g., a note or a recording) as an indicator of suicide. We might look for more than one such indicator and would need to figure out what to do if some of the evidence was contradictory. In fact, determining whether a death is a suicide can be elusive, as Rockett and Kapusta's focal research, below, suggests. Whatever decisions we make along the way would be part of our measurement strategy—one that usually starts with the broader issues of how to conceptualize something and then goes to the more specific issue of how to operationalize it.

CONCEPTUALIZATION

Researchers who start by creating a research question the way Hoffnung did in the study discussed in the focal research section of Chapter 4 are working with the building blocks of all scientific concepts—words or signs that refer to phenomena that share common characteristics. Hoffnung asked questions about what would happen to young women who had graduated from college in 1993: Would they "have it all"? Would they establish long-term relationships, become mothers, share family responsibilities, and have careers?

Whenever we refer to ideas such as "sharing family responsibilities," "long-term relationships," or "careers," we are using concepts. But let's notice something important: We each develop our basic understandings of concepts through our own idiosyncratic experiences. You initially might have learned what an "accident" is when, as a child, you spilled some milk, and a parent assured you that the spilling was an accident. Or you might have heard of a friend's bicycle "accident." Someone else might have had a baby brother who had an "accident" in his diaper. We might all have reasonably similar ideas of what an accident is, but we come to these ideas in ways that permit a fair amount of variation in just what we mean by the term.

BOX 6.1 How Do You Feel?

Take a moment to answer the following questions suggested by Diener, Emmons, Larsen, and Griffin (1985). Later in the chapter, we'll talk about what these questions are intended to measure.

Following are five statements with which you may agree or disagree. Using the 1–7 scale shown, indicate your agreement with each item by placing the appropriate number on the line preceding that item. Please be open and honest in responding.

The 7-point scale is as follows:

7—Strongly agree
6—Agree

5—Slightly agree
4—Neither agree nor disagree
3—Slightly disagree
2—Disagree
1—Strongly disagree

____ In most ways my life is close to my ideal.
____ The conditions of my life are excellent.
____ I am satisfied with my life.
____ So far I have gotten the important things I want in life.
____ If I could live my life over, I would change almost nothing.

So beware: Concepts are not phenomena; they are only symbols we use to refer to (sometimes quite different) phenomena. Concepts are abstractions that permit us to communicate with one another, but they do not, in themselves, guarantee precise and lucid communication—a goal of science. In fact, in non-scientific settings, concepts are frequently used to evoke multiple meanings and allusions, as when poets write or politicians speak. If, however, concepts are to be useful for social science purposes, we must use them with clarity and precision. The two related processes that help us with clarity and precision are called conceptualization and operationalization.

conceptualization, the process of clarifying just what we mean by a concept.

conceptual definition, a definition of a concept through other concepts.

Conceptualization is the process of clarifying just what we mean by a concept. This process usually involves providing a theoretical or **conceptual definition** of the concept. Such a definition describes the concept through other concepts.

🛑 **STOP & THINK**   *What concepts are part of your definition of suicide?*

If we use Webster's dictionary definition (or conceptualization) of "suicide" as "the act of killing oneself intentionally," this definition employs the concepts of "killing" and "intention." Or, if we define "satisfaction with life" using the definition of Diener, Emmons, Larsen, and Griffin (1985) as "a cognitive judgment of one's life as a whole," we'd have to understand what "cognitive judgment" and "life as a whole" mean. We could define "power," using Max Weber's classic formulation, as the capacity to make others do what we want them to do despite their opposition—a definition that refers, among other things, to the concepts of "making others do" and "opposition." In each case, we would be defining a concept through other concepts: ideas (like "intention," "opposition," and "cognitive judgment") that themselves might deserve further clarification, but which nonetheless substantially narrow the possible range of meanings a concept might entail. Thus, for instance, in Diener's conceptual definition, he is implicitly suggesting that life satisfaction involves the conscious overall evaluative judgment of life using the person's own criteria. He implies that the concept doesn't involve the degree of frequency that the person feels joy, sadness, or anxiety—things that others might associate with life satisfaction. In doing so, he narrows and clarifies the meaning of his concept. Being specific and clear about a concept are the goals of conceptualization.

As part of creating and considering a conceptual definition, we should ask if the concept is so broad and inclusive that it could profitably be divided up into aspects or **dimensions** to further clarify it. A concept that refers to a phenomenon that is expected to have different ways of showing up or manifesting itself is said to be **multidimensional**. For example, we often hear about the "fear of crime," which can be conceptually defined as the fear of victimization by criminal activity. However, we could follow Mark Warr and Christopher Ellison's (2000) lead and divide it up into two dimensions or two kinds of fear of crime: a "personal fear" of crime, which is when people worry about their own personal safety, and an "altruistic fear" of crime, when they worry about the safety of others, such as spouses, children, and friends.

dimensions, aspects or parts of a larger concept.

multidimensionality, the degree to which a concept has more than one discernible aspect.

OPERATIONALIZATION

As we've said, we often can't perceive the real-world phenomena to which our concepts refer, even after we've provided relatively unambiguous conceptual definitions of what we mean by the concepts. "Satisfaction with life" and "fear of crime" and "religion" are very hard to observe directly, and even such a "concrete" concept as "suicide" can elude our direct perception. Someone found dead in a body of water might have been drowned (died by homicide), or accidentally drowned (died by accident), or died as result of his or her own effort (died by suicide)—or even died of natural causes. When the phenomena to which concepts refer can't be observed directly, we might make inferences about their occurrence (or lack or degree thereof).

The process of specifying what particular indicator(s) one will use for a variable is called **operationalization**. **Indicators** are observations that we think reflect the presence or absence of the phenomenon to which a concept refers. Indicators may be answers to questions or comments, things we observe, or material from documents and other existing sources. For an example, let's take a concept that we're all familiar with—age. Sounds simple enough, right? However, reading the gerontology literature (see Atchley, 2003: 6–8), we could notice that there are several ways to conceptualize age: as chronological age (time since birth), functional age (the way people look or the things they can do), and life stages (adolescence, young adulthood, and so on).

How we conceptualize age will make a huge difference in what we do when we operationalize the concept. Let's say we conceptualize age in a particular way—as chronological age; we'll still have to decide on an operational definition. We could, for example, operationalize chronological age by using as our indicator the answer to a variety of questions, some of which follow:

1. Are you old?
2. How old are you? Circle the letter of your age category:
   a. 25 or younger   b. 26 to 64   c. 65 or older
3. How old are you?

And many others. Each question is a possible indicator of age. Nevertheless, some of these might be much more useful than others are.

operationalization, the process of specifying what particular indicator(s) one will use for a variable.

indicators, observations that we think reflect the presence or absence of the phenomenon to which a concept refers.

An Example of Conceptualization and Operationalization

Stevanovic and Rupert (2009) were interested in how personal feelings of accomplishment at work affected feelings of family-related stress and, in turn, life satisfaction among professional psychologists. To study the topic, they sent out a questionnaire to practicing psychologists, receiving completed questionnaires from 485 of them. To examine their three variables of interest—personal feelings of accomplishment at work, family-related stress, and life satisfaction, they used well-established conceptualizations and operationalizations available in the social-psychological literature. Importantly, like all other researchers, they conceptualized and operationalized each of these variables separately. When it came to measuring "life satisfaction," Stevanovic and Rupert employed the tools that Diener et al. (1985) had provided. Like Diener et al., they conceptualized life satisfaction as "a cognitive judgment of one's life as a whole." Note again that this definition of "life satisfaction" not only provides the reader with an idea of what Stevanovic and Rupert mean by "life satisfaction," it also eliminates other possible meanings—like the degree to which one feels (or lacks) anxiety on a daily basis. Furthermore, when they wanted to operationalize "life satisfaction," Stevanovic and Rupert used the five questions that Diener et al. developed and that we've reproduced in Box 6.1. The **index**—a composite measure that is constructed by adding scores of individual items—for life satisfaction is created by adding the scores of the five items so that the totals range from 5 (low satisfaction) to 35 (high satisfaction). In Figure 6.1 we summarize the process researchers go through to measure each variable using the Stevanovic and Rupert measurement strategy for life satisfaction as our example.

index, a composite measure that is constructed by adding scores from several indicators.

🛑 **STOP & THINK**   *Just for fun, why don't you use your answers to the five statements in Box 6.1 to calculate a "life satisfaction" score for yourself.*

Stevanovic and Rupert used their measure of life satisfaction to show that feelings of accomplishment at work are associated with reduced feelings of family-related stress that, in turn, correlate with increased life satisfaction among professional psychologists. Other researchers have used this life satisfaction measure to determine that middle-aged women who have a strong sense of generativity (i.e., who feel they've made a difference in other people's lives) have greater life satisfaction than women who don't (Rittenour and Colaner, 2012) and that if the elderly feel younger than their actual age, they have a higher life satisfaction than if they don't (Stephan, Caudroit, and Chalabaev, 2011). Researchers also used the measure in a variety of countries, including the United States, Japan, Australia, Northern Ireland, Portugal, Cape Verde,

FIGURE **6.1** An Illustration of Measurement in Quantitative Research Based on Stevanovic and Rupert (2009)

Korea, England, and France (Kim and Hatfield, 2004; Lewis, Dorahy, and Schumaker, 1999; Neto and Barros, 2000; Schumaker and Shea, 1993; Smith, Sim, Scharf, and Phillipson, 2004; Stephan, Caudroit, and Chalabaev, 2011).

STOP & THINK

Hamarat et al. (2001: 189) studied life satisfaction and found that in their sample of 192 people, the older adults experienced the highest satisfaction and the younger adults the lowest. The average scores were 21.9 for those 18 to 40, 24.2 for those 41 to 65, and 26.4 for those 66 and over. How did your score compare to those in your age group?

Composite Measures

composite measures, measures with more than one indicator.

Indexes like the one of life satisfaction, and other **composite measures**, abound in the social sciences. All composite measures, or measures composed of more than one indicator, are designed to solve the problem of ambiguity that is sometimes associated with single indicators by including several indicators of a variable in one measure. The effort to confirm the presence (or absence or the degree of presence) of a phenomenon by using multiple indicators occurs in both qualitative and quantitative research.

In qualitative research (see especially Chapters 10 and 11), a researcher might or might not need to prepare measurement strategies in advance of, say, an interview or observation. The careful qualitative researcher is likely to use multiple indicators of key concepts. A researcher can judge a study participant's "emotional tone" and body language and pay attention to the actual answer to a question like "How are you today?" Also, a question can be followed up with other questions. A little later in this chapter, in the section on "Measurement in Visual Analysis and Qualitative Research," we'll give you an idea of how qualitative researchers frequently do measurement, not in advance of, but after making their observations or collecting their data.

In quantitative research, there has been much sophisticated thinking about the preparation of composite measures in advance of data collection. A full discussion of this thinking is beyond the scope of this chapter, but we would like to introduce you to it briefly. All strategies for making composite measures of concepts using questions involve "pooling" the information gleaned from more than one item or question. Two such strategies, the index and the scale, can be distinguished by the rigor of their construction. An index is constructed by adding scores given to individual items, as occurs in the computation of the life satisfaction score. The items don't have to "go together" in any systematic way because indexes (sometimes called indices) are usually constructed with the explicit understanding that the given phenomenon, say "life satisfaction," can be indicated in different ways in different elements of the population. (Incidentally, we define a **scale** as an index in which some items can be, or are, weighed more than others and will illustrate scales a bit later in the chapter.)

scale, an index in which some items are given more weight than others in determining the final measure of a concept.

If a concept is multidimensional—that is, has different ways of showing up or manifesting itself—then an index can sum up all the dimensions. The Consumer Price Index, for instance, summarizes the price of a variety of goods and services (e.g., food, housing, clothing, transportation, recreation) in the United States as a whole in different times and for different cities at a given time. It is an index (of a multidimensional concept) because one wouldn't necessarily expect the price of individual items (say, the price of food and the price of recreation) to go up and down together. The Consumer Price Index is, however, a good way of gauging changes in the cost of living over time or space.

Other examples of indexes are evident in this chapter's first focal research on Gross National Happiness, by Emily. Emily became interested in the concept of Gross National Happiness when she read an article about "the happiest country in the world." She had to go and check where Bhutan (the country so described) was and found that it was a small Buddhist country in the Himalayas. As a methodologist, she'd been dissatisfied with how concepts of life satisfaction or happiness have been measured and was attracted to what the Bhutanese were doing. While you read Emily's essay, see if you can answer this question: How does Bhutan conceptualize and operationalize the concept of Gross National Happiness?

FOCAL RESEARCH

"How's Life?" and other ways to measure happiness

By Emily Stier Adler*

Suppose I were to ask you, "Taken altogether, how would you say things are these days—would you say that you are very happy, pretty happy or not too happy?" What would you answer? This question is one that the General Social Survey has asked US respondents for about 40 years. Tabulations of respondents' answers find little variation in responses over the years. In 1972, 30% of Americans said that they were very happy, 53% reported being pretty happy and 17% said they were not too happy. The results for the categories in 2010 were 29%, 57% and 14% (Social Data Archive, 2012). How do you compare?

The degree to which a person is happy is clearly important, at least to that person. But is it an important attribute of a country as a whole? How important is it to know that 86% of Americans see themselves as either very or pretty happy? Does it tell us that we are on the right track as a society?

For a long time Western societies have been much more interested in measuring economic productivity than the happiness of their citizens. Nations started using economic measures like gross national product as early as the 1930s. Only recently have they begun to think about the usefulness of overall measures of wellbeing or happiness.

Jigme Singye Wangchuck, the King of Bhutan, began this international discussion in 1972. Soon after his enthronement, he declared that Gross National Happiness (GDH), rather than Gross Domestic Product (GDP), would be his country's measure of success (Thinley, 2005). He believed that happiness, not productivity, was a better way to make decisions about his country's development. Since his announcement, numerous countries and international organizations have responded to Wangchuck's challenge. France commissioned economists to come up with a measure of quality of life; British and American experts came up with measures of subjective well-being. The United Nations actually generated a happiness index. But it was an organization in Bhutan itself, The Centre for Bhutan Studies that, after 7 years of work, came up with the world's first Gross National Happiness Index.

*Article prepared for this book.

Although happiness can be experienced personally, for Bhutan, its pursuit is a collective goal (Centre for Bhutan Studies, 2012a: 1). Prime Minister of Bhutan, Jigmi Y. Thinley (2012) argues that "Gross national happiness is the philosophy or development paradigm that has guided Bhutan's development in a way that has been holistic and based on the belief that the aspiration, the ultimate goal of every human individual, is happiness, so then it must be the responsibility of the state, or the government, to create those conditions that will enable citizens to pursue this value, this goal." Unlike the General Social Survey's concept of happiness, this concept of happiness is not focused only on a person's subjective well-being at one point in time, but is multidimensional and explicitly includes responsibility and other-regarding motivations such as altruism, concern for others and living in harmony with nature (Centre for Bhutan Studies 2012a: 7)

Courstesy of Gross National Happiness

For the Gross National Happiness Index, the researchers conceptualized happiness as having nine domains or dimensions, with each having between two and four subdimensions, for a total of 33 subdimensions (Centre for Bhutan Studies, 2012b). These dimensions and subdimensions are:

1. psychological well-being (*life satisfaction, emotional balance* [positive and negative emotions], *spirituality*)
2. health (*mental health, self-reported physical health status, healthy days, long-term disability*)
3. time use (*working hours, sleeping hours*)
4. education (*literacy, educational qualifications, knowledge, values*)
5. culture (*speak native language, artisan skills, cultural participation, "the way of harmony"*)
6. good governance (*political participation, fundamental rights, service delivery, government performance*)
7. community vitality (*social support or the donating of time and money to others, community relationships, family, safety versus victim of crime*)
8. ecological diversity and resilience (*pollution, environmental responsibility, wildlife damage, urbanization issues*)
9. living standard (*household income, assets, housing quality*)

After a pilot test of an instrument in 2006, national surveys of the Bhutan population were conducted in 2008 and 2010. The indicators, selected in accordance with the values and culture of Bhutan, were deemed to be understandable by ordinary people and to reflect their life experiences.

The 2010 survey had a sample of 7,142 people representative of the 738,000 people in Bhutan with the appropriate representation from the 20 districts and the proportion of urban and rural households (Centre for Bhutan Studies, 2012a).

The composite measure of happiness was based on more than 100 indicators, each of which was a question asked by an interviewer in the respondent's home. For example, the indicators for the *service delivery aspect* of good governance were seven questions that asked respondents to rate the performance of

(continues)

FOCAL RESEARCH *(continued)*

the government in the past 12 months on these major objectives: employment, education, equality, health, anti-corruption, environment and culture. For each, the respondent ranked the services on a 5-point scale from "very good" to "very poor" and the individual scores were added up for totals ranging between 7 and 35 (Centre for Bhutan Studies, 2012b: 159). The internal consistency among the seven indicators was high (Cronbach's alpha = .89).

The indicators of *community relationships* were two questions, one on a sense of belonging (with answers ranging from very strong to very weak) and the other trust in neighbors (with answers from trust most of them to trust none of them) (Centre for Bhutan Studies, 2012b: 163).

Life satisfaction was measured by multiple items so as to combine individuals' subjective assessment of contentment levels with respect to health, occupation, family, work-life balance and standard of living (Centre for Bhutan Studies, 2012b: 126). For each item, respondents were asked to rate themselves on a 5-point scale ranging from 1 "very dissatisfied" to 5 "very satisfied." (Cronbach's alpha for these indicators was .70).

After each dimension was scored, individuals were categorized as "happy" if they had sufficient levels of happiness on at least six of the nine dimensions. Otherwise, they were categorized as "not yet happy." Some results from the 2010 survey (Centre for Bhutan Studies, 2012a) include the following:

- Forty one percent of Bhutanese have sufficient happiness in six or more of the nine dimensions and are considered 'happy'.
- On average, men are happier than women.
- In urban areas, 50% of people are happy; in rural areas it is 37%.
- Of the nine dimensions, Bhutanese have the most sufficiency in health, then ecology, psychological wellbeing, and community vitality.
- Urban areas do better in health, living standards and education. Rural areas do better in community vitality, cultural resilience, and good governance.
- Happiness is higher among people with a primary education or above than among those with no formal education, but higher education does not affect GNH very much.
- The happiest people by occupation include civil servants, government officials and monks. Interestingly, the unemployed are happier than corporate employees, housewives and farmers.
- Unmarried people and young people are among the happiest.

The advantages of a multidimensional measure of happiness over the one-dimensional measure used by the General Social Survey, especially for a government, or any agency, interested in improving overall happiness, may now be obvious. On the one hand, and at the individual level, it enables insight into the aspects of life that produce happiness or unhappiness. What matters most—one's health, community, living standard—or do all count equally? On the other hand, at the national level, Gross National Happiness facilitates an understanding of the areas in which a country is successfully creating conditions which lead to well-being of their citizens and points to areas needing improvement.

Bhutan does particularly well at promoting a sense of good health and ecological diversity and resilience, but less well on the dimensions of community vitality and psychological wellbeing in urban areas and education, living standard and time use in rural areas. With information like this, public policy initiatives can be created so that more people become happy, something that Bhutan's leaders consider an inalienable human right (Thinley, 2012).

References

Centre for Bhutan Studies. 2012a. A short guide to GNH index 2012. Thimphu, Bhutan. Online at http://www.grossnationalhappiness.com/wp-content/uploads/2012/04/Short-GNH-Index-edited.pdf

Centre for Bhutan Studies. 2012b. Extensive Analysis of GHP Index. On-line at http://www .grossnationalhappiness.com/wp-content/uploads/2012/10/An%20Extensive%20Analysis%20of %20GNH%20Index.pdf

Social Science Data Archive. 2012. Online at http://dataarchives.ss.ucla.edu/

Thinley, Jigmi Y. 2005. What is Gross National Happiness? Presented at the Second International Conference on GNH in Halifax, Canada in 2005. Online at http://www.bhutanstudies.org.bt /pubFiles/8-Rethinking.pdf

Thinley, Jigmi Y. 2012. Interview with Prime Minister of Bhutan, Jigmi Y. Thinley. UN News Centre, April 2, 2012. Online at http://www.un.org/apps/news/newsmakers.asp?NewsID=49

STOP & THINK *How is Gross National Happiness conceptualized in Bhutan, according to Emily? Does this definition make sense to you?*

Bhutan's researchers conceptualize Gross National Happiness as positive feelings on nine dimensions of possible concern, from psychological wellbeing to living standard. Like all scientific conceptualizations, this one winnows away many possible connotations of the word "happiness" in favor of one with relative precision. It suggests, for instance, that happiness isn't just a sense of personal wellbeing at one point in time. This is, in fact, just one of nine dimensions of happiness. Some of the others continue to focus on personal conditions like health, while some are concerned with more communal concerns like community vitality. The researchers further clarify what they mean by each of these dimensions by telling us about subdimensions they will investigate for each of the dimensions. Thus, the community vitality dimension, we're told, entails four subdimensions: social support or the donating of time and money to others, community relationships, family, and safety. Finally, we're told that each of these subdimensions is operationalized in terms of particular questions. Thus, Emily tells us that the community relationships subdimension of the community vitality dimension is measured in terms of answers to two questions: one on a sense of belonging (with answers ranging from very strong to very weak) and the other trust in neighbors (with answers from *trust most of them* to *trust none of them*).

STOP & THINK *The Life Satisfaction subdimension of the Bhutanese Psychlogical Wellbeing dimension is measured by the sum of answers on five questions, Emily reports. Respondents were asked to rate themselves on a five-point scale ranging from 1, "very dissatisfied," to 5, "very satisfied," in five areas: health, occupation, family, work–life balance and standard of living. How does this measure of Life Satisfaction differ, would you say, from the one created by Diener et al. (1985) (see Box 6.1)?*

Bhutan's researchers used composite measures to tap each of the subdimensions of their nine major dimensions of Gross National Happiness. To measure Life Satisfaction they used the five indicators mentioned in the Stop & Think. Respondents could score anywhere from 5 to 25 when these individual indicators were added up (as opposed to the range of 7 to 35 on the indicator used by Diener et al., 1985). They were also providing information about their satisfaction with regard to five different concerns of life, from their health to their standard of living (as opposed to the measure used by Diener et al., 1985, a measure that doesn't focus on respondents' feelings about any of these particular concerns).

Indexes can be used to measure multidimensional concepts. We often compute an index score, as Bhutan's researchers did for Gross National Happiness, by adding all the scores on over 100 individual items. Methodologists often call indexes in which some items can be, or are, weighed more than others *scales*. One kind of scale, the Bogardus social distance scale (designed by Emory Bogardus), was created to ascertain how much people were willing to interact with other kinds of people. Suppose you wanted to find out how accepting people are of Martians. You could ask them questions like:

1.   Would you be inclined to let Martians live in the world?
2.   Would you be inclined to let Martians live in your state?
3.   Would you be inclined to let Martians live in your neighborhood?
4.   Would you be inclined to let a Martian be your next-door neighbor?
5.   Would you be inclined to let a Martian marry your child?

Because you might expect that a person who answered yes to Question 5 would also answer yes to all the other questions, you could reasonably give a higher score to a yes for Question 5 than the others. Similarly, you could reasonably give a higher score to a yes for Question 4 than for yeses to Questions 3, 2, and 1.

Indexes and scales are not hard to make up, though assuring yourself of their validity and reliability can be challenging. (See the discussion of these topics that follows.) You're frequently well advised to find an index whose validity and reliability has already been tested if one exists for the variable you're trying to measure. (See Touliatos, Perlmutter, and Straus, 2001, or Miller and Salkind, 2002, and see also the journal *Social Indicators Research* for indexes that are often used in the social sciences.) Keep your eyes peeled during your literature review and you might see ways (or references to ways) that others have operationalized the concepts you're interested in. If you think the same approach might work for you, try it out (and remember to cite your source when you write up your results). There's no sense in reinventing the wheel.

Measurement in Visual Analysis and Qualitative Research

Before turning to some finer detail associated with quantitative research measures, we'd like to introduce you briefly to qualitative measures and, particularly, to a specialized version of such measures: those used in visual analysis, an exciting new field in the social sciences. **Visual analysis** refers to a set of techniques used to analyze images. As John Grady (2001: 83) has observed,

visual analysis, a set of techniques used to analyze images.

"visual media and images increasingly dominate mass communications in contemporary society" and not to study how such images are produced and what they tell us is a serious oversight.

Images should be treated as data—about the image-maker's view of the world, if nothing else. We encourage you, for instance, to take a critical look at the cover image of our book. Consider, for instance, what it tells you about what we want you to think about social research methods. Here you couldn't begin, as Bhutan's researchers did in their study of happiness, with predetermined variables and predetermined measurement strategies. You'd have to see what themes emerged as you examined the picture. In the case of the cover picture, you might notice, for instance, that it is actually made up of multiple images, each having something to convey about our message about research. Then, using what qualitative researchers call **coding**, or the assigning of your observations to categories, you could begin to create your own variables. You might notice, for instance, how many of the images include women and how many include men. In doing so, you'd not only be "measuring" our presentation of social researchers by "gender," but you might also begin to suspect that we (Emily and Roger) are "suggesting," through the images we've chosen, that social research can be and is done by both women and men.

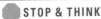

coding, assigning observations to categories.

🛑 **STOP & THINK**   *Can you code the images on our cover for what they seem to say about researchers' ethnic backgrounds?*

Grady (2007a) used a very similar process for measuring the messages of 590 advertisements containing images of African Americans in *Life* magazine from 1936 to 2000. His approach, called "grounded theory" coding, involved successive observations of the advertisements, seeing what codes occurred to him, and then assigning each advertisement to categories of emerging variables. One thing he discovered was a "steadily growing white commitment to racial integration" (Grady, 2007a: 211) expressed through the magazine editors' choice of published advertisements. Notice that the creation of variables for Grady, as in our example of analyzing the cover of this book, came after the collection of data (the images) and is comparatively ad hoc and idiosyncratic to the individual researcher. Another researcher might have chosen different codes and different variables.

We have more to tell you about measurement in qualitative research in Chapter 15. For now, however, we'd like to direct you to Figure 6.2, an image meant to suggest that, in qualitative research, unlike quantitative research (see Figure 6.1), measurement often begins after observations have been made and involves the coding of those observations until one begins to imagine "variables" that make meaning of the codings.

The kind of measurement we've attributed to versions of visual analysis is actually fairly typical of qualitative research generally. Another example of qualitative coding comes from the author of the focal research in Chapter 10, Sandra Enos. She was curious about how incarcerated women "mothered" from prison. Early on, she took copious notes on her observations at a parenting program in a prison in the Northeast. She then coded the notes, finding

FIGURE **6.2** An Illustration of Measurement in Qualitative Research from the Examination of This Book's Cover Image

that some of the women's children were under the care of their fathers, some were under the care of another relative, and some were in foster care. She eventually noticed that she made a variable of the various residences of the children and that this variable included the following categories: grandparent, husband, other relative, and nonrelative.

Exhaustive and Mutually Exclusive Categories

Measuring a concept (e.g., life satisfaction or age) means providing categories for classifying units of analysis as well as identifying indicators. For "life satisfaction," for example, Diener (2001) suggests creating seven categories by taking the answers to the five questions, adding the individual scores together (for a total of between 5 and 35), and classifying the score as follows:

31 to 35	extremely satisfied
26 to 30	satisfied
21 to 25	slightly satisfied
20	neutral
15 to 19	slightly dissatisfied
10 to 14	dissatisfied
5 to 9	extremely dissatisfied

But what of the nature of these categories? In this section, we'd like to emphasize two rules that should stand as self-checks any time you engage in measuring a variable concept.

The first rule for constructing variable categories is that they should be **exhaustive**. That is, there should be a sufficient number of categories to enable you to classify every subject in your study. The aim of creating exhaustive categories doesn't require any particular number of categories, although the minimum number of categories for a variable concept is two. As an example, consider measuring political party affiliation with the question, "What is

exhaustive, the capacity of a variable's categories to permit the classification of every unit of analysis.

your political party affiliation?" followed by answer categories: (1) Republican, (2) Democrat, and (3) other.

 STOP & THINK *Can you think of any likely instance that couldn't be "captured" by the net offered by this question's answer categories? If you can't, the categories are exhaustive.*

mutually exclusive, the capacity of a variable's categories to permit the classification of each unit of analysis into one and only one category.

The second rule for variable construction is that the categories should be **mutually exclusive.** This means you should be able to classify every unit of analysis into one and only one category—that is, your categories should have no overlap. A woman can give birth as the result of a planned pregnancy, but then it won't be unplanned. A man can fall into the category "65 or older," but if he does, he can't also fall into the "26 to 64" category.

 STOP & THINK *Suppose you asked the question "How old are you?" and then told the person to circle the number associated with his or her age category, giving them the answer categories of (1) 25 or less, (2) 25 to 65, and (3) 65 and older. Which of the rules for variable construction would you be offending? Why?*

The example in the "Stop & Think" is of a variable whose categories are not mutually exclusive because they overlap. A person who was 25 years of age, for instance, could put himself or herself in both categories (1) and (2) and a person who was 65 could fall into both categories (2) and (3). But if we asked the question "How old are you?" and told the respondents to circle the number associated with his or her age category, giving them the answer categories of (1) 25 or less and (2) 26 to 65, then we would be violating the rule of exhaustiveness for categories of variables because there are people, 66 and older, who wouldn't fit into a category.

measurement error, the kind of error that occurs when the measurement we obtain is not an accurate portrayal of what we tried to measure.

Measurement error, unlike the kinds of survey error we discussed in Chapter 5—coverage error, nonresponse error, and sampling error—refers to the kind of error that occurs when the measurement we obtain is not an accurate portrayal of what we tried to measure. In a survey using a questionnaire, two possible sources of measurement error are that question categories are not exhaustive or mutually exclusive. In the following section, we'll look at two other qualities of measurement that must exist to avoid measurement error.

QUALITY OF MEASUREMENT

reliability, the degree to which a measure yields consistent results.

validity, the degree to which a measure taps what we think it's measuring.

We'd now like to introduce two other important qualities of measurement: **reliability** and **validity.** A measurement strategy is said to be reliable if its application yields consistent results time after time. If you were to use a scale to measure how much a rock weighs, the scale would provide you with reasonably reliable measurements of its weight if it tells you that the rock weighs the same today, tomorrow, and next week. A measurement strategy is valid if it measures what you think it is measuring. If it's the weight of a rock you want to measure, a scale works pretty well. But, however reliable (and valid) your scale is when measuring the rock's weight, it cannot validly measure its shape or texture—it's just not supposed to.

STOP & THINK *Suppose you attempted to measure how prejudiced people are with the question "What is your sex?" Which quality of measurement—reliability or validity—is most apt to be offended by this measurement strategy? Why?*

Checking Reliability

The reliability of a measurement strategy is its capacity to produce consistent results when applied time after time. A yardstick can consistently tell you that your desk is 30 inches wide, so it's pretty reliable (even if it's possibly inaccurate—we'd have to check whether the yardstick's first inch or so has been sawed off). There are many ways to check the reliability of a measurement strategy. One is to measure the same object(s)—like your desk—two or more times in close succession. This **test–retest method** has been used by the researchers who created the life satisfaction measure. Diener et al. (1985) gave the life satisfaction questions to 76 students twice, two months apart. They found that students who scored high on the index the first time around were much more likely to do so the second time than students who had not scored high the first time.

test–retest method, a method of checking the reliability of a test that involves comparing its results at one time with results, using the same subjects, at a later time.

The major problems with test–retest methods of reliability assessment are twofold. First, the phenomenon under investigation might actually change between the test and the retest. Your desk won't grow, but a person's life satisfaction might actually change in two months. So if people's categories change because there really have been changes in what's being measured, this does not mean that the measurement strategy is unreliable. Second, earlier test results can influence the results of the second. Do you think that if you completed a life satisfaction questionnaire today, it might influence your answers on the same questionnaire next month?

internal consistency method, a method that relies on making more than one measure of a phenomenon at essentially the same time.

The **internal consistency method** is one way to deal with the problems of the test–retest method. It relies on making more than one measure of a phenomenon at essentially the same time. That's in fact what Diener et al.'s five life satisfaction questions (in their life satisfaction index) do. And some researchers, including Vera et al. (2008) in their study of subjective wellbeing among urban adolescents of color, have checked the degree to which the answers to these questions are correlated with one another. Among their sample, Vera et al. found that individual indicators of the life satisfaction measure were highly correlated with one another. In other words, they all seem to give consistent reports about adolescents' satisfaction. Emily reports that the service delivery subdimension of Bhutan's good governance dimension contained seven indicators. These asked for ratings of the government's delivery of particular services, from education to health (all measured on a five-point scale). She reports that the internal consistency of these seven indicators was high, as suggested by a Cronbach's alpha of .89. Cronbach's alpha is actually a way of estimating the internal consistency of items. An alpha of 0.70 or greater suggests that items in a scale "go together" very well—that is, that they are internally consistent, so an alpha of 0.89 indicates high internal consistency.

STOP & THINK

Go back to the focal research piece and look at the alphas for the Life Satisfaction subdimension of the Gross National Happiness's Psychological Wellbeing dimension. Does this measure have high internal consistency?

---

interobserver (or inter-rater) reliability method, a way of checking the reliability of a measurement strategy by comparing results obtained by one observer with results obtained by another using exactly the same method.

alternate-forms (or parallel-forms) method, a way of checking the reliability of a measurement strategy by comparing its results with those obtained using a measurement strategy meant to measure the same variable.

Another common way to check the reliability of a measurement strategy is to compare results obtained by one observer with results obtained by another using exactly the same method. We'll see this **interobserver reliability or inter-rater reliability method** figure extensively in later chapters. In Chapter 13, on content analysis, Roger and two colleagues report that their counts of single-gender human and animal pictures were correlated at .93 or more. The fact that the correlation of each of these numbers for the three raters was never less than 0.93 means that the three raters used the same measurement strategy to come up with virtually the same results all the time. In other words, the measurement scheme yielded consistent results when different raters used it.

A fourth way to check the reliability of a measurement approach is to compare one measure of a phenomenon with another measure of the same phenomenon. This is sometimes called an **alternate-forms procedure** and sometimes a **parallel-forms procedure.** King and King (1990) made two "short" questionnaires meant to measure respondents' attitudes toward equality between the sexes by taking 25 questions from the previously existing 95-question Sex-Role Egalitarianism Scale. They administered both questionnaires to 608 undergraduate students and found that the two "short" measures yielded very similar results.

 STOP & THINK

Suppose that you found a way of measuring homophobia with 10 questionnaire items and you wanted to use those questions in a questionnaire to see if there's a connection between homophobia and knowledge of AIDS. Describe a way in which you might check the reliability of the homophobia measure you were using.

Checking Validity

Validity refers to how well a measurement strategy taps what it intends to measure. You might wonder how to tell whether a strategy is valid. After all, most measurement in the social sciences is indirect. Asking people "All things considered, how satisfied are you with your life as a whole these days?" isn't exactly the most direct way of determining personal wellbeing, even though this what is used in the World Values Survey in 81 countries (Kahneman and Krueger, 2006).

Ultimately the validity of a measure cannot be proven, but we can lend credence to it. Perhaps the most intuitively satisfying check on validity is this question: Does this measurement strategy feel as if it's getting at what it's supposed to? If a measure seems as if it measures what you and I (and maybe a whole lot

face validity, a test for validity that involves the judgment of everyday people, like you and me.

of others) believe it's supposed to be measuring, we say it has **face validity,** and we can feel pretty comfortable with it. The answer to the question, "How old are you?" seems like a pretty good measure of age and, therefore, has high face validity. On the other hand, the question, "What, in your opinion, is the proper role of the military in foreign affairs?" would yield a poor measure of age and, therefore, has low face validity.

content validity, a test for validity that involves the judgment of experts in a field.

Content validity is similar to face validity, except that it generally involves the assessment of experts in a field, rather than just regular people, like you and me. Oransky and Fisher (2009), for instance, developed a 34-item Meanings of Adolescent Masculinity Scale (MAMS) by first doing a qualitative analysis of boys' own conceptions of what it means to be masculine in today's society. Oransky and Fisher used this analysis to come up with a tentative list of 60 items for their index and then asked a series of focus groups, made up of diverse adolescent males (the real experts, in the view of Oransky and Fisher, in what it takes to possess adolescent masculinity), to pick the items they believed were most descriptive of what they viewed as "being a man"—and then to add some others. The resulting 34-item masculinity scale, then, had been tested and approved by a group of experts—in this case, adolescent males.

predictive criterion validity, a method that involves establishing how well the measure predicts future behaviors you'd expect it to be associated with.

Establishing **predictive criterion validity** is different from establishing face validity and content validity, inasmuch as it doesn't directly involve the judgment of people about the measure. Predictive criterion validity involves establishing how well the measure predicts future behaviors you'd expect it to be associated with. Srivastava, Tamir, McGonigal, John, and Gross (2009), for instance, measured the suppression of emotion (the inhibiting of the outward expression of emotion) in a sample of students entering college. They predicted that such suppression would lead to both lower social support for students and less closeness to others within a short period after the start of college. When they found that this was true, they inferred that their measure of emotional suppression had predictive criterion validity, because it did, in fact, go together with the predicted behavioral outcomes. Importantly, the reason a predictive criterion validity test involves the adjective criterion is that, implicitly, it's the behavior against which the new measure is being tested that is the variable of greatest interest to the researchers. In this case, Srivastava et al. were crucially interested in a condition they felt would predict or help them understand the capacity of students to develop satisfying relationships, early on, in college.

concurrent criterion validity, how well a measure is associated with behaviors it should be associated at the present time.

A similar test for validity is embodied in **concurrent criterion validity** tests. Here the concern isn't so much with behaviors a measure should predict, but with things with which it should happen at the same time. Salmela-Aro, Kiuru, Leskinen, and Nurmi (2009), for instance, devised a measure of school burnout, consisting of dimensions related to exhaustion with school, cynicism toward the purpose of school, and a sense of inadequacy in relation to school. The researchers figured that something that might go along with school burnout would be poor academic achievement. When they found, in fact, that their measure of burnout was related to poor school performance, they concluded that their measure was valid because it was associated with a behavior it should be associated with.

construct validity, how well a measure of a concept is associated with a measure of another concept that some theory says the concept should be associated with.

Again, the distinguishing characteristic of predictive and concurrent validity checks is that the criterion (e.g., how one interacts with fellow students at school or school performance) is the researcher's focus, or concern, rather than the variable that's being measured (e.g., emotional suppression or school burnout). The variable that's being measured (e.g., burnout) can been seen as being designed as a tool for predicting or understanding the behavior (e.g., school performance) of interest. **Construct validity,** on the other hand, is used when the trait or quality being measured itself is of central importance, rather than the measures used to check its validity (Cronbach and Meehl, 1955). Construct validity refers to how much a measure of one concept is associated with a measure of another concept that some theory says it should be associated with. Howe, Cate, Brown, and Hadwin (2008), for instance, developed a test of empathy in preschool children. They checked its construct validity by seeing how it was correlated with something their theory suggested it should be associated with: pro-social behavior (as evaluated by their teachers). Now, one might say that the main reason for being concerned with preschool children's capacity for empathy would be its association with pro-social behavior (and that Howe et al. were therefore checking their measures through concurrent criterion validity). But that's not the way Howe et al. saw it. They believed such empathy was crucial in and of itself and so were measuring its construct validity when they correlated it with teacher assessment of pro-social behavior.

Our second focal research piece of the chapter focuses on strategies to check the reliability and validity of suicide classification. Ian Rockett, one of the authors (along with and Nestor Kapusta) of the following article, is an epidemiologist, with training in sociology and demography. Epidemiologists study the distribution and causes of diseases and health conditions. Rockett developed an interest in the determinants of suicides, but almost immediately found reason to doubt the basic data used by researchers since Emile Durkheim published his (1897) *Suicide*: official suicide reports. He observed that suicides are often misclassified in the categories labeled "injury of undetermined intent" or "ill-defined or unknown causes." He and Nestor Kapusta use this observation to generate the following reliability and validity tests.

FOCAL RESEARCH

Measuring Suicide

By Ian R.H. Rockett and Nestor D. Kapusta*

Loss of a loved one to suicide exerts a devastating emotional toll upon families and close friends. Suicide can trigger profound second-guessing, guilt, shame, and bewilderment among these "survivors." Suicide is likely also the most poorly measured major cause of death in the United States. The problem is undercounting rather than overcounting, because suicide is not a default category for the decision-makers in determining cause of death, namely, medical examiners and coroners. Detection of suicides draws much less attention and fewer resources than that other cause of intentional injury mortality, homicide.

Officially about twice as many people die by suicide nationally as by homicide and yet the official counts undoubtedly underestimate this difference. The means for adequate counting don't exist, but neither does the motive: examiners and coroners are aware of the insurance penalties against suicide deaths as well as the stigma for surviving family members. For these reasons actual suicides are often misclassified. They are frequently labeled as *injury of undetermined intent*. In fact, this category (injury of undetermined intent) is the most prone to

(continues)

_____

*Article prepared for this book.

FOCAL RESEARCH (continued)

hiding suicides. However, many more actual suicides are probably misclassified under *ill-defined or unknown causes* because this is a much larger cause-of-death category than the undetermined category.

If the deceased person hasn't left a note (only about 30% do so) or the death was unwitnessed, medical examiners and coroners have to go out of their way to determine that a death was a suicide. If a true suicide is by shooting or hanging, relatively active methods, suspicions of suicide may be raised more easily for obvious reasons. If, however, a true suicide is by poisoning or drowning, relatively passive methods, it is more likely to fall under the radar.

Two key concerns about any measurement procedure are reliability and validity. Reliability refers to the consistency of measurement across time and space. Validity refers to whether a measure measures what we think it is measuring. The validity of a suicide measurement system is affected by two attributes: whether it correctly classifies true suicides as suicides and whether it correctly classifies true nonsuicides as nonsuicides.

STOP & THINK   *Given our previous discussion, which do you think is the greater threat to accurate suicide measurement? The classification of true suicides as suicides or the classification of true nonsuicides as nonsuicides?*

If you said the classification of true suicides as suicides is the greater threat, you're right. It's sometimes very hard to figure out whether a death is a suicide and the decision-makers often miss the signs.

The proportion of true suicides that are classified as suicides by a suicide measurement system is called the measure's **sensitivity**, one aspect of validity. Our colleagues and we believe most suicide classification systems do not have a high degree of sensitivity. The proportion of true nonsuicides that are classified in some other category of death besides suicide is called the measure's specificity. We believe that most suicide classification systems have a high degree of **specificity**, because of the tendency to classify all deaths, even many actual suicides, as nonsuicides. The standards for ruling a death a suicide are extremely stringent. Authorities have to show that a decedent both deliberately injured him- or herself and intended to die.

Our research shows that suicide classification at the national level across countries has a low degree of both reliability and validity. Let us give you a sense of what we've found.

The way we've examined the **reliability** of national-level suicide classification is through a version of the alternate-forms procedure. Here we've correlated the official suicide rates in countries with a composite measure, namely with what the suicide rates would be if all the deaths in the other suicide-obscuring categories of death were suicides. We'd expect, for instance, that the official suicide rate would be highly correlated with a rate that combined official suicides with deaths by injuries of undetermined intent. The latter, in fact, might be considered the lower limit of the actual suicide rate in a country since, even if all deaths by injuries of undetermined intent were really suicides, they wouldn't add very much to the official number of suicides. (Relatively few deaths are placed into the injuries of undetermined intent category is most countries.) And, in fact, the correlation coefficient of the official suicide rates in countries with the lower limit rate is 0.93 out of 1.00, about as high as could be expected (see Table 6.1). But to be truly confident about an official suicide rate, we'd also like it to be highly correlated with an upper limit for that actual suicide rate, something we estimate by combining official suicides with both deaths by injuries of undetermined intent and deaths by ill-defined and unknown causes. The correlation of this rate and the official suicide rate is not so high, only 0.40 (see Table 6.1) and this leads us to believe that scientists who use official suicide rates in cross-national studies may be using unreliable data.

TABLE **6.1**
Official suicide rates and estimated lower and upper rate limits for selected countries, 2006*

Country	(a) Official suicide rate per 100,000	(b) Lower estimated suicide rate (combined official rates per 100,000 of suicide and undetermined injury intent mortality)	(c) Upper estimated suicide rate (combined official rates per 100,000 of suicide, undetermined injury intent mortality, and ill-defined and unknown cause mortality)
Lithuania	28.9	42.2	58.7
Hungary	23.2	24.1	25.5
Slovenia	22.8	26.4	41.0
Latvia	19.3	26.2	97.8
Finland	19.0	20.5	25.1
Estonia	18.7	29.0	66.9
France	15.4	16.0	50.8
Switzerland	14.9	15.8	34.3
Poland	14.3	20.4	77.9
Austria	13.4	15.0	21.2
Luxembourg	13.1	14.3	32.5
Czech Republic	12.2	16.3	24.0
Sweden	12.0	14.8	30.8
Slovakia	11.9	16.7	29.7
Romania	11.9	13.2	17.2
Norway	11.2	11.2	34.7
USA	11.2	12.9	17.7
Denmark	10.6	13.46	41.7
Bulgaria	10.5	12.4	45.8
Germany	9.8	11.9	28.7
Portugal	9.6	15.4	77.5

(continues)

FOCAL RESEARCH *(continued)*

TABLE **6.1** *(continued)*

Country	(a) Official suicide rate per 100,000	(b) Lower estimated suicide rate (combined official rates per 100,000 of suicide and undetermined injury intent mortality)	(c) Upper estimated suicide rate (combined official rates per 100,000 of suicide, undetermined injury intent mortality, and ill-defined and unknown cause mortality)
Ireland	9.1	10.5	12.7
Netherlands	8.7	9.2	34.9
Spain	6.6	6.7	22.6
United Kingdom	6.5	9.9	20.3
Malta	6.0	6.4	13.2
Italy	5.2	5.3	14.3
Cyprus	2.4	2.6	46.2

Correlation between (a) & (b) Spearman's rho = 0.93

Correlation between (a) & (c) Spearman's rho = 0.40

*Data for Estonia, Hungary, Slovakia, and Spain are for 2005, and for Portugal 2004.

Data Sources: World Health Organization, *European Health for All Database (HFA-DB), and the* United States Centers for Disease Control and Prevention, CDC WISQARS (Web-based Injury Statistics Query and Reporting System) and CDC WONDER (Wide-ranging Online Data for Epidemiologic Research).

We now turn to **validity**. We've already noted that the specificity of suicide classification—or the percent of true nonsuicides that are classified as nonsuicides—approximates 100%. If the sensitivity of suicide classification—the degree to which true suicides are classified as suicides—were consistently as good, we could declare it to be valid. But our tests of sensitivity cast grave doubts. These tests involve comparing the number of official suicides in 28 countries to their corresponding estimated lower and upper limits for the number of true suicides. The lower-limit estimate, again, means adding the number of official suicides in each country to the number of deaths from injury of undetermined intent. We then created our upper-limit measure of the sensitivity of suicide classification by dividing the official number of suicides by this lower-limit estimate for the number of true suicides and multiplying by 100, to get our first estimate of the percentage of true suicides in each country represented by the official number. (Note: this is an upper-limit, or relatively high percentage because it results from dividing a relatively small number, the lower-limit estimate of true suicides, into the official suicide number.) Our upper-limit estimate is the number of official suicides plus the number of deaths from both injury of undetermined intent and ill-defined and unknown causes. We then created our lower-limit measure of the sensitivity of suicide classification by dividing the official number of suicides by this larger number to get our second estimate of the percentage of true suicides represented by the official number. (This is a lower-limit, or relatively low, percentage, because it results from dividing a relatively high number, the upper-limit estimate of true suicides, into the official suicide number.)

Our estimated measures of the sensitivity of suicide classification for 28 countries are shown in Table 6.2. Our upper-limit measures vary between 62% and 100%, while our lower-level measures vary from 5% to 92%. If the lower-level measures are apt, they indicate that, in quite a few countries, many true suicides are currently missed by the existing system of suicide classification. And research suggests that even this measure of the missed true suicides may be conservative. This is because it neglects two other categories of death classification that are likely to hide suicides: unintentional drowning and unintentional poisoning.

TABLE **6.2**

Estimated percentage sensitivity of suicide certification for selected countries, circa 2006*

Country	Lower limit	Upper limit
Lithuania	49.3	68.5
Hungary	90.9	96.1
Slovenia	55.5	86.2
Latvia	19.7	73.5
Finland	75.8	92.6
Estonia	27.9	64.5
France	30.3	96.1
Switzerland	43.5	94.3
Poland	18.3	69.9
Austria	63.3	89.3
Luxembourg	40.3	91.7
Czech Republic	50.8	74.6
Sweden	38.9	81.3
Slovakia	40.0	71.4
Romania	69.0	90.1
Norway	32.3	100.0
USA	63.3	87.0
Denmark	25.4	79.4
Bulgaria	22.9	84.7
Germany	34.1	82.0
Portugal	12.4	62.5
Ireland	71.4	86.2
Netherlands	24.9	94.3

(continues)

FOCAL RESEARCH *(continued)*
TABLE **6.2** *(continued)*

Country	Lower limit	Upper limit
Spain	29.2	98.0
United Kingdom	31.9	65.4
Malta	45.4	93.5
Italy	36.4	97.1
Cyprus	5.2	90.9

*Data for Estonia, Hungary, Slovakia, and Spain are for 2005, and for Portugal 2004.
Data Sources: Same as for Table 6.1.

THINKING ABOUT ETHICS

Rockett and Kapusta use aggregated data rather than data about individuals. Their main ethical responsibility, then, is to report their findings honestly.

Rockett and Kapusta illustrate a creative use of a standard check on reliability (the alternate-forms procedure) and some useful new ways of thinking about validity. In their check on reliability, they argue that the official suicide rate is but one estimate of the true suicide rate in each country. They claim that two alternate forms of measuring the suicide rate would be to create a rate that added official suicides to all injuries of undetermined intent, on the one hand, and another that added these two to all deaths by ill-defined or unknown causes. If the official suicides rates are to be considered reliable, they argue, they should be highly correlated with each of these other estimates of the true suicide rate. But they aren't. In fact, the official rate is correlated only at a moderate 0.40 level with the second rate. They therefore infer, using the logic of the alternate-forms procedure, that the official suicide rate is an unreliable measure of the true suicide rate for cross-national research.

 STOP & THINK

Do Rockett and Kapusta claim that the validity of suicide measurement systems is hampered by a lack of sensitivity or specificity?

sensitivity, the ability of a system of measurement to identify correctly those who have a condition of concern.

specificity, the ability of a system of measurement to identify correctly those who do not have a condition of concern.

In discussing validity, Rockett and Kapusta introduce us to a couple of new concepts: **sensitivity** and **specificity**. These are both related to our concept of validity, or the degree to which a measure measures what we think it is measuring. When we are dealing with a measurement that involves deciding whether an individual case falls into a category or not (such as whether a death is a suicide or not), we can redefine **validity** as the ability of the system to indicate which individuals fall into the category and which do not. Sensitivity, then, refers to the ability of a measurement scheme to identify correctly those who should fall into that category, while specificity refers to its ability to identify correctly those who should not fall into that category. Rockett and Kapusta make the believable case that most suicide classification systems get it right about non-suicides and puts them into some category other than suicide (i.e., the systems have high levels of specificity). But they show that, at the national level, systems of classification don't always get it right when putting suicides into the suicide category (i.e., they have low levels of sensitivity). They therefore conclude that, for the purposes of cross-national research, suicide measures are neither valid nor reliable.

One last quality of measurement is its precision or exactness. Some measurement strategies allow greater precision than others do. If we ask people how old they were at their last birthday, we can be more precise in our categorization than if we ask them to tell us their ages using broad categories such as "26 to 64." If we use the seven categories of life satisfaction (from extremely satisfied to extremely dissatisfied) that Diener (2001) suggests, we can be more precise than if we use only two categories (satisfied and dissatisfied), though less precise than using the 5- to 35-point scale on which the categories are based.

LEVEL OF MEASUREMENT

Consideration of precision leads us to consider another way that variable categories can be related to one another—through their level of measurement or their type of categorization. Let's look at four levels of measurement: nominal, ordinal, interval, and ratio, and the implications of these levels.

Nominal Level Variables

nominal level variables, variables whose categories have names.

Nominal (from the Latin, *nomen*, or name) **level variables** are variables whose categories have names or labels. Any variable worth the name is nominal. If your variable is "cause of death" and your categories are "suicide," "homicide," "natural causes," "accident," and "undetermined," your variable is a nominal level variable. If your variable is "sex" and your categories are "male" and "female," your variable is nominally scaled. Nominal level variables permit us to sort our data into categories that are mutually exclusive, as the variable "religion" (with categories of "Christian," "Jewish," "Muslim," "none," and "other") would. Another example of a nominal level variable is marital status, if its categories are "never married," "married," "divorced," "widowed," and "other."

A nominal level of measurement is said to be the lowest level of measurement, not because nominal level variables are themselves low or base, but because it's virtually impossible to create a variable that isn't nominally scaled.

Ordinal Level Variables

ordinal level variables, variables whose categories have names and whose categories can be rank-ordered in some sensible way.

Ordinal (from the Latin, *ordinalis*, or order or place in a series) **level variables** are variables whose categories have names or labels and whose categories can be rank-ordered in some sensible fashion. The different categories of an ordinal level variable represent more or less of a variable. If racism were measured using categories like "very racist," "moderately racist," and "not racist," the "very racist" category could be said to indicate more racism than either of the other two. Similarly, if one measured social class using categories like "upper class," "middle class," and "lower class," one category (say "upper class") could be seen to be "higher" than the others.

All ordinal level variables can be treated as nominal level variables, but not all nominal level variables are ordinal. If your variable is "sex" and your categories are "male" and "female," you'd have to be pretty sexist to say that one category could be ranked ahead of the other. Sex is a nominal level variable, but not an ordinal level variable. On the other hand, if your categories of age are "less than 25 years old," "between 25 and 64," and "65 or older," you can say that one category is higher or lower than another. Age, in this case, is ordinal, though it can also be treated as nominal. Ordinal level variables guarantee that categories are mutually exclusive and that the categories can be ranked, as the variable "team standing" (with the categories of "first place," "second place," and so on) does in many sports.

Interval Level Variables

interval level variables, variables whose categories have names, whose categories can be rank-ordered in some sensible way, and whose adjacent categories are a standard distance from one another.

Interval level variables are those whose categories have names, whose categories can be rank-ordered in some sensible way, and whose adjacent categories are a standard distance from one another. Because of the constant distance criterion, the categories of interval scale variables can be meaningfully added and subtracted. SAT scores constitute an interval scale variable, because the difference between 550 and 600 on the math aptitude test can be seen as the same as the difference between 500 and 550. As a result, it is meaningful to add a score of 600 to a score of 500 and say that the sum of the scores is 1100. Similarly, the Fahrenheit temperature scale is an interval level variable because the difference between 45 and 46 degrees can be seen to be the same as the difference between 46 and 47 degrees. All interval level variables can be treated as ordinal and nominal level variables, even though not all ordinal and nominal level variables are interval level.

Ratio Level Variables

ratio level variables, variables whose categories have names, whose categories may be rank-ordered in some sensible way, whose adjacent categories are a standard distance from one another, and one of whose categories is an absolute zero point—a point at which there is a complete absence of the phenomenon in question.

Ratio level variables are those whose categories have names, whose categories can be rank-ordered in some sensible way, whose adjacent categories are a standard distance from one another, and one of whose categories is an absolute zero point—a point at which there is a complete absence of the phenomenon in question. Age, in years since birth, is often measured as a ratio level variable (with categories like one, two, three, or four years, and so on) because one of its possible categories is zero. Other variables such as income, weight, length, and area can also have an absolute zero point and are therefore ratio level variables. They are also interval, ordinal, and nominal level variables. For practical purposes, however, interval and ratio level variables are similar, and we will refer to them later as either interval ratio or simply interval level variables. This raises the question of what practical purposes there are for learning about levels of measurement in the first place, however, and that's the topic of our next section.

STOP & THINK   *When medical examiners must decide whether a death is due to suicide or not, what level of measurement are they using? What makes you think so? When the World Health organization reports that the U.S. has a suicide rate of 11.2 per 100,000 people, what level is it ascribing to suicide rates? What makes you think so?*

The Practical Significance of Level of Measurement

Perhaps the most practical reason for being concerned with levels of measurement is that frequently researchers find themselves in the position of wanting to summarize the information they've collected about their subjects, and to do that they usually use statistics (see Chapter 15). An interesting thing about statistics, however, is that they've all been designed with a particular level of measurement in mind. Thus, for instance, statisticians have given us three ways of describing the average or central tendency of a variable: the mean, the median, and the mode. The mean is defined as the sum of a set of values divided by the number of values. The mean of the ages 3 years, 4 years, and 5 years is $(3 + 4 + 5)/3 = 4$ years. Thus, the mean has been designed to summarize interval level variables. If age were measured only by categories "under 25 years old," "25 to 64," and "65 and older," you couldn't hope to calculate a mean age for your subjects. The median, however, has been designed to describe ordinal level variables. It's defined as the middle value, when all values are arranged in order. You could calculate a median age if your (five) respondents reported ages of, say, 3, 4, 5, 4, and 4 years. (It would be 4, because the middle value would be 4, when these values were arranged in order—3, 4, 4, 4, 5). You could calculate a median because all interval level variables (as age would be in this case) are ordinal level variables. But you could also calculate a median age even if your categories were "under 25 years old," "25 to 64," and "65 and older."

STOP & THINK   *Suppose you had five respondents and three reported their ages to be "under 25 years old," one reported it to be "25 to 64," and one reported it to be "65 and older." What would the median age for this group be?*

A third measure of central tendency or average, the mode, has been designed for nominal level variables. The mode is simply the category that occurs most frequently. The modal age in the previous "Stop & Think" problem would be "under 25 years old" because more of the five respondents fell into this category (three) than into either of the other categories. The modal age for a group with ages 3, 4, 5, 4, 4, would be 4, for the same reason. You can also calculate the modal sex if, of five participants, three reported they were female and two reported they were male. (It would be "female.") And, you wouldn't have been able to calculate either a median or a mean sex.

The upshot is that the scale of measurement used to collect information can have important implications for what we can do with the data once they've been collected. In general, the higher the level of measurement, the greater the variety of statistical procedures that can be performed. As it turns out, all statistical procedures that scientists use are applicable to interval (and, of course, ratio) data; a smaller subset of such procedures is applicable to ordinal data; a smaller subset still to nominal data.

SUMMARY

Measurement means classifying units of analysis by categories to represent a variable concept. Measuring concepts in quantitative research involves conceptualizing, or clarifying, what we mean by them, and then operationalizing them, or defining the specific ways we will make observations about the concepts in the real world. Measurement in qualitative research often involves making observations, coding those observations into categories, and then looking for relationships among the categories that might constitute usable variables.

The first step of quantitative measurement is to construct a conceptual definition of a concept. The second step is to operationalize the concept by identifying indicators with which to classify units of analysis into two or more categories. The categories should be exhaustive, permitting you to classify every subject in the study, and mutually exclusive, permitting you to classify every subject into one and only one category. Depending on the complexity of the concept involved, you can use simple or complex measures of it and be more or less precise in your measurement. Indexes, like those used by the Bhutanese researchers in their study of Gross National Happiness, frequently involve the summation of scores on a variety of items used to measure a concept.

The best measurement strategies are reliable and valid. Reliability refers to whether a measurement strategy yields consistent results. We can examine the reliability of a measurement strategy with a test–retest, internal consistency, inter-rater reliability, or alternate-forms check. Rockett and Kapusta used an alternate-forms check to determine that official suicide rates have low levels of reliability for cross-national research. Validity refers to whether a measurement strategy measures what you think you are measuring. We can examine the validity of a measurement by considering its face validity (by seeing how we and others feel about its validity), by considering its content validity (by seeing how experts evaluate its various elements), by testing its predictive or concurrent validity (by seeing how it's correlated with future or current behaviors of interest), or by testing its construct validity (by seeing how it's correlated with some other variable that theory predicts it will be associated with). Rockett and Kapusta point to two concerns about the validity of death classification systems: 1) whether they correctly classify people who have died of a particular cause of death as having died of that cause (i.e., whether they have sensitivity), and 2) whether they correctly classify people who have not died of a particular cause as not having died of that cause (i.e., whether they have specificity). They find that suicide classification systems lack sensitivity.

There are three significant levels of measurement (nominal, ordinal, and interval ratio) in the social sciences. In general, the data analyses that can be used with a given set of variables depend on the level of measurement of the variables.

EXERCISE 6.1

Putting Measurement to Work

This exercise gives you the opportunity to develop a measurement strategy of your own.

1. Identify an area of the social sciences you know something about (aging, marriage and the family, crime and criminal justice, social problems, social class, and so on).
2. Identify a concept (e.g., attitudes toward the elderly, marital satisfaction, delinquency, social class) that receives attention in this area and that interests you.
3. Describe how you might conceptualize this concept.
4. Describe one or more measurement strategies you might use to operationalize this concept. List at least one or more indicators and two or more categories.
5. Identify the level of measurement of this concept given the operationalization you suggested. Give support for your answer.
6. Discuss the reliability of your measurement strategy. How might you check the reliability of this strategy?
7. Discuss the validity of your measurement strategy. How might you check the validity of this strategy?

EXERCISE 6.2

Comparing Indicators of Life Satisfaction: Global Evaluation versus Areas of Life

Some researchers have asked people to evaluate how satisfied they are in various domains of their lives rather than asking for an overall evaluation. A Gallup poll, for example, asks a series of questions about different areas of life as one of the ways it measures life satisfaction. In this exercise, you're asked to compare two alternative sets of indicators for life satisfaction—the one suggested by Diener and his associates and one used by the Gallup poll.

1. Make two copies of the five questions on life satisfaction suggested by Diener et al. at the beginning of the chapter and find two people to answer the questionnaires.
2. After the people complete the questionnaire, interview them and ask the following questions adapted from a Gallup poll. Read the questions to them and write in the answers they give. You could also answer the questions yourself to see how you feel about the issues.

   Using the answers "very satisfied," "somewhat satisfied," "somewhat dissatis-fied," or "very dissatisfied," how satisfied are you with each of the following aspects of your life?
   a. Your community as a place to live in _____
   b. Your current housing _____
   c. Your education _____
   d. Your family life _____
   e. Your financial situation _____
   f. Your personal health _____
   g. Your safety from physical harm or violence _____
   h. The opportunities you have had to succeed in life _____
   i. Your job, or the work you do _____

3. Americans responded to these questions in a survey conducted on June 11–17, 2001. Review Table 6.3 from Gallup.
4. Discuss the measurement of "life satisfaction" by answering the following questions:
   a. Which set of questions—Diener's or Gallup's—do you think results in a more valid measure of life satisfaction? Why?
   b. Which set of questions do you think results in a more reliable measure of life satisfaction? Why?

TABLE **6.3**

Percentage Answering "Very Satisfied" to Questions

Adults	Men	Women	Non-Hispanic	Whites	Blacks	Hispanics
Your family life	69%	67%	72%	72%	52%	73%
Your current housing	63	63	63	66	41	45
Your community as a place to live in	58	59	57	61	42	53
Your education	45	48	44	48	38	38
Your financial situation	26	28	25	29	15	23
Your personal health	45	57	52	54	54	57
Your safety from physical harm or violence	55	61	50	59	33	47

c. What would be an advantage of using both the Gallup and the Diener measures of satisfaction in a study?

d. Can you suggest an index that could be constructed based on the nine questions asked by the Gallup organization?

e. Suggest one or more ways of measuring "life satisfaction" in addition to the two sets of questions presented here. Describe one or more additional questions that could be asked, one or more observations that could be made, or one or more pieces of information that could be obtained from records or documents.

Cross-Sectional, Longitudinal, and Case Study Designs

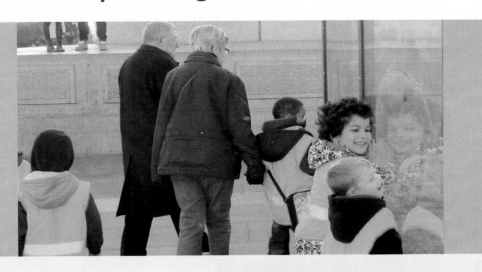

CHAPTER OBJECTIVES

This chapter is designed to help students

- **Understand study design**
- **Identify and compare cross-sectional, panel, trend, cohort, and case study designs**
- **Recognize advantages and disadvantages of each nonexperimental design**
- **Know the limitations of cross-sectional designs for testing causal hypotheses**
- **Design cross-sectional, panel, trend, cohort, and case studies for specific research purposes**

CHAPTER OUTLINE

Designing a Study

Study Design Choices

Summary

Exercises

DESIGNING A STUDY

You might be thinking about career choices you'll make in the near future. What factors have influenced your career goals? Do you remember what you wanted to be when you were younger? What would you have answered to the question, "What do you want to be when you grow up?" Have you changed your mind in recent years?

RESEARCH IN THE NEWS

The British Broadcasting Corporation (BBC) reported the results of a 2012 study conducted by researchers at London University which asked more than 11,000 seven-year-olds what they wanted to be when they grew up. The study found that the 12 most popular occupations were teacher, scientist, hairdresser, sports player, firefighter, police officer, artist, entertainer, animal carer, vet, doctor, and builder. Girls were more interested in careers involving helping or caring for people or animals while boys were more likely to be motivated by dreams of becoming powerful, popular, or wealthy. Do you think this study's findings would be different if conducted in other countries?

Source: http://www.bbc.co.uk/news/education-19586824

If you were to create a hypothesis about career aspirations, what would it be? What kind of research could you do to test it? Based on what you've learned so far, you could plan almost all parts of a study to test your hypothesis including using theory in some way, knowing how to be ethical in conducting your study, making decisions about how to select a sample, considering how to measure the concepts, and figuring out whether your study is feasible in terms of time, money, and access. In this and the next chapters, we'll cover the remaining research decisions you'd need to make: the selection of a study design and the choice of one or more methods of data collection.

A **study design** is a research plan that results from several interconnected decisions made on ethical, practical, and methodological bases. It is obvious that data must be collected at least once from at least one unit of analysis, but that is the minimum. The study design choices that must be made are as follows:

> **study design,** a research strategy specifying the number of cases to be studied, the number of times data will be collected, the number of samples that will be used, and whether the researcher will try to control or manipulate the independent variable in some way.

1. Is it appropriate to study only one or a few cases or is a larger sample needed?
2. Is it appropriate to collect data only one time or more than once?
3. Is it appropriate to collect data about only one sample or about more than one sample?
4. Is it appropriate to measure the independent variable or should it be manipulated or "controlled" in some way?

> **causal hypothesis,** a testable expectation about an independent variable's effect on a dependent variable.

For a study with a **causal hypothesis,** where a researcher speculates that one variable, the independent variable, is the *cause* of another variable, the dependent variable, there is an additional decision. As discussed in Chapter 2, it is difficult to establish causality in part because to show that change in the independent causes the change in the dependent variable, we must know that the independent variable comes first. One way to determine the time order would be to *control or manipulate the independent variable* in some way. In a study you'll read about in Chapter 8, for instance, Chris Caldeira wanted to know if using film clips in class is an effective way to help students comprehend core concepts in sociology. The independent variable (whether or not students saw film clips) could be manipulated so that some students saw film clips and some did not. The study is an experiment—a design in which the manipulation of an independent variable is integral.

Researchers must balance methodological concerns against practical matters and ethical responsibilities when designing a study. Because it is almost always easier, less time consuming, and less expensive to collect data once, from one sample, and to measure rather than control or manipulate independent variable(s), making decisions that will cost more in time or money must be justified. In some cases, it is just not possible to control the independent variable; in other cases, it might be unethical to do so. Researchers know that controlling or manipulating something is an important responsibility because of the potential effect on people's lives. In addition, studying a sample more than once means we must keep confidential the names and addresses of study participants.

Connections between the Uses of Research, Theory, and Study Design

Decisions about the design are based partly on the purposes of the research, the researcher's interest in testing causal hypotheses, and the use of theory. You might remember that in earlier chapters we talked about using research for purposes of exploration (breaking new ground), description (describing groups, situations, activities, events, and the like), explanation (explaining why things are the way they are), and evaluation (assessing the impacts of programs, policies, or legal changes). We described theory as an explanation about how and why something is as it is and compared the deductive and inductive approaches to theory—seeing that theory can be used to deduce a testable hypothesis or can be generated after data are collected and analyzed. Our point in recalling the purposes of research and the deductive and inductive approaches here is that these purposes and approaches have implications for research design decisions.

STUDY DESIGN CHOICES

STOP & THINK   *Let's say you were interested in studying what people do with their time after they retire from their career jobs. How might you go about planning this study? Would you decide to talk to people who have already retired or those who are planning to retire? Would you observe people at Senior Centers? Ask for information from local organizations that serve older members of the community? How might you pick your population and sample? Would you need information about many people or just a small group? Would you want to find about their plans or their actual behavior? Would you need to collect data more than once?*

Cross-Sectional Study Design

Given a projected shortage of primary care doctors, researchers studied the retirement plans of older physicians. Phillip Sloane and his colleagues thought the shortage might be partially alleviated by physicians working in part-time or volunteer capacities after retirement. They mailed a four-page questionnaire to a random sample of North Carolina physicians in direct care–related specialties who were 55 or older and received 910 responses (a 59 percent response rate) (Sloane et al., 2008). They asked about current work status, age, health, hours working for pay, retirement plans, current volunteer activities, and interest in volunteerism in retirement. Fifty-seven percent of the respondents were working full-time, 22 percent were working part-time, and 21 percent were currently not working for pay. Forty percent were doing some kind of volunteer work within medicine (such as working in a free medical clinic, providing disaster relief services, or doing volunteer medical teaching), with those employed having the highest rate (54 percent) and those not working for pay having the lowest (37 percent). Interest in volunteering in retirement was high: 43 percent of those working full-time, 38 percent of those not working for pay, and 33 percent of those working part-time expressed this interest (Sloane et al., 2008: 320). Based on their data, Sloane and his colleagues can *describe* the current behavior and future plans of the sample and some of the relationships among the variables. Of course, they can't know whether the doctors actually follow through with their plans after retirement.

cross-sectional study, a study design in which data are collected for all the variables of interest using one sample at one time.

The research by Sloane and his colleagues is an example of the **cross-sectional study,** the most frequently used design in social science research. In the cross-sectional design, data are collected about one sample at one point in time, even if that "one time" lasts for hours, days, months, or years. It might have taken the research team weeks or even months to obtain all of the completed questionnaires, but each physician answered the questions only once. Another example of the cross-sectional design is Desirée Ciambrone's study of women living with HIV or AIDS (the focal research in Chapter 3).

Cross-sectional designs are widely used because they enable the description of samples or populations and because it is least expensive to collect data once from one sample. Cross-sectional studies often have large samples and usually have data that lend themselves to statistical analyses. In looking for patterns of relationships among variables, the sample can be divided into two or more categories of an independent variable to identify differences and similarities among the groups on some dependent variable. Sloane et al. (2008), for example, could analyze the connection between age and employment status by comparing the percentages of each age group that are employed full-time and part-time. Cross-sectional studies like Sloane et al.'s that have at least one independent and one dependent variable (age and employment status, e.g.) can be drawn as a diagram[1] as shown in Figure 7.1.

---

[1]This diagram and the others used in this chapter and Chapter 8 have been influenced by those presented by Samuel Stouffer (1950) in his classic article "Some Observations on Study Design."

		time 1
one sample (divided up into categories of the independent variable during analysis)	category 1 of independent variable *ages 55 to 59*	measure of dependent variable *83% are employed full-time* *13% are employed part-time*
	category 2 of independent variable *ages 60 to 64*	measure of dependent variable *61% are employed full-time* *28% are employed part-time*
	category 3 of independent variable *ages 65 to 69*	measure of dependent variable *35% are employed full-time* *28% are employed part-time*

FIGURE **7.1** The Cross-Sectional Study Design

 STOP & THINK   *Here are four variables that might "go together" in one or more patterns:*

1. student's gender
2. parental annual income
3. student's perceptions of peers' alcohol use
4. student's alcohol use

Use at least two of these variables and construct one or more hypotheses. Think about the time order and speculate about which of these variables might "go before" others. State your hypothesis and think about how to do a cross-sectional study to test your hypothesis.

Cross-Sectional Studies and Causal Relationships

Cross-sectional designs are sometimes used to examine causal hypotheses. You might remember from our discussion in Chapter 2 that just because two variables "go together" doesn't mean that one variable *causes* change in another. Let's take a simple example of a causal hypothesis and hypothesize that the sight of a bone makes your dog salivate.

      Our example would need to meet three conditions before we could say a **causal relationship** exists. In a causal relationship, there must be an *empirical association* between the independent and dependent variables. In the case of the dog and the bone, first, we'd need to see the dog generally salivating in the presence of a bone and generally failing to salivate in its absence. Second, we'd need to make sure that the *temporal precedence* or *time order* was correct—that is, we'd need to make sure that the bone appeared before, rather than after, salivation. (This is to make sure that the bone causes salivation, rather than salivation "causing," in some sense, the bone.)

causal relationship, a nonspurious relationship between an independent and a dependent variable with the independent variable occurring before the dependent variable.

spurious relationship, a non-causal relationship between two variables.

Finally, we'd want to make sure that it is not a **spurious relationship,** or caused by the action of some third variable, sometimes called *antecedent variable,* that comes before, and is actually responsible for, making the independent variable and the dependent variable vary together. We'll illustrate spuriousness with a different example. Consider, for instance, that in some regions there is a relationship between two autumnal events: leaves falling from deciduous trees and the shortening of days. One might be tempted after years of observing leaves falling while days shorten (or, put another way, days shortening while leaves fall) to hypothesize that one of these changes causes the other. But the relationship between the two events is really spurious, or non-causal. Both are caused by an antecedent factor: the change of seasons from summer to winter. It is perfectly true that, in some regions, leaves fall while days shorten, but it is also true that both of these events are due to the changing of the seasons, and so neither event causes the other. Therefore, the two events are spuriously related.

requirements for supporting causality, the requirements needed to support a causal relationship include a pattern or relationship between the independent and dependent variables, determination that the independent variable occurs first, and support for the conclusion that the apparent relationship is not caused by the effect of one or more third variables.

The reason we mention the criteria for determining causation in the context of research designs is simple: Some research designs are better than others for examining the **requirements for supporting causality.** Thus, for instance, although it is possible to use a cross-sectional design to test causal hypotheses, this design might be less than ideal for such tests in some circumstances. We can determine the relationships between variables and sometimes figure out the time order when collecting data for two or more variables at one point in time. For example, several hypotheses were tested in a cross-sectional study that compared the career aspirations of ethnic minority and European American psychology graduate students. The findings included some significant differences between groups: Latino and Latina students were much more likely than any other group to report direct service rather than career aspirations and women were somewhat more likely than men to report interest in direct service (Maton et al., 2011: 72). While the effects of third variables still need to be considered, the time order is clear: ethnicity and gender come before adult career aspirations. In other studies it may be difficult to disentangle the time order. It is always possible to ask questions about past behavior, events, or attitudes, but such retrospective data can involve distortions and inaccuracies of memory.

⬤ **STOP & THINK**   *You may remember that the research on happiness in Bhutan includes the finding of greater happiness in urban rather than rural areas. Can you think of a problem with concluding that being living in a city **causes** a person to be happier?*

While there is a relationship between the two variables and perhaps a theoretical reason to assume that happiness is a dependent variable, we don't really know which comes first (and is a cause)—being happy or where a person lives—since both variables are measured at the same time. Nor do we know if a third, antecedent, variable came first—perhaps a lower fear of crime among those choosing to live in cites—which contributed to the relationship between the variables. Time order can be established by measuring by using an alternative study design, one that enables the collection of data more than once. That way the researchers can determine the time order of the variables.

So, be cautious in interpreting the results of studies that use a cross-sectional design for explanatory purposes. The cross-sectional study design is very appropriate for describing a sample of one or more variables and for seeing connections between the variables. It can also be useful for studying causal hypotheses when the time order between the variables is easy to determine and when sophisticated statistical analyses can be used to control for possible antecedent variables.

Longitudinal Designs: Panel, Trend, and Cohort Designs

In the focal research that follows, Emily and Roger studied the attitudes and experiences of workers, including feelings about current jobs, plans for retirement, and actual experiences in retirement but, unlike the study by Sloane and his colleagues, we did not do a cross-sectional study.

FOCAL RESEARCH

Moving On? Continuity and Change after Retirement

By Emily Stier Adler and Roger Clark*

Retirement is an important life transition. In addition to the issues of income and identity, retirement means changes in daily routines and typically provides the retiree with more discretionary time. Before a planned and voluntary retirement, workers think about post-retirement activities. Often their plans include a list of things they won't be doing— such as not setting an alarm clock, not eating lunch at the desk, and not having a schedule for every hour during the week—as well as a list of added activities—such as volunteering, getting exercise, doing some work for pay, spending time with grandchildren, reading novels, and the like. With the first of the 76 million "Baby Boomers"—people born between 1946 and 1964—starting to retire (U.S. Census Bureau, 2006), it's especially important to know how people will spend their time after retirement.

Labor Force Participation, Retirement, and Volunteerism

The National Health and Retirement Study (HRS), a biannual panel study of a nationally representative sample of more than 20,000 Americans over the age of 50, finds that the majority of men and women in their fifties work for pay (70% and 60%, respectively), most on a full-time basis, but that labor force activity declines after age 60 (HRS, 2007, p. 41). By age 62, approximately half of men and more than 40% of women are working for pay, at least two thirds of them full-time; by age 65, half of those employed are working part-time and the labor force participation of both women and men is close to half of the participation rate of people in their fifties (HRS, 2007, p. 41). Analysis of HRS data also finds increasing proportions of those in their early to mid-fifties who *expect* to work full-time after age 65. For example, among college educated men, the proportion increased from just over 30 percent in 1992 to almost 50 percent in 2004, while for college educated women it went from under 30 percent to almost 40 percent (HRS, 2007, p. 49). A Pew Research Center (2006) survey finds that

(continues)

*This article was written for this text and is published with permission.

FOCAL RESEARCH *(continued)*

77% of workers expect to do some kind of work for pay after retirement. The same survey, however, found that only 12 of those retired were currently working for pay either full-time or part-time (Pew Research Center, 2006, p. 2).

Retirement may be a gradual process. One analysis of the HRS data finds that at least half of those moving out of full-time career jobs don't leave the labor force, but at some point take short-duration, part-time jobs or transitional jobs (often called bridge jobs) after retirement (Cahill, Giandrea, and Quinn, 2006). In their analysis of the HRS data, Gustman and Steinmeier (2000, p. 63) found that from 1992 to 1998, 71% of the sample reported no change in status, 17% moved from greater to less labor force participation, and 6% moved from less to more participation.

There are other ways to be productive beyond paid employment, including volunteering for community organizations. Such activities may replace paid work and can provide the individual with an identity, a social role, friendship networks, a sense of being productive, and daily or weekly structure for activities. In 2008, 26% of the American population and 28% of those 55 to 64 years of age did volunteer work, spending about an hour a week on these activities (U.S. Department of Labor, 2009). A survey found that 62% of Baby Boomers who currently volunteer expect to devote more time to service in retirement while 36% who are currently not actively engaged in volunteering expect to do so (AARP, 2004). While it might be expected that those employed in the labor force would have less time to volunteer, those who are employed part-time have the highest rate of volunteering (34%), followed by those employed full-time (28%) and those who are unemployed or not in the labor force have the lowest rate of volunteering (22%) (U.S. Department of Labor, 2009).

As workers retire in the next few decades, how will they spend their time? Role theory and continuity theory can help us predict role constellations in the immediate post-retirement period. *Role theory* (e.g., Ashforth, 2001; Wang, 2007) suggests that to the extent people are highly invested in a particular role, their feelings of self-worth are related to the ability to carry out that role. Role theorists also argue that the loss of roles can lead to depression or anxiety. We posit, then, that after retirement, to the extent possible, people will retain previous roles because doing so enables them to maintain feelings of self-worth and avoid feelings of anxiety and depression. Moreover, we posit that pre-retirement expressed *desires* to perform new roles (like new volunteer roles) after retirement are less likely to be based upon the same emotional calculus of self-worth, depression, and anxiety that informs the desire to continue performing a pre-retirement role. In fact, we expect that all workers have at least some cognitive access to the emotional and psychological needs that work roles have satisfied for them and that their plans to do some kind of work after retirement will therefore be related to whether they actually work for pay after retirement. More specifically, our research is guided by three hypotheses, generated from role theory:

Hypothesis 1: People who express a desire for work after retirement are more likely to actually engage in post-retirement work than others.

Hypothesis 2: People who have performed volunteer roles before retirement will be much more likely to perform such roles after retirement than people who have not performed such roles.

Hypothesis 3: People who before retirement express a desire to perform roles (such as volunteer roles) that they have not previously performed are not more likely to actually perform those roles after retirement than people who do not express this preretirement desire.

Role theory alone cannot explain which people will be attracted to working for pay after retirement because, by definition, everyone who retires is retiring from a work role. *Continuity theory*, on the other hand, suggests reasons why there may be a *selective investment* in roles (including occupational roles), such as differential feedback on the performance of roles (Atchley, 1994). Within a given occupational or professional level one may expect that those who have worked for more years to derive greater satisfaction from the work role. The longer people participate in a role, we posit, the longer they will have had to organize their ideas and feelings about themselves in relationship to those roles and the more likely they will be to seek the structure of organized roles after retirement. More specifically, our research is guided by one additional hypothesis derived from continuity theory:

Hypothesis 4: The longer a person has been in the workforce, the more likely she/he is to pursue a workplace role after retirement.

Methodology

In order to study the *transition* to retirement, it is necessary to use a longitudinal design. We used a panel study with two data collections, one before and one after retirement. Because there are no lists of workers who plan to retire from a variety of occupations, we selected a convenience sample of middle- and upper-middle-class workers who had definite plans to retire. A call for participants, describing the project, was distributed. We requested participants from those who were college educated or working as professionals, currently working at least 25 hours a week and had specific plans to retire within the next 18 months. Using e-mail, letters, Internet discussion group notices, announcements in a college bulletin, and personal contacts, we asked that the call for participants be passed along to others that might fit the sample and estimate that approximately 1,000 people received information about the study.

The final sample included 44 North American college-educated and/ or professional workers, currently working for pay 25 hours or more per week with plans to retire from full-time career employment. At time 1, the participants lived in 15 different states and worked in a variety of occupations: six college professors, four managers or administrators for large organizations, five public school teachers, four administrative assistants, three engineers, three computer information technology managers, three nurses working in hospital administration, three state or city department heads, two physicians, two school administrators, a psychologist, a dentist, an accountant, a podiatrist, a clergyman, a small business owner, an elementary school staff member, a social researcher, and a social worker. All of them identified their retirements as voluntary. They'd worked for pay between 16 and 51 years with a mean of 35.5 years in the labor force. Twenty-seven (61%) were female and 17 (39%) were male; all were white; 77% were married; and the rest were never married, divorced, widowed, or cohabitating with a same or opposite sex partner. Their anticipated ages at retirement ranged between 52 and 75, with an average age of 61 at planned retirement. The actual mean age at retirement of the 44 respondents was 61.4 years.

(continues)

FOCAL RESEARCH (continued)

All of the study participants signed informed consent forms and were interviewed by the same member of the research team twice. The pre-retirement interviews lasted between one and three hours and were conducted between November 2001 and July 2002 using a semi-structured format. The interviews were tape-recorded and transcribed. As the period immediately after retirement is often seen as a vacation or "honeymoon" and is frequently a transitional period before settling into new patterns, the second interview was conducted between 18 and 20 months after retirement. The participants were contacted within weeks of their expected retirement to see if they had in fact retired as planned. If not, they were re-contacted every few months until retirement. There was no panel attrition; all 44 post-retirement interviews were completed by 2007.

Findings

Post-Retirement Paid Work

Three quarters of the study participants did retire within a few weeks of their planned dates. The 25% who did not retire as anticipated postponed retirement from two months to four years with a median of one year. The major reason for postponing retirement was financial— either incentives to continue to work or worries about financial readiness for retirement—but a few retirees waited because of worries about how they would spend their time once retired.

Before retirement, participants were asked about their plans to work for pay after retirement, and if they planned to work, their expectations about the kind of work and the hours/days per week they'd thought they'd work. Ten (23%) of the sample did not expect to work after retirement, 14 (32%) were unsure of their plans to work, and 20 (46%) thought they probably would work after retiring. None expected to work full-time after retirement. In reality, at the time of the second interview, 15 (34%) were doing no paid work, 11 (25%) were working eight hours or fewer per week for pay, 13 (30%) were working more than 8 hours but less than 24 hours a week for pay, and 5 (11%) were working at least 24 hours a week for pay.

Our first hypothesis received support, as expectations of working after retirement were positively associated with both whether respondents were actually working after retirement and the number of hours they were employed. The Spearman's rho for the association between expectation of work and work after retirement is .46, while that between expectation of work and hours of work after retirement is .47 (Table 7.1). In fact, 90% of those planning to work after retirement were actually employed in the labor force at the time of the second interview, while 40% of those with no plans to work and 50% who had been unsure were employed.

When we looked at the number of years in the labor force, we found support for hypothesis 4 as years of paid work was correlated with pursuing a role in the workplace after retirement (rho is .37) and with number of hours employed (rho is .39) (see Table 7.1).

The correlation between years in the labor force and expectations of work at retirement was positive but weak (rho is .13). Tests not shown here suggest that both years in the labor force and expectations of work at retirement have independent effects on both whether people will work after retirement and how much they will work.

TABLE **7.1**

Associations of Work after Retirement and Hours of Work after Retirement with Expectation of Work, Age at Retirement, Years in the Labor Force, and Gender using Spearman's Rho (N = 44)

	Work after Retirement (1)	Hours of Work after Retirement
Expectation of Work (2)	.46**	.47***
Age at Retirement	−.08	.04
Years in Labor Force	.37*	.39**
Gender (3)	−.18	−.14

*Indicates significant at .01 level; **indicates significant at .001 level; ***indicates significant at .05 level.

(1) Work after retirement coded Yes = 1, No = 0; (2) Expectation of work coded No = 1, Maybe = 2, Yes = 3; (3) Gender coded female = 1, Male = 0.

Table 7.1 also shows that a person's age at retirement has almost no association with his or her chances of working after retirement (rho is .08) and that gender is not much associated with either of these two variables, even though males are slightly more likely than females to have been engaged in work for pay after they retired (rho is .18).

Some people seem to be able to predict their post-retirement needs for work very well before retirement. Joe, a high school math teacher planning to retire after 36 years, believed that he would have too much time on his hands if he did not work. He thought he would teach math and coach in some capacity after retirement because of his need for some structure and the advantages of extra income. And Joe was, in fact, teaching part-time at a local college and coaching track and field at a nearby high school two years after retirement. Joy, a high school biology and chemistry teacher for 30 years, on the other hand, had no plans to work after her retirement and, in fact, did not work afterward. Although she had been offered several jobs, including part-time teaching, Joy turned them down. She had suffered and recovered from a major health crisis, felt financially secure, and enjoyed being away from the stresses associated with work.

But some people do not foresee the extent to which work has become central to their lives or the extent to which they'll enjoy their lives without paid employment. Michael, an engineer for 42 years before he retired, did not expect to work afterward. In the years before retirement, he disliked the substantial amounts of travel his job required. He was able to separate his pleasure in his work from his dislike of the travel, asserting that "if you took away the travel, I would say I love the job," but he could not foresee wanting to work again. He spent the month after retirement hunting, hiking in the woods, and painting and repairing his house. "And then it became boredom time," Michael said. He found he did not like the lack of structure and the lack of contact with other people in retirement. By the time of the second interview, he had returned to his old job, minus the travel, on a contract basis, full-time for about half the year.

(continues)

FOCAL RESEARCH (continued)

On the other hand, some people found that, despite their pre-retirement intention to do some kind of remunerative work after retirement, they changed their minds. Before retirement, Anne, a department head in a state agency, had planned on post-retirement work as a consultant one or two days a week. But after a very few short-term consulting projects, Anne was a bit surprised to find she found excuses to turn down jobs until she realized, as she noted, "Wait a minute. I've done all this. There are other things I want to do with my life."

We wondered whether our data could help us determine when a career became long enough to have the relationship outlined above. We found that 30 years of labor force participation seemed to maximize the differences. Thus while 9 (64%) of 14 respondents with 30 or fewer years on the job were not engaged in work for pay two years after retiring, only 6 (20%) of the 30 respondents with 31 or more years of work were not so engaged. Moreover, all five of our respondents who were working at least 24 hours per week two years after retirement had worked 31 or more years before retiring.

Volunteer Work

Based on role theory, we hypothesized that those who had taken a volunteer role in the community before retirement were more likely to be volunteering after retirement. Before retirement, 19 (43%) reported no formal volunteer work, 21 (48%) reported up to four hours of formal volunteer work each week, and 4 (9%) reported five or more hours of volunteer work a week. After retirement, 43% still reported no formal volunteer work while 39% percent reported up to four hours of formal volunteer work and 18% reported five or more hours a week of formal volunteer work. Pre-retirement formal volunteer work is very strongly associated with post-retirement formal volunteer work (rho = .63).

Before retirement many participants told us that they wanted to spend time volunteering but we find no association between the expressed hope to do more volunteer work and the reported level of post-retirement volunteer work (rho = .02). But what about the hope to do more volunteer work after retirement with the level of pre-retirement volunteer work controlled? (After all, one might reasonably claim that the hope for doing more volunteer work after retirement should only be associated with *change* in levels of volunteer work, not with absolute levels of volunteer work after retirement.) Additional tests, not shown here, suggest that the pre-retirement desire to do more volunteer work after retirement still has no association with reported levels of post-retirement volunteer work, even after pre-retirement levels are controlled. Moreover, we find the number of hours participants work for pay after retirement, age at retirement, or years in the labor force are not significantly associated with post-retirement volunteer work.

Discussion

People on the cusp of retirement have expectations or hopes about their lives after retirement. They may plan to work for pay part-time or hope to become more engaged in their communities. In this paper, we have considered hypotheses generated from role and continuity theory to test whether people's expectations about these activities or their past behaviors are better predictors of post-retirement activity and found support for a continuation of activities. Among our participants, retirement seems a time of continuity and only moderate change. Before retirement, for instance, all of our participants expressed their expectations about how much they would work in the early years after retirement. They realized these expectations to a significant, but imperfect degree. However, an equally powerful predictor of their work after retirement was the number of years they had spent in the labor force. The longer people had channeled their energies into work, the more likely they were to be employed and employed for more hours two years after retirement, no matter what their plans were before retirement. For volunteer work, the major predictor of post-retirement behavior was not expectation but previous degree of engagement in volunteer work.

Interestingly, post-retirement working for pay and involvement in volunteer work was not substantially related in our sample. It is not true that people who are engaged in substantial amounts of post-retirement paid work are less likely than others to be involved in their communities.

Ours is not a representative sample and our causal conclusions are speculative. Our data are supportive of a causal connection between the length of one's career and the likelihood of working after retirement. We cannot be sure that the continuities we find in pre- and post-retirement activities are not spurious— perhaps the result of some personality need for external structure and community involvement pre-dating both work years and retirement. This said, among our participants, retirement does not seem to be a time in which people reinvent themselves. When it comes to both paid and volunteer work, the habits of a lifetime seem to be good predictors of post-retirement activity.

There are several implications of our study for members of a Baby Boomer generation about to retire. If people hope to engage in activities like volunteer work after retirement, they should begin these activities in the years prior to retirement and increase their time commitments once their paid work responsibilities have diminished. If they haven't had the chance to join organizations or volunteer prior to retirement, they may need to make special efforts to begin. On the other hand, if they have put in a full lifetime of work, they may be surprised how hard it is to withdraw from the workforce cold turkey or, conversely, how necessary it may feel to continue working for pay, at least to some degree. Given the enormous size of this generation, as well as its work experience, one macro-level implication is that it may be a supply of productive energy for years to come.

References

AARP. 2004. Baby Boomers Envision Retirement II: Survey of Baby Boomers' expectations for retirement. Retrieved February 15, 2009, from http://assets.aarp.org/rgcenter/ econ/boomers _envision.pdf

Ashforth, B. 2001. Role transitions in organizational life. Mahwah, NJ: Erlbaum.

Atchley, R. C. 1994. *Social forces and aging.* Belmont, CA: Wadsworth Publishing Company.

Cahill, Kevin E., Michael D. Giandrea, and Joseph F. Quinn. 2006. Retirement patterns from career employment. *The Gerontologist* 46: 514–523.

Gustman, A. L., and T. L. Steinmeier. 2000. Retirement outcomes in the health and retirement study. *Social Security Bulletin* 63(4): 57–71.

HRS. 2007. *Growing older in America: The health and retirement study.* Retrieved February 15, 2009, from http://hrsonline.isr.umich.edu//docs/databook/HRS_ Text_WEB_Ch2.pdf

(continues)

FOCAL RESEARCH (continued)

Pew Research Center. 2006. Working after Retirement: The Gap between Expectations and Reality. Retrieved February 15, 2009, from http://pewresearch.org/pubs/320/ working-after -retirement-the-gap-between-expectations-and-reality

U.S. Census Bureau. 2006. *Facts for Features Special Edition Oldest Baby Boomers.*
Turn 60! Retrieved February 15, 2009, from http://www.census.gov/Press-Release/ www/2006 /cb06ffse01-2.pdf

U.S. Department of Labor. 2009. Volunteering in the United States, 2008. Retrieved February 15, 2009, from http://www.bls.gov/news.release/pdf/ volun.pdf

Wang, M. 2007. Profiling retirees in the retirement transition and adjustment process: Examining the longitudinal change patterns of retirees' psychological well-being. *Journal of Applied Psychology* 92(2): 455–474.

THINKING ABOUT ETHICS

We submitted our research to our university's Institutional Review Board (IRB) for approval. The participants volunteered to be part of the study by responding to the request for participation; they also voluntarily agreed to the second interview and each time gave informed consent before being interviewed. Furthermore, we believe that no harm came to any of the participants as a result of being part of the study, we have kept all names and information confidential, and, in all reports, we have identified none of the participants by their real names.

 STOP & THINK

In our study we collected data twice from the same sample and therefore went beyond the cross-sectional design. Why do you think we didn't use a cross-sectional study design?

internal validity, agreement between a study's conclusions about causal connections and what is actually true.

longitudinal research, a research design in which data are collected at least two different times, such as a panel, trend, or cohort study.

We could have talked to a group of retired people and, using a cross-sectional design, asked these participants questions about what they used to feel about their jobs, what their activities outside of work had been, what they thought they might do in retirement, and what they were actually doing in retirement. However, a cross-sectional design with retrospective questions would have been problematic for the **internal validity** or accuracy of the study's conclusions. Validity concerns would result, in part, because of reliance on people's answers about what they felt or did years before. Memories fade, and current realities affect recollections and interpretations of past experiences and attitudes. We instead chose an over-time or a **longitudinal research** approach by collecting data twice—once before and once after retirement from career jobs.

STOP & THINK

*We found no relationship between people's expectation to do volunteer work after retirement and the degree to which they actually did volunteer work after retirement. Do you see how people who, say, weren't doing volunteer work after retirement might have (often falsely) remembered that they **didn't** expect to do such work before retirement if we'd waited until then to ask them about those expectations?*

	time 1 *2001–2002*	time 2 *2003–2006*
one sample of professional or college educated workers	All variables are measured *Expectations of retirement; current activities, including number of hours working for pay and doing volunteer work*	All variables are measured *Views of retirement; current activities, including number of hours working for pay and doing volunteer work*

FIGURE **7.2** The Minimum Panel Study Design

The Panel Study

panel study, a study design with data collected about one sample at least two times where the independent variable is not controlled by the researcher.

The design that we used is a **panel study**, a design that follows one sample over time with two or more data collections. (See Figure 7.2 for a representation of the minimum panel design using the Adler and Clark study as an example.) The study by Michele Hoffnung, described in Chapter 4, is also a panel study—one in which 200 women who graduated from college in 1992 were followed until 2009 with many repeated data collections.

A cross-sectional study would not have worked for us as we needed more than a snapshot of people's lives at one time. The longitudinal or over-time nature of a panel study allowed us to document patterns of change and to establish time order sequences. We wanted to find out about people's expectations of retirement and learn what they did in their time away from work while they were still employed. We also needed post-retirement information about what they actually were doing and their views of retirement. Remembering accurately what one did or felt years before is problematic. Sometimes people make the past "fit" the present, which leads to misremembering actual behavior and previous attitudes.

A key issue for panel studies is deciding when to collect data. In the retirement study, we were interested in the initial transition after leaving full-time careers so our second data collection was approximately 18 months after retirement. Results might have been different if we'd waited three or more years to follow up. As the research described in Box 7.1 demonstrates, selecting the time periods affects the findings.

STOP & THINK
Although contacting each person in the sample two or more times has advantages, it also has disadvantages. What do you think are some of the difficulties facing a researcher who collects data more than once from the same sample?

The ability to see change over time is very important, but there are offsetting costs and difficulties in doing a panel study. The most obvious concerns are time and money. Assuming identical samples and no inflation, a study with

BOX **7.1**	**A Question of Timing**

Studying children's development calls for longitudinal research to see changes over time. One important study has answered many questions about development while raising methodological questions about the *timing* of data collection. The National Institute of Child Health and Human Development (NICHD) Study of Early Child Care (SECC) is a comprehensive longitudinal study on the connections among child care quantity, care quality, and types of care on children's developmental outcomes. The 1,364 infants in the original sample were followed from birth to age 15 with 14 separate assessments (Vandell et al. 2010). The research results about the effects of child care have changed over time (Barry, 2002). One of the first findings was no relationship between attachment to mothers and kinds of care: At 15 months, children in day care were as strongly attached to their mothers as children reared at home. But at age two, day care seemed to have negative effects: Those spending many hours in day care showed less social competence and more behavior problems. At age three, the news was different:

Those in day care at least 10 hours a week had better cognitive and language skills. But a year later, the researchers announced that children who spent more than 30 hours a week in day care displayed higher incidence of disobedience, aggression, and conflict with caregivers. In 2008, researchers reported that the *size* of the groups counted: Children who had spent more time in small-sized peer groups in child care when younger were seen as more sociable and cooperative in third grade but were rated by teachers as more aggressive; children who had spent more hours in medium-sized groups also received higher ratings for peer aggression by their teachers (NICHD ECCRN, 2008). The age 15 follow-up found children who experienced high quality care had better academic-cognitive outcomes and that those who had spent many hours in day care exhibited more risk taking and impulsivity (Vandell et al. 2010: 751). While the study's findings are important to parents, educators, and policy makers, as methodologists we note that *timing matters* when doing repeated data collections.

three data collections takes much longer and costs three times as much as a one-time study. With the same resources but three data collections, a researcher could select a sample only one-third as large as for a study using a cross-sectional design. It's important to consider the need for a longitudinal approach to establish time order or improve data validity before incurring extra expenses and time or decreasing sample size. The expenses connected to research—in this case, travel and the costs of transcription—were two of the reasons for our small sample size.

It's essential to keep track of study participants and encourage them to continue participation in panel studies. Data must be kept confidential as anonymity cannot be promised to participants because keeping information, including names and addresses, is necessary to re-contact them. In our highly mobile society, keeping track of respondents is difficult, especially in long-term studies, even when the participants are highly motivated. **Panel attrition**, the loss of subjects from the study because of disinterest, death, illness, or moving without leaving forwarding information, can become a significant issue. In a study of the career and family goals of academically talented individuals, for example, the original sample in 1989 was 1,724 high school seniors (Perrone, Civiletto, Webb, and Fitch, 2004). The participants were mailed annual questionnaires with self-addressed, stamped response envelopes, but many were returned unopened and marked "cannot forward." For the thirteenth survey, Perrone and

panel attrition, the loss of subjects from a study because of disinterest, death, illness, or inability to locate them.

his colleagues (2004) report that they received only 113 responses—a response rate of about 6 percent! In contrast, Michele Hoffnung has done a much better job of keeping in touch with the participants in her long-term panel study. In the focal research in Chapter 4, she notes that the response rate for the fourteenth data collection was 76 percent.

In our research on retirement, we had no panel attrition. Our sample was motivated to participate; we had phone numbers, e-mail addresses, and/or home addresses for all of the participants and information about a contact person, such as a son or daughter; and typically less than two years passed between contacts with them.

panel conditioning, the effect of repeatedly measuring variables on sample members in a panel study.

Another concern for panel studies is **panel conditioning**, the effect of repeated measurement of variables. Participants tend to become more conscious of their attitudes, emotions, and behavior with repeated data collections which might be partially responsible for differences in attitudes, emotions, or behavior over time. At the second interview, some of our participants mentioned that talking about their feelings about work and their expectations about retirement during the first interview was useful to them and helped them sort out some of their feelings and think about their post-retirement plans. This is an example of panel conditioning.

Finally, the issue of changes due to time period needs to be considered, as "while each generation's path shares many milestones and processes in common, each generation's journey is unique" (Smith, 2005: 177). The events of September 11, 2001, occurred right before we began our study; the economic crisis of 2008 and beyond occurred afterward. Those events and many others can affect how people see retirement. Historical events of each time period are intertwined with the experiences of individual transition, and unless multiple samples of individuals who are experiencing the same transition in different decades are followed over time, a panel study like ours cannot sort out the effects of historical era from other variables.

🛑 **STOP & THINK**   *Can you think of any situations where you'd want to study changes over time and therefore would want to do a longitudinal study, but would prefer to select a new sample each time rather than contact the original sample multiple times?*

The Trend Study

Selecting new samples for longitudinal studies can be very useful when researchers are interested in identifying changes over time in large populations, such as registered voters in one country or college students around the world. It can be very time consuming and expensive to re-locate the same individuals to track changes in attitudes, opinions, and behaviors of a very large sample over a number of months or years. It is much more practical to select a new **trend study,** a study design in which data are collected at least two times with a new sample selected from the same population each time. probability sample each time. In addition, selecting new samples allows for anonymous data collection, which can be a more valid way to find out about embarrassing, illegal, or deviant behavior. A **trend study** is a longitudinal design that selects a new probability sample from the same population for two or more data collections.

STOP & THINK *If we were interested in doing a study on attitudes about legalizing medical marijuana, we could do a trend study by studying the attitudes of a population two times, selecting a new sample each time. What would be the advantages and disadvantages of this strategy?*

Trend studies avoid panel attrition and panel conditioning, save the expense of finding the original participants, and enable the researcher to collect data anonymously. Furthermore, trend studies are useful for describing changes in a population over time and in studying the impact of time periods. For these reasons, studies of changes in attitudes often use the trend study design. A good example is the General Social Survey (GSS), a study conducted periodically in a wide variety of countries, including Japan, Germany, Great Britain, and Australia. In the American GSSs, probability samples of more than 1,000 adults living in households in the United States are interviewed with some topics rotating in and out and others continuing (Smith, Kim, Koch, and Park, 2005). Examples of using the GSS as trend data include research on the impact of race and gender on abortion attitudes over time (Carter et al., 2009) and a study on happiness which concluded that happier people are more active in social activities, in religion, and in newspaper reading but spend less time watching television (Robinson and Martin, 2008).

Another example of a trend study is the annual survey on religion and public life conducted by the Pew Research Center. Each year since 2001, a new random sample of 2,000 or more adult Americans are interviewed. Based on comparisons of data from 2007 and 2012, the researchers conclude that "[t]he number of Americans who do not identify with any religion continues to grow at a rapid pace.... In the last five years alone, the unaffiliated have increased from just over 15 percent to just under 20 percent of all U.S. adults. Their ranks now include more than 13 million self-described atheists and agnostics (nearly 6 percent of the U.S. public), as well as nearly 33 million people who say they have no particular religious affiliation (14 percent)" (Pew Research Forum on Religion & Public Life, 2012: 9).

Trend studies can also include a cross-sectional aspect when comparisons are made *between* groups *within* each time period. The Pew Center compared the percentages of unaffiliated among different demographic groups at each time. In Figure 7.3, a diagram of the trend study design using two surveys on religious affiliation, demonstrates that there is a consistent pattern between variables with men and younger people more likely to be unaffiliated than women or older people.

Trend studies have limitations. Because a trend study does not compare a specific sample's responses at two or more times, the trend study can only identify *aggregated* changes, not changes in individuals. People move in and out of populations, such as the population of college students or American adults, so identified changes in attitudes, opinions, or characteristics over time could reflect changes in the composition of the population or in the sample rather than changes in the individuals. If a trend study finds a change over time, such as the percentage of college students who say they voted for Obama in 2008 and 2012, we don't know if there would have been no change, the same change, or

	Time 1 2007	Time 2 2012
First sample Adult Americans	Measure all variables *18% of men are religiously unaffiliated* *13% of women are religiously unaffiliated* *25% of 18–29 year old are religiously unaffiliated* *17% of 30–49 year olds are religiously unaffiliated* *12% of 50–69 year olds are religiously unaffiliated*	
Second sample Adult Americans		Measure all variables *23% of men are religiously unaffiliated* *17% of women are religiously unaffiliated* *35% of 18–29 year old are religiously unaffiliated* *37% of 30–49 year olds are religiously unaffiliated* *21% of 50–64 year olds are religiously unaffiliated*

FIGURE **7.3** The Trend Study Design

Source: Pew Research Forum on Religion & Public Life 2008 and 2012.

a more extreme change if the sample had included the *same* people (i.e., if a panel study had been done). In addition, it is possible for "no change" to result for the aggregate even if individual members of the population change in opposite directions between data collections. True to its name, the trend study design is best for describing trends in a population; it cannot identify how individuals have changed, nor can it pinpoint the causes of change.

The Cohort Study

cohort study, a study that follows a cohort over time.

cohort, a group of people born within a given time frame or experiencing a life event, such as marriage or graduation from high school, in the same time period.

A **cohort study** is longitudinal study design that follows a cohort over time. In social science research, a **cohort** is a group of people who have experienced the same significant life event at the same time. The most frequently used are birth cohorts, that is, people born in the same year. People in birth cohorts are the same chronological age and it is often assumed that they have experienced similar social and cultural influences as each other in the past and continue to do so. Cohorts can also be groups that experience a life event at the same time, such as graduating from college, retiring from the labor force, or getting divorced. Although people can "exit" from a cohort (e.g., by dying), no one can join after its formation.

In a cohort study, the same population is followed at least two times using *either* a panel or a trend study design. In the panel study approach, a sample that is a cohort is followed with at least two data collections. Since factors like the specific years that people were born, life course events, and/or time periods can influence behavior and attitudes, cohort studies can help to sort out these factors.

A cohort study can examine a small cohort or an entire generation, such as the "Baby Boomers," "Generation X" or "The Millennials." The focal research study on the transition to retirement is a cohort study in addition to being a panel study because it followed a cohort of people who retired within a few years of each other. Michele Hoffnung's study of 200 women who graduated from college in the same year also a panel *and* a cohort study.

A very ambitious cohort study is the Early Childhood Longitudinal Study (ECLS), which includes three panel studies of three different cohorts: a sample of approximately 14,000 children born in 2001 who were followed from birth through entering kindergarten, another sample of more than 12,000 children followed from kindergarten through the eighth grade and the kindergarten class of 2010–11, and a sample of approximately 18,000 children who will be followed from kindergarten through the fifth grade (ECLS, 2013). These three cohort studies will include many variables including child development, school readiness, and early school experiences.

A trend study approach can also be used in cohort studies. In a regular trend study, if we study the same population over time (e.g., all married men in Rhode Island, all bus drivers in France, all registered voters in Canada), the population itself changes over time as people move in and out of the population. But in a cohort study using the trend design, the same population is followed over time. If, for example, we were studying the birth cohort of Americans born in 1980 over time, we could ask new probability samples of this cohort questions about their attitudes toward cohabitation, careers, marriage, and divorce several times over several decades. We would then have the ability to describe changes in attitudes as the cohort gets older. We might find the cohort becoming more liberal, becoming more conservative, or remaining constant in their attitudes as they moved through their teens and into their twenties and thirties.

An interesting cohort study looked Americans at the other end of the age spectrum. Using the responses of older Americans who participated in the National Health Interview Survey between 1982 and 2009, Lin and colleagues (2012) described the health and disability trends among those aged 70 and older concluded that older American adults have become more disabled in recent years.

The Case Study

case study, a research strategy that focuses on one case (an individual, a group, an organization, and so on) within its social context at one point in time, even if that one time spans months or years.

The last non-experimental study design is the **case study,** a design with a long and respected history. In a case study, the researcher purposively selects one or a few individuals, groups, organizations, communities, events, or the like and analyzes the selected case(s) within their social context(s). Each case might be studied over a brief period of time or for months or years. The distinguishing feature of this study design is its *holistic* approach. A case study is distinguished from a cross-sectional study that uses a small sample because in a case study design, the researcher does *not* analyze relationships among variables but rather makes sense of the case or cases as a whole. The case study design is especially useful when the research has a "how" or "why" question and the researcher is studying contemporary events that he or she cannot control (Yin, 2009: 13).

Case study analysis is somewhat paradoxical. On the one hand, case studies "frankly imply particularity—cases are situationally grounded, limited views of social life. On the other hand, they are something more ... they invest the study of a particular social setting with some sense of generality" (Walton, 1992: 121). While case studies describe specific people and places, the best of them help us understand more general categories of the social world. Typically, the case study provides an in-depth understanding because it relies on several data sources, such as observations, interviews, photographs, media accounts, and other materials.

Many classic studies have used the case study approach. For example, "Goffman (1961) tells us what goes on in mental institutions, Sykes (1958) explains the operation of prisons, Whyte (1943) and Leibow (1967) reveal the attractions of street corner gangs, and Thompson (1971) makes food riots sensible" (Walton, 1992: 125).

Robert Stake (2003: 136–138) identifies three types of case studies: the *intrinsic*, undertaken because the particular case itself, with its particularity and ordinariness, is of interest; the *instrumental*, where a typical case is selected to provide insight into and understanding of, or a generalization about, an issue that it illustrates; and the *collective*, where the instrumental approach is extended to several cases selected to lead to a better understanding or theory about a larger collection of cases.

In Box 7.2, Susan Chase describes her instrumental case study of one university. Using a variety of data sources, her work focuses on how undergraduates in one university environment engage with each other across social differences.

Other projects use different units of analyses, kinds of data, and approaches. Leigh Culver (2004) investigated the impact of new immigration patterns on police–Latino community relations in the rural Midwest, using as her cases three communities in central Missouri that had recently experienced substantial increases in the Latino population. She collected a variety of data—including observations of uniformed patrol officers and formal and informal interviews with officers, community, and government leaders. Noting the language barriers and the amount of fear and distrust in interactions, Culver (2004) was able to describe the difficulties in developing cooperative, working relationships between the Hispanic community and the police.

Catherine Neimetz (2011) used interviews and observations in her case study of how family frameworks are integrated into one privately operated orphanage in China. She concluded that the staff took on family roles similar to those of mothers and fathers and organized daily family-like routines in order to provide for the children's emotional needs (Neimetz, 2011).

A major advantage of the case study is that it allows a "close reading" of social life and gives attention to the broader social context (Feagin, Orum, and Sjoberg, 1991: 274). Using this approach can be appropriate when working inductively, that is, moving from observations to theory, and can be particularly useful for exploratory and descriptive purposes. A disadvantage of the case study is its limited generalizability. Focusing on a sample of one or a

BOX 7.2 A Case Study of City University
By Susan E. Chase*

My research uses a case study approach to understanding undergraduates' engagement with diversity issues. I selected City University (CU, a pseudonym), a predominantly white institution, because I knew from an earlier comparative study that something was working well on this campus. I studied CU to see how students speak and listen to each other across social differences and to discover how the campus environment can encourage that kind of interaction.

I collected a variety of data. I did individual and group interviews: with faculty, staff, and administrators to get their perspectives on CU as an institution with a commitment to diversity; with a variety of students of different backgrounds and interests; and with leaders of student organizations to get their perspectives on how diversity issues are addressed on campus. With the help of Jessie Finch and Misti Sterling, two sociology majors at my university, I also studied documents. We analyzed every issue of the student newspaper for two years, the minutes of every student government meeting for three years, the university's monthly calendar of events for two years, and the undergraduate course schedule for three years. We counted how often specific diversity issues (race, class, gender, sexual orientation, ability, and global/international) were addressed in these documents. I found that in all the documents (newspaper, minutes, events, and curriculum) there was considerable discourse about diversity with race the most prominent diversity category in three contexts (newspaper, minutes, and events).

I also did a qualitative analysis of CU's student newspaper and student government minutes for 2004–2005, the year I did the interviews. Focusing on which diversity issues got play and how they were addressed, I found that specific racial issues were not only more prominent but also more contentious. For example, issues related to sexual orientation were usually presented in educational or supportive tones in both the newspaper and the student government minutes, while race-related issues evoked heated exchange in both contexts. This analysis provided strong corroboration of the understandings I had gained from the interviews.

Using a case study design with multiple data sources allows me to understand the "how" and "why" of CU's environment where diversity issues of all types are frequently on the table and in which race is the most prominent and contentious diversity issue, especially in comparison to sexual orientation. I have insight into the reflexive relationship between CU's narrative environment and students' narratives about how they have learned to speak and listen across social differences.

*Susan E. Chase is a professor of Sociology at the University of Tulsa. Her most recent book is *Learning to Speak, Learning to Listen: How Diversity Works on Campus* (Cornell University Press, 2010). This essay is published with permission.

few, with a specific phenomenon in its specific context, we cannot know whether what is observed is unique or typical. In addition, sometimes doing a study at one point in time makes it hard to disentangle causal connections.

STOP & THINK *Two researchers, Baker and Miller, are interested in studying changes in college students' attitudes about attending graduate school. Baker decides to do a cross-sectional study with a sample of freshmen and seniors at one university. Based on the interviews, Baker notes that the seniors have more positive opinions toward graduate school than the freshmen. In contrast, Miller selects a panel design and interviews the same sample of students at one university twice—once when they are freshman and three years later when most are* *seniors. Miller finds that the students' attitudes as seniors are more positive toward graduate school than they were as freshmen. Baker and Miller both conclude that students' attitudes toward graduate education change, becoming more positive as students progress through the undergraduate years. In whose conclusions do you have more confidence, Baker's or Miller's? Why?*

SUMMARY

Researchers begin with specific questions and purposes. They might be interested in exploring new areas, generating theories, describing samples and populations, seeing patterns between variables, testing a causal hypothesis, evaluating a program, or, as in the examples of Baker and Miller, documenting changes over time.

The purposes or reasons for the research influence the choice of study design. If, for example, we are interested in *describing* students' attitudes toward graduate school, selecting a cross-sectional design would be effective and sufficient. A cross-sectional design, like the study by Baker, allows us to describe the samples' attitudes toward graduate school and to *compare* the attitudes of freshmen and seniors. A panel design like that used by Miller, although more expensive and time consuming, is better if we want to study *changes* in attitudes. A panel study allows us ask about current attitudes at two times rather than depend on retrospective data, documents the time order of the variables, and lets us analyze changes in the attitudes of individuals, not just aggregated changes.

Judging the appropriateness of a design for a given project involves a series of decisions. The answers to the following questions provide a basis for selecting one of the five study designs covered in this chapter and the experimental designs discussed in Chapter 8:

- Is the goal to analyze only one case or a small number of cases holistically or is it to focus on variables and the connections between them?
- Is it possible and important to collect data more than one time from the same sample?
- Is there a cohort that can be studied to answer the research question?
- If it is important to collect data more than one time, is it useful to select a new sample for each data collection?
- If there is a causal hypothesis, is it possible, useful, and ethical to try to control or manipulate the independent variable in some way?

The focus of this chapter has been on the strengths and weaknesses of non-experimental designs as related to internal validity and generalizability. When selecting a study design, methodological concerns need to be balanced along with ethical considerations and practical needs. Although the summary shown in Figure 7.4 focuses on the methodological issues, ethical and practical criteria must also be considered when selecting a research design.

Summary of Cross-Sectional, Panel, and Trend Study Designs

Study Design	Design Features	Design Uses
CROSS-SECTIONAL STUDY One Sample (divided up into categories of the independent variable during analysis)	One sample of cases studied one time with data about one or more variables	Useful for describing samples and the relationship between variables Useful for explanatory purposes especially if the time order of variables is known and sophisticated statistical analyses possible

	time 1
category 1 of an independent variable	measure of a dependent variable
category 2 of an independent variable	measure of a dependent variable

Study Design	Design Features	Design Uses
PANEL STUDY one sample	One sample of cases is studied at least two times	Useful for describing changes in individuals and groups over time (including developmental issues) Useful for explanatory purposes

	time 1	time 2
one sample	variables measured	variables measured

Study Design	Design Features	Design Uses
TREND STUDIES first sample second sample	Different samples from the same population are selected	Useful for describing changes in populations and for analyzing the effect of time period

	time 1	time 2
first sample	measure several variables	
second sample		measure the same variables

FIGURE **7.4** Summary of Cross-Sectional, Panel, and Trend Study Designs

EXERCISE 7.1

Identifying Study Designs

Find an article in a social science journal that reports the results of actual research. (Be sure it is not an article that only reviews and summarizes other research.) Answer the following questions about the article and the research.

1. List the author(s), title of article, journal, month and year of publication, and pages.
2. Did the researcher(s) begin with at least one hypothesis? If so, specify one hypothesis, identifying the independent and dependent variables. If there was no hypothesis, but there was a research question, include it.
3. Identify the main purpose(s) of this study as exploratory, descriptive, explanatory, or applied.

4. Describe the population and the sample in this study. Is the population a cohort?
5. If this study used only one or a few cases or analyzed the cases as a whole rather than by focusing on variables, then it might be a *case study*. If you believe it is a case study, give support for this conclusion and skip Questions 6–9. If it's not a case study, skip this question and answer the rest.
6. How many times were data collected in this study?
7. If data were collected more than once, how many samples were selected?
8. Did the researchers try to control or manipulate the independent variable in some way?

9. Did the researchers use a case, cross-sectional, panel, trend, or cohort study design? If so, which one? Give support for your answer.

10. Was the design useful for the researcher's purposes and goals or would another design have been better? Give support for your answer.

EXERCISE 7.2

Planning a Case Study of Your Town or City

In his book *Code of the Street*, Elijah Anderson introduces his study by describing the 8.5-mile major artery in Philadelphia that links the northwest suburbs with the heart of the inner city. Anderson (1999: 15) says,

> Germantown Avenue provides an excellent cross section of the social ecology of a major American city. Along this artery live the well-to-do, the middle classes, the working poor, and the very poor—the diverse segments of urban society. The story of Germantown Avenue can therefore serve in many respects as a metaphor for the whole city.

In a case study of a geographic entity such as a city or a town, one or more streets, areas, organizations, groups, or individuals can be selected purposively. Think about the town or city where you live or one where you have lived in the recent past. Describe who or what your focus would be if you wanted to use a case study approach to study this town or city. Think of a research question you could answer with your case study.

1. Name the geographic area you want to focus on and what research question you would like to answer about it.
2. Describe who or what you would study. (You can identify one unit of analysis or you can propose to study more than one.)
3. Give support for your choices by discussing why it would be useful to do a case study of your city or town to answer your research question.

EXERCISE 7.3

Evaluating Samples and Study Designs

Janice Jones, a graduate student, designs a research project to find out if women who are very career-oriented are more likely to be mothers than women who are less career-oriented. Jones selects a random sample of women who work for large corporations and hold jobs as president or vice-president. Interviewing the women she finds that their average age is 45 and that 35 percent are mothers, 55 percent are childless by choice, and 10 percent are childless involuntarily. She concludes that strong career orientations cause women to choose childlessness.

1. Discuss some of the challenges to the internal validity of Jones's conclusions.
2. Think of a different way this study could have been done. Specifically describe a sample and a study design that you think would be more useful to answer Jones's question.

Experimental Research

CHAPTER OBJECTIVES

This chapter is designed to help students

- **Review independent and dependent variables**
- **Reexamine criteria for demonstrating a causal relationship**
- **Appreciate the experiment as a study design for evaluating causal experiments**
- **Compare pretest–posttest and quasi-experimental designs**
- **Understand the differences between field and laboratory experiments**
- **Evaluate internal validity in experiments**
- **Understand the natural experiment**
- **Compare experiments to non-experimental designs**

CHAPTER OUTLINE

Causal Hypotheses and Experimental Designs

Experimental Designs

Comparing Experiments to Other Designs

Summary

Exercises

© Andersen Ross/Jupiter Images

CAUSAL HYPOTHESES AND EXPERIMENTAL DESIGNS

explanatory research, research designed to explain why subjects vary in one way or another.

causal hypothesis, a testable expectation about an independent variable's effect on a dependent variable.

We'll continue our discussion of study designs by considering those that are especially useful for research with an explanatory purpose—that is, research seeking to explain *why* cases vary on one or more variables or *why* something is the way it is. **Explanatory research** begins with a **causal hypothesis** or a statement that one variable, the independent variable, *causes* another, the dependent variable. Frequently the researcher has deduced the hypothesis from an existing theory.

 STOP & THINK   *Explanatory research can focus on any area of social life. If you were doing a study on some aspect of education, there are many research questions and hypotheses you could consider. Create one causal hypothesis that uses level of education or kind of education as a variable.*

There are many causal hypotheses that can be constructed using an aspect of education as an independent, dependent, antecedent, or intervening variable. Doing a literature review of research, we found many studies with causal hypotheses about one or more aspects of education. Here are three, each with the *independent variable* in italics.

1. *Level of education* affects the extent to which people rely on the Bible and the teachings of their place of worship for guidance in their decision making (Schieman, 2011).
2. *Pre-school education* can help to close social class disparities in children's vocabulary (Becker, 2011).
3. *Participation in structured extracurricular activities* in high school affects the self-esteem of adolescents (Kort-Butler and Hagawen, 2011).

As these three are causal hypotheses, to test them, it is important to consider the requirements for demonstrating causality. As noted in Chapters 2 and 7, causality cannot be directly observed but, instead, must be inferred from other criteria. You may remember that the factors that support the existence of a causal relationship include the following:

1. An independent and dependent variable that are associated, or vary in some way, with each other. This is the condition of *empirical association*.
2. An independent variable that occurs *before* the dependent variable. This is the condition of *temporal precedence* or *time order*.
3. The relationship between the independent and dependent variable is not explained by the action of a third variable (sometimes called an antecedent variable) that makes the first two vary together. That is, the relationship between the independent and dependent variables is not *spurious* or non-causal.

For any study, it is necessary to make the series of choices we've discussed in the previous chapters. We'd need to find a way to measure the variables, select a population and sampling strategy, determine the number of times to collect data, and establish the number of samples to use. We might decide that it is appropriate to use one of the study designs that we presented in Chapter 7 and design a panel, trend, cohort, case, or cross-sectional study. Before making a final decision, however, we'd want to consider the possibility of using an **experimental design**. Although other study designs usually allow the researcher to determine if there is a pattern or relationship between independent and dependent variables (and, in the case of longitudinal designs, temporal sequencing), experimental designs are also especially helpful for determining the time order of variables and for minimizing the third-variable effects.

experimental design, a study design in which the independent variable is controlled, manipulated, or introduced in some way by the researcher.

The Experiment: Data Collection Technique or Study Design?

As we discussed in Chapter 6, a critical set of decisions focuses on how to measure variables. That is, for each variable, the researcher has to figure out how to put every unit of analysis into one of two or more categories. The uniqueness of the experimental design becomes apparent when we think about the issue of measurement because, in a traditional experiment, the independent variable is not measured at all! Instead, in an experiment, the independent variable is "introduced," "manipulated," or "controlled" by the researcher. That is, the independent variable does not occur naturally but is the result of an action taken by the researcher.

In the next few chapters, we will be describing methods of data collection, including asking questions, observing, and using available data. Because of the unique way that independent variables are handled in experiments, some textbooks define the experiment as a method of data collection or as a mode of scientific observation. Although we recognize the usefulness of this approach, we want to highlight the experiment's special treatment of the independent variable as a unique *design* element. For this reason, we include experiments in our discussion of designs and see this chapter as a natural transition between the consideration of study designs and the presentations on data collection techniques.

In the first hypothesis listed above, level of education is the independent variable and importance of religion in decision making is the dependent variable. Schieman (2011) used a cross-sectional design and collected information one time by asking questions of a large sample of people. Education was measured by asking how much education was completed with "less than high school," "high school graduate," "some college," college graduate," and "postgraduate or professional school after college" as categories (Schieman, 2011: 576). Use of religion in decision making was measured by asking about the importance of the Bible and about the teachings of the respondents' church or synagogue in helping them make decisions about their lives (Schieman, 2011: 576.) Analyzing the pattern, Schieman found an association: those with more years of formal education were less likely to say that the Bible or their church or synagogue were important in helping them make decisions. However, it's not clear which of the variables came first—education or importance of religion—and while Schieman does statistical analyses to control for variables like income, it is possible that the relationship between education and importance of religion is explained by other variables.

One useful strategy to determine causality is to actually manipulate, introduce, or control the independent variable in some way rather than measuring it. If *practical* and *ethical*, a study can be designed so that the dependent variable is measured first, then an independent variable is introduced or manipulated, and then the dependent variable is measured again. In this way, one can see whether the introduction of the independent variable comes *before* change in the dependent variable.

While it's not possible to do an experiment for Schieman's hypothesis because we can't manipulate the number of years of education that a person completes, there are many variables that can be manipulated such as teaching techniques in college classrooms or rehabilitation programs for cardiac patients. By manipulating the independent variable (for example, using different techniques or programs) and measuring changes in the dependent variable (such as learning or heart health), we can see the effect of an independent variable.

EXPERIMENTAL DESIGNS

stimulus, the experimental condition of the independent variable that is controlled or "introduced" by the researcher in an experiment.

placebo, a simulated treatment of the control group that is designed to appear authentic.

internal validity, agreement between a study's conclusions about causal connections and what is actually true.

pretest–posttest control group experiment, an experimental design with two or more randomly selected groups (an experimental and a control group) in which the researcher controls or "introduces" the independent variable and measures the dependent variable at least two times (pretest and posttest measurement).

Internal Validity and Experiments

Determining who receives which condition of the independent variable and controlling its timing are central elements in the experimental design. The control group is exposed to *all* the influences that the experimental group is exposed to *except* for the **stimulus** (the experimental condition of the independent variable). The researcher tries to have the two groups treated exactly alike, except that instead of a stimulus, the control group receives no treatment, an alternative treatment, or a **placebo** (a simulated treatment that is supposed to appear authentic). For maximum internal validity, the experimental design tries to eliminate any systematic differences between the experimental and control group beyond the introduction of the independent variable.

Internal validity is concerned with factors that affect the internal links between the independent and dependent variables, specifically factors that support alternative explanations for variations in the dependent variable. In experiments, internal validity focuses on whether the independent variable or other factors, such as prior differences between the experimental and control groups, are explanations for the observed differences in the dependent variable.

Pretest–Posttest Control Group Experiment

The gold standard for experiments is the **pretest–posttest control group experiment,** which is also called the classic controlled experiment. In this kind of experiment, there are at least two samples or groups (although there may be more than two groups) with two measurements of the dependent variable—one before and one after the introduction of the independent variable. These

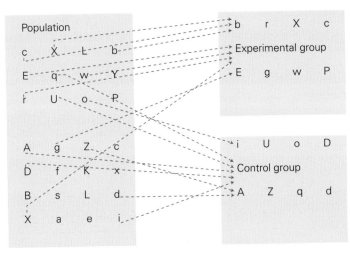

Population

FIGURE **8.1** Use of Probability Sampling to Select Groups

pretest, the measurement of the dependent variable that occurs before the introduction of the stimulus or the independent variable.

posttest, the measurement of the dependent variable that occurs after the introduction of the stimulus or the independent variable.

probability sampling, a sample drawn in such a way as to give every member of the population a known (nonzero) chance of inclusion.

random assignment, a technique for assigning members of the sample to experimental and control groups by chance to maximize the likelihood that the groups are similar at the beginning of the experiment.

measurements are called the **pretest** and the **posttest**, respectively. The samples or groups are the experimental and the control group, groups that can be selected in one of several ways. The preferred method of picking the groups is to identify an appropriate population and, using **probability sampling,** to select two random samples from the population. Figure 8.1 illustrates the use of probability sampling to pick experimental and control groups. This method gives the greatest generalizability of results because selecting the samples randomly from a larger population increases the chances that the samples represent the population.

However, participants in experiments are rarely selected from a larger population usually because it's usually not practical and might not be ethical. Instead, participants are more typically a group of people who have volunteered or have been selected by someone in the setting. In such cases, the sample is really a convenience sample and might not be representative of a larger population.

There are two ways to sort a sample into experimental and control groups to minimize preexisting, systematic differences between the groups. By minimizing such differences, experiments can offer support for the conclusion that no third variable is making the independent and dependent variables appear to "go together"—the third requirement of causality.

One way to assign members to experimental and control groups is **random assignment,** the process of randomly selecting experimental and control groups from all study participants. This can be done by flipping a coin to determine which subject is assigned to which group or by assigning each subject a number and randomly selecting the numbers (and members) at random for each group (experimental or control). Figure 8.2 illustrates the use of random assignment to create two groups.

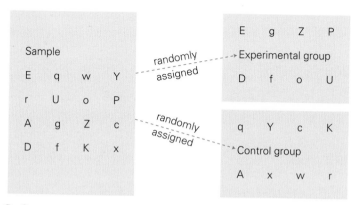

FIGURE **8.2** Use of Random Assignment to Select Groups

matching, assigning sample members to groups by matching the sample members on one or more characteristics and then separating the pairs into two groups with one group randomly selected to become the experimental group.

Another way to assign members of the sample to experimental and control groups is through **matching**. Using this method, the sample's members are matched on as many characteristics as possible so that the whole group is "paired up." The pairs are then split, forming two matched groups (see Figure 8.3). A concern is that typically several characteristics in the sample are matched while others are not. After the sample members are matched into groups, one group is randomly selected to become the experimental group and the other becomes the control group.

The pretest-posttest control group experiment has several elements that distinguish it from other study designs:

1. The study uses at least one experimental and one control group, selected using a strategy to make the groups as similar as possible (probability sampling, random assignment, or matching). Experimental groups are sometimes called treatment groups because an experiment will have at least one and possibly more experimental treatments.

2. One or more dependent variables are measured at least two times for the experimental and control groups. The first measurement is before and the

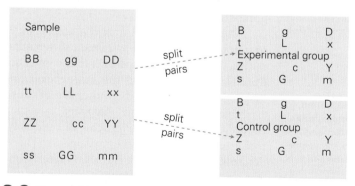

FIGURE **8.3** Use of Matching to Select Groups

second is after the independent variable is introduced. (These measurements are called the pretest and the posttest, respectively.)

independent variable, a variable that is seen as affecting or influencing another variable.

3. The **independent variable** is introduced, manipulated, or controlled by the researcher between the two measurements of the dependent variable. While the experimental group receives the stimulus, the control group gets nothing, an alternative treatment, or a placebo.

dependent variable, a variable that is seen as being affected or influenced by another variable.

4. The differences in the **dependent variable**(s) between the pretest and posttest are calculated for the experimental group(s) and for the control group. The differences in the dependent variable(s) for the experimental and control groups are compared.

Sociology has a long history of trying to sort out cause-and-effect relationships, especially those concerned with social problems. For more than a century, research has documented links between neighborhood poverty and outcomes such as crime, economic dependency, poor physical health, dropping out of school, and teenage pregnancy. One issue of concern to researchers has been whether the characteristics of the *individuals* living in deprived neighborhoods or those of the *neighborhood* itself best explain these connections (Sampson, 2012: 1364).

 STOP & THINK   *If you were interested in studying a causal relationship between neigbhorhood poverty and one outcome, such as the physical health of residents, could you design an experiment to study this? Think about how you'd get an experimental and a control group, how you'd measure a dependent variable, and how you'd introduce the independent variable.*

 ### RESEARCH IN THE NEWS

In 2012, *The New York Times* reported on results of a vast social experiment to determine the impact on low-income families of moving to more prosperous communities. While the project found little evidence that new neighborhoods made a difference in either annual income or level of education, it did find that participants who had moved had lower rates of obesity, diabetes, and depression than those who had not. Beyond that, there was a big change in happiness. Jens Ludwig, a professor of public policy at the University of Chicago and the lead author of the study, said that moving to neighborhoods that were less poor caused families making $20,000 a year to feel as happy as families making $33,000 a year.

Source: http://www.nytimes.com/2012/09/21/health/intangible-dividend-in-an-anti-poverty-experiment-happiness.html

The news story above reported on the results of a long-term study that began in the 1990s as a result of attention to the plight of inner cities. While very costly and time consuming, the researchers conducted Moving to Opportunity

(MTO) to test causal hypotheses about neighborhood poverty. From 1994 to 1998, the research project enrolled 4,604 low-income households living in public housing in Baltimore, Boston, Chicago, Los Angeles, and New York. Families were randomized into three groups: the experimental group, (a Low-Poverty Voucher group), which received housing vouchers that subsidize private-market rents but could only be used in census tracts with 1990 poverty rates below 10 percent; an alternative treatment group (a Traditional Voucher group), which received regular housing vouchers without any MTO relocation constraint; and a control group, which received no assistance through MTO (Ludwig et al., 2012: 1506). Data were collected about all three groups both before and after the experimental group was offered the opportunity to move to a better neighborhood and the alternative treatment group was offered the chance to move but no assistance with moving to better neighborhoods. The experimental design was a pretest–posttest experiment to determine the long-term effects of neighborhood poverty by studying three groups of households ten to fifteen years apart.

The advantages of the experimental design are evident. The researchers could see whether or not there was association between the experimental condition of the independent variable (offering housing vouchers to allow families living in public housing projects in high-poverty neighborhoods to move to lower-poverty neighborhoods) and any of the dependent variables (including economic status, mental and physical health). In addition, the researchers could demonstrate time order as the dependent variables were measured both *before* and *after* the independent variable was introduced.

These two conditions are *necessary* but not *sufficient* to support a cause-and-effect relationship as other factors could have caused the changes in the dependent variables. The final element of the design was the selection of several groups using random assignment to minimize the chances of preexisting systematic differences between the groups. Figure 8.4 diagrams a simplified version of this experiment. For a more complete discussion of the Moving to Opportunity experiment, see Housing and Urban Development (2012).

The researchers conclude that the chance to move to less distressed neighborhoods has no detectable long-term effects on adult economic self-sufficiency but does have an impact on two important aspects of physical health (extreme obesity and diabetes) as well as self-assessment of well-being (Ludwig et al., 2012: 1508–1509). In this pretest–posttest control group experiment, the researchers have satisfied the three conditions that support causality cited at the beginning of this chapter.

selection bias, a bias in the way the experimental and control or comparison groups are selected that is responsible for preexisting differences between the groups.

maturation, the biological and psychological processes that cause people to change over time.

Advantages of Pretest–Posttest Experiments

In addition to handling the internal validity concern about **selection bias** or preexisting differences between the groups, some other advantages of these kinds of experiments noted by Donald Campbell and Julian Stanley (1963) are as follows:

1. The experiment is useful for separating the effects of the independent variable from those of **maturation,** the biological and psychological

Groups	Time 1 (pretest) Mid 1990s	Independent variable	Time 2 (posttest) 10 to 15 years later
Experimental group Vouchers to move to better neighborhoods	Measure dependent variables including Mental health Physical health Employment Receiving public assistance Happiness	Stimulus	Measure dependent variables including Mental health Physical health Employment Receiving public assistance Happiness
Alternative treatment group Traditional vouchers	Mental health Physical health Employment Receiving public assistance Happiness	Alternative treatment	Mental health Physical health Employment Receiving public assistance Happiness
Control group No vouchers or assistance	Mental health Physical health Employment Receiving public assistance Happiness	No treatment	Mental health Physical health Employment Receiving public assistance Happiness

FIGURE **8.4** Diagram of the Pretest–Posttest Control Group Experiment Illustrating Moving the Opportunity Research Project

processes that cause people to change over time, because, in experiments, both the experimental and control groups are maturing for the same amount of time.

testing effect, the sensitizing effect on subjects of the pretest.

2. The experiment can control for some aspects of the **testing effect,** the separate effects on subjects that result from taking the pretest, because both groups take the pretest.

history, the effects of general historical events on study participants.

3. The experiment can control for some aspects of **history**, the effects of general historical events not connected to the independent variable that influence the dependent variable, because both groups are influenced by the same historical events.

 STOP & THINK   *Do you play video games? Have you ever thought about what effect they can have on behavior? Construct a hypothesis to test something about the effect of playing video games on some aspect of behavior and think of a study design that could test it.*

Posttest-Only Control Group Experiment

posttest-only control group experiment, an experimental design with no pretest.

Another experimental design, the **posttest-only control group experiment,** has no Time 1 measurement (no pretest), either because it is not possible to do a pretest or because of a concern that using a pretest would sensitize the experimental group to the stimulus (the testing effect). Otherwise, it has the same design elements as other experiments: control or manipulation of the stimulus and two or more groups using random selection or assignment.

An experiment using posttest-only designs tested the desensitization hypothesis that links playing violent video games to decreased helping behavior. Using a posttest experiment, as illustrated in Figure 8.5, a sample of 320 college students, with equal numbers of men and women, was recruited to be part of a study (Bushman and Anderson, 2009). The students were told that the research's purpose was to see what types of people liked what types of games. The sample was randomly assigned to experimental and control groups. Those in the experimental group played one of four violent video games while those in the control group played one of four nonviolent video games. When the game was completed, each student was handed a long questionnaire on video games to complete while the experimenter left the room. The actual purpose of the questionnaire was to keep the student busy while a recording of a staged fight was played outside the room. The six-minute recording had actors of the

Groups	Stimulus	Posttest
Experimental group (played violent video game)		Measure dependent variable (Helping behavior)
Control group (played nonviolent video game)	*	Measure dependent variable (Helping behavior)

*The control group receives no stimulus, receives an alternative treatment, or receives a placebo.

FIGURE **8.5** Posttest-Only Control Groups Experiment Illustrated Using Video Game Research same gender as the participant sounding like they were fighting about one of them stealing the other's boyfriend or girlfriend—complete with lots of shouting and scuffling. At one point in the recording, the experimenter threw a chair on the floor outside the closed door of the lab and kicked the door twice. After one "person" in the fight left, the "other" groaned in pain. The measures of the dependent variable were whether the participant tried to help the person in pain and, if so, the length of time he or she waited before trying to help. Bushman and Anderson (2009: 276) found that "Although in the predicted direction, there was no significant difference in helping rates between violent and nonviolent video game players, 21 percent and 25 percent.... When people who played a violent game did decide to help, they took significantly longer (M = 73.3 seconds) to help the victim than those who played a not violent game (M = 16.2 seconds)."

A major challenge to the internal validity of the posttest-only experimental design is the unverifiable assumption that a difference between the experimental and control groups on the dependent variable represents a *change* in the experimental group. Without a pretest, it is difficult to determine whether the difference between groups on the dependent variable is a change in the experimental group because of the stimulus or is a continuation of a preexisting difference between the groups. After all, though unlikely, it is possible in the example of the posttest-only experiment that the students in the control group were just nicer, more helpful people than those in the experimental group. We need to be cautious in reaching the conclusion that a causal hypothesis has been supported if there is no Time 1 measurement. However, when it is not practical to measure the dependent variable before the introduction of the dependent variable, as was the case in the video game experiment, the posttest-only experiment provides many of the advantages of the pretest and posttest design.

Extended Experimental Design

Solomon four-group design, a controlled experiment with an additional experimental and control group, with each receiving a posttest only.

Experimental designs that add groups can be useful in increasing internal validity. The **Solomon four-group design** has four groups—two experimental and two control groups, with only one set of groups being given a pretest. It is most useful when there is a concern about a possible interaction effect between the first measurement of the dependent variable and the introduction of the stimulus. The compensation for the extra effort, time, and expense of this design is the ability to determine the main effects of testing, the interaction effects of testing, and the combined effect of maturation and history (Campbell and Stanley, 1963: 25).

🛑 **STOP & THINK**   *Let's say you want to evaluate a six-month "quit smoking" program. You could evaluate its effectiveness by taking a sample of smokers who volunteer to join the program, divide them into two groups at random, and enroll only the experimental group in the program. At the end of six months, you could see how many in each group have quit smoking and what health changes, if any, have occurred in each group. But what about the ethical concern of depriving potential "quitters" of the chance to stop smoking and the practical problem that members of the* *control group might get help elsewhere if not accepted into the program? What other study design could you use to study the impact of the "quit smoking" program?*

Quasi-Experimental Design

In situations where it is not ethical or practical to do a true controlled experiment, the researcher can do a **quasi-experiment.** In this design, the researcher does not have full control over all aspects of the experiment. The missing element(s) could be control over timing, the selection of subjects, or the ability to randomize. Most frequently missing is random assignment into experimental and control groups. In the comparison group quasi-experiment, for example, there are two groups, but the groups are not selected by random selection, randomization, or matching to have pre-experimental equivalence. In this design, there is an experimental group and a comparison group in place of the control group (see Figure 8.6).

So, for example, if we were interested in the causal hypothesis that an antismoking program is effective in reducing smoking and improving health, we could do a comparison group quasi-experiment. We'd compare a group of volunteers who, on their own, decided to enroll in a reputable stop smoking program and find other similar smokers, for example, from a health center, who weren't enrolled in the program and collect information from both groups over the same time period. Because we didn't pick or assign the groups randomly, they'd serve as our experimental and comparison groups.

An example of a quasi-experiment is the study of a probation supervision program designed to reduce criminal recidivism. One county in the UK implemented a new county-wide program (Citizenship, a motivational program individually tailored to the offenders' risk level and needs) for all offenders (Pearson et al., 2011). The researchers used this policy change to conduct an experiment. The experimental group was the more than 3,000 offenders who

Groups	Time		
	1 (pretest)		2 (posttest)
Experimental	Measure dependent variable	Stimulus	Measure dependent variable again
Comparison	Measure dependent variable	*	Measure dependent variable again

*The comparison group receives no stimulus, receives an alternative treatment, or receives a placebo.

FIGURE **8.6** Comparison Group Quasi-Experiment

received the new program and were released from prison between 2005 and 2007; the comparison group was the more than 2,000 offenders who had received traditional probation supervision from 2004 to 2005. Comparing the two groups on a dependent variable (percentage reoffending within two years of being released from prison), the researchers found a statistically significant difference: 41 percent of the experimental group and 50 percent of the comparison group had reoffended within two years (Pearson et al., 2011: 88).

There are limitations to the internal validity of this and all quasi-experiments. Although the researchers doing the probation study were able to compare the characteristics of the two groups at baseline, they cannot entirely rule out the possibility that unknown factors could have been responsible for the differences in outcome. For example, they note that during the time period between the comparison and experimental groups' release, the national rate of reconvictions was reduced (Pearson et al., 2011:96).

● **STOP & THINK**   *In some experiments, the subjects do not know they are part of an experiment. What are some of the advantages and disadvantages of this strategy?*

Experiments in the Field

field experiment, an experiment done in the "real world" of classrooms, offices, factories, homes, playgrounds, and the like.

The probation quasi-experiment and the Moving to Opportunity (MTO) pretest–posttest control group experiment described in this chapter are also **field experiment**, experiments that take place in "real-life" settings where people live or congregate naturally—schools, supermarkets, hospitals, prisons, workplaces, neighborhoods, and the like. One of the biggest field experiments was discussed in Chapter 4—an experiment involving 61 million Facebook users. In that study, to assess the impact of social influence on political mobilization, an experimental group received a "get out the vote" message from Facebook with pictures of friends while a control group received the identical message without the pictures.

While all research is conducted within social settings, there is variability between settings and the degree to which the researcher has control over them. Some field experiments are actually built into the setting so the subjects do not know they are part of an experiment. The Facebook study and probation research were these kinds of studies. Facebook users saw the "get out the vote" message as just another message. In the probation study, being part of a specific probation program upon release from prison was just a normal part of judicial processing.

In the focal research that follows, we'll read about a field experiment on student learning that Chris Caldeira conducted in two university classes that were regular classes as far as the students were concerned. Having been in many classrooms over the years, with dozens of teachers, using a wide variety of styles and approaches, you may have ideas about effective teaching strategies. Like Caldeira, you could plan a study to find out if some teaching techniques are more effective than others. Here Caldeira describes her field experiment to determine if using film clips in introductory sociology enhances student learning.

Student Learning: A Field Experiment in the Classroom

By Chris Caldeira*

I was searching for a topic to study in order to complete my M.A. in Sociology while I was working as a senior acquisitions editor for sociology at Wadsworth/ Cengage Learning. In gathering feedback about textbooks, I was in contact with many faculty members and heard about the struggle that students often had in understanding and retaining sociology's key concepts. Some instructors talked about supplementing their classroom instruction with film and documentary clips in order to present the concepts in real-world or simulated contexts and make them more easily remembered.

While few instructors expressed doubts that films are useful, I realized that there was little systematic research addressing the effectiveness of film in enhancing students' understanding and comprehension of sociological concepts. Because I was fortunate to find a faculty member who was willing to allow me to use her introductory sociology classes as the research setting, I decided to do a field experiment to test the effectiveness of using film and documentary clips in conjunction with text- and lecture-based explanations of concepts. The hypothesis I tested is that using film clips to teach sociological concepts in conjunction with the textbook and lecture is more effective in helping students learn core concepts than the lecture and textbook without the clips.

Working in consultation with a professor in a university setting made it possible for me to easily get access to data about student achievement and be able to complete my study over the course of one semester. Working in the field also meant limitations on my research. I could specify the independent variable (the intervention) and select the actual film clips to be used. On the other hand, I could not control many aspects of the research, including the composition of the classes or which students would be in the professor's afternoon or morning classes. I was, however, able to adapt my research strategy to the field requirements and use a quasi-experimental design and one of the course requirement results as the indicator of my dependent variable.

Methodology

The professor I collaborated with was teaching two introductory sociology classes at a mid-sized, public, open-enrollment university. As the professor used the same lectures, textbook, assignments, and examinations in both classes, I selected the students in them as my sample. Each class had approximately 50 undergraduate students who were enrolled to satisfy a general education or major requirement. The slight differences between the classes in their year at the university or average GPA were not statistically significant. The only statistically significant difference was the gender breakdown of the classes: 40% of students in the morning class were women as compared to 68% of the afternoon class. While studies have noted sex-related differences in learning and test performance, the experimental design described below compensates for this difference between the classes.

---

*Chris Caldeira conducted this research as part of the requirements for an MA in Sociology from San Francisco State University. This article is published with permission.

The independent variable was the set of film clips that was selected and showed. We selected a group of 12 concepts of that are typically covered in an introductory sociology course, including mechanical solidarity, organic solidarity, race as a social construction, culture shock, and stigma and labeling theory, and found clips for each of them. The clips were chosen not only to stimulate interest but because each presented a clear visual representations of a sociological concept. For example, for the concept "culture shock," a term used to describe the physical and emotional discomfort that occurs when people reorient themselves to a new culture or social environment, we selected 3 scenes from Lost Boys of Sudan, a film about a group of boys from Sudan who move to the U.S. as part of an immigrant resettlement program. The focus in the scenes is the difficult process the boys face in a culture that is foreign to them.

I selected a quasi-experimental design with two classes alternating as experimental and comparison groups. Over the course of a semester, the professor showed six clips on six concepts to the morning class and six different clips on six other concepts to the afternoon class. In other words, the independent variable (the film clips) was manipulated or controlled so that one class saw them while the other did not. The morning class functioned as the experimental group for the six clips shown in that class while the afternoon class functioned as the comparison group (and vice versa).

Before showing each film clip to the experimental group, the instructor introduced the related concept, gave some general examples, and prepared students to view the clip with the concept in mind. After showing the clip, the instructor asked the students to write a paragraph describing how the film illustrated the concept, collected their writings, summarized the film's connection to the concept, and asked for questions or comments. For the comparison group, the instructor introduced the related concept with related examples and asked them to write a paragraph describing their understanding of that concept, collected their writings, and asked for questions or comments.

The dependent variable was student learning. The concept was operationally defined as student performance on multiple choice tests, the method by which student learning is actually assessed in most introductory sociology classes. The indicators were approximately five multiple choice questions for each sociological concept covered. The test questions were those that accompany the assigned textbook with additional multiple choice questions written by the professor. I received the grades on the tests with student names removed.

Findings

If film clips are an effective teaching tool, then the experimental group will perform better on the examination questions related to the concepts to which the film applied. The hypothesis was supported by the data. In 11 out of 12 cases, the experimental group did better on exams than the comparison group by between 2.6 and 16.8 points and by an average of 7.4 points (better by more than a half letter grade). For the one film clip where the experimental group did not score better, the experimental group scored lower than the comparison group by 1.1 points, a very small difference. For 6 of 12 concepts, the mean difference between the experimental and comparison group test scores was large enough to be statistically significant (using .05 level of significance). Of these six, the experimental group was the morning class for two and the afternoon class was the experimental group for other four.

(continues)

FOCAL RESEARCH *(continued)*

Conclusion

The hypothesis for this study was that using film clips as an instructional supplement to traditional lectures and readings will help students learn core concepts. Carefully chosen scenes from films and documentaries were used to illustrate specific (rather than broad) sociological concepts. The quasi-experimental design was used with two introductory sociology classes, with the classes alternating as experimental and comparison groups. The students in each class saw six films that those in the other class did not. The experimental group tended to perform better on tests than those in the comparison group. Differences in sex composition of two classes had no significant impact on the results. This research offers support for the conclusion that film clips are effective in teaching core concepts in sociology.

In terms of this study's generalizability, it is important to note that this experiment involved a professor well known for effective teaching as measured by student evaluations and teaching awards. Would film clips have the same impact on student achievement in other classrooms? Other research will need to be done to determine the impact of this kind of media when the instructor is less experienced and/or less effective in the classroom.

 THINKING ABOUT ETHICS *The project was not given a full ethics review because of the exemption for research conducted in educational settings involving normal education practices such as research on instructional strategies or classroom management methods. However, the research does meet ethical standards. The study design was dictated in part by concern for not harming students. If, as the researcher hypothesized, watching film clips would in fact improve grades, those who saw the films would have received an advantage over the comparison group. By having students alternate between being in the experimental and the comparison groups, no group of students had an extra advantage and therefore there was no academic harm. In addition, all student data, including examination grades, were kept confidential.*

When field experiments seem like part of real life to participants instead of part of a social experiment, they typically have good internal validity. The introduction of the independent variable (such as the teaching technique, opportunity to move, administrative decision, marketing strategy, medical treatment, or another variable) usually seems quite natural. In Caldeira's study, since the students didn't need to give informed consent, she was able to introduce an independent variable, obtain data on the dependent variable, and complete an experiment in the real world. For the subjects, it was not a research project but just another class.

generalizability, the ability to apply the results of a study to groups or situations beyond those actually studied.

Field experiments can also have good **generalizability** (also called external validity). In real-world settings, the researchers can use samples that are more like (or actually may be) the actual students, residents of a community, patients, workers, customers, and other groups to whom they want to generalize their results. Although working in real settings has advantages, it can present difficulties. Researchers doing a field experiment are typically guests of the institutions or community settings and might not be able to control as much of the experiment as would be ideal. Flay and Collins (2005) looked at school-based evaluations of prevention and intervention programs for problem behavior (such as smoking) and found that although study designs have improved over the past 20 years, often there are difficulties in doing such experiments: school personnel resist randomization, appropriate control schools might not be used, the intervention (the stimulus) might not be fully implemented in some of the experimental group schools, permission for student participation is difficult to obtain, and biases can be introduced by data collection when the same people deliver the intervention and measure its outcome.

Experiments in the Laboratory

Another, perhaps more common, setting for experiments is the "laboratory," which is a setting controlled by the researcher. The literature of psychology, social psychology, education, medical sociology, delinquency and corrections, political sociology, and organizational sociology is replete with reports of **laboratory research,** much of it conducted in universities using students as participants. For example, while both kinds of studies use experimental and control groups, a field experiment testing a hypothesis about the effect of gender on job offers might have had matched "applicants" of both genders apply for real jobs in real companies, while a laboratory experiment on the same topic might have undergraduate subjects rate hypothetical job applicants.

laboratory research, research done in settings that allow the researcher control over the conditions, such as in a university or medical setting.

The Bushman and Anderson (2009) study on the desensitizing effects of violent media on helping others described earlier is an example of a laboratory experiment. The sample in this study was composed of groups of students who had volunteered to be part of a research project and participated in a lab on a college campus. Bushman and Anderson (2009) had control over the setting and could easily randomly assign sample members to groups.

Although realistic situations can be created in labs and subjects can be kept in the dark about the real purpose of the study, the participants in laboratory experiments know that they are involved in research. If the setting and staging of the experiment, including the way the independent variable is introduced, do not feel real, the study's internal validity will be hampered. For example, Bushman and Anderson (2009: 275) used a pilot or trial study with 50 participants to see if the students found the "fight" realistic. Only five of the first 10 students thought it was genuine so they added throwing the chair and kicking the door. The next 40 students reported believing that the fight was real.

It's important to evaluate a study's external validity or the ability to generalize the results from the lab to the "real world" by considering the following issues:

1. Was the situation very artificial, or did it approximate "real life?"
2. How different were study participants from other populations?
3. To what extent did the participants believe that they were up for inspection, serving as guinea pigs or play-acting, or have other feelings that would affect responses to the stimulus (Campbell and Stanley, 1963: 21)?
4. To what extent did the researcher communicate his or her expectations for results to the subjects with verbal or nonverbal cues?

experimenter expectations, when expected behaviors or outcomes are communicated to subjects by the researcher.

double-blind experiment, an experiment in which neither the subjects nor the research staff who interact with them knows the memberships of the experimental or control groups.

One way to handle the issue of **experimenter expectations** is to use a design called the **double-blind experiment,** which involves keeping both the subjects and the research staff who directly interact with them from knowing which members of the sample are in the experimental or the control group. In this way, the researcher's expectations about the experimental group are less likely to affect the outcome.

Natural Experiments

STOP & THINK   *What if you were interested in finding out if returning "home" after release from prison to the same neighborhoods with the same criminal opportunites and criminal peers has a negative impact on ex-prisoners? Could you design an experiment to test this?*

natural experiment, a study using real-world phenomena that attempts to approximate an experimental design even though the independent variable is not controlled, manipulated, or introduced by the researcher.

There are many situations where it's not possible to control the independent variable for practical and/or ethical reasons. So, even if we had an explanatory purpose, such as trying to find out if a change in residence has an impact on recidivism, we couldn't do a traditional experiment because it's not possible for a researcher to control where people live. However, sometimes real-world phenomena approximate the experimental design in a **natural experiment** even though the independent variable is not controlled, manipulated, or introduced by the researcher. Sometimes events like legal changes, wars, economic crises, natural disasters, and other phenomena approximate the independent variable and allow testing of a causal hypothesis.

A study on recidivism conducted by David Kirk used a natural experiment. Kirk noted (2009: 485) that studying the impact of "neighborhood effects" on ex-prisoners means considering the possibility of selection bias as those who move "home" following incarceration may be fundamentally different on unobserved characteristics compared to those who move elsewhere. The tragedy of Hurricane Katrina, damaging or destroying so many housing units in New Orleans, presented an opportunity to do a natural experiment on the effect of residential change on recidivism. With fewer local housing options, more prisoners had to move to new communities—allowing comparison of crime rates of those who moved and those who did not. Kirk found that prior to Katrina, 75 percent of parolees returned to their parish upon release from prison but only 50 percent of those released post-Katrina did so.

Moreover, he found that "individuals who migrated to a different parish were significantly and substantially *less* likely to be re-incarcerated within one year of release from prison" (Kirk, 2009: 496).

COMPARING EXPERIMENTS TO OTHER DESIGNS

Although the experimental study design is best suited for explanatory research and is a useful choice when considering causal hypotheses, many research questions cannot and should not be studied experimentally. Researchers should consider non-experimental designs if they have descriptive or exploratory research purposes, need large samples, or if it is not practical, possible, or ethical to control or manipulate the independent variable.

STOP & THINK   *Can you think of a hypothesis that does not lend itself to being studied with an experiment? What design could be used to test your hypothesis?*

Social scientists often construct hypotheses using background variables like gender, race, age, or level of education as independent variables. Let's say that we hypothesize that level of education affects the extent to which young adults are involved in community service and political activity; more specifically, that an increase in education, the independent variable, causes increases in the dependent variables (amounts of community service and political activity). We can't control how much education an individual receives, so we'd need to use a non-experimental design to examine these relationships. For example, we could do a panel study and observe changes in the variables over time by starting with a sample of 18-year-olds. Following the sample for a number of years, we could measure the independent variable (years of schooling completed) and the dependent variables (involvement in community service and political activities) at least twice. We'd be able to see the pattern or relationship between the variables and their time order. We could use statistical analyses as a way to disentangle possible "third variables" and spuriousness.

Panel studies have been widely used to study the cause-and-effect relationship between behaviors such as drinking alcohol, smoking tobacco, and eating a diet rich in red meat and the onset and progression of illnesses such as heart disease and cancer. Health researchers can't and wouldn't randomly select groups of people and force them to drink alcohol, eat red meat, or smoke cigarettes for a number of years while prohibiting the members of other randomly selected groups from doing so, but they can follow people who make these choices themselves.

Panel studies like the Nurses' Health Study (NHS) have followed people for decades. Started in 1976 when the youngest participants were 25 and expanded in 1989, the nurses' study is world's largest, longest-running study of women's health with information provided by 238,000 nurse participants (NHS, 2013). Every two years the nurses enrolled in the study are asked for medical histories, questioned about daily diet and major life events, and periodically provide physical samples to the researchers. The response rate for the questionnaires is typically 90 percent for each two-year cycle. The study is ethical as researchers do not attempt to intervene in subjects' behavior and keep all data confidential.

The NHS produced landmark data on cancer prevention, cardiovascular disease, diabetes, and many other conditions. As a panel study can take a long-term view, there is strong support for the validity of the many analyses of the NHS data focusing on how diet, physical activity, and other lifestyle factors affect health. This panel study has been used to test many explanatory hypotheses.

Let's consider another non-experimental design, the cross-sectional study, for its use in explanation. Assume we hypothesize that gender affects the degree of social isolation among senior citizens. While we can't "control" or randomly assign gender, we could measure this independent variable and any number of dependent variables, including social isolation in a sample of senior citizens. Because the time order of these independent and dependent variables is clear (the gender of a senior citizen obviously occurs before his or her social isolation), if we found an association between the variables, we could then look for other support for causality. We could, for example, use statistical analysis techniques to eliminate the effect of other variables that might make the original relationship spurious.

SUMMARY

This chapter adds to the researcher's toolbox by describing the experimental design options. Because experiments allow researchers to control the time order of the independent and dependent variables and to identify changes in the dependent variable, they are very useful for testing causal hypotheses.

Experiments can be done as field experiments in the "real world," or they can be conducted in laboratories. Although the field experiment might be less artificial, the laboratory experiment can provide a greater ability to regulate the situation. Another option is the natural experiment, which is really not an experiment in the traditional sense, but which attempts to approximate an experimental design.

Experimental designs can be extended by adding extra experimental and control groups or can be truncated into a quasi-experiment by eliminating one or more aspects of the classic design. Some design choices are more practical than others, whereas others have greater internal validity. Experiments typically use small, non-random samples and therefore usually have limited generalizability.

Ethical considerations and the practical concerns of time, cost, and the feasibility of controlling the independent variable are important issues. Cross-sectional studies are typically the most practical, and when a large and representative sample is selected, the study will have good generalizability. However, the tradeoff for selecting a non-experimental design may be questions about internal validity if the time order is not clear or if there are preexisting differences between groups.

Each study design has strengths and weaknesses. As always, it's important to consider a project's goals against each design's costs and benefits and to balance methodological, ethical, and practical considerations when selecting a research strategy. Both experimental and non-experimental designs should be considered when evaluating options (see Table 8.1).

TABLE **8.1**
Summary of Design Studies

Study Design	Design Features	Design Uses
Pretest–Posttest Control Group Experiment	This has at least one experimental and one control group; the independent variable is manipulated, controlled, or introduced by the researcher; the dependent variable is measured at least two times in both groups (before and after the independent variable is introduced).	Useful for explanatory research testing causal hypotheses.
Posttest-Only Control Group Experiment	Same as above except that the dependent variable is measured only after the independent variable is introduced.	Same as above.
Extended Experimental Design	Same as the pretest–posttest control group experiment with an additional experimental and control group measured only at the posttest.	Same as above but can also sort out the effects of pretesting.
Quasi-Experiment	The researcher does not have complete control of experimental stimuli. The independent variable is introduced, controlled, or manipulated and a comparison group rather than a control group is used with an experimental group.	Same as experiment but with increased concerns about internal validity. May be more ethical or practical than a control group experiment.
Cross-Sectional Study	One sample of cases studied one time with data about one or more variables. Typically, the sample is divided into categories of the independent variable and compared on the dependent variable during analysis.	Useful for describing samples and the relationship between variables; useful for explanatory purposes especially if the time order of variables is known and sophisticated statistical analysis is possible.
Panel Study	One sample of cases is studied at least two times.	Useful for describing changes in individuals and groups over time (including developmental issues); useful for explanatory purposes.
Trend or Cohort Study	Two or more different samples are selected from the same population at least two times.	Useful for describing changes in populations.
Case Study	In-depth analysis of one or a few cases (including individuals, groups, and organizations) within their social context.	Useful for exploratory, descriptive, and, with caution, explanatory purposes; useful for generating theory and developing concepts; very limited generalizability.

EXERCISE 8.1

Reading about Experiments

Find a research article in a psychology or social science journal that reports the results of an experiment. Answer the following questions about the study.

1. List the author(s), title of article, journal, month and year of publication, and pages.
2. What is one hypothesis that the research was designed to test? Identify the independent and dependent variables.
3. Describe the sample.
4. How many groups were used and how were they selected? Is there a control group?
5. Was there a pretest? If so, describe it.
6. How was the independent variable (or the stimulus) manipulated, controlled, or introduced?
7. Was there a posttest? If so, describe it.
8. What kind of experimental design was used—pretest and posttest experiment, a posttest-only experiment, a Solomon four-group, a comparison group quasi-experiment, or some other kind? Give support for your answer.
9. Comment on the issues of internal validity and generalizability as they apply to this study.

EXERCISE 8.2

Conducting a Quasi-Experiment

What to do:

First, check with your professor about the need to seek approval from the Human Subjects Committee, the institutional review board, or the committee at your college or university concerned with evaluating ethical issues in research. If necessary, obtain approval before completing this exercise.

In this quasi-experiment, you will be testing the hypothesis that when there is a "conversational icebreaker," people are more likely to make conversation with strangers than when there is no icebreaker.

Select a situation with which you are familiar from your daily life where strangers typically come into contact with each other. Some examples of appropriate situations include riding in an elevator, being in a doctor's waiting room, waiting for a bus or a train, and standing in the check-out line in a supermarket.

Select one item to use as a conversational icebreaker. (Make sure it is something that could be commented on but is not something that could be perceived as threatening, hostile, or dangerous.) Some examples of icebreakers that you can use are carrying a plant or a bouquet of flowers, holding a cake with birthday candles on it, pushing an infant in a stroller, carrying a fishbowl with water and tropical fish in it, or wearing a very unusual item of clothing or jewelry.

Over the course of several hours or days, put yourself in the same situation six times: three times with your icebreaker and three times without it. To the extent that you can, make sure that you keep everything else under your control the same each time (such as how you dress, the extent to which you make eye contact with strangers, and whether you are alone or with friends). But recognize that the groups of people you interact with, with and without the icebreaker, are two convenience samples—with one serving as your experimental and the other as your comparison group.

Immediately after leaving each situation, record your field notes (detailed descriptions of what happened). In your field notes, record information about the number of strangers you interacted with, what they were like, and the way they interacted with you, specifically recording any comments that were made.

What to write:

After completing your experiment, answer the following questions:

1. What situation did you use for your experiment?
2. What icebreaker did you use?
3. Describe the dependent variable in the experimental group—that is, the kind of interactions that you had and any comments that

strangers made in the three situations where you had an icebreaker.

4. Describe the dependent variable in the comparison group—that is, the kind of interactions that you had and any comments that strangers made in the three situations where you did not have an icebreaker.
5. Does your data support the hypothesis that an icebreaker is more likely to foster conversation and interaction with strangers?
6. Comment on the internal validity and generalizability of your study.
7. Comment on the experience of doing this exercise.

EXERCISE 8.3

Considering Non-Experimental Study Designs

In some situations, it won't be practical, ethical, or feasible to do a controlled experiment. Think about what you would do if, like Chris Caldeira, whose study you read in this chapter, you had a hypothesis that one teaching technique is more effective than others, but you think it's not practical or ethical to do an experiment. Construct a specific hypothesis that has a teaching technique or curriculum approach as the independent variable and design a study using a non-experimental design to test it (see Figure 7.4, in Chapter 7, for these designs).

1. State your hypothesis, identifying the specific independent and dependent variables you have selected.
2. Identify, describe, and draw a diagram of a study design you could use to examine your hypothesis.
3. Describe how you would measure your independent and dependent variables in this study.
4. Comment on the issues of internal validity and generalizability for the study you propose.
5. Compare the internal validity, generalizability, and practicality of your proposed study design with the kind of study conducted by Caldeira.

Questionnaires and Structured Interviews

SURVEYS

RESEARCH IN THE NEWS

In 2013, *The Washington Post* reported that according to a survey of teachers, a gap in access to the Internet between the rich and poor is leading to troubling disparities in education. Half of the teachers of high-income students say that all or almost all of their students have home access to needed digital tools; only 20% of teachers of middle-income students and just 3% of teachers of the poorest students say this. Three-quarters of teachers surveyed said Google and other search engines have conditioned students to expect to find information quickly and easily and discourages them from using a wide range of sources for research.

For the complete article see: http://articles.washingtonpost.com/2013-02-27/business/37338686_1_teachers-internet-access-poorest-students

---

STOP & THINK   *What do you think are the most important issues for policy makers to focus on this year? Each year the Pew Research Center for People & the Press does a national poll of Americans to find out what they think should be on the public agenda. In 2013, the researchers reported that strengthening the economy and improving the job situation were at the top of the list of priorities; reducing the budget deficit, defending against terrorism, making Social Security financially sound, and improving education followed closely behind (Pew Research Center, 2013). How do you think the Pew researchers obtained these poll results?*

---

questionnaire, a data collection instrument with questions and statements that are designed to solicit information from respondents.

structured interview, a data collection method in which an interviewer reads a standardized list of questions to the respondent and records the respondent's answers.

survey, a study in which the same data, usually in the form of answers to questions, are collected from one or more samples.

Have you ever received a phone call asking you to participate in a survey? Do you remember someone in your household completing a questionnaire from the U.S. Census Bureau? Have you heard or read about recent polls on current issues or policies? Most of us are familiar with interviews and questionnaires. In fact, it's hard to avoid them. **Questionnaires** and their first cousins, **structured interviews,** are the most widely used methods of data collection in the social sciences. Research projects that use structured interviews and questionnaires are sometimes referred to as surveys, because although these methods can be used in experiments or case studies, they are most often used in surveys.

A **survey** is a study in which the same data (most often a specified set of questions, but possibly observations or available data) are collected from one or more samples (usually individuals, but possibly other units of analysis) and analyzed statistically. Surveys typically use large probability samples and the cross-sectional study design, which is when data are collected about many variables at one time from one sample. However, surveys can also employ nonprobability samples and trend and panel study designs. The Research in the News section and Box 9.1 provide examples of recent

BOX **9.1** Surveys and Polls

Surveys are all around us. If we want to know what people think about something, we usually can find a survey that's asked questions on the topic. Two sources of interesting and methodologically careful surveys are Public Agenda (www.publicagenda.org) and the Pew Research Center for the People & the Press (http://people-press.org). Both of these nonprofit organizations conduct many studies annually. The work of Public Agenda has included studies on families, children, crime, corrections, education, health care, and the economy. The Pew Research Center focuses on attitudes toward the press, politics, and public policy.

STOP & THINK   *What are your educational goals and career aspirations? Do you think it's important to graduate from college? Do you think your feelings are typical for those your age?*

A national survey conducted by the Public Agenda (2012) of a national random sample of 1,000 Americans ages 18 through 25 found that most believed strongly in the value of going to college, seeing it as both a way to earn society's respect and to ensure career advancement and financial security. See the details of this study at http://www.publicagenda.org/files/life_after_high_school.pdf

respondents, the participants in a survey who complete a questionnaire or interview.

self-report method, another name for questionnaires and interviews because respondents are most often asked to report their own characteristics, behaviors, and attitudes.

surveys. Although these are from the U.S., survey research is conducted around the world.

The various modes of data collection that use communication between respondents and researchers fall along a continuum from least interactive to most interactive—that is, from self-administered questionnaires to in-person interviews (Singleton and Straits, 2001). In all of the modes of data collection covered in this chapter, data are collected by asking study participants, typically called **respondents**, a set of predetermined questions. This technique is also called the **self-report method** because respondents are usually asked to provide information about themselves. The researcher can take the answers as facts or interpret them in some way, but the basic data consists of information provided by respondents.

Answering questions in a survey is a social act in the same way that attending a class or browsing in a shopping mall is. The social context of the data collection and the views of those we question are influenced by the surrounding culture. When we meet new acquaintances in our everyday lives, we often ask questions. In the same way, researchers use questions to elicit information about the backgrounds, behavior, and opinions of respondents. And, just as in our personal conversations, researchers try to figure out if the answers they've received are honest and accurate.

In the study described in the focal research that follows, Brandon Lang and Christopher Podeschi asked college students questions about environmental issues. They were interested in whether attitudes were connected to background characteristics and behavior patterns. This study was conducted for both descriptive and explanatory purposes.

Environmentalism among College Students: A Sociological Investigation

By K. Brandon Lang and Christopher W. Podeschi*

Introduction

In recent decades, Americans have become more aware of environmental issues such as air and water pollution, waste management, ozone depletion and global warming. Dozens of studies have shown, however, that environmental attitudes and behaviors vary according to a number of social and cultural variables. In this paper, we build upon previous work that considers these social and cultural predictors by examining the "social bases" of environmentalism using a sample of first-year students at a comprehensive public university in Pennsylvania. We also explore the relationship between the environmental worldview and behaviors of these students.

Previous Research

For decades, social scientists have examined environmental attitudes and people's willingness to engage in environmentally friendly behaviors. Researchers have also studied people's fundamental values and beliefs with regard to society's place in relationship to nature. Dunlap and Van Liere (1978) first examined adherence to either the "New Ecological Paradigm" or the "Dominant Social Paradigm." They identify the former as an ecologically-informed worldview, the latter a human-centered orientation that fails to acknowledge the need for ecological limits.

Early evidence and theory led scholars to predict a correlation between higher socio-economic positions and environmentalism (see Van Liere and Dunlap 1980). The "post-materialist" thesis says that people without economic concerns move on to considerations like beauty or ecology (Inglehart 1995). Recent studies, however, find mixed results concerning the relationship between environmentalism and social class (see Jones and Dunlap 1992; Klineberg et al. 1998; Marquart-Pyatt 2008).

Research on the relationship between environmentalism and residence also yields mixed results. An early hypothesis asserts that rural culture, because of dependence on occupations involving natural resources (e.g., farming, mining, logging), is more utilitarian than ecologically-inclined. In addition, urbanites have become concerned because they experience environmental problems like air pollution (see Van Liere and Dunlap 1980). While there is some support for this position (see Jones and Dunlap 1992; Tremblay and Dunlap 1978), research by Jones et al. (1999) and others yields contradictory findings about residence (also see Jones et al. 2003). Perhaps environmental concern "diffused" or perhaps, as Bennett and McBeth (1998) suggest, economic diversification in rural areas has weakened the utilitarian aspect of rural culture.

*K. Brandon Lang and Christopher W. Podeschi are associate professors of Sociology at Bloomsburg University of Pennsylvania. This article is published with permission.

More consistent predictors of environmentalism include political orientation and gender. Research on environmental attitudes and politics considers the effect of party identification and political ideology (e.g., attitudes about the government's role in the economy). Dunlap et al. (2001) provide an extensive review of this literature. They note that in the early stages of the modern environmental movement, an attitude distinction based on politics was not as apparent as it is today when Democrats and liberals are more pro-environment than Republicans and conservatives (see Dunlap et al. 2001). Also notable is the fact that while party identification and political ideology both predict environmental attitudes, the former has been a weaker predictor.

Much recent research finds women to be more environmentally oriented than men in terms of both attitudes and behaviors (see Zelezny et al. 2000; Davidson and Freudenburg 1996). Scholars feel that women tend to be more environmentally oriented because their socialization to be nurturing attunes them to risks to others and altruism (see Dietz et al. 2002). Furthermore, concern for others and consequently for collective goods like environmental health is thought to be fostered by women's disadvantaged position which is seen as leading to empathy (Kalof et al. 2002). Kalof et al. (2002) find that this "disadvantage" hypothesis is supported by data on both gender *and* race/ethnicity.

Finally, there is the relationship between environmental attitudes and behavior. While we might expect that these things would come together, research finds only modest or weak correlations (see Theodori and Luloff 2002; Tarrant and Cordell 1997). Why might the connection not be stronger? Bell (2004) identifies how the "social constitution of daily life" constrains our choices. For example, people drive despite their environmental concerns because we have developed car-centered communities. Similarly, Diekmann and Priesendorfer (2003) find strong support for the idea that attitudes and behaviors are more congruent when costs are low, but that relationships weakens as costs increase. For example, in a car-centered society, alternative transportation may have high time costs with the result that transport choices are less likely to be congruent with people's pro-environment views.

Research Questions and Hypotheses

Our study asks about the predictors of environmental worldview and about the link between attitudes and behavior. We test hypotheses in the settled and unsettled areas of research discussed above.

In the unsettled areas, we test two hypotheses:

1. Despite mixed results in the literature, higher class position will be associated with a pro-environment worldview because of "post-material" conditions and because parents of these students are more likely to have higher levels of education, a consistent predictor of environmentalism.
2. The difference in environmental worldview between rural and other will be small or non-existent because of the diffusion of environmentalism and changes in the rural occupation structure that has weakened the utilitarian aspect of rural cultures. Notably, coal mining has declined in rural Pennsylvania, and like the rest of the country, the population is shifting away from farming occupations.

(continues)

FOCAL RESEARCH *(continued)*

In the more settled areas, we test three hypotheses:

3. Female students will have a more pro-environment worldview than male students.
4. Students that identify as Democrats or liberals will have a more pro-environment worldview than college students that identify as Republicans or conservatives, but party will be a weaker predictor than political ideology (i.e., liberal vs. conservative will be a stronger predictor).
5. Since recycling behavior and willingness to pay increased student fees for campus environmental programs are easy ways to enact one's environmental worldview, there will be a solid correlation between environmental worldview and these behaviors.

Methodology

The data were collected in August, 2008, at the end of orientation for new first-year students at a comprehensive public university in rural Pennsylvania. A short self-administered questionnaire was distributed to groups of approximately thirty students that had been grouped by major. Completed questionnaires were obtained from 1,226 out of the population of approximately 1,500 new students.

The sample is sixty-one percent female. Forty percent are politically liberal, 60.5 percent identify as Democrat, 52.9 percent are from a rural area or a small city, and seventy-one percent characterize themselves as wealthy or comfortable.

We measure environmental worldview with an index of ten items from the "New Ecological Paradigm" scale (NEP hereafter; see Dunlap et al. 2000). The appendix lists these and all other indicators used. In hypotheses one through four, environmental worldview is the dependent variable. For hypothesis five, it serves as an independent variable.

We measure environmental behavior in two ways: recycling and willingness to pay increased fees for campus environmental programs. Recycling behavior is measured with a three item index (alpha.755). Six items are indexed for the "willingness to pay" variable (alpha.706).

The independent variables are class, place of residence, gender, political party identification and political ideology. Rather than rely on multiple dummy variables, variables with multiple response categories have been recoded into dichotomies.

Results

The possible range of scores for NEP adherence extends from 10 to 40. The mean score on this measure for our sample is 27.8 (s.d. = 3.81). The possible range of scores for the recycle scale is 3 to 12 with an average score in the sample of 7.1 (s.d. = 2.98). The possible range of scores for the fees scale is 6 to 12 with an average score of 8.46 (s.d. = 1.79).

Eighty percent of those surveyed answered that they always or usually recycle bottles and cans. Roughly one third answered that they always or usually recycle paper while about half said they always or usually recycle cardboard. About half of the respondents answered that they would be willing to pay higher student fees in order for older buildings on campus to be renovated, for biodiesel to be used in campus buses and for recycled paper to be more widely used on campus. However, students were much less willing to pay higher fees for organic food, natural pesticides and low-flow shower heads.

We use means testing and OLS regression to test hypotheses one through four. Means tests reveal that students who consider themselves wealthy or comfortable do not score significantly higher on the NEP scale. The difference between rural/small city students and urban/suburban students is also not significant. By contrast, means tests show that females, liberals and those that identify as Democrats do have significantly higher NEP scores than males, conservatives and those that identify as Republicans, respectively.

In the regressions, we consider the effects of gender, politics (as ideology or party), social class and place of residence together upon NEP adherence. With regard to gender, female students score, on average, 1.50 units higher on the NEP than male students ($p < .001$). Our results also show that liberal students score 1.52 units higher on the NEP than conservative students ($p < .001$). In a second model, we substitute party identification for political ideology and find that the effect of considering oneself a Democrat is stronger than the effect of being liberal, producing a 2 unit increase in NEP ($p < .001$). Neither class nor residence has a significant impact upon NEP adherence.

We also use OLS regression to test hypothesis five. We consider the effect of NEP adherence upon two behaviors while controlling for all the above independent variables. Our results show a positive and significant correlation between NEP score and recycling behavior. Every unit increase in NEP adherence produces a .086 unit increase in recycling ($p < .001$). Similarly, each unit increase in NEP adherence produces a .132 unit increase in willingness to pay fees for campus environmental programs ($p < .001$). Simple bivariate correlations between NEP adherence and these behaviors are stronger than the standardized regression coefficients that show the relation between NEP adherence and these behaviors, controlling for demographics and politics.

Conclusions

We find support for all but one of our hypotheses. No significant differences in NEP adherence are found between rural/small city and urban/suburban students, lending support to the notion that environmentalism has diffused into rural areas and/or that the utilitarian worldview that comes with reliance on extractive industries has weakened because of economic changes. We find female students stronger adherents to the NEP than males, something that fits the existing literature. Our finding that students with a liberal political ideology and those that identify as Democrats have higher NEP scores as well also supports the findings of previous research. Contrary to findings in the literature, however, we find that party is a stronger predictor than ideology. These political findings should be interpreted with caution given the possibility that first-year college students may not be well-enough versed in politics to provide valid data in this area.

As expected, we also find the NEP is a solid predictor of both recycling behavior and willingness to pay fees for campus environmental programs. This is supported by both simple bivariate correlations and regressions that control for demographic and political factors. The fact that the bivariate correlations are stronger than the regression coefficients suggest that demographics and politics

(continues)

FOCAL RESEARCH *(continued)*

play a role in the link between environmental attitudes and behavior, something also found by Tarrant and Cordell (1997).

In addition, we find that reported recycling is significantly higher among urban and suburban students than among rural students. This suggests that ease of access plays a role since we expect that urban and suburban areas are more likely to have recycling programs.

The only unsupported hypothesis focuses on class differences. The literature is mixed, but we expected more comfortable classes to be more "post-materialist" and that those with more affluent parents would be more supportive of environmentalism. This is not the case, lending support to arguments about the diffusion of environmentalism in the class system or to contentions that there's been a dampening of environmentalism among those higher on the class ladder. As with the political variables, however, we are aware of validity concerns about our measure of class position.

References

Bell, Michael. 2004. *An Invitation to Environmental Sociology.* Thousand Oaks, CA: Sage.

Bennett, Keith and Mark McBeth. 1998. "Contemporary Western Rural USA Economic Composition: Potential Implications for Environmental Policy and Research." *Environmental Management.* 22,3:371–381.

Davidson, Debra and William Freudenburg. 1996. "Gender and Environmental Risk Concerns: A Review and Analysis of Available Research." *Environment and Behavior.* 28,3:302–339.

Diekmann, Andreas and Peter Priesendorfer. 2003. "The Behavioral Effects of Environmental Attitudes in Low-Cost and High-Cost Situations." *Rationality and Society.* 15,4:441–472.

Dietz, Thomas, Linda Kalof and Paul Stern. 2002. "Gender, Values and Environmentalism." *Social Science Quarterly.* 83,1:353–364.

Dunlap, Riley and Kent Van Liere. 1978. "The 'New Environmental Paradigm': A Proposed Measuring Instrument and Preliminary Results." *Journal of Environmental Education.* 9:10–19.

Dunlap, Riley, Kent Van Liere, Angela Mertig and Robert Jones. 2000. "Measuring Endorsement of the New Ecological Paradigm: A Revised NEP Scale." *Journal of Social Issues.* 56,3:425–442.

Dunlap, Riley, Chenyang Xiao and Aaron McCright. 2001. "Politics and Environment in America: Partisan and Ideological Cleavages in Public Support for Environmentalism." *Environmental Politics.* 10,4:23–48.

Inglehart, Ronald. 1995. "Public Support for Environmental Protection: Objective Problems and Subjective Values in 43 Societies." *PS: Political Science and Politics.* 28,1:57–72.

Jones, Robert and Riley Dunlap. 1992. "The Social Bases of Environmental Concern: Have They Changed Over Time?" *Rural Sociology.* 57:28–47.

Jones, Robert, Mark Fly and Ken Cordell. 1999. "How Green Is My Valley? Tracking Rural and Urban Environmentalism in the Southern Appalachian Ecoregion." *Rural Sociology.* 64,3:482–499.

Jones, Robert, Mark Fly, James Talley and Ken Cordell. 2003. "Green Migration into Rural America: The New Frontier of Environmentalism?" *Society and Natural Resources.* 16:221–238.

Kalof, Linda, Thomas Dietz, Gregory Guagnano and Paul Stern. 2002. "Race, Gender and Environmentalism: The Atypical Values and Beliefs of White Men." *Race, Gender and Class.* 9,2:112–130.

Klineberg, Stephen, Matthew McKeever and Bert Rothenbach. 1998. "Demographic Predictors of Environmental Concern: It Does Make a Difference How It's Measured." *Social Science Quarterly.* 79,4:734–753.

Marquart-Pyatt, Sandra. 2008. "Are There Similar Sources of Environmental Concern? Comparing Industrialized Countries." *Social Science Quarterly.* 89,5:1312–1335.

Tarrant, Michael and Ken Cordell. 1997. "The Effect of Respondent Characteristics on General Environmental Attitude-Behavior Correspondence." *Environment and Behavior.* 29,5:618–637.

Theodori, Gene and A.E. Luloff. 2002. "Position on Environmental Issues and Engagement in Proenvironmental Behaviors." *Society and Natural Resources.* 15:471–482.

Tremblay, Kenneth and Riley Dunlap. 1978. "Rural-Urban Residence and Concern with Environmental Quality: A Replication and Extension." *Rural Sociology.* 43,3:474–491.

Van Liere, Kent and Riley Dunlap. 1980. "The Social Bases of Environmental Concern: A Review of Hypotheses, Explanations and Empirical Evidence." *Public Opinion Quarterly.* 44:181–197.

Zelezny, Lynette, Poh-Pheng Chua and Christine Aldrich. 2000. "Elaborating on Gender Differences in Environmentalism." *Journal of Social Issues.* 56,3:443–457.

Appendix A: Group-Administered Questionnaire

Instructions read to respondents by the orientation workshop leader:

The University Campus Green Initiative has prepared a survey for the in-coming students to fill out during their orientation week. The survey should take about ten minutes to complete. Every student will be given a survey and an answer sheet. Please answer all of the questions on the answer sheet with a blue pen, black pen or pencil. Upon completing the survey, please return both sheets to the survey administrator. For the last questions, please write the answers directly on the answer sheet.

Please note that the surveys are anonymous. The information provided will not be linked with any individuals. Also, fully completed surveys are very helpful, but you may skip any questions that you would prefer not to answer. Thank you very much for completing these surveys. The members of the Campus Green Initiative value your opinions and appreciate that you are taking the time to answer these questions.

Thanks very much,
Dr. Brandon Lang, Department of Sociology, Social Work and Criminal Justice
Dr. Chris Podeschi, Department of Sociology, Social Work and Criminal Justice

2008 First-Year Student Environmental Survey (abridged)

Section 1:

With regard to what is said in the following statements, please answer whether you…

A: strongly agree   B: agree   C: disagree   D: strongly disagree

1. Humankind is severely abusing the environment.
2. We are approaching the limit of the number of people the earth can support.
3. The earth has plenty of natural resources if we just learn how to develop them.
4. Plants and animals have just as much right as humans to exist.
5. The balance of nature is strong enough to cope with the impacts of modern industrial nations.
6. Despite our special abilities, humans are still subject to the laws of nature like other species.

(continues)

FOCAL RESEARCH *(continued)*

7. The so-called 'ecological crisis' facing humankind has been greatly exaggerated.
8. Humans were meant to rule over the rest of nature.
9. Humans will eventually learn enough about how nature works to be able to control it.
10. If things continue on their present course, we will soon experience a major ecological catastrophe.

Section 2:

Would you be willing to pay increased student fees in order to…

1. Have organic food options on campus? A: yes   B: no
2. Have 100% recycled paper for on-campus printing? A: yes   B: no
3. Have only biodiesel and other low-emission fuels used in campus vehicles and buses? A: yes   B: no
4. Have low-flow showerheads installed in dormitory bathrooms? A: yes   B: no
5. Support the use of organic pesticides and fertilizers on campus? A: yes   B: no
6. Renovate older buildings so they're more environmentally-friendly? A: yes   B: no

Section 3:

How frequently do you recycle the following types of items…

1. Paper/Magazines/Newspapers? A: always   B: usually   C: sometimes   D: never
2. Cardboard? A: always   B: usually   C: sometimes   D: never
3. Bottles and cans? A: always   B: usually   C: sometimes   D: never

Section 4:

1. I am from a… A: rural area   B: small city   C: large city   D: suburb
2. I am… A: male   B: female
3. Politically speaking, I am… A: very conservative   B: conservative   C: liberal   D: very liberal
4. I support the… A: Republicans   B: Democrats   C: Green Party   D: none (I'm an independent)
5. Growing up, my family was… A: very wealthy   B: comfortable   C: struggling   D: poor
6. For work, my parents… PLEASE WRITE YOUR PARENTS' JOBS ON THE ANSWER SHEET
7. I was born in… PLEASE WRITE YOUR YEAR OF BIRTH ON THE ANSWER SHEET

THINKING ABOUT ETHICS *The researchers submitted a proposal to their university's Institutional Review Board (IRB), which approved it. The respondents were informed about the study's sponsor, were assured that the surveys were anonymous, and were told that they could skip any of the questions.*

THE USES OF QUESTIONNAIRES AND INTERVIEWS

Once researchers decide on the kind of information required, they'll typically consider whether asking questions is an appropriate way to collect this information and then think about ways of administering those questions. Brandon Lang and Chris Podeschi selected a self-report method because it is very useful for collecting information about *attitudes, beliefs, values, goals,* and *expectations,* although data on these topics can sometimes be inferred from observed behavior. They asked the respondents to indicate their level of agreement with statements such as "Humankind is severely abusing the environment" and "The earth has plenty of natural resources if we just learn how to develop them."

Lang and Podeschi also asked questions about respondents' background characteristics and behavior. Although data about *social characteristics and past experiences of individuals* and information about *organizations or institutions* might be accessible through records and documents, they are suitable topics for questions. Questions are frequently employed to obtain information about *behavior,* even though observational methods will often yield more valid data. If you are interested in attitudes or behavior that typically occurs without observers (such as sexual victimization), or if you want to find out about behavior that would be unethical, impractical, or illegal to observe (such as voting behavior or getting tested for HIV), then a self-report method might be the only data collection option. In addition, questions can be used to measure *knowledge* about a topic. In Box 9.2, you'll see some examples of questions using the self-report method on a variety of topics.

STOP & THINK   *In addition to this chapter's focal research, several other focal research studies have collected data by asking questions. Can you remember any of the studies that have used this method? Do you have any concerns about the validity of the self-report data collected in any of the focal research studies?*

The self-report method has a wide variety of applications, including public opinion polls, election forecasts, market research, and publicly funded surveys. In this chapter, we'll consider structured ways of asking questions; in the next chapter, we'll consider less structured formats.

More structured ways of asking questions facilitate the collection of the same information from large samples, especially when the samples have been selected from regional, national, or multinational populations. The resulting analyses are often used by local, state, federal, and international organizations and are frequently the focus of articles in the mass media. For example, the National Crime Victimization Survey (NCVS) collects interview data each year about people's experiences with criminal victimization from a nationally representative sample of 76,000 households. This data enables the Bureau of Justice Statistics (BJS) to estimate the numbers of rapes, sexual assaults, robberies, assaults, thefts, and burglaries for the population as a whole as well as for segments such as women, the elderly, members of various racial groups, and others. Numerous surveys on a wide variety of topics make it possible to gather large amounts of information from mass publics that can be used descriptively and to inform public policy.

BOX **9.2**   Topics and Examples of Questions

Questions can be asked about attitudes.

An example: Employers should be allowed to require drug testing of employees.

☐ agree strongly

☐ agree somewhat

☐ undecided

☐ disagree somewhat

☐ disagree strongly

Questions can be asked about organizations.

An example: Does your current place of employment have a policy concerning sexual harassment?

☐ no

☐ yes

Questions can be asked about an expectation.

An example: What is your best guess as to whether you will graduate with honors from college?

☐ I have no chance

☐ I have very little chance

☐ I have some chance

☐ I have a very good chance

Questions can be asked about behavior.

An example: During a typical week, how much time do you spend exercising or playing sports?

☐ none

☐ less than an hour

☐ one to two hours

☐ three to five hours

☐ six or more hours

Questions can be asked about a social characteristic or attribute.

An example: How old will you be on December 31 of this year?

☐ under 18

☐ 18–20

☐ 21–23

☐ 24–26

☐ 27–30

☐ 31 or over

Although surveys are very widely used, there are important concerns about the *validity* of the information that they generate. Self-report methods assume that people have the information being asked about and that they will answer based on their core values and beliefs. As Sudman and Bradburn (1982) have argued, the accuracy or validity of survey data is affected by memory, motivation, communication, and knowledge. It is problematic when the respondent wants to present a good image or tries to conform to what are perceived to be the researcher's expectations. It is critical to be aware of respondents' motivations to be perceived in positive ways.

Reliability concerns are that responses might represent casual thoughts rather than deep feelings and can reflect what the respondent heard on the news most recently rather than personally salient answers (Geer, 1991; Kane and Schuman, 1991: 82). In addition, those in positive moods have been found to give more favorable answers to questions than people feeling more negative (Martin, Abend, Sedikides, and Green, 1997).

It is fairly simple to design questions that ask respondents to describe their attitudes toward abortion, to share their views on laws banning assault weapons, or to recall the number of sex partners they had in the past 12 months.

measurement error, the kind of error that occurs when the measurement we obtain is not an accurate portrayal of what we tried to measure.

However, as noted in Chapter 6, we must consider **measurement error,** the giving of inaccurate answers to questions. In addition to the issue of question wording, which we will discuss later in this chapter, measurement error is affected by the way questions are administered, the respondents' level of knowledge, whether they have an opinions about the topic, whether they can recall events accurately from the past, and whether they have the intent to answer honestly.

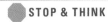 **STOP & THINK**

You agree to participate in a survey on driving behavior that includes the following questions: "Have you been pulled over by the police anywhere in this state in the past year?" "How many times in the last year were you pulled over?" "How many of these pull-overs were for speeding?" Will your answers be accurate? What's your estimate about the accuracy of other respondents' answers?

Recall of prior events by respondents has been found to be influenced by the time that has elapsed between the event and the data collection, the importance of the event, the amount of detail required to be reported, and the social psychological motivation of the respondents (Groves, 1989: 422) and the way that questions are asked (Glasner and van der Vaart, 2009).

Researchers have found that some survey respondents report having voted when they did not actually do so probably because of embarrassment at not voting (Holbrook and Krosnick, 2010). Respondents give socially desirable responses on many topics such as church attendance. For example, while a sizable percentage of respondents reported having attended religious services the previous week, studies have found that when the numbers of service attendees were checked, significant numbers had overreported their attendance (Brenner, 2011).

Factors such as gender and race can also influence responses. In a classic study of children's play activities, Janet Lever (1981) compared answers to the question "Where do you usually play after school, indoors or outdoors?" with activity logs that the fifth-graders kept and found that boys overestimated and girls underestimated their outdoor play. In order to estimate the accuracy of respondents' answers, Tomaskovic-Devey, Wright, Czaja, and Miller (2006) surveyed a sample of drivers and asked them about having been stopped by the police while driving. They found that 23 percent of the white respondents and 29 percent of the African Americans, who according to official records had received one or more speeding citations in the previous six months, did not report any stops. Although studies that generate data by asking questions can be valuable, it's important to consider the specific questions carefully before accepting the results.

PARTICIPANT INVOLVEMENT

response rate, the percentage of the sample contacted that actually participates in a study.

A critical issue for all studies is the **response rate,** the percentage of the sample that is contacted and, once contacted, agrees to participate and completes the survey. The response rate is affected by the number of people who cannot be reached (non-contacts), the number who choose not to participate (refusals), and the number who are incapable of performing the tasks required of them (e.g., due to illness or language barriers). The response rates of studies using self-report methods vary greatly. An analysis of the response rates of 490 organizational research studies published between 2000 and 2005 found that the average response rate was 53 percent in studies where data was collected from individuals and 36 percent where it was collected from organizations (Baruch and Holtom, 2008).

In-person interviews tend to have high response rates, often over 80 percent, while phone surveys typically have lower response rates. For example, the California Health Interview Survey, the largest health survey conducted in any state, uses phone interviews. In 2007–2008, the response rate for the California study was 35 percent for landlines and 22 percent for cell phones (CDC, 2013a).

STOP & THINK   *What do you think has been happening to the response rates of surveys in recent years? Why do you think this is such a source of methodological concern?*

Refusal rates have grown in the past few decades, in the United States and internationally. Yehuda Baruch (1999) examined 175 published studies over three decades and found that the average participation rate reported in 1975 was 64 percent but only 48 percent in 1995. Examination of the response rate in four large surveys over time provides clear evidence that nonresponse rates have continued to increase, especially for those using random-digit dialing (Brick and Williams, 2013). The nonresponse rate for sampled adults from 1996 to 2007 in the National Health Interview Survey, which uses in-person interviews, went from 20 percent to 30 percent; nonresponse to the National Household Education Survey, conducted by phone, went from 40 percent to over 60 percent in the same time frame (Brick and Williams, 2013). There are many reasons for the rise in nonresponse, including increases in non-contact (often attributed to caller ID), increased numbers of cell phone–only households and awareness of publicized identity scams and misuse of personal data. An analysis of verbal and nonverbal interactions by interviewers for the National Health Interview Study found that the most common reasons for not responding were being too busy, having privacy concerns, not being interested or not wanting to be bothered, or feeling that the interview would take too much time (Bates, Dahlhamer, and Singer, 2008).

nonresponse error, an error that results from differences between non-responders and responders to a survey.

A major reason for concern about response rates is **nonresponse error**, which, as discussed in Chapter 5, is the error or bias that results from the differences between those who participate and those who do not. If participants and nonparticipants differ in social characteristics, opinions, values, attitudes, or behavior, then generalization from the sample to the larger population becomes problematic. Comparing response rates, some studies have found higher refusal rates among different groups, such as young people, older persons, those with low income, or those with less than a high school education (Groves, 1989; Krysan, Schuman, Scott, and Beatty, 1994; Fitzgerald, 2011).

Recently, however, some have questioned whether higher rates of refusal mean more bias (Groves, 2006; Keeter et al., 2006). Moreover, Groves (2006) and Groves and Peytcheva (2008) demonstrated that nonresponse errors are only indirectly related to nonresponse rates and are particularly likely to exist when the survey's topic is strongly associated with the likelihood of responding. Thus, if people are simply more averse than they once were to *surveys of all kinds*, lower response rates, generally, don't necessarily mean greater levels of nonresponse bias or error.

STOP & THINK *Have you ever been asked to complete a questionnaire or an interview? What factors did you consider when deciding whether to participate? What you would do if you were on the other side and were trying to convince someone to participate in a survey?*

Participation in studies can best be understood within a social exchange context. That is, once potential sample members are contacted, they must decide about cooperation after thinking about the costs and benefits. Respondents are asked to give up their time, engage in interactions controlled by the interviewer, think about issues or topics that might cause discomfort, and take the risk of being asked to reveal embarrassing information. Even though they are usually assured that the information they provide is anonymous or will be confidential, participants might worry about privacy and lack of control over the information. On the other hand, others might want to participate because of an interest in the topic, a desire to share their views, or knowledge that their information will be useful to science and society.

Advance mailings or other forms of contact may be used to combat the decline in response rates. However, because pre-contact tends to disproportionately raise participation rates among some segments of the population, potential bias needs to be considered (Link and Mokdad, 2005). Another method is to use incentives. For example, in the focal research in Chapter 5, Keeter and his colleagues at the Pew Research Center gave $10 to each cell phone respondent to defray the respondent's cost of the cell phone time. Singer and Cong (2013) systematically reviewed journal articles on surveys since 2002 and analyzed the findings about egotistical versus altruistic reasons for participation. Their analysis concludes that while other factors do matter, incentives, specifically monetary and prepaid ones, increase survey response rates (Singer and Cong, 2013: 134).

In addition, focusing on the interesting aspects of the study, re-contacting possible participants who have not responded, making it clear that the research is legitimate with a bona fide sponsor, and minimizing the costs of participation (such as time and possible embarrassment) can all be effective in increasing participation. See Box 9.3 for a discussion of one researcher's efforts to recruit a sample for her study.

Lang and Podeschi had a high response rate (82 percent of all first-year students at one university) because of the way they administered their questionnaire. Both researchers are members of the Campus Green Initiative, a college environmental group, and were interested in how much student support there was for initiatives such as recycling, low-flow shower heads, and keeping rooms at lower temperatures in the winter. They approached the college's orientation coordinator, who agreed to have all student orientation leaders administer the questionnaires to each group of entering students as part of the

BOX **9.3**   **Recruiting a Sample**

Australian sociologist Jo Lindsay found that it wasn't easy to find a sample of nonprofessional workers for a survey on sexual health, alcohol, and drug consumption. Once she gained access to workplaces, she put up fliers about the project but had trouble finding respondents to volunteer to complete an individually administered questionnaire and return it to her (Lindsay, personal communication). She was more successful when she made in-person contact with potential participants at lunchtime. She says, "In the pitch I made to the young workers, I presented myself as their advocate. I told them it was their right to have good health information and that I would give them a voice in the health issues affecting them. I emphasized the low impact of the research again. I talked about the limited time the survey would take and the confidentiality of the research" (Lindsay, 2005: 124).

university's summer orientation program (Lang and Podeschi, personal communication). The distribution of the questionnaire during an official university event by the orientation leaders, the fact that the respondents were a "captive audience," the relative brevity of the questionnaire, and the anonymity of the responses are all likely to have contributed to the high response rate.

SELF-ADMINISTERED QUESTIONNAIRES

self-administered questionnaires, questionnaires that the respondent completes by himself or herself.

interview, a data collection method in which respondents answer questions asked by an interviewer.

There are different methods or modes of data collection that ask questions of respondents. A variety of new technologies have resulted in more ways of asking and answering surveys. **Self-administered questionnaires** can be offered in paper-and-pencil versions or in computer-assisted formats (CASI) such as web-based designs, versions that use touch-screen responses, or ones that have respondents answer computer-controlled questions by pressing keys on phones. An **interview** can be done by phone or in person, with the interviewer recording the responses for later data entry and transcription, or by using computer-assisted technology to enter data as the interview proceeds. Time, money, population, kind of sample, and response rate are issues to consider when selecting a self-report mode.

While one analysis of 48 studies that compared respondents' answers on various methods of administering questions found that, typically, computer administration did not have an overall effect on how respondents answer questions (Richman, Kiesler, Weisband, and Drasgow, 1999), other studies have found differences due to method. One study that focused on risk-taking behavior among adolescents found more reports of health-risk behaviors in self-administered computer questionnaires than in either face-to-face interviews or written questionnaires (Rew, Horner, Riesch, and Cauvin, 2004). Similarly, Chang and Krosnick (2009, 2010) compared interviewing with computerized questionnaires and Internet surveys and found that mode shift affects answers. The absence of an interviewer allowed respondents to provide answers that were not socially admirable (and probably more honest because of fewer concerns with impression management). On the other hand, a study of child abuse with a sample of female college students compared three modes—paper-and-pencil questionnaire, computer-based assessment, and face-to-face interview. The study raises ethical issues because although there was little variation in reported victimization rates among the different modes, the abuse victims experienced more emotional distress after taking the computerized version of the survey than after the other versions (DiLillo, DeGue, Kras, Di Loreto-Colgan, and Nash, 2006: 416).

The setting for taking a survey can also influence answers. One study of self-reported cigarette smoking found that 20 percent of youths who had reported having smoked in the past year when answering a school survey reported never having smoked when completing their household survey (Griesler, Kandel, Schaffran, Hu, and Davies, 2008).

Group-Administered Questionnaires

Lang and Podeschi used a paper questionnaire administered in group settings— the classrooms where the orientation workshops for new students were held. If you want to conduct a survey using a questionnaire and if, like Lang and Podeschi, you could locate a group setting for an appropriate sample (such as church-goers, club members, or students) and were able to obtain permission to recruit participants, you could use a **group-administered questionnaire**. Administering a questionnaire to a group of people who have congregated gives the researcher the opportunity to explain the study and answer respondents' questions, provides the researcher with some control over the setting in which the questionnaire is completed, allows the respondents to participate anonymously, and usually results in a good response rate.

Group-administered questionnaires are typically inexpensive. Lang and Podeschi volunteered their time to develop the instrument and analyze the data and did not have to pay the orientation leaders for administering the questionnaire. Once approved by the university, they did not consider any other method. They were pleased with their response rate and kept their expenses to a minimum: $90 to print the double-sided, one-page questionnaire. The cost of 7 cents per completed questionnaire demonstrates the bargain that this method of data collection can be.

On the other hand, group-administered questionnaires have drawbacks, beyond the fact that there might be no group setting for the population the researcher wants to study. The extra pressure to participate that people might feel when in a group setting raises an ethical concern about voluntary participation, and the limit to the length of time that groups will allot to data collection mandates using a relatively short questionnaire. Lang and Podeschi (personal communication) feel that they obtained enough data for their purposes using many short questions. They estimate that their questionnaire took about 10 minutes to complete.

Mailed and Individually Administered Questionnaires

If mailing addresses are available, another way to administer questionnaires is to send them to respondents' homes or, less commonly, to their workplaces.

group-administered questionnaire, questionnaire administered to respondents in a group setting.

mailed questionnaire,
questionnaire mailed to the respondent's residence or workplace.

The **mailed questionnaire** remains a commonly used method, with advantages to recommend it. Michele Hoffnung, in the panel study of college graduates described in the focal research in Chapter 4, used a brief mailed questionnaire for some data collection.

Mailed questionnaires have a long history. The biggest U.S. survey is the U.S. Census. The census of the population is required by the U.S. Constitution and has been conducted every 10 years since 1790. For the most recent version, census employees went door-to-door in 2009 to update address lists nationwide, and then between February and March 2010, every American household received the Census questionnaire, most by mail (U.S. Census Bureau, 2009b). The estimated cost of the 2010 Census is $11.3 billion over its life cycle, making it the most expensive in the nation's history (GAO, 2007).

Mailed questionnaires are widely used for several reasons: people are familiar with them, they are fairly inexpensive, and the researcher can cover a wide geographic area. There is a wide range in the response rates of mailed questionnaires, with one meta-analysis finding that individual studies have response rates from 10 to 89 percent (Shih and Fan, 2008: 265). Some research suggests that mailed questionnaires can get better response rates than phone interviews (Link, Battaglia, Frankel, Osborn, and Mokdad, 2008).

A new option for researchers is to use database technology, such as the Delivery Sequence File (DSF) used by the U.S. Postal Service, a computerized file containing all delivery-point addresses with the exception of General Delivery. This offers the potential of covering 97 percent of all U.S. households for selecting national and local probability samples (Link et al., 2008).

If respondents can be encouraged to complete and return the questionnaire, the mailed questionnaire can be useful because filling it out at home or at the workplace without time deadlines permits respondents to check their records and not feel rushed. One way to increase the response rate is to personalize a mailed survey, such as by including names and addresses on letters and the use of salutations and real signatures or postscripts. An analysis of 9 studies that contained 17 experimental comparisons of personalization found that response rates increased from 3 percent to 12 percent with increased personalization (Dillman et al., 2007).

Those who receive mailed questionnaires can answer them in private with fewer social pressures to conform to the expectations of others and eliminating the worry about trying to please an interviewer. This might explain some of the research that compares the same question using different modes of data collection. Krysan and her colleagues (1994) found that mail respondents were more likely than phone respondents to indicate that they were having problems in their neighborhoods and less likely to express support for affirmative action (both for women and for blacks) or for open housing laws. A study of attitudes in Los Angeles found that phone respondents were more likely than those who completed a mailed questionnaire to voice approval for the police and express fear of crime but were less likely to perceive social cohesion in their neighborhoods (Hennigan, Maxson, Sloane, and Ranney, 2002).

individually administered questionnaires,
questionnaires that are hand-delivered to a respondent and picked up after completion.

There are also **individually administered questionnaires**. In this method, a research associate hands out or delivers the instrument and makes sure that it is returned. Approaching respondents in person is similar to using a mailed questionnaire, although it is more expensive and often has a better response rate. For example, in a study of the attitudes that gay, lesbian, bisexual, and transgendered persons have about the police and community policing tactics, Wayne Gillespie (2008) used rented space at a three-day Atlanta Pride Festival Market to obtain completed individually administered questionnaires from a convenience sample of 179 passersby.

STOP & THINK *What do you think of posting questionnaires on websites or e-mailing them to respondents? Are they likely to be equally effective with all populations? Better for some topics than others?*

Internet and Web Surveys

web or Internet survey, survey that is sent by e-mail or posted on a website.

The rapid rise in the use of computers has created an explosion in the surveys using the Internet. In 2011, 78 percent of American adults had Internet access (Zickuhr and Smith, 2012) with other countries not far behind. So many people can be reached electronically that many questionnaires can be e-mailed or put on websites for respondents to complete. Doing a **web or Internet survey** is becoming more widespread as researchers seek survey cooperation from respondents, a trend in line with the more general culture of the web where people are encouraged to be content producers and not just passive recipients of information (Couper and Miller, 2008). The Internet can be very useful if you need a large sample, don't have much money, and can't travel to the participants. In addition, the web-based survey can be interactive and can be programmed to vary the questions and keep respondents from skipping questions. (See Box 9.4 for ideas about how to conduct your own Internet survey.)

When a percentage of a population does not have Internet access, an issue for Internet surveys is coverage error, the kind of sampling error that arises when the sampling frame is different from the intended population. However, probability samples can be obtained by using other methods to sample and recruit respondents and Internet access can be provided to those without it (Couper and Miller, 2008).

If a list of e-mail addresses of a population is available, such as from a workplace, a school, or an organization, then a random sample or the whole population of those with access can be invited to participate. Five e-mail surveys of members of the Evangelical Lutheran Church, for example, were

BOX 9.4   On the Web

It's easy to create and distribute your own web survey. Check out www.surveymonkey.com, www.zapsurvey.com, www.formsite.com, or www.zoomerang.com. Sites like these allow you to create a basic survey without any financial expense. They have survey templates and questionnaires that can be adapted to your needs and offer a choice of languages, a wide variety of question formats, and advice to increase response rate. If you do a web survey, consider the issues of the validity, reliability, and generalizability of your data.

conducted with response rates ranging from 10 percent to 58 percent (Sims, 2007). Convenience samples of self-selected Internet users can be recruited by posting invitations to participate on websites or sending them to Internet discussion groups, social networking sites, chat rooms, e-mail lists, and other venues.

One concern is that Internet surveys tend to have low response rates, especially in comparison to other methods. Diment and Garrett-Jones (2007) found that when given the choice of completing either a web- or a paper-based survey, the great majority of their respondents selected the paper-based mode. Comparing 39 recent studies, Shih and Fan (2008: 264) found considerable variation between studies, but that overall, web surveys had, on average, a 10 percent lower response rate than mailed questionnaires. Follow-ups, such as e-mail reminders to complete the Internet questionnaire, have been found to increase response rates (Klofstad, Boulianne, and Basson, 2008). However, Couper and Miller (2008: 833) note that response rate issues affect "all types of Web surveys, from list-based samples to pre-recruited probability panels and opt-in or volunteer panels … [as] an ever increasing number of panel vendors and surveys appear to be chasing an ever-dwindling group of willing respondents, raising future concerns about the selectivity of respondents in such surveys."

Some researchers have found relatively high rates of abandonment (i.e., when the participants begin but do not complete the study) for web surveys (Crawford, Couper, and Lamias, 2001), but other studies have found their completion rates to be acceptable (Denscombe, 2006).

Cover Letters and Follow-Ups

cover letter, the letter accompanying a questionnaire that explains the research and invites participation.

Whether distributed in person, by mail, or electronically, self-administered questionnaires require reading and language skills and are less likely to be completed and returned than interviews. Without an interviewer, a researcher using a questionnaire often depends on an introductory or **cover letter** to persuade respondents to participate. The cover letter provides a context for the research and information about the legitimacy of the project, the researcher, and the sponsoring organization (see Box 9.5). The physical appearance of the questionnaire and the difficulty of completing it can affect the response rate. The format should be neat and easy to follow, with clear instructions, including

---

| BOX **9.5** | **Desirable Qualities of a Cover Letter** |

A good cover letter should

- Explain that participation is voluntary
- Look professional and be brief
- Describe the confidentiality or anonymity of responses
- State the purpose of the study
- Tell how the respondent was selected

- Include information about the researcher or a contact person in case of questions or concerns
- Describe the benefits of participation
- Explain how to return the questionnaire
- Contain appeals for participation including both the positive impacts of participation and the negative effects of nonparticipation

statements about whether one or more than one answer is appropriate, and sufficient room to answer each question. The question order should be easy to follow and should begin with interesting and non-threatening questions, leaving difficult or "sensitive" questions for the middle or end.

Response rates can be improved with multiple requests for participation, and for mailed questionnaires, including a return envelope with first class postage. Wolfe, Huang, and Oswald (2008) compared first responses and those received after follow-ups with various combinations of mail, e-mail, and web-based questionnaires with a sample of teachers. When used initially, the mailed-based questionnaire was more effective than an e-mail-directing participants to a questionnaire on the web (81 versus 70 percent response rate), but a follow-up using the other method brought in additional responses for both versions.

When Hoffnung used questionnaires in her longitudinal study described in Chapter 4, she found that about 50 percent were returned within a month, another 35 percent came in after follow-up letters, e-mails, or phone calls, and a few additional responses were returned months later. Each year, she'd give her study participants a choice of a regular mailed questionnaire or one via e-mail.

Unless everyone in the sample will be getting a follow-up letter, the researcher needs to know who has not returned their questionnaires. This can be accomplished by using some sort of identification on the instrument. Identification means that replies will not be anonymous. Although respondents can be assured of confidentiality, some respondents might be less willing to participate when the survey is not anonymous.

● STOP & THINK   *Given the choice, would you rather complete a questionnaire yourself or have the questions read to you by an interviewer? Would it matter what the topic was? How about the gender, race, and age of the interviewer—would these affect your responses?*

INTERVIEWS

An alternate way of getting answers to questions is by having an interviewer ask them. In a structured interview, an interviewer reads a standardized set of questions and response options for closed-ended questions. The interview has some similarities to a conversation, except that the interviewer controls the topic, asks the questions, and does not share opinions or experiences. In addition, the respondent is often given relatively few options for responses. The set of instructions to the interviewer, the list of questions, and the answer categories make up the **interview schedule**, but it is basically a questionnaire that is read to respondents.

interview schedule, the list of questions and answer categories read to a respondent in a structured or semi-structured interview.

Interviewers are supposed to be neutral and are trained to use their interpersonal skills to encourage the expression of attitudes and information but not to help *construct* them (Gubrium and Holstein, 2001: 14). The goal is to make the questioning of all respondents as similar as possible. The assumption is that differences in answers are the result of real differences among respondents rather than because of differences in the instrument (Denzin, 1989: 104).

The use of a structured interview allows for some flexibility in administration, clarification of questions, and the use of follow-up questions. Interviews can be useful for respondents with limited literacy and education. Although rates have dropped for all surveys in recent years, interviews, especially in-person interviews, tend to have good response rates.

On the other hand, interviews are more expensive than questionnaires because interviewers need to be hired and trained. In addition, using an interviewer adds another factor to the data collection process—the **interviewer effect**, or the changes in respondents' behaviors or answers that result from some aspect of the interview situation. Some studies have found that answers to the same question vary depending on how the question is administered (Chang and Krosnick, 2010). The interviewer effect can be the result of the respondents' desires to make good impressions, the interviewers' personal and social characteristics, or the specific way each interviewer presents questions. Moreover, the same interviewer might present questions differently on different days or in different settings.

interviewer effect, the change in a respondent's behavior or answers that is the result of being interviewed by a specific interviewer.

The majority of paid interviewers in many large surveys are women, and interviewers and respondents frequently differ from each other on important background traits including age, ethnicity, gender, and/or race. Characteristics such as race and gender have been found in some studies to affect responses for at least some topics. For example, in a study of attitudes on race and race-related issues, using both African American and white respondents and interviewers, the results varied by question topic with race-of-interviewer effects for only some questions. In this study by Krysan and Couper (2003), African Americans provided less liberal racial attitudes to white interviewers, while white interviewers seemed to reduce the negative attitudes of white respondents. In a parallel result, Fuchs (2009) found gender-of-interviewer effects most common on questions concerning gender-related topics or sexual behavior.

In-Person Interviews

in-person interview, an interview conducted face-to-face.

The **in-person interview** was the dominant mode of survey data collection in this country from 1940 to 1970 (Groves, 1989: 501). Depending on the sampling procedure—as, for example, when the researcher has a list of home addresses—and when there is a need for privacy in conducting the interview, interviews can be conducted in respondents' homes. Interviews can also be conducted in public places, such as an employee cafeteria or an office. The in-person interview is a good choice for questions involving complex reports of behavior, for groups difficult to reach by phone, for respondents who need to see materials or to consult records as part of the data collection (Bradburn and Sudman, 1988: 102), when the interview is long, and when high response rates are essential. The General Social Survey (GSS), for example, collects data by means of in-person interviews of approximately 3,000 people every other year; in 2006, the GSS conducted 4,500 interviews. The GSS response rate has stayed fairly constant: for example, 76 percent in 1975 and around 71 percent during data collections from 2000 to 2006 (Davis, Smith, and Marsden, 2007).

rapport, a sense of interpersonal harmony, connection, or compatibility between an interviewer and a respondent.

Another benefit of the in-person interview is **rapport**, or sense of interpersonal harmony, connection, or compatibility between the interviewer and the respondent. Increased rapport might be why people are more willing to provide information on topics such as income in face-to-face interviews than on the phone (Weisberg, Krosnick, and Bowen, 1989: 100). Furthermore, the in-person interviewer can obtain visual information about respondents and their surroundings and a sense of the honesty of the answers, especially if the interviewer pays attention to verbal and nonverbal cues.

On the other hand, beyond the interviewer effect, a disadvantage of in-person interviewing is that sometimes others, most typically a spouse, is present despite the goal of interviewing in private. The lack of privacy has been found to affect some responses (Zipp and Toth, 2002). In addition, in-person interviews are the most expensive of the self-report methods because they involve locating and contacting respondents, scheduling interviews ahead of time, the cost of interviewers' salaries, and travel expenses to do interviews. Computer-assisted personal interviewing (CAPI) methods are now available which make in-person interviewing more cost effective. However, some research indicates that adding a computer to the interview process affects the interviewer–respondent interaction in terms of the way the interviewer asks questions and enters responses as well as how the respondent answers (Couper and Hansen, 2001). On the other hand, Childs and Landreth (2006) found that interviewers using handheld computers obtained responses of acceptable quality for the most part, especially when short questions tailored to this mode of delivery were used.

Phone Interviews

telephone interview, an interview conducted over the telephone.

Before the 1960s, the proportion of households with telephones in the United States was too small to do interviews by phone for national probability samples. By the 1970s, 90 percent of U.S. households had telephones, and the phone interview was commonplace (Lavrakas, 1987: 10). Today, the **telephone interview** is a dominant mode of survey data collection in the United States (Holbrook, Green, and Krosnick, 2003). It became the preferred approach because it can yield substantially the same information as an interview conducted face-to-face at about half the cost (Groves, 1989; Herzog and Rogers, 1988), interviewers can be closely supervised in order to obtain interview standardization, it eliminates interviewer travel time and, using random-digit dialing, it can be conducted without names and addresses, which allows for anonymity. In addition, newer technology makes it cost effective to do computer-assisted telephone interviews (CATI), in which data are collected, stored, and transmitted during the interview.

On the other hand, disadvantages of phone interviews include the previously mentioned lower response rates as compared to other formats as well as greater difficulty locating those without landline phones. In 2012, 36 percent of all American households had only wireless phones (Blumberg and Luke, 2012), with significant variation by state: the percentage of wireless-only households ranged from a high of 45 percent in Idaho to a low of 15 percent in Rhode Island (Blumberg et al., 2012). This is of concern because cell phone interviews typically cost more and have higher refusal rates (Brick, Edwards, and Lee, 2007).

Qualitative interviews that asked people about their willingness to participate in telephone interviews indicate that potential respondents base their decisions to participate on the time they have available, their interest in the interview topic, and their evaluation of the purpose and sponsor of the study (Kolar and Kolar, 2008). Phone interviews are shorter and have a quicker pace than in-person interviews as it is unlikely for respondents to stay on the phone for more than 20 or 30 minutes; in-person interviews often last 40 or more minutes. Complicated questions are more difficult to explain over the phone than in person and it's harder to judge the respondent's comprehension.

Phone interviews are more likely to generate higher proportions of "no opinion" answers (Holbrook et al., 2003). "Sensitive" or threatening questions (questions that are perceived as intrusive or raise fears about the potential ramifications of disclosure) are answered less frequently and less accurately on the phone (Kreuter, Presser, and Tourangeau, 2008).

Because the request for a phone interview is frequently the first contact with a respondent, and requests for interviews are often mistaken for solicitations or telemarketing, the interviewer's introductory statement is crucial. Interviewers must make it clear that the research is legitimate and convince potential respondents that they aren't soliciting contributions or selling anything.

In interviews where an appointment is scheduled, the rest of life is "put on hold" to some extent while the interview takes place. But, in most surveys, interviewers call or ring doorbells while respondents are in the midst of daily life and must expect interruptions as part of the interview process. For phone interviews, the interviewer must often call back to contact the household or find the person specified by sampling decisions. From 1979 to 1996, for example, the interviewers for the Survey of Consumer attitudes had to double the number of call-backs to obtain the same response rate (Curtin, Presser, and Singer, 2000). Trying to reach potential respondents over a period of weeks rather than over a few days can result in a much higher response rate (Keeter et al., 2000).

Whether they are volunteers or paid staff, interviewers are trained so that they understand the general purpose of the study and are familiar with the introductory statement, the questions, and the possible answers. For the 2010 Census, classes for interviewers typically used a verbatim approach with a focus on reading interview scripts during a week's training session (GAO, 2007). When interviewing, the interviewer should be polite, open, and neutral, neither critical nor overly friendly, and establish some sense of rapport or connection with the respondent. In keeping with ethical standards of research, all interviewers should enter into an agreement not to violate the confidentiality of the respondents.

Interesting new developments are blurring the line between the interview and the questionnaire as researchers experiment with a variety of options, including using virtual interviewers, adding human-like elements in questionnaires such as computer-generated faces that ask questions, and interviewing using instant messaging (IM). For example, Fontes and O'Mahony (2008) did interviews about online social networks using IM and obtained a 60 percent response rate. They note that on a practical level, IM interviewing is a low-cost option with respondents from diverse locations that results in transcribed data. Gerich (2008) compared questions delivered in video clips to some respondents to those that were self-administered and those asked by a real interviewer face-to-face. He found few differences by mode of administration, with some evidence that the "virtual" interviewers produced responses least affected by social desirability issues. As technology changes, it's likely that we'll see new modes of interviewing as well.

CONSTRUCTING QUESTIONS

Researchers make a series of decisions based on assumptions about how respondents read or hear questions. Theories of answering questions assume that the stages are understanding and interpreting the question, retrieving or constructing an answer, and then reporting the answer using the appropriate format (Schaeffer and Presser, 2003: 67).

Types of Questions

open-ended questions, questions that allow respondents to answer in their own words.

Questions can be asked with or without answer categories, as in the examples in Box 9.6. **Open-ended questions** allow the respondents to answer in their own words. On a questionnaire, one or more blank spaces or lines can be used to indicate an approximate length for the answer as Lang and Podeschi did when they asked "I was born in …. PLEASE WRITE YOUR YEAR OF BIRTH ON THE ANSWER SHEET." In an interview, the interviewer waits for the response and can ask for an expanded answer or specific details.

closed-ended questions, questions that include a list of predetermined answers.

Closed-ended questions use a multiple-choice format and ask the respondent to pick from a list of answer categories. One common practice is to use open-ended questions in a preliminary draft and then develop closed-ended choices from the responses. Answer categories for closed-ended questions must be exhaustive and, if only one answer is wanted, mutually exclusive. Exhaustive answers should have sufficient answer choices so that every respondent can find one that represents or is similar to what he or she would have said had the question been open-ended. (Sometimes "other" is included as one of the answers to make the list exhaustive.) Answer categories should be mutually exclusive so that each respondent can find *only one* appropriate answer. For example, some surveys offer the less useful choice of "single" rather than "never married" when asking about marital status on their questionnaire. Some divorced and widowed respondents might define themselves as single in

| BOX **9.6** | **One Question—Two Formats** |

Open-Ended Version:

Thinking about your job and your spouse's, whose would you say is considered more important in your family?

Probe Questions for Follow-Up:

Could you tell me a little more about what you mean? Could you give me an example?

Closed-Ended Version:

Whose job is considered more important in your family? Would you say your spouse's job is considered more important, both of your jobs are considered equally important, or your job is considered more important?

addition to divorced or widowed and be tempted to check two answers. Another example is the common error made by beginning researchers of listing overlapping categories, such as when age groups are used as closed-ended answers to a question about age. If ages are listed as 15 to 20, 20 to 25, and 25 to 30, the category set will not be mutually exclusive.

STOP & THINK

Lang and Podeschi used almost all closed-ended questions in their survey. They included four answer categories, from "strongly agree" to "strongly disagree," as responses to each of the statements on the environment and from "always" to "never" on questions about recycling behavior. These answer categories are mutually exclusive and exhaustive. Why do you think they used closed-ended rather than open-ended questions?

Selecting whether to use open- or closed-ended questions involves several issues. Answer choices can provide a context for the question, and they can make the completion and coding of questionnaires and interviews easier. Surveys that use closed-ended questions are typically less expensive to administer. On the other hand, respondents might not find the response that best fits what they want to say, and possible answer categories (such as "often" or "few") can be interpreted differently by different respondents. Although a small percentage of respondents will not answer open-ended questions (Geer, 1988), and other respondents give incomplete or vague answers, leaving the answers open-ended does permit respondents to use their own words and frames of reference. As a result, open-ended questions typically encourage fuller and more complex answers. One time-saving and cost-effective alternative is to use mostly closed-ended questions and include a few open-ended ones where necessary.

coding, assigning observations to categories.

In questionnaires and interviews, all open-ended responses must be categorized before doing statistical analyses. A limited number of answer categories must first be created for each question so that the data can be coded. **Coding,** the process of assigning data to categories, is an expensive and time-consuming task that can result in the loss of some of the data's richness. In the focal research, Lange and Podeschi restricted their instrument to mostly closed-ended questions because they felt closed-ended questions adequately measured the variables of interest while minimizing the time necessary to complete the questionnaire and code the data.

screening question, a question that asks for information before asking the question of interest.

A useful type of question is the **screening question,** a question that asks for information before asking the question of interest. It's important to ask if someone voted in November, for example, before asking for whom they voted. Screening questions can help reduce the problem of "nonattitudes" (when people with no genuine attitudes respond to questions anyway). By asking "Do you have an opinion on this or not?" or "Have you thought much about this issue?" the interviewer can make it socially acceptable for respondents to say they are unfamiliar with a topic; when this type of question is used, some studies have found a sizable number saying they have no opinion (Asher, 1992). A

contingency questions, questions that depend on the answers to previous questions.

screening question is typically followed by one or more **contingency questions,** which are the questions that are based on (or are contingent upon) the answer to the previous question (see Box 9.7). For example, in an interview, those who answer "yes" to the question, "Do you work for pay outside the home?"

BOX **9.7**   Using a Contingency Question

Have you applied to a graduate or professional program?

☐ no

☐ yes →

If yes, please list the program(s)
_____
_____
_____

can then be asked the contingency question, "How many hours per week are you employed?" On a questionnaire, the respondent can be directed to applicable contingency questions with instructions or arrows.

Most self-report methods use direct questions to ask for the information desired, such as "What is your opinion about whether people convicted of murder should be subject to the death penalty?" Other self-report techniques are less direct.

 STOP & THINK   *How would you feel about keeping a 24-hour time diary or a weekly calendar to record your activities? Do you think the data would be valid and reliable?*

time diary or calendar, a self-report method that asks about amount of time spent on particular activities in time blocks.

One way to find out how people spend their time is to ask respondents to keep a **time diary or calendar** of daily events or create an event history by listing information reconstructing their lives in specific time increments. Researchers using this method have studied a variety of topics. For example, Hiromi Ono and Hsin-Jen Tsai (2008) focused on computer use time among young school-age children. A sample of 6- to 11-year-olds or their caregivers completed a 24-hour chronology for a randomly selected weekday and a weekend day in a study using the calendar method. The researchers coded the data and aggregated them into 20 categories including computer activities, excluding school use, and found that of all activities, home computer use time increased the most in percentage terms among 6- to 11-year-olds between 1997 and 2003 (Ono and Tsai, 2008). Diary and calendar instruments often take longer to administer than traditional questions and show mixed results with regard to data quality, but they are often preferred by respondents and increase the completeness of respondents' accounts (Glasner and van der Vaart, 2009).

Occasionally researchers use indirect questions, especially when there is a socially desirable answer. One way to do this is to ask what others think as an indicator of the person's own opinion. For example, in 2005, polls found that 72 percent of Americans said that they would be willing to vote for a woman for president, but only 42 percent said that most of *their family, friends,* and *co-workers* would be willing to vote for a woman for president (Rasmussen Reports, 2005). Streb, Burrell, Frederick, and Genovese (2008) used another indirect way to measure the same attitude. Asking a respondent which of a list of things makes them "angry or upset," they found that 26 percent expressed anger over the statement "A woman serving as president."

vignettes, scenarios about people or situations that the researcher creates to use as part of the data collection method.

Another approach is to create **vignettes**—scenarios or stories about people or situations—which may be less threatening for sensitive issues than direct questions. In a study of friendship with a sample of 9- to 11-year olds, MacEvoy and Asher (2012) used vignettes describing situations that violate friendship expectations such as "You're really upset because your pet is sick. You tell your friend about how sick your pet is but your friend won't listen to you talk about your pet and instead says, 'It's no big deal—it's just a pet.'" After each vignette, the children rated themselves on three scales (angry, sad, and "okay") in answer the question, "How would you feel if this really happened to you?"

How to Ask Questions

Keep questionnaires and interviews as short as possible and ask only the questions necessary for the planned analysis. Each concept or variable is measured by one or more questions so consider the issues of validity and reliability to make sure that questions asked are accurately and consistently measuring what they are supposed to be measuring. Lang and Podeschi, for example, measured students' social class backgrounds with two questions, one open-ended ("For work, my parents … PLEASE WRITE YOUR PARENTS' JOBS ON THE ANSWER SHEET"), and one closed-ended ("Growing up, my family was… A: very wealthy B: comfortable C: struggling D: poor"). But after analyzing their data, they concluded that these were not adequate. They intend to do similar surveys in the future and will ask additional questions about social class (Lang and Podeschi, personal communication).

If you will be constructing scales or indices from answers to a series of questions, consider the approach Lang and Podeschi used in constructing their New Ecological Paradigm (NEP) index. They varied the ends of the four-point answer categories so that "strongly agree" was not always the more environmentally concerned answer. This meant that respondents couldn't just find a pattern—like choosing "strongly agree"—and follow it but had to think about each question. In calculating the NEP scores, the researchers had to reverse the coding for some of the statements. This meant that, for example, for the statement "Humans were meant to rule over the rest of nature," the answer "strongly disagree" was the most environmentally concerned answer and received a code of "4," but for the statement "Plants and animals have just as much right as humans to exist," the answer "strongly agree" was coded as "4."

pilot test, a preliminary draft of a set of questions that is tested before the actual data collection.

If you plan on doing a survey, it's useful to start by making a list of the variables that are of interest and then do a literature search to see if another study's measurement strategy can be used or adapted. A preliminary draft or **pilot test** should be used with a small sample of respondents similar to those who will be selected for the actual study. Giving the interview or questionnaire in this kind of pre-test will help determine if the questions and instructions are clear, if the open-ended questions generate the kinds of information wanted, and if the closed-ended questions have sufficient choices and elicit useful responses. Lang and Podeschi tried out drafts of their questionnaire and made numerous changes based on the feedback from the pilot tests.

STOP & THINK   *An interviewer was thinking of asking respondents, "Do you think that men's and women's attitudes toward marriage have changed in recent years?" and "Do you feel that women police officers are unfeminine?" Try to answer these questions. Do you see anything wrong with the wording?*

Guidelines for Question Wording

Interviews and questionnaires are based on language, so it's important to remember that words can have multiple meanings and can be interpreted differently by different respondents. Keeping questions concise and using carefully selected words is important, as the specific choice of words can affect the results. There will always be advantages and disadvantages in asking questions in specific ways. One study found 45 percent saying they favored physician-assisted suicide when the question was "Do you favor or oppose physician-assisted suicide," while another found 61 percent saying "yes" to the question "When a person has a disease that cannot be cured and is living in severe pain, do you think doctors should be allowed by law to assist the patient to commit suicide if the patient requests it, or not?" (Public Agenda, 2002). Similarly, researchers found a very significant difference when a question about courts and criminals was worded "In general, do you think the courts in this area deal too harshly or not harshly enough with criminals?' than when it was worded "In general, do you think the courts in this area are not lenient enough or too lenient with criminals?" (Applegate and Sanborn, 2011).

In cross-national studies and within countries with large immigrant populations, translation becomes important. The choice of words is particularly significant because there might be no functional equivalents of some words in other languages. In one study, for example, Japanese researchers concluded that no appropriate word or phrase in Japanese approximated the Judeo-Christian-Islamic concept of God (Jowell, 1998).

Our suggestions for question construction include the following:

1. Avoid *loaded words*, words that trigger an emotional response or strong association by their use.
2. Avoid *ambiguous* words, words that can be interpreted in more than one way.
3. Don't use *double negative* questions, questions that ask people to disagree with a negative statement.
4. Don't use *leading questions*, questions that encourage the respondent to answer in a certain way, typically by indicating which is the "right" or "correct" answer.
5. Avoid *threatening* questions, questions that make respondents afraid or embarrassed to give an honest answer.
6. Don't use *double-barreled* or *compound* questions, questions that ask two or more questions in one.
7. Ask questions in the language of your respondents, using the idioms and vernacular appropriate to the sample's level of education, vocabulary of the region, and so on.

BOX 9.8   Examples of Poorly Worded Questions and Suggested Revisions

Ambiguous question:

In the past year, about how many times have you seen or talked with a doctor or other health professionals about your health?

Revised:

We are going to ask about visits to doctors and getting medical advice from them. How many office visits have you had this year with all professional personnel who have M.D. degrees or those who work directly for an M.D. in the office, such as a nurse or medical assistant?

Double negative question:

Do you disagree with the view that the President has done a poor job in dealing with foreign policy issues this past year?

Revised:

What do you think of the job that the President has done in dealing with foreign policy issues this past year?

Double-barreled question:

What is your opinion of the state's current economic situation and the measures the governor has taken recently?

Revised:

What is your opinion of the economic situation in the state at present?

What do you think of the measures that the governor has taken in the past few months?

Leading question:

Like most former presidents, the President faces opposition as well as support for his economic policies. How good a job do you think that the President is doing in leading the country out of the economic crisis?

Revised:

Are you aware of the President's economic policies?

If so, what is your opinion of them?

Threatening question:

Have you been arrested for driving under the influence of alcohol in the past year?

Revised:

In the past year, have you driven a vehicle after having had too much to drink?

Review the poorly worded questions and the suggested revisions in Box 9.8 to see how the general guidelines are applied to specific question wordings.

 STOP & THINK   *Did you identify what was wrong with the questions in the previous "Stop & Think"? Did you see that the first question is double-barreled in that it asks two questions about marriage, one about men's attitudes and one about women's? Did you notice that the word "unfeminine" in the question about women police officers is really a loaded word?*

Response Categories Guidelines

When response categories are used, a large enough range of answers is needed. For some questions, it's better to use a continuum rather than give two answer choices. Lee, Mathiowetz, and Tourangeau (2007) compared the yes–no format to continuums and found the rates of disability to be higher when respondents had answer choices with a range (such as "no difficulty at all" to "completely unable") for questions like "How much difficulty do you have performing basic physical activities such as walking or climbing stairs?" Some common closed-ended responses are listings of answers from strongly agree to strongly disagree, excellent to poor, very satisfied to very dissatisfied, or a numerical rating scale with only the end numbers given as descriptors. Researchers interested in the impact of 10-point answer categories found that questions with answers going from 1 to 10 had much higher levels of nonresponse than those going from 0 to 10 and that those presenting "5" as a midpoint consistently had the least (Courser and Lavrakas, 2012).

For questions with more than one possible answer, a list can be used and the respondent directed to identify all that apply, by asking, for example, "Which of the following television shows have you watched in the past week? Check all that apply." Some, however, argue against a "mark all that apply" approach. Research finds that, although it takes more time, deeper processing and more options are selected when questions are broken into separate questions that are followed by a forced choice format such as "yes" or "no" (Smyth, Dillman, Christian, and Stern, 2006).

Another debate is whether a "no opinion," "don't know," or "middle category" should be offered. (For example, "Should we spend less money for defense, more money for defense, or continue at the present level?") Some argue that it is better to offer just two polar alternatives, whereas others claim it is more valid to offer a middle choice. Studies have found people to be much more likely to select a middle response if it is offered to them (Asher, 1992; Schuman and Presser, 1981) and that listing the "middle" option last rather than in the middle makes it more likely to be selected (Bishop, 1987). Lang and Podeschi decided to use four answer categories ranging from "strongly agree" to "strongly disagree" for one set of questions after "really grappling with the inclusion of a fifth response to our Likert scales. We considered putting in a 'neutral' or 'neither agree nor disagree' response. In the end, we did not include that choice because we were worried that most respondents would gravitate towards it as an answer" (Lang and Podeschi, personal communication).

Question Order and Context

Responses to questions can be affected by the question order as earlier questions provide a context for later ones and people may try to be consistent in their answers to questions on the same topic. For example, a study found that support for gender-target affirmative action decreased when the questions followed those about race-targeted affirmative action; however, race-targeted affirmative action support increased when the questions followed those about affirmative action based on gender (Wilson, Moore, McKay, and Avery, 2008). Some research on question order in surveys of attitudes on privacy issues indicates that when order effects are found, they are not especially large and may be relatively local, that is, restricted to a few nearby items that are related conceptually (Tourangeau, Singer, and Presser, 2003).

There are no easy solutions to the issue of question order because each choice involves tradeoffs. Consider a logical order that makes participation easier for respondents. To encourage respondent participation, start with interesting, non-threatening questions, and save questions about sensitive topics for the middle or near the end. If questions about emotionally difficult topics are included, it's preferable to have some "cool down" questions follow them so as to minimize psychological discomfort.

SUMMARY

Self-report methods are widely used in social research. They can be used profitably to collect information about many different kinds of variables, including attitudes, opinions, knowledge, social characteristics, and some kinds of behavior. Clear introductory statements and instructions and the judicious use of carefully worded open- and closed-ended questions are critical if questionnaires and interviews are to provide useful results. As we saw in the focal research article on environmentalism, the researchers were able to collect enough data inexpensively with clearly worded questions and answer choices and a sample of respondents who were willing to participate.

Questionnaires and structured interviews are less useful when the topic under study is difficult to talk about or when asking questions about a topic is unlikely to generate honest or accurate responses.

In choosing between the various modes of questionnaires and interviews, factors like cost, response rate, anonymity, and interviewer effect must be considered. Questionnaires are typically less expensive than interviews. In addition, they can usually be completed anonymously and privately, which makes them a very good choice for sensitive topics. The group-administered questionnaire, like the one used by Lang and Podeschi, is among the least expensive and typically has a good response rate. If the questionnaire is delivered on the web, by mail, or by e-mail, the researcher will need to think of ways to encourage participation, as these methods tend to have lower response rates than the group-administered questionnaire.

Interviews, especially in-person ones, usually have good response rates and little missing data. They can be especially helpful with populations that have difficulty reading or writing, when topics are complex, and when rapport with respondents is important. Phone interviews have lower response rates but allow researchers to collect self-reports from national samples in a less costly way than in-person interviews. For both kinds of interviews, interviewer effect should be considered.

EXERCISE 9.1

Open- and Closed-Ended Questions

Check with your instructor about the need to seek approval from the Human Subjects Committee, the IRB, or the committee at your college or university concerned with evaluating ethical issues in research. If necessary, obtain approval before doing the exercise.

1. Make five copies of the list of questions that follows. Find five people who are working for pay and are willing to be interviewed. Introduce yourself and your project, assure potential participants of the confidentiality of their answers, and ask them if they are willing to participate in a brief study on occupations. Write down their answers to these open-ended questions as completely as possible. At the end of each interview, thank the respondent for his or her time.

    a. Are you currently employed? (If yes, continue. If no, end the interview politely.)
    b. What is your occupation?
    c. On average, how many hours per week do you work?
    d. Overall, how satisfied are you with your current job?
    e. In general, how does your current job compare to the job you expected when you took this job?
    f. [*If the job is different from what was expected*]
    In what ways is the job different from what you expected?
    g. If it were up to you, what parts of your job would you change?
    h. What parts of your job do you find most satisfying?

2. Before data can be analyzed statistically, it must be coded. Coding involves creating a limited number of categories for each question so that each of the responses can be put into one of the categories. From your interview data, select the answers to three questions from d to h and for each respondent, *code* the data using *no more than five* response categories. We've started a table below that you can use as a model for your coding. You can add up to one additional coding response for question d if appropriate or feel free to create your own coding strategy.

Respondent	Answers to Question d coded	Answers to Question __ coded	Answers to Question __ coded
1	She is very satisfied with her job		
2	He thinks that his job is better than most so is somewhat satisfied		
3	She thinks her job is awful and is very dissatisfied		
4	He thinks there is an equal number of good and bad things about this job		
5	_____		

3. Based on your respondents' answers, your coding strategy, and other possible answers to the questions, rewrite and include the three interview questions you selected for Question 2 and make them closed-ended by including the answer categories.

4. Describe your response rate by noting the number of people you asked to participate to get five respondents. Based on your experience, comment on the reasons you think some people agree to participate in a short in-person interview and others decline.

5. Attach the five completed interviews to your exercise.

EXERCISE 9.2

Asking Questions (or, Have I Got a Survey for You!)

Check with your instructor about the need to seek approval from the Human Subjects Committee, the IRB, or the committee at your college or university concerned with evaluating ethical issues in research. If necessary, obtain approval before doing the exercise.

1. Select between two and four concepts or variables to study. Do not choose "sensitive" or threatening topics. (Appropriate topics include educational goals, career aspirations, involvement in community service, marriage and family expectations, social characteristics, and so forth.) List the variables you have selected.
2. Construct between one and five questions to measure each variable. Feel free to use or adapt questions from studies you've read or those discussed in this chapter. A good source of questions can be found at the end of the study "Life After High School" at http://www.publicagenda.org/files/life _after_high_school.pdf. Develop a list of no more than 20 questions.
3. Make six copies of the interview questions, leaving enough space to write down respondents' answers. Contact potential respondents to find six people who are willing to be interviewed. Introduce yourself and your project, assure potential participants of the confidentiality of their answers, and ask them to participate. Do three of the interviews over the phone and three in person.
4. Turn the list of interview questions into a questionnaire, complete with instructions and a cover sheet explaining the research and its purpose. (Introducing yourself as a student and the survey as a part of the requirements of your course would be appropriate.) Give the questionnaire to three new people to complete.
5. Compare the information you obtained using the three methods of data collection. Comment on the validity and reliability of the information you received using each method. Which did you feel most comfortable using? Which was the most time-consuming method? If you were planning an actual study using these questions, which of the three methods would you select? Why?
6. Based on your experience, select one of the methods, and make any changes to the questionnaire or list of questions that you think would make the answers more valid indicators of your variables in an actual study.
7. Attach the six completed interviews and the three completed questionnaires to this exercise as an appendix.

EXERCISE 9.3

Direct and Indirect Questions

Check with your instructor about the need to seek approval from the Human Subjects Committee, the IRB, or the committee at your college or university concerned with evaluating ethical issues in research. If necessary, obtain approval before doing the exercise. In this exercise, you'll be comparing asking direct questions versus asking indirect questions using a vignette.

1. Make five copies of the following questionnaire, leaving enough space after each of the first four questions for the respondent to write a few sentences. Get five envelopes to give with the questionnaires.
2. Using a convenience sample, find five people who are willing to answer this questionnaire anonymously. Have them complete the questionnaire while you wait, and ask them

to seal it in the envelope before returning it to you.

3. Analyze the information on the completed questionnaires. Specifically, compare the information from the first four direct questions to the information from Question 5, an indirect question that employs a vignette. Comment on the validity and reliability of the information you received using each way of asking questions. Which way of asking questions do you think is better? Why?

4. Attach the five completed questionnaires to this exercise as an appendix.

Questionnaire on Family Roles

Thank you for agreeing to participate in a study on family roles for a course on research methods. Please don't put your name on this questionnaire so that your responses will be anonymous. Answer the first four questions with a few sentences. For the last set of questions, place an X in the space on the scale that best reflects your point of view.

When you have completed your responses, please put this questionnaire in the envelope provided, seal it, and return it to me.

1. All couples in long-term relationships need to make decisions about family roles. How do you think married couples should divide household and economic responsibilities?

2. If there are children in a two-parent household, do you think that the mother's and father's roles should be different in some ways when it comes to caring for the children?

3. Do you think husbands should be expected to make career sacrifices to accommodate their families? If so, what kinds of sacrifices?

4. Do you think wives should be expected to make career sacrifices to accommodate their families? If so, what kinds of sacrifices?

5. Susan and Steve have been married for five years and both have worked full-time during this time. Susan makes a slightly higher salary than Steve and is expecting a promotion soon. Steve's company has been bought by another company, and he has been told that his job will disappear after the two companies merge. Susan is pregnant with their first child, and both she and Steve agree that it will be better for their child if one parent stays home for the first few years of the child's life. After discussing their situations, Steve and Susan agree that she will continue to work full-time while he will stay home with the baby.

What's your impression of Susan and Steve? For each of the six questions that follow, put an X in the answer category that best reflects your viewpoint.

Susan is

[ ]   [ ]   [ ]   [ ]   [ ]   [ ]
 0   1   2   3   4   5
Very dedicated to   Not dedicated
her family   to her family

Susan is

[ ]   [ ]   [ ]   [ ]   [ ]   [ ]
 0   1   2   3   4   5
A very caring   Not a very
person   caring person

Susan is

[ ]   [ ]   [ ]   [ ]   [ ]   [ ]
 0   1   2   3   4   5
Very ambitious   Not ambitious

Steve is

[ ]   [ ]   [ ]   [ ]   [ ]   [ ]
 0   1   2   3   4   5
Very dedicated to   Not dedicated
his family   to his family

Steve is

[ ]   [ ]   [ ]   [ ]   [ ]   [ ]
 0   1   2   3   4   5
A very caring   Not a very
person   caring person

Steve is

[ ]   [ ]   [ ]   [ ]   [ ]   [ ]
 0   1   2   3   4   5
Very ambitious   Not ambitious

EXERCISE 9.4

Using Structured Interview Questions from the GSS

The GSS, conducted by the National Opinion Research Center (NORC), uses a structured interview schedule. The GSS was administered annually from 1972 and then biennially from 1994 to a national probability sample. Some questions are asked on every survey, other questions have rotated, and some have been used only in a single survey.

The GSS measures a great many variables using answers to questions as indicators. The variables included on past surveys cover a wide range of behaviors, among them drinking behavior, membership in voluntary associations, and the practice of religion, and numerous attitudes, such as opinions about abortion, affirmative action, capital punishment, family roles, and confidence in the military.

Browse through the archives of the GSS to find questions asked between 1972 and 2010, which is available at http://sda.berkeley.edu/archive.htm and select a few variables of interest.

1. List the names of the variables and the questions on the GSS that are the indicator(s) of those variables.
2. Comment on the specific questions that were used. Do you think they are valid indicators for the variables? Are the questions clear and easy to understand? What do you think of the answer categories?

EXERCISE 9.5

Doing a Web Survey

There are many websites that allow users to create and distribute their own web-based surveys. Some of them are free of charge. Check out this process by using a website to construct and implement a questionnaire. Afterward, evaluate the data you obtained. Some sites that you can check out are

www.zapsurvey.com
www.zoomerang.com
www.formsite.com
www.surveymonkey.com

When you have completed the survey, write a brief report about this experience, including the following:

1. What was the topic you were interested in?
2. Print a copy of your questionnaire.
3. Evaluate the quantity and quality of the data you obtained using the Internet survey method. Which website did you use?
4. Based on your experience, what do you think of web-based surveys?

Qualitative Interviewing

CHAPTER OBJECTIVES

This chapter is designed to help students

- **Understand the uses of qualitative interviews**
- **Identify the differences between qualitative and structured interviews**
- **Compare semi-structured, unstructured, and life story interviews**
- **Consider ways to use visual methods in interviewing**
- **Explain the uses of group interviews such as focus group interviews**
- **Learn about locating participants and encouraging participation**
- **Evaluate the impact on interviewing of race, gender, class, and other characteristics**
- **Begin analysis of qualitative interview data**

CHAPTER OUTLINE

QUALITATIVE INTERVIEWS

 Suppose a researcher who was about your age asked you to spend an hour or two talking about your college experiences and your plans for the future. Would you agree to be interviewed? Would you feel flattered that someone wanted to hear some of your life story or annoyed that you were being asked to spend the hour or two? Would you feel intimidated by having your views recorded? On the other hand, consider being the interviewer. How would you feel about conducting a series of interviews with your peers?

Qualitative interviewing has been an important social science tool for more than a century—and a method whose uses have been debated for almost as long (Lazarsfeld, 1944; Thurlow, 1935). These less structured interviews are opportunities to learn a wide variety of things, among them people's backgrounds and experiences; their attitudes, expectations, and perceptions of themselves; their life histories and the meaning they ascribe to events; and their views about groups of which they are a part and organizations with which they interact.

qualitative interview, a data collection method in which an interviewer adapts and modifies the interview for each interviewee.

The **qualitative interview,** also called the **in-depth or intensive interview,** has much in common with the **structured interview.** Both kinds of interviews rely on self-reports and anticipate that the interviewer will do more of the asking and the respondent will do more of the answering. Both kinds, "far from being a kind of snapshot or tape-recording … in which the interviewer is a neutral agent who simply trips the shutter or triggers the response" (Kuhn, 1962: 194), are opportunities for data to emerge *from the social interaction* between the interviewer and the interviewee.

structured interview, a data collection method in which an interviewer reads a standardized list of questions to the respondent and records the respondent's answers.

Despite these commonalties, the two kinds of interviews are distinct. In the structured interview, the interviewer uses a standardized set of questions, heavily weighted toward closed-ended questions. The questions and closed-ended answer choices are delivered to each respondent with as little variation as possible to achieve uniformity in questioning. In addition, some kinds of structured interviews, such as phone interviews using random-digit dialing, can be done anonymously without identifying the respondent.

In contrast, the qualitative interview is designed to allow interviewees to structure and control much more of the interaction. Some researchers use the interview to obtain information about lives and social worlds; others describe it as an opportunity for participants to tell their stories or construct narratives. In the qualitative interview, as Box 10.1 illustrates, the interviewer asks open-ended questions (either preformulated or constructed during the interview), frequently modifies the order and the wording of the questions, and typically asks respondents to elaborate or clarify their answers. One of the major advantages of the more conversational tone of the qualitative interview is that it is useful for allowing **rapport** or a sense of connection to develop between the interviewer and interviewee.

rapport, a sense of interpersonal harmony, connection, or compatibility between an interviewer and an interviewee.

Before the interview, the interviewer typically knows the topics the interview is intended to cover and the kind of information that is desired, but has the opportunity to tailor each interview to fit the participant and situation. In

You're seated across from the interviewer and she continues the interview by asking, "So, how has your family been of help as you've pursued your undergraduate education?" Perhaps while you think about what to say, you wonder exactly whom you should talk about. You might think she means for you to talk about your mother or father, who have saved over the years to help pay your tuition. But you wonder if she also means for you to talk about your great-aunt, whom you rarely see but who sometimes sends you little notes of encouragement. Then you consider whether she can possibly mean for you to tell about your "cousin," who is not a relative at all, but rather a family friend who has helped you find a summer job each year. You might even consider mentioning your roommate, the person who gives you emotional support when you most need it. If you're like most participants in an interview, you have your own unique understanding of what is meant by each word and question. Fortunately, the in-depth interview can provide an opportunity to foster the kind of give-and-take where the participants can ask for clarifications about the questions and discuss their meanings.

in-depth interviews, the answers are usually longer; the interviewees can impose their own structure, ask for clarification about questions, and answer in their own terms; and the interviewers can follow up by asking additional questions. Because these interviews are almost always scheduled in advance and usually occur face-to-face, the interviewer typically knows the identity of the interviewee and, ethically, must keep confidential all information obtained.

Qualitative interviewing has been used as the sole data collection technique in a wide variety of studies, including research on friendship, trust, and love in Latino marriages (Harris, Skogrand, and Hatch, 2008) and in a study of how working class people in their 20s and 30s construct adult identities (Silva, 2012). Qualitative interviewing has also been used in conjunction with **observational techniques** (collecting data by observing people in their natural settings) and with other methods of data collection. For example, Gayle Sulik (2007) completed 60 interviews and did participant observation while working for six months as an administrative assistant and volunteer with a community-based breast cancer organization to learn how women with breast cancer balance illness with caring for themselves and others. To study how adoptive parents validate their children's ethno-cultural origins, Amy Traver (2007) conducted semi-structured interviews with 91 Americans adopting children from China, read first-person accounts of adoption from China, attended cultural celebrations with a group called Families with Children from China, and spent weekends at Chinese culture camps.

observational techniques, methods of collecting data by observing people, most typically in their natural settings.

Selecting an interviewing method means deciding on the amount of structure. More structured interviews can be useful when the researcher wants to collect some data on many topics and when surveying representative samples. Structured interviews can also be appropriate when researchers focus on people who are similar to themselves in terms of social characteristics, when the topics lend themselves to standardized questions, and when large samples and cost containment are necessary.

On the other hand, structure in an interview can limit the researcher's ability to obtain in-depth information. Furthermore, using a standardized format implicitly assumes that all respondents understand and interpret questions in the same way. Some researchers don't want the very asymmetric and hierarchical interaction produced by structured interviews with interviewers having significant control of the interaction and not sharing their own views. Some critics have gone so far as to argue that the approach used by the structured interview "breaks the living connection with subjects so that it is forever engaged in the dissection of corpses" (Mies, 1991: 66).

For researchers who are less interested in measuring variables and more interested in understanding how individuals subjectively see the world and make sense of their lives, the less structured approaches are more appropriate. In qualitative interviewing, the interviewer can "break the frame of the interview script" and shift her or his perspective to that of the interviewee and adapt the questions for the person being interviewed rather than using standardized wordings.

Encouragement to elaborate can increase an interviewee's investment in the interview and decrease the sense of being a machine producing acceptable answers to questions (Suchman and Jordan, 1992: 251). Qualitative interviews can provide insight into the meanings of participants' everyday lives and explore behavior, roles, feelings, and attitudes by allowing interviewees to tell their stories in their own words and in ways that are meaningful to them. Although the interviewer and interviewee are usually strangers, the qualitative interview can provide vibrant data.

In their book *Premarital Sex in America,* discussed in the Research in the News section that follows, Mark Regnerus and Jeremy Uecker (2011) drew upon data from four comprehensive surveys that used structured approaches but also wanted to obtain meaningful information through deep conversations with real people. To better understand the behavior of 18- to 23-year-olds in heterosexual relationships, Regnerus and Uecker (2011: 11) did 40 qualitative interviews with undergraduates at the University of Texas, focusing on the ways the students pursued or refrained from sexual involvement with peers.

 RESEARCH IN THE NEWS

In the *New York Times,* op-ed columnist Ross Douthat discussed a study by sociologists Mark Regnerus and Jeremy Uecker on heterosexual relationships among young adults which reported significant correlations between sexual restraint and emotional well-being, between monogamy and happiness, and between promiscuity and depression. Douthat noted that the correlations among these variables was much stronger for women than for men and that the happiest women were those with a current sexual partner and only one or two partners in their lifetime. Douthat also described the study as concluding that women who were virgins were almost as happy as the minimally sexually active women and that a young woman's likelihood of depression rose steadily as the number of partners increased.

Whether or not Regnerus and Uecker would agree, Douthat draws this conclusion: "When social conservatives talk about restoring the link between sex, monogamy and marriage, they often have these kinds of realities in mind. The point isn't that we should aspire to some Arcadia of perfect chastity. Rather, it's that a high sexual ideal can shape how quickly and casually people pair off, even when they aren't living up to its exacting demands. The ultimate goal is a sexual culture that makes it easier for young people to achieve romantic happiness—by encouraging them to wait a little longer, choose more carefully and judge their sex lives against a strong moral standard."

http://www.nytimes.com/2011/03/07/opinion/07douthat.html?_r=0

Perhaps because of its engaging nature, the qualitative interview typically has a high response rate. For example, in a study of heroin and methadone users, 90 percent of those approached in two methadone clinics agreed to complete a one- to two-hour interview (Friedman and Alicea, 1995: 435). In the focal research that follows, most of the prison inmates asked agreed to participate in a qualitative interview.

Sandra Enos, a sociologist whose research interests included families and corrections, combined these interests in a study on the social processes of mothering when mothers are in prison. She first collected data while working as a volunteer in the prison parenting program, observing and talking informally to the participants. In the excerpt that follows, she draws on her qualitative interviews for her analyses of "mothering" while in prison.

FOCAL RESEARCH

Managing Motherhood in Prison
By Sandra Enos*

Introduction

The United States has experienced an incarceration boom. By mid-year 2000, the population behind bars had grown from 732,000 in 1985 to just under 2 million prisoners (Beck and Karberg, 2001). The United States has one of the highest incarceration rates in the world, which means that proportionately we send more individuals to prison than does almost any other country in the world (Mauer, 1997). Although males are more likely to be imprisoned, a growing number of women are finding themselves serving terms behind bars. While the "imprisonment boom" increased the male population behind bars by 165 percent, the number of sentenced women and those awaiting trial increased by over 300 percent during the same time (Beck and Karberg, 2001).

(continues)

---

*This article has been revised by Sandra Enos from her paper of the same name that appeared in the first edition of this book and is published with permission.

FOCAL RESEARCH *(continued)*

The children of these inmates are also affected by imprisonment. It is estimated that there were 1.5 million children with parents in prison in 1999 (Mumola, 2000). Research has shown that children of inmates are likely to follow their parents' criminal careers; their grades are likely to suffer in school; they display behavioral problems; and they become early users of drugs and alcohol (Johnston, 1995). The impact of having either parent in prison is likely to be detrimental, but because mothers are more often the primary caretakers of children, the incarceration of a female inmate is especially burdensome for children and for families. Most women in prison are mothers, with approximately 65 to 80 percent of them having children under the age of 18 (Baunach, 1985). Because arrest, pretrial holding, and court processing are usually unplanned, the placement of children may be haphazard; imprisonment means that existing family and extended family units must reorganize to adjust to life without the offender. Nationally, 90 percent of children who have fathers in prison live with their mothers, while only 25 percent of those with mothers in prison reside with their fathers (Snell, 1994). The children of inmate mothers are more likely to experience changes in their living arrangements when their mothers come to prison and may, in fact, be significantly more affected by the incarceration of their mothers than by that of their fathers.

Inmates as Mothers

Mothering under normal conditions is challenging and often stressful. A variety of myths and ideologies about the "family" and about mothering carry enormous weight in influencing our values, norms, and behavior patterns. These ideas about what it means to be a family and what it means to be a mother have also affected the research that social scientists have done (Baca Zinn and Eitzen, 1993). Traditional sociological work has proposed one normative family form and defines other forms as deviant or underdeveloped (O'Barr, Pope, and Wyer, 1990).

Some of the most important myths that underlie contemporary American views of mothering include the following: mothers are totally absorbed in mothering, the mother is the only figure who can give the child what s/he needs, the mother has resources to give the child what s/he needs, and the mother should be able to manipulate the child's behavior toward successful adulthood (Caplan and Hall-McCorquodale, 1985; Chodorow and Contratto, 1982). These assumptions do not make room for mothering styles outside the middle class (Dill, 1988; Maher, 1992) or consider that motherhood is experienced differently by women in different race/ethnic categories or life cycle positions. There is only a small body of work (Collins, 1990; Rubin, 1994; Stack, 1974; Young and Wilmott, 1957) that has investigated how mothers under strain and economic stress "do mothering." There is little room for models of mothering where caretakers share mothering responsibilities, such as "other mothers" identified by Collins (1990). Few have studied mothering in a way that makes room for interpretations of the mothers themselves. It is important to analyze mothering, including race and class differences, in a way that goes beyond overly idealistic images or negative categorizations of mothers as evil or neglectful caretakers. In other words, it is important to "unpack" motherhood if we wish to learn what is really involved in this work.

Because images of mothers are idealized, we are likely to look harshly upon women who violate our expectations. Women's involvement in crime and drug use is viewed through a gendered lens. Women are strongly condemned for these behaviors because they suggest ignoring the welfare of their children. These expectations seldom come into play for male offenders. When women who are mothers are sent to prison, this brings the case not only to the attention of criminal justice agencies but also to the attention of child welfare authorities. In some instances, a sentence of longer than one year can be grounds to move to terminate a mother's right to a child. So, the importance of motherhood carries more weight for a female inmate than does fatherhood for a male inmate.

Several research questions guided this project: How mothers remain connected to their children, how they maintain a place and position in their families, how they arrange and evaluate child-care options and the impact on their long-term careers as mothers are elements of the social process of "managing motherhood" in prison.

The Sample and Method

A women's prison in the Northeast served as the study site. The facility is small in size, holding a maximum of 250 women awaiting trial or serving sentences. Approximately half of the mothers in the prison were involved with the state's child welfare agency as a result of complaints of child abuse, neglect, or the presence of drugs in infants at birth. In other cases, the women sought help from child welfare to place children when they felt they had no other viable option.

The parenting program in this prison is operated off-site and allows considerable freedom from usual security constraints. The program facility is filled with toys and more closely resembles a school than a correctional site. Child welfare authorities view participation in the program as an indication of inmate mothers' interest in rehabilitation. Participation in the program is a way for inmate mothers to demonstrate their "fitness" as parents. This may help them to regain custody and resume care of their children upon release. For these reasons, participation in the program is a welcome opportunity.

To participate in the parenting program, the women had to be mothers or guardians of minor children, have either minimum or work release security status, and be free from work obligations at the time of the program. The participants in the parenting program were serving sentences similar to those of the general population—ranging from six months to a few years for nonviolent offenses such as embezzlement, fraud, soliciting for prostitution, drug offenses, breaking, and entering.

I was granted access to the parenting program by the Warden and the Director of the parenting program in 1993. I assumed two roles— observer and volunteer/program aide. The mothers gave me permission to sit in on the parenting group (an hour or more discussion group designed to help mothers work out issues related to reunification with children) and to observe the children's visits. Attendance at the program varied from week to week, attracting from 7 to 12 women and between 12 and 25 children. There was turnover in the population; approximately 20 women participated during the four months I was an observer.

I attended 16 sessions and held informal conversations with both the women and their children during the parent–child visits. The conversations were typically about the child and the women's path to prison. Although I was able to interview some women fairly completely during this time, the informal

(continues)

conversations with the women and children were typically limited to a few minutes at each session so that my research interests did not detract from the real purposes of the parenting program— to allow contact and enhance communication between parents and their children. During this time, I took extensive field notes, and generated methodological and analytical memos, a method that supports the grounded theory approach (Charmaz, 1983).

I observed two differences during this preliminary work: Both the placement of children and their mothers' paths to prison were different for African Americans and whites. An interest in the patterns and possible connections among them led me to select qualitative interviewing as my main method of data collection. I conducted 25 additional interviews, most with women who had participated in the parenting program. I gathered a purposive sample consisting of women serving short sentences and those with longer terms; those with involvement with child welfare and those without; and with women who were recidivists and those serving their first terms in prison. Because I was interested in the effect race and ethnicity played on the management of motherhood in prison, the sample reflected the racial/ethnic composition of the prison population. Finally, because I was investigating where children were placed, I located mothers whose children were living in a variety of placements: with grandmothers, with fathers, in foster care, and with other relatives. The 25 women had given birth to 77 children, 18 of whom had been adopted by relatives or other individuals.

I identified myself as a researcher and told the mothers that the interview was about the challenges female inmates faced in maintaining relationships with their children while in prison. I made it clear that unlike other conversations about their children with representatives from child welfare, counseling agencies, drug treatment facilities, school departments, and so on, their conversations with me would not directly affect their status at the prison. All signed informed consent forms (see Appendix B) and were assured that our conversations were confidential. Most were curious about my study and wanted to share their stories, hoping it might help others or lead to changes in the parenting program.

In the lower security levels, the interviews were held in a hallway or program area. In the higher security level, a correctional official stood outside the room where I was allowed to conduct interviews. Some of the participants allowed me to tape-record the interview and all but one of the rest agreed to my taking notes. One woman insisted on neither notes nor tape, which required reconstructing the interview from memory immediately after leaving the site. The interviews ranged from one to two and one-half hours.

The Emotional Content of Interviewing

I was concerned about how forthcoming the women would be. Given the dynamics related to the presentation of self and to the pressures for women to conform to idealized conceptions of mothering, I was concerned about the fruitfulness of certain lines of questioning. For the most part, I directed conversations along paths initially suggested by women themselves. In later interviews, I occasionally paraphrased the responses from previous interviews as a way of giving the women something to respond to. In very few instances did I ask questions that would result in yes or no answers.

My aims were to strike a balance between protecting each woman's privacy, getting an accurate picture of how she viewed her role and place as a mother, and hearing her story in her own voice. To establish this balance, I asked very broad questions in a semi-structured interview format, such as "What brought you to prison?" (See Appendix A for the interview schedule.) The answers often required additional clarification. Some women had been in prison several times and were confused about whether I was referring to their first incarceration or their most recent term. The question about where their children were living was too vague, as some women had been caring for children who were not their biological children and others had already lost custody of their children to the state or other caretakers before imprisonment. Many issues were brought into the conversation by the women themselves. For instance, several women, more typically the middle-class women or those in prison for the first time, brought up problems with the conditions of confinement, such as the lack of meaningful programming and the arbitrary nature of the rules, although I did not ask about them.

In the interviews, the women were direct and forthcoming. Some interviewees stated very simply that they were not emotionally bonded to certain of their children; others expressed overly optimistic views of the possibility of reestablishing good and easy relations with their children; others were not in touch with their children because current caretakers prevented contact; some discussed in detail strategies for maintaining a drug habit while care taking their children. In other words, while the emotional context of the interviews was rich and the conversation occasionally difficult, the mothers rarely presented an idealized view of themselves as mothers or a view of mothering as nonproblematic.

The women expressed an appreciation for the time spent speaking about their children and their hopes for the future. Another rewarding aspect for the mothers was that I offered to photograph their children, giving them both the photographs and the negatives. Since many of the women didn't have recent pictures of themselves or their children, this opportunity was appreciated.

I transcribed the interviews and coded them using a software program called HyperRESEARCH. The social situations that affect inmate mothers' careers as mothers that emerged from the analysis include arranging care and managing caretakers, demonstrating fitness, negotiating ownership of children, constructing and managing motherhood, and balancing crime, drugs, and motherhood. Because of space limitations, my discussion of results will be brief.**

Findings

Arranging Care and Managing Caretakers

A recent survey has reported that 35 percent of state and 15 percent of federal women inmates have not lived with their children prior to their incarceration (Mumola, 2000). For those who co-reside with their children, child placement becomes a challenge upon imprisonment. As in previous research (Snell, 1994), there were differences in white, African American, and Hispanic child placements (see Table 10.1). While children most frequently reside with grandparents, white

(continues)

**For the complete findings, see S. Enos, *Mothering from the Inside: Parenting in a Women's Prison*, Albany: State University of New York Press, 2001.

FOCAL RESEARCH (continued)

and Hispanic mothers are much more likely to leave their children with husbands; African American women are more likely to place their children with their mothers or other relatives. Making a decision about placing a child is not easy. Stacey,*** a white woman in her twenties, wanted to protect her child but also wished to minimize her involvement with child welfare. Her four-year-old son was living with his father, and she had heard from friends that the father had resumed a drug habit and the selling of narcotics. Stacey knew that contacting the child welfare authorities would prompt an investigation of the father. However, this could also mean having an agency remove the child from the home and place him in foster care. As she noted, "Most of the time, trying to get help means getting into trouble."

TABLE **10.1**
Race/Ethnicity of the Sample of Female Inmates (N = 25) and Current Residences (N = 29) of Their Children*

	Grandparent	Foster	Husband	Other Relative	Total
White	4	1	5	–	10
African American	7	2	–	5	14
Hispanic	2	–	2	–	4
Native American	–	–	1	–	1
Total	13	3	8	5	29

*Several siblings do not reside in the same residence.

Constructing and Managing Motherhood

Maintaining presence as a mother is difficult when a woman is absent from the home and is even more problematic when she has few resources. Bernice, a pregnant African American mother of a five-year-old son who had been in many foster homes, echoed Stacey's concern. She said, "I feel like I'm having babies for the state. Once they [the child welfare agency] get in your life, they're always there. They stick with you forever." This mother remarked that foster care was the worst place for an African American child to be. Her child was placed with a foster family only because her family and friends wouldn't (or couldn't) respond to her need for a place for him to live during her incarceration.

Other women were more positive about putting their children in the care of foster parents. Some white women characterized it as a "life saver." Louise talked about her satisfaction with foster care, and she was careful to make a distinction between being a child's mother and the care taking of the child. Her comments indicate that she was able to separate the biological identity of mother and the social and psychological work of "mothering."

***To protect confidentiality, Stacey and all the other names of interviewees are pseudonyms.

Enos:	So, you had a baby just before you came to jail. Where is he now?
Louise:	He's with foster parents who really love him a lot. They are just great with him.
Enos:	Does he visit you here?
Louise:	Sure he does, a little, the foster parents bring him to me.
Enos:	And does he know you?
Louise:	Well, he knows who I am. But, he doesn't know me know me.
Enos:	Can you explain that?
Louise:	Well, he knows I'm his mother, but he doesn't know me that way.

The distinction Louise makes may imply special challenges when the mother resumes care taking of the child or when she reflects upon how she may handle the tasks associated with care taking. Pam, a white middle-class woman, noted that her husband had taken up much of the responsibility for their developmentally delayed son. She felt that some activities can be delegated, while others are better for mothers to do. As an example, she described what happened when the father took the son shopping for school shoes. She was dismayed at his choice, saying "I looked at what they picked out and thought to myself, 'My God, I would have never bought such a thing for him. No mother would have.' There is a difference in what a mother would have done."

Unlike other women, Pam had no difficulty maintaining her position in her family. She was still defined as the family's major caretaker. She reported that her husband visited often and that they maintained the close working relationship that parenting a special child required.

Negotiating Ownership

Family members who assume caretaking responsibilities immediately after a child's birth were viewed differently than those who became involved later in the child's life. Vanessa, an African American woman, said, "Just after I had the baby, my mom took him. Then, I got a short bid. But ever since he was born, he was her baby. I never had a chance." Vanessa is suggesting she has a marginal role in the child's life as his mother—either because others have asserted ownership in her absence or because of her lack of effort.

Other women noted the same development, but suggested that this may have turned out for the best. Kate, a white woman, said, "I don't know what was wrong with me then. I just never had any feeling for my first. I just had it and then my mom took over. We used to do drugs together and party all the time and then she went straight and was preaching to me all the time. Then, the baby came and she took over." Kate conveyed a simpler delineation of responsibility by putting her child in her mother's hands with little regret.

Significant differences were also revealed as women described their paths to prison. White women were much more likely to attribute blame to their families of origin for their problems, suggesting that they were pushed out of their homes or subjected to early sexual and physical abuse and neglect. Few white women relied on their families to take care of their children, and most were wary of incurring "unpayable debts" if they utilized family resources to help them with child care.

(continues)

FOCAL RESEARCH *(continued)*

On the other hand, African American women tended to trace their paths to prison to their attraction to the "fast life," arguing that their families had tried unsuccessfully to divert them from a life of crime. Placing children with family and friends was expected and acceptable. In many cases, the mothers had previously prevailed upon relatives to care for their children when they were engaged in criminal activities. These children were cared for by a variety of adults before the women's imprisonment, and most African American women expected to continue using these shared child caring arrangements after release.

There were exceptions to these patterns. Not all children of white women were in the care of husbands, the state, or other non-kin, and not all children of African American mothers lived with relatives or other kin. White women with access to deviant life styles had placements similar to the African American inmates. That is, white women with parents or relatives involved in law-breaking behavior, or who lived in areas where entry into this lifestyle was easy, did place children with their own mothers. On the other hand, African American women who traced their imprisonment to deficient families of origin and were estranged from them placed their children in foster care.

Family "expectations of trouble" occur in white and African American families. Histories of conflict with the law and with other institutions create capacities within families to extend resources when needed. Some women suggested that racial and ethnic differences characterized families of white women and those of color. They stated that expectations for girls in white families were more rigid and punitive, with young women expected to "stay out of trouble," and to follow gendered prescriptions for behavior. Most of the white women did report that their families had expected them to follow the rules, and handled deviations by constricting the family's resources to protect the other members of the family and to keep the offender isolated.

Balancing Crime, Drugs, and Motherhood

Relationships with family before and during incarceration, the quality of those relationships, whether they were supportive or undermined the relationship between mothers and their children, the extent of the involvement of women in crime and drugs, all affect the long-term career of inmates as mothers. Research indicates that about one-third of these mothers will not maintain custody of their children (Martin, 1997). Some mothers maintained that their criminal involvement (such as larceny, selling drugs, or assault) supported the work they did as mothers by supplementing their meager incomes or by protecting children from physical danger. Others stated that they managed to keep their long-term involvement in drugs and crime from their partners and from their small children. There was disagreement among the women, however, on the issue of balancing drug use and motherhood. Some inmate mothers claimed that one could separate what one did as a drug user or an offender and what one did as a mother by confining each activity to a separate part of the day and making certain that the drug use did not grow out of control. Other women challenged this perception, as demonstrated by the comments of Tee and Margaret.

You think you are doing right, but you are really not. You are doing so many things, you really don't know what you're doing really. You think you're being good, but you're not. You're neglecting them because you're taking a chance by going stealing and risking yourself. You might not come back to them knowing that they love you.

Can't nobody say they didn't do it [use drugs] in front of their kids, because they're lying. That addiction is always there.

There were a variety of mother-careers among incarcerated mothers. Some were primary caretakers of children before incarceration; some shared care with others; some did not live with children or assume major responsibilities for their care. Upon release, some inmate mothers will resume the major responsibility of caretaker. Others will assume this role for the first time as other caretakers terminate their roles as the child's primary care giver. Still others will lose their rights to children in formal judicial hearings. Finally, some will continue to share child care with others.

Summary and Conclusion

Imprisonment provides an important vantage point for sociologists to examine family responses to women in trouble and how these responses are affected by larger structural forces. The availability of families to assist women in prison is very much influenced by normative expectations. The financial resources and capabilities of families to lend a hand—here evidenced by their taking care of children—appear to be less important than cultural beliefs about what members do for each other in crisis. Race and ethnicity have a powerful impact on the resources women inmates have in arranging care for their children. Few child welfare organizations recognize these differences and few provide support that matches the needs of families from different racial and ethnic groups.

This examination has also focused on the complex nature and meanings of motherhood. Motherhood might mean a biological connection to child, taking care of a child, maintaining overall responsibility for a child's welfare (even if not directly engaged in the care), or sustaining a unique and irreplaceable relationship with a child. We gain insight into the management of motherhood by seeing how women in prison understand motherhood and meet its challenges. While parenting programs may provide opportunities to visit with children, the variety of mother careers supports the need for a range of programs to help inmate mothers with the challenges they'll face upon release from prison.

References

Baca Zinn, M. and D. S. Eitzen. 1993. *Diversity in families*, 3rd ed. New York: HarperCollins College.

Baunach, P. J. 1985. *Mothers in prison*. New Brunswick, NJ: Transaction.

Beck, A. J. and J. Karberg. 2001. *Prison and jail inmates at midyear 2000*. Bureau of Justice Statistics: Washington, D.C.: U.S. Department of Justice.

Caplan, P. J. and I. Hall-McCorquodale. 1985. Mother-blaming in major clinical journals. *American Journal of Orthopsychiatry* 55: 345–353.

Charmaz, J. 1983. The grounded theory method: An explication and interpretation. In *Contemporary field research: A collection of readings*, edited by R. M. Emerson, pp. 109–126. Boston: Little, Brown.

Chodorow, N. and S. Contratto. 1982. The fantasy of the perfect mother. In *Rethinking the family*, edited by B. Thorne and M. Yalom. New York: Longman.

(continues)

FOCAL RESEARCH *(continued)*

Collins, P. Hill. 1990. *Black feminist thought: Knowledge, consciousness, and the politics of empowerment*. New York: Routledge.

Dill, B. T. 1988. Our mothers' grief: Racial ethnic women and the maintenance of families. In *Race, class and gender: An anthology*, edited by M. L. Anderson and P. H. Collins, pp. 215–237. Newbury Park, CA: Sage.

Johnston, D. 1995. The effects of parental incarceration. In *Children of incarcerated parents*, edited by K. Gabel and D. Johnston, pp. 59–88. New York: Lexington.

Maher, L. 1992. Punishment and welfare: Crack cocaine and the regulation of mothering. In *The criminalization of a woman's body*, edited by C. Feinman, pp. 157–192. New York: Harrington Park.

Martin, M. 1997. Connected mothers: A follow-up study of incarcerated mothers and their children. *Women and Criminal Justice* 8(4): 1–23.

Mauer, M. 1997. *Americans behind bars: U.S. and the international use of incarceration*. The Sentencing Project. Washington, D.C.

Mumola, C. J. 2000. *Incarcerated parents and their children*. Bureau of Justice Statistics: Washington, D.C.: U.S. Department of Justice.

O'Barr, J., D. Pope and M. Wyer. 1990. Introduction. In *Ties that bind: Essays on mothering and patriarchy*, edited by J. O'Barr, D. Pope, and M. Wyer, pp. 1–14. Chicago: University of Illinois Press.

Rubin, L. B. 1994. *Families on the fault line*. New York: HarperCollins.

Snell, T. L. 1994. *Women in prison: Survey of state inmates, 1991*. Bureau of Justice Statistics. Washington, D.C.: U.S. Department of Justice.

Stack, C. 1974. *All our kin: Strategies for survival in the Black community*. New York: Harper & Row.

Young, M. and P. Wilmott. 1957. *Family and kinship in East London*. London: Routledge & Kegan Paul.

Appendix A: Interview Questions

Information was first collected from the correctional staff and official records and checked in the course of the interview, as appropriate. This information included the woman's race and ethnicity, living arrangements of child(ren), length of sentence, recidivism status, and involvement of child welfare.

Paths to Prison

Can you tell me what brought you to prison?
How did you get involved in drugs/crime?
Did your family and friends provide an entry? Did a boyfriend? How would you describe your family when you were growing up? What has been the hardest thing about being in prison?

Children

Where are your children living right now?
Were you living with your children before you came to prison? If not, who was taking care of them?
How old are your children?
Is DCF (child welfare) involved? If so, in what way?
What are your plans after you are released? Will you live with your children? Immediately? Eventually?
Have any of your children been adopted?

People say that crime and getting into trouble with the law might be passed down through generations. Are you concerned about your children getting into trouble? What can be done, if anything, to prevent this?

Caretaker Characteristics

How did you decide where to place your children?

Did you feel you had some good options?

What are the pros and cons of placing your children with your parents?/foster care?/your husband?/relatives?

Do you think your child's caretaker is doing a good job? A better job than you can just because of your situation?

Are you involved in making decisions about your children, like where they will go to school? If they need to go to the doctor?

What kind of burden do you think it is for other people to take care of your children? Do you think it is hard for them?

What kinds of obligations do you think families have for each other? What are family members supposed to do if somebody in the family needs help?

Are you comfortable asking your family for help?

Management of Motherhood

Women who are serving a long sentence must have to make lots of arrangements for their children. What are some of these and how is that different from women who are serving a short sentence?

Are there things that only mothers can give and do for their children? If so, what are these?

Are there things that you think your children are missing because someone else is taking care of them?

Do your children understand that you are in prison? What do you think they think about this?

How often do you see your children? How often are you in touch with them?

Do your children understand that you are their mother even though someone else is taking care of them?

Have you heard of instances where children call women other than their mother "Mom?"

What are some things mothers are supposed to do for their children? How do you think women in prison try to make up for the fact that they are in prison?

What are some of the things that make it hard to be a good mother to your children?

In terms of your family and friends, do you need to prove to them that you are a good mother? How do you do that?

In terms of child welfare, do you need to prove to them that you are a good mother? How do you do that?

Women's Understanding of Other Mothers

There are some racial differences in where children live when their mothers come to prison. African American and Hispanic children seem to be more likely to live with relatives while white kids seem more likely to go into foster care or live with husbands. Can you explain why you think that happens?

(continues)

FOCAL RESEARCH *(continued)*

Can you tell when women in prison are ready to make a change in their lives and go straight? What are some of the signs?

There has been a lot of talk about making it easier for the state child welfare agency to terminate parental rights if there has been a child death in the family or if the children have been exposed to drugs. What do you think about these new laws?

Do you think it is possible to tell if someone is a good mother by seeing how she acts with her children in prison?

Do you think the courts are easier on women who have children or harder? Why does this happen?

Appendix B: Excerpts from the approved Informed Consent Form Used by Sandra Enos

What will be done: If I decide to take part in the project, I will be asked about my children, their living arrangements, what brought me to prison, and how imprisonment is affecting my relationship with children and other family members. I may also be asked about involvement with the child welfare agency. My part in the study will involve an interview which will last about 1–2 hours and which will be tape recorded. My name will not be on the tape and after the interview is typed, the tape recording will be destroyed.

Risks: The possible risks in the study are small. I may feel some discomfort as I talk about the past. There is a chance that some of the interview questions may result in my feeling uncomfortable or anxious. If that is the case, the interview may be suspended at that point if I wish. The researcher will ask several times during the interview if I want to stop. The decision to participate in the study is up to me. I may terminate the interview at any time. Whatever I decide will not be held against me. I understand that the researcher is not affiliated with the Department of Corrections and that my participation in the interview will not have an impact on my treatment, criminal processing, or any other matter.

Reportable child abuse: If, while talking with Sandra Enos, I tell her about some abuse or neglect of a child that I say has not been reported to DCYF, the researcher will inform me that (1) she is required by law to report the abuse/neglect to DCYF and (2) that we must terminate the interview. The tape recording of our interview will be immediately destroyed. I understand that the purpose of the study is not to track reportable incidents of abuse or neglect.

Benefits: There are no guarantees that my being in this research will provide any direct benefit to me. I understand that taking part in this research will have no effect on my parole, classification status, and/or inmate record. My taking part will provide important information for people who are trying to understand the impact of prison on women and their families.

Confidentiality: My participation in the study is confidential to the extent permitted by law. None of the information collected will identify me by name. All information provided by me will be confidential. Within two weeks after the interview, the tape will be transcribed, and the recording will be destroyed. No information that is traceable to me will be on the transcript. Transcripts will be maintained in a locked cabinet in a secure location available only to the researcher. No information collected by the research that identifies me will be given to the Department of Corrections.

**THINKING
ABOUT
ETHICS** *Enos submitted her proposal to her university's Institutional Review Board
(IRB), which approved it. All of the women interviewed in this research
volunteered and gave informed consent before being interviewed. All information was kept confidential, although each participant was cautioned about the
legal limits of confidentiality, and no participant's actual name is used by Enos.*

VARIATIONS IN QUALITATIVE INTERVIEWS

Number and Length of Interviews

Qualitative interviews can vary in several ways: the degree of structure, the
length of each interview, the number of times each member of the sample is
interviewed, the number of interviewees and interviewers participating in the
session, and whether interviews are used alone or with other techniques. In
any given study, there is typically variation, sometimes quite a lot, in how long
each interview takes. The interviews in Emily and Roger's study on retirement
typically took between one hour and 90 minutes; the Enos interviews lasted
between one and two-and-a-half hours.

Interviews can be used in longitudinal research, with multiple interviews as
in Hoffnung's long-term study of college graduates (in Chapter 4) and in Emily
and Roger's panel study of the transition to retirement (in Chapter 7). However, in studies like Enos's, where participants are not followed over time,
each member of the sample is typically interviewed only once. Despite the additional cost, some researchers suggest dividing the material into sequential interviews. Irving Seidman (1998: 11–12), for example, suggests a three-session
series, with the first focused on the past and background to provide a context
for the participant's experience, the second to cover the concrete details of the
participant's present experience in the topic area, and the third directed to the
participant's reflections on the meaning of the experience. Other researchers
advocate multiple interviews because they feel that the additional contact will
give interviewees more confidence in the procedure and increase their willingness to report fully. In one study, Robert Weiss (1994) found that only in a
fourth interview did one man talk about his wife's alcoholism.

 STOP & THINK

**semi-structured
interview,** an interview
with an interview guide
containing primarily
open-ended questions that
can be modified for each
interview.

interview guide, the list
of topics to cover and the
order in which to cover
them that can be used to
guide less structured
interviews.

*Sometimes an interviewer doing a structured interview feels foolish reading the
questions almost rigidly, not being able to modify, add, or delete questions.
How do you think you'd feel in the opposite situation—without a "script," and
needing to spontaneously construct questions as the interview proceeds?*

Degree of Structure

Qualitative interviews can vary from unstructured to semi-structured interactions. **Semi-structured interviews** are designed ahead of time but are modified
as appropriate for each participant. They begin either with a list of interview
questions that can be modified or with an **interview guide**, the list of topics to
cover in a suggested order. Although some semi-structured interviews use at
least a few closed-ended questions, more typical is a core set of open-ended
questions, supplemented liberally with probes and questions tailored for each

interview. Such an approach will generate some quantifiable data and some material that allows for in-depth analysis. Constructing questions ahead of time makes the interviewer's job easier because there is a "script" to ensure coverage of all the topics. The interviewer must judge whether the questions are appropriate or not, re-order and re-word them if necessary, and use follow-up questions and encouragement to help the interviewee answer fully.

In her study of college graduates, described in Chapter 4, Hoffnung chose the semi-structured interview for two of her data collections to make the interview more like a conversation, to be able to identify new issues as they came up, and, given her heterogeneous sample, to have the ability to re-word questions for each individual participant. Similarly, Emily and Roger, in their study on the retirement transition, decided that this method was best suited to understanding the changes in people's lives as they moved from full-time employment to other activities. Sandra Enos used semi-structured interviews because pre-formulated questions helped her organize her thoughts and paths of investigation, but left things open enough to pursue "surprises"—unexpected but interesting avenues of investigation (Enos, personal communication).

The semi-structured approach is most useful if you know in advance the kinds of questions to ask, feel fairly sure that you and the interviewees "speak the same language," and plan an analysis that requires the same information from each participant. It allows the study participants to be actively engaged and have an impact on the direction of the interview. In addition, the questions for a semi-structured interview can be made available to other researchers, allowing them the opportunity to evaluate the questions and to replicate the interview in other settings.

unstructured interview, a data collection method in which the interviewer starts with only a general sense of the topics to be discussed and creates questions as the interaction proceeds.

At the far end of the qualitative interview continuum is the **unstructured interview.** Researchers doing totally unstructured interviews start only with a sense of the information needed and formulate questions as the interview unfolds. Although some interviewers develop a guide or a list of topics, they are not bound by it. In all unstructured interviews, interviewee "digression" tends to be valued as much as core information. Flexibility in questioning can provide insight into the participant's viewpoint and the meaning behind statements. In their study of drug use among street-working London prostitutes, Brenda Mosedale, Christos Kouimtsidis, and Martina Reynolds (2009) did not use an interview guide, but instead began interviews by asking participants to tell their story, starting from wherever they wanted. This approach allowed the researchers to hear about the women's early lives, which often involved abuse and violence, and about how they first started using drugs and the current vicious cycle between their work and taking drugs, with the work generating funds for drugs and drugs facilitating their work.

life story interview, a short account of an interviewee's life created in collaboration with an interviewer.

The unstructured or narrative interview lends itself to situations where the researcher wants people to tell about their lives as a whole from their own perspectives. For example, the life history or **life story interview** includes a focus on the participant's social context to help the researcher develop hypotheses or theories during data collection. Roma Hanks and Nicole Carr (2008) used this approach in their study of turning points in the lives of women currently in jail. They began each interview by asking the women to talk about their lives and, as they shared their life stories, used a visual aid—a sheet of paper with a single line printed across the page with the word "Birth" on the left and "Now" on the right—to help participants identify important life events (Hanks and Carr, 2008).

As understandings or hypotheses are formulated, additional questions can be added to the interview. Sulik (2007: 862) was interested in the worldviews of women with breast cancer and also in constructing theories of coping. As she developed models of the women's coping strategies, she modified the interview questions to incorporate and evaluate her understandings of ways the interviewees accept and ask for help and the extent to which they feel entitled to care.

Shulamit Reinharz (1992) labels the unstructured approach an "interviewee-guided" interview because the focus is more on understanding the interviewee than on getting specific questions answered. This method may be most useful when the researcher does not know in advance which questions are important, when interviewees are expected to find different meanings in questions or possess different vocabularies (Berg, 1989: 16), or when the researcher wants to work inductively. Wanting to work collaboratively with the study's participants, Westervelt and Cook (2007) used this approach to talk to people who had been on death row before being exonerated. They felt their primary research role was to facilitate the participants' stories, allowing them to be heard as authentically as possible to "explore and understand the flesh-and-blood realism of their experiences" rather than reducing them to statistical patterns (Westervelt and Cook, 2007: 28). The less structured interview can also be used first to develop themes and topics that are then covered in more structured ways or used in addition to surveys to obtain more complex narratives—the strategy used by Regnerus and Uecker (2011) in their research on heterosexual relationships among young adults.

STOP & THINK   *The interview methods we've discussed so far have relied solely on written or verbal responses. How about more visual data? Can you see a way that drawings or photographs can be used to collect data?*

Interviews Using Visual Methods

Interviewing may be improved by the addition of visual methodologies. In these methodologies, the data are the images and the explanations that the interviewee gives about them. One visual method uses drawings made by interviewees. For example, in a study of women recovering from postnatal depression, after the individual interviews, the women were handed a set of blank cards and colored pens. They were asked to draw two pictures—one illustrating what it was like for them during the illness and the other depicting what it is like for them now—and then asked to discuss and explain the drawings (Guillemen and Drew, 2010).

photo elicitation, a technique in which photographs help reveal information and encourage discussion of events and meanings in the participant's life.

Other visual approaches involve photographs. The most common of these is **photo elicitation,** a technique where photographs help reveal information and encourage discussion of events and meanings in the participant's life. Participants typically take or bring photographs of themselves or their surroundings and are asked to talk about them. For example, in Coralee McLaren's (2009) study of the experiences of hospitalized children, research assistants escorted the study participants on a walking tour of the hospital's lobby atrium. The children were asked to take photographs of architectural and design features that interested them and these photographs were used to generate thoughts about how the patients perceived and navigated the hospital space. Anamika Majumdar's (2007) primary method was the life history approach in her study of marriage in South Asian communities, but she also asked the women to bring photographs of the houses and areas where they lived before marriage as memory triggers for discussions of emotionally close premarital relationships. The photographs grounded the women's narratives in specific places and helped the participants discuss personal experiences (Majumdar, 2007).

These visual methods can be used in combination with interviews or with the observational methods described in the next chapter. Dave Snell and Darrin Hodgetts (2007) combined observational, visual, and verbal methods to document the efforts of a group of heavy metal fans to negotiate a sense of community with a shared history, identity, and sense of belonging. The researchers used direct observations at a heavy metal bar, a series of individual interviews with patrons and bar staff, and a photo-voice project where the fans were asked to take and discuss photographs that represented their experiences of the music and community.

photo-interviewing, a data collection technique using photographs to elicit information and encourage discussion usually in conjunction with qualitative interviewing.

Photo-interviewing and other visual methods have limitations, such as the ethical concerns of privacy, questions of probability sampling, and the fact that they are literally a snapshot of reality, not a complete picture. However, advocates of these methods believe they are very effective for a number of reasons: images are ways to gain insight into another's perspective and make it possible for participants to introduce new and possibly contentious topics they might not talk about in a purely verbal interview (Chronghan et al., 2008: 355); images can make experiences and feelings more accessible to research participants and the study becomes more collaborative when participants create the images (Guillemen and Drew, 2010).

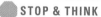 **STOP & THINK** *So far, all the interviewing we've discussed has been one-to-one. Are there any situations you can think of when it might be better to have more than one interviewer or more than one interviewee?*

Joint Interviewers

The use of more than one interviewer is fairly uncommon. When used, it works best when the person being interviewed is not easily intimidated, and the perspective of two or more interviewers is important. For example, in a study of the women who served in the Rhode Island legislature (Adler and Lemons, 1990), one of us, Emily, interviewed several of the more than 50 women jointly with her co-researcher, historian J. Stanley Lemons. This enabled the researchers to develop similar interviewing styles and to elicit information that was of interest both sociologically and historically. After the joint interviews, we moved to one-to-one interviewing to be able to complete the interviews (each lasting two to three hours) in a timely fashion.

Group and Focus Group Interviews

group interview, a data collection method with one interviewer and two or more interviewees.

It is more usual to have more than one interviewee as in the joint or **group interview,** where one interviewer or moderator directs the inquiry or interaction in an interview with at least two respondents. The individuals in the group are selected because they have something in common. They might know each other (such as a married couple or members of the same church) or be strangers (such as teachers from schools in different towns or patients in a given hospital). Such interviews can be based on a predetermined set of questions or can use an unstructured format. Group sessions are less time consuming than interviewing the same number of participants individually and have "the advantages of being inexpensive, data rich, flexible, stimulating to respondents, recall aiding, and cumulative and elaborative over and above individual responses" (Fontana and Frey, 1994: 365).

Group interviews are useful when the group is the unit of analysis. It can provide either a shared or a disparate view of people's experiences and allow the researcher to observe interaction among participants while the interview is in progress. Racher, Kaufert, and Havens (2000: 367) interviewed frail, rural, elderly couples jointly and, to maximize understanding each couple as a unit, used the interview itself as an opportunity to observe each couple's verbal and nonverbal interaction while focusing on the content of their answers. When the unit of interest is not the group itself, the group interview can have the advantage of releasing the inhibitions of individuals who are otherwise reluctant to disclose what for them are private matters. Because some "are more willing than the others to speak of personal experiences and responses, they tend to be among the first to take active part in the discussion. As one ventilates his experiences, this can encourage others to ventilate theirs … and establishes a standard for the rest who progressively report more personalized responses" (Merton, Fiske, and Kendall, 1956: 142–143). In addition, the interaction in the group might remind each individual of details that would otherwise not be recalled (Merton et al., 1956: 146).

Group interview data can suppress negative attitudes and may be biased toward the views of one participant. William Aquilino (1993: 372) found that even though the absolute magnitude of effects was small, the presence of a spouse influenced subjective assessment of marriage in the direction of more positive assessment. Some argue that it is better to interview husbands and wives separately to enable them to talk more freely about their feelings and views of each other (Adler, 1981; Rubin, 1983) and when the researcher is interested in the individual's rather than the couple's experience. In one study of married couples, women's perspectives were found to be so prominent in the joint interviews that the researchers suggested interviewing men on their own to find out about the husbands' experiences of illness and fatherhood (Seale et al., 2008). Even among those who are strangers, some participants might take over the group, monopolize the interaction, or inhibit other members.

focus group interview, a type of group interview where participants converse with each other and have minimal interaction with a moderator.

The most commonly used group interview is the **focus group interview** where the emphasis is on obtaining data through group interaction. To hear many points of view, focus groups use a moderator or facilitator who doesn't primarily ask questions but instead encourages group participants to talk to and ask questions of each other. Focus groups were originally used in social sciences in the 1940s and 1950s (Merton et al., 1956), but then became much more widely used in market research. Rediscovered by social scientists in recent decades, the focus group is now a common method in their fields.

Group interviews can be used alone to provide useful qualitative data or in combination with other methods. Some focus groups are planned, with the moderator being more involved in controlling the group dynamics and keeping it on the researcher's interests; others are less planned and become more conversational and self-managed (Morgan, 2001: 147).

Some argue that focus group interaction highlights issues and concerns that would have been neglected by questionnaires (Powell, Single, and Lloyd, 1996) and encourages participation by those who would be more likely to decline one-to-one interviews (Madriz, 2000). But, the use of focus groups is not without problems. Like other studies that use small and nonrandom samples, focus group studies will have limited generalizability. In addition, in all group interviews, including the focus group, it is very important ethically for participants to keep confidential the information provided by others. An agreement to maintain confidentiality can be included on an informed consent form, but it is hard for the researcher to enforce compliance.

Examples of focus group research include an exploration of Koreans' perceptions of what makes them happy (Kim, Kim, Cha, and Lim, 2007) and a study on the effect of the University of Nebraska-Lincoln football team on which found local residents believing that having the team in their community bolstered their sense of pride in Nebraska's values and lifestyle (Satterfield and Godfrey, 2011). Focus groups can use visual material as conversation starters. For example, an Australian study used photos of hospice care activities that the workers had taken to generate discussion in the workers' focus groups (Horsfall, Noonan, and Leonard, 2012).

Focus groups have between 3 and 12 participants, selected for homogeneity on the characteristic for which the researcher recruited them, such as people who have been recently widowed or have specific health concerns. The participants usually don't know each other before the group interaction. Projects will use anywhere from just one to dozens of groups.

The groups are usually conducted with participants seated around a table with a moderator who starts the interaction with one or more questions. In the study on Korean happiness, only three questions were asked: "What makes you happy? What could make you happier than now? In general, who is a happy person?" (Kim et al., 2007). See Box 10.2 for the questions Erica Chito Childs (2005) used in her study of black college women's views of interracial dating and her thoughts on focus groups.

While face-to-face interaction is an important feature of the focus group, some researchers have begun to use "virtual focus groups" to be able to include those who are far apart geographically or unable to travel. One video-conferenced group was conducted as part of a study on parenting support systems for new mothers. One mother said she enjoyed participating in the virtual focus group because "It's better than a Q [questionnaire] as sometimes someone else's answer can make you think of something you wouldn't have thought of by yourself" (Madge and O'Connor, 2006: 203).

BOX **10.2**   A Focus Group on Interracial Dating

Erica Chito Childs used focus groups as one source of data in a study of black college students' attitudes toward interracial dating. The participants were women active in black/African American student organizations on three college campuses who were between the ages of 18 and 23. Chito Childs used focus groups because of her interest in the collective responses of the women and the attitudes toward interracial dating that they would discuss within a group. The focus group discussion allowed respondents to debate and to build upon the comments of others, either adding to what another said or offering a different perspective. Chito Childs (personal communication) notes that while one person's comments can affect what others say, focus groups can give the researcher potential access to collective attitudes, experiences, and beliefs.

Chito Childs has very generously provided a description of her interview questions. Note that she began with general questions, moved to more specific ones, and then used follow-up questions to encourage participants to expand their answers.

I began by asking about their general views on race relations with the question "How would you characterize race relations between blacks and whites?" I started with very general questions about societal views to encourage the discussion to flow, and then would follow up with more specific questions about their own views/experiences, encouraging them to talk. I next asked "Tell me about interracial dating. Do you have certain ideas about these relationships?" I followed this up with questions about their own views. I also asked them about the idea that black women as a group are opposed to interracial dating and if they agreed with that view. After the women talked about their own feelings, some discussed their views of how other black women felt. I followed up with questions, such as "Where do these views come from?", "Explain that a little more", "Why do you think that is?", or "Could you give me some examples of a situation like that?" Since much of the discussion was on black men dating white women, I also asked "What about black women dating white men?" and followed up with questions to have them explain in more depth. Finally, I asked "How do you think whites feel about interracial dating?" I found using informal, conversational questions that did not steer the discussion too much worked best for me. (Chito Childs, personal communication)

ISSUES OF VALIDITY AND PRACTICALITY

As with other self-report methods, we must consider validity concerns when using data generated by qualitative interviews. Inaccurate memories, misunderstandings, and miscommunications and misrepresentations must be considered in evaluating the information obtained. As early as 1935, researchers debated the "questionable value" of the interview for securing reliable data when people become defensive concerning their private and personal lives (Thurlow, 1935). But Enos and others who have used the qualitative interview believe that, overall, interviewees tell the truth as they understand it, rarely offering false information knowingly.

The qualitative interviewer is not a passive listener. The way in which an interviewer questions, responds, and acts can affect participants' responses. Different interviewers or even the same interviewer using different wording might be told different accounts, raising concerns about reliability as well. Furthermore, it's important to consider whether different researchers would interpret responses in the same ways.

Specific validity concerns for focus groups are that some voices might not be heard and that there is pressure to conform to group attitudes. We also should be aware that status within the group can affect responses and that with "group think" participants may echo what the first speakers say rather than giving their own points of view. A one-to-one interview might be selected to elicit individual answers about specific behaviors and experiences and focus groups used to find out about a community. Researchers may find it useful to compare data from group and individual interviews. For example, in a study of boyhood in London, researchers conducted individual and focus group interviews with 14-year-old boys and obtained different data. In focus groups, the boys' comments were affected by hierarchies among the boys, demonstrated the boys' humor and naughtiness, and contained a great deal of ridicule of girls. In individual interviews, the boys were quieter and more serious, and talked about their close relationships with parents, grandparents, pets, children, and girls. Rather than privileging one account over another, the researchers concluded that "What the very different and contradictory accounts of boys in the different modes of research suggested was that boys were both powerful and vulnerable" (Pattman, Frosh, and Phoenix, 2005: 560).

The counterpoint to concerns about qualitative interviews is that standardizing might be no better at producing comparable data because different respondents can hear the same question in different ways. The ability to reword and restate questions in the more qualitative method can give the researcher a chance to present them in ways that are better suited to each participant. In addition, a relatively natural interactional style allows interviewers to better judge the information obtained. Qualitative researchers can try out what they think they have come to know as a result of interviewing others, so that interviewees can say "That's what I meant" or "No. What I meant was …" to validate or correct understandings (Gold, 1997).

Conducting and transcribing interviews is a time- and labor-intensive task. As a result, qualitative interviewing is an expensive method that can sometimes mean a small sample with limited generalizability. As in the focal research for this chapter, Enos conducted only 25 semi-structured interviews in her study; Emily and Roger did 88 in theirs. However, other researchers have been able to use larger samples. Lillian Rubin (1994), for example, conducted 388 separate in-depth interviews in her study of 162 working-class and lower-middle-class families. Because of the benefits and the costs, some argue that the qualitative interview is most appropriate for exploratory research, for research that seeks an understanding of interviewees' worlds and meanings, and when the researcher is interested in generating "grounded theory"—that is, theory developed in intimate relationship with data, without preconceived notions, and with researchers aware of themselves as instruments for developing theory (Strauss, 1987: 6).

Locating Interviewees, Presenting the Project, and Encouraging Participation

In all interviews, researchers select a population and sampling method before contacting potential interviewees. If names and contact information are available, the approach can be a letter, a phone call, or an e-mail. Hoffnung, whose panel study of women college graduates was discussed in Chapter 4, began with a list of seniors who had been selected at random from each of five colleges. She called the women, introduced herself, and asked them to participate. Researchers interested in the workplace experiences of lesbian, gay, bisexual, and transgender physicians used a snowball sample by contacting all physicians listed in the Gay and Lesbian Medical Association database and asking them to participate and to forward the request to other GLBT physicians (Eliason, Dibble, and Robertson, 2011).

STOP & THINK   *What about groups of people for whom there are no lists? For example, what would you do if you wanted to talk to homeless people, those planning to retire in the next few years, or working-class young adults? How could you locate your sample? How could you encourage them to participate?*

gatekeeper, someone who can get a researcher into a setting or facilitate access to participants.

Qualitative researchers are frequently interested in people for whom there are no lists—such as mothers in prison, fathers who do not live with their children, or grandfathers raising grandchildren—so selecting a random sample might not be possible. One strategy is to contact participants through **gatekeepers**, that is, people who control access to others. Gatekeepers include the parents and guardians of children under 18 and the heads of institutions, community organizations, agencies, or groups whose members are the desired sample. Some researchers use their personal networks, notices on bulletin boards, announcements at meetings, e-mails, and the like to recruit participants. Sulik (2007) began with one community-based breast cancer organization and then added cancer support groups, two treatment centers, cancer-related community events, bulletin boards, and snowball sampling to recruit her sample of 60 breast cancer survivors.

The more "political" or "deviant" one's topic is, the more difficult it is to get access to and participation of potential respondents. For example, Terrell Hayes (2010) was denied access to meetings of a chapter of Debtors Anonymous when he revealed that he was conducting research.

Sometimes researchers encounter unexpected difficulties. Interested in the implications of the decline in traditional cultural pathways to adulthood, Jennifer Silva (2012) sought out working-class people in their 20s and 30s to participate in a two-hour semi-structured interview on adult identities. In Box 10.3, she describes her experience.

Some researchers encounter participant concerns that their information will be used against them or that the researcher is not who she says she is. Enos presented herself as someone who was "outside the system"—not part of the prison administration or the child welfare system. Her interviewees believed her and her pledge of confidentiality with limitations, including her promise not to use their real names. Some researchers find that potential participants view them with suspicion—perhaps as vice officers or narcotics agents in disguise (Miller, 1986) or as investigative reporters, union organizers, or industrial spies (Harkess and Warren, 1993: 324). Others note that some will interview the interviewer before agreeing to participate so as to ensure professional competence (Gonzalez-Lopez, 2011: 452).

| BOX **10.3** | **What Does it Mean to Grow Up Today?** |

By Jennifer Silva*

When I set out to interview working-class young adults about their struggles to grow up—to get good jobs, graduate from college, or find a committed romantic partner—I was completely unaware of what a challenge it would be. In one sense, working-class young people are all around us, whether serving coffee at a local coffee shop, changing oil at the mechanic's, or working check-out at the grocery store. I started out by spending long days driving around Lowell, MA, and Richmond, VA, hitting every service workplace and community college I could find. My strategy included approaching young adults in their various places of employment, such as retail chains and fast food joints, and asking, "Would you like to participate in a study of what it's like to grow up today?" While it was sometimes awkward to approach people cold, and I often worried about getting employees in trouble with their managers, I was excited that most people agreed to speak with me. Problems arose, however, when it came to scheduling and following through with an interview. Working-class young adults have high rates of job turnover and residential mobility; many "bounce around," crashing on couches when they can't make ends meet and working temporary jobs that start and end according to the demands of the capricious labor market. In addition, when they can't pay their bills, their cell phones are often shut off, and they get a new number with a pre-paid cell phone—making staying in touch exceedingly difficult. On top of that, many are balancing the demands of work and parenting and are not sure if they can count on childcare or if they will be called to work a shift.

I approached this dilemma by seeing these challenges as data—indeed, when I was repeatedly stood up for interviews or texted at the last minute to reschedule, I experienced firsthand how insecure and unstable these young people's lives really were. Doing an interview with a child running around and demanding constant attention showed me how difficult it is to be both a good parent and a good employee without support networks to fall back on. One morning, I drove through a blizzard to make it to National Guard training by 6 A.M. because I knew the soldiers were required to be there. I also learned to conduct the interview "on the spot" whenever possible, sometimes sitting down in the community college cafeteria or taking a respondent to a nearby Arby's or Subway so that they could take their lunch break while speaking with me.

Over the course of my research, I was continually forced to think about the potential difficulties in forging connections across lines of power and identity—whether class, race, gender, or sexual orientation. John, a twenty-seven-year-old black man who was studying for an accounting exam at a Richmond community college, stopped halfway through the interview to tell me point-blank, "You know, the average white woman won't even look me in the eye. They look away. I was so shocked when you sat down and talked to me and would even look me in the face. . . . Then I learned that you wanted something from me and it made sense."

Being close in age to my respondents helped me to bridge the gap between our disparate social locations. In my typical graduate school uniform of jeans, a sweater, boots, and a ponytail, I could blend in at the community college or at Walmart. When I laughed about my own prolonged transition to adulthood, talked about the fourteen-hour days my mom pulls as a small business owner, or recalled my experiences of apprehension as a first-generation college student, respondents would often visibly relax and respond with stories or even advice of their own. Some, only partially jokingly, referred to the interview as "free therapy," seeming to savor the opportunity to communicate their difficult emotions to someone who would listen.

*Jennifer Silva is a postdoctoral fellow at Harvard University. This essay is published with permission.

Some studies provide interviewees with incentives to encourage participation. Ciambrone (Chapter 3) gave the HIV participants in her study a small gift certificate to a local store. Campbell and Adams (2009) recruited rape victims from Chicago neighborhoods by advertising their study through posters, fliers, and in-person presentations, noting that there would be financial compensation of $30 for participating. In response to the question about how they decided to participate, only 14 percent of the 102 women they interviewed specifically mentioned the $30 compensation as a primary motivating factor. Much more common was the desire to help other women (38 percent), the feeling that being part of the study would be helpful to themselves and their recovery (34 percent), and a general interest in supporting research (25 percent) (Campbell and Adams, 2009: 399–402).

Many doing qualitative interviewing don't offer financial incentives yet still find participants. In Enos's study, the women appreciated her offer of photographs of their children, but participated because they enjoyed being listened to with interest and respect and wanted to help others in their position. Emily and Roger had many interviewees comment on their desire to talk about their retirement plans, noting that doing so helped them to clarify their thinking. Although researchers try to explain the difference between therapy and the research interview, Silva and others note that some participants see the interview as "free therapy."

PLANNING THE INTERVIEW

Using Consent Forms

informed consent form, a statement that describes the study and the researcher and formally requests participation.

Anonymous surveys like those discussed in the previous chapter are often exempted from the requirements of using **informed consent forms**. In some studies using face-to-face interviews, researchers rely on verbal rather than written permission because it is possible in an interview to explain the purpose of a study, offer confidentiality of answers, and assume that the decision to participate implicitly means giving consent. However, as discussed in Chapter 3, IRBs, the groups explicitly designed to consider ethical issues, usually require interviewers to use a written informed consent form. The researcher provides a document that informs potential participants about the purpose of the study, its sponsor, the benefits to participants and others, the identity and affiliation of the researcher, the nature and likelihood of risks for participants (such as the possibility of raising sensitive issues), and an account of who will have access to the study's records and for what purposes. With such information, potential participants can weigh the benefits and risks in deciding whether to participate. Chapter 3 has the approved informed consent form that Ciambrone used in her study of women's experiences with HIV/AIDS and excerpts from Enos's form are in this chapter. Both researchers gave a signed copy to each interviewee. Both offered confidentiality to the extent provided by law, although some researchers question the impact of this limitation on the validity and reliability of the data (Palys and Lowman, 2001).

Constructing an Interview Guide or Schedule

Once the study's objectives are set, the researcher determines what information is essential. If the interview is unstructured, using a list of topics, an introductory statement and having a general sense of the order of discussion is sufficient. The interviewer typically starts with general questions and follows up on the participants' comments. More information can be sought to provide the context, chronology, or additional meanings of the answer. Usually, it's better to wait until rapport is developed before asking questions about sensitive issues.

If doing a semi-structured interview, a list of questions, both basic and follow-up, are needed. Sometimes "filler questions" are used to provide a transition from one topic to another. In less structured interviews, starting with broad, interesting questions is important. The guidelines for question construction provided in Chapter 9—such as staying away from double-barreled, double-negative, or threatening questions and not using wording that is ambiguous or leading—also apply to qualitative interviews. Doing practice interviews helps to prepare for the actual experience and allows the interviewer to gain experience covering issues and developing conversation generators. Questions that might elicit more information include "Can you tell me more about that?", "When did that happen?", "Who did you go with?", and "What happened next?" Noticing not only what is mentioned, but what is *not* mentioned, can lead to additional questions.

The intended analysis affects the choice of specific questions. Reviewing Sandra Enos's questions at the end of the focal research article, you'll see that they invite the participants to tell their stories. Susan Chase took a similar approach in her case study of one university (discussed in Chapter 7). Although she was interested in the issue of diversity on campus, she didn't start with direct questions on that topic. Instead she first asked about the past (such as "Tell me a bit about your high school years. What were your interests then?"), before moving to the college career ("What kinds of transitions did you go through during your first semester and first year at here?" "Who has been most influential in your life during your college years?"). She notes that most students bring up diversity issues "naturally" as they tell their stories about transitions to college and through their college years. At that point, she asked a series of questions about diversity (such as "Do you have friends who are 'different' from you in important ways?") (Chase, personal communication).

Once an interview guide or schedule is constructed, pilot testing with people similar to those who will be interviewed during the actual data collection is helpful. Trying out questions will allow the researcher to select those most useful for focusing the conversation on desired topics. The interviewees in these preliminary interviews can be asked about the interview experience—what was effective and what was difficult to understand. In a qualitative study, the list of questions can continue to change during the course of data collection, especially as the focus of the study evolves.

STOP & THINK   *As a college student, you use words and language in certain ways. Are there any groups that you think might use language in ways that are so different from how you use words that you might initially have difficulty communicating with them?*

Speaking the Same Language

Obviously it is essential for the interviewer and interviewee to literally speak the same language, but for participants who speak more than one language, language of preference should be considered. For example, Jeong-eun Rhee found some of her interviewees wanted to speak Korean while others used English or a combination. In her interviews, she came to see that some participants thought that English was the appropriate language for public and academic settings and others saw their language choice as a statement about identity (Subedi and Rhee, 2008).

It is necessary to be familiar with the cultural milieu of study participants and to understand what they mean by terms. Umaña-Taylor and Bámaca (2004: 265), in their research with Latino parents, for example, note their participants sometimes switched between English and Spanish, perhaps because some words had no direct translation and others, like "abuelita," had more sentimental value than the English equivalent, "grandmother." They note that while Latinos share Spanish as a common language, there are expressions and words whose meanings are unique to specific nationalities. Enos found that she'd picked up enough "prison lingo" during the observation part of her study that she knew what the interviewees meant when they reported "getting jammed up" (getting in trouble), "catching a bid" (getting sentenced), and the like. If she didn't understand something, she'd usually ask that it be repeated or explained. For example, one woman reminisced about her life, talking about things "back in the day." Enos wondered about those words until she asked another inmate mother about the phrase and was told that it was when you were young and free or out on the streets—getting high, having fun, or whatever. In the context of her study, it isn't surprising that Enos sometimes had to ask the women to explain family relationships because some of the children they talked about "mothering" were not their biological children, but those for whom they were informal guardians.

CONDUCTING THE INTERVIEW

Where and How to Interview

Interviews can be held in offices, in the interviewee's home, or elsewhere. Ching Yoon Louie (2001: 8) preferred to conduct her interviews of Chinese, Korean, and Mexican sweatshop workers who had become leaders in the movement to improve work conditions wherever they were working: "camped out at movement offices, sandwiched between pickets and workshops, while driving across the state to demonstrations, leafleting at factory gates, and fighting with government officials."

If possible, individual interviews are conducted in private. But sometimes others are present. In a study of married couples, when Emily was interviewing the wives in their homes (while her associate was interviewing the husbands), she found that, on occasion, children would be present for part of the interview—sometimes commenting on the woman's comments! If privacy is important for an interview, it's important to consider that when scheduling the time and place.

Although the interview might not follow all the conventions of conversation, it *is* a social interaction. The interviewer should be aware of all that the interviewee is communicating: verbal and nonverbal messages and any inconsistencies in them. The interviewer should strive to be nonjudgmental in voice tone, inflection, phrasing of questions, and body language. A goal is to communicate openness to whatever the interviewee wants to share, rather than conveying expectations about his or her answers. Following up answers, the interviewer can ask for details, clarification, and additional information. It's useful to know when to keep quiet as interested and active listening is a good way to obtain in-depth responses.

Recording the Interview

If the interview covers a large number of topics, a recording—either audio or video—is invaluable; the more unstructured the interview, the more necessary recording becomes. If there are no pre-set questions and probing is an important part of the process, the interviewer may find it difficult to attend to what the interviewee is saying while trying to write. Recording will also allow later review of specific words, voice tone, pauses, rephrasing of answers, and other subtle information. But the impact of recording an interview is hard to assess. On the one hand, it can help rapport by allowing the interviewer to make eye contact, nod, show interest, and concentrate on follow-up questions. On the other hand, some interviewees refuse permission to record the interview and others are intimidated or inhibited by the recording process.

STOP & THINK *After describing his relationship with his father, the person you are interviewing asks you how you get along with your father. What would you do? Would you tell him that what you think really isn't important because he's the one being interviewed, answer that you'll tell about your father after the interview is over, describe your relationship with your father honestly but briefly, or launch into a detailed description of how you get along with your father?*

Being "Real" in the Interview

In more traditional qualitative interviewing, the interviewer maintains social distance from the interviewee. This means using a style that gives evidence of interest and understanding in what is being said (nodding, smiling, murmuring "uh-huh"), but that prohibits judgment, reciprocal sharing, or "real conversation." The interviewer is advised not to share opinions or any personal information with respondents because it can increase the chance of leading subjects to say what they think the interviewer wants to hear and shift attention from the interviewee.

Critics dispute the view that the traditional interviewer response is really neutral. Ruth Frankenberg (1993: 31), interviewing people about issues involving race, felt that "evasive or vague responses mark one as something specific by interviewees, be it 'close-mouthed,' 'scientific,' 'rude,' 'mainstream,' 'moderate,' or perhaps 'strange'—and many of those are negative characterizations in some or all of the communities in which I was interviewing." Fontana and Frey (2000: 658) believe that there is more support now for researchers' attempts to minimize status differences and show their human side by answering questions and expressing feelings, especially as this can allow for the development of a closer relationship or greater rapport between the two interview participants.

Although Enos was asked only an occasional personal question about herself or her opinions, quite a few interviewers report interviewees asking numerous questions—about the study, about their personal lives, or for advice on the topic under study. For the research on women living with HIV/AIDS (Chapter 3), Desiree Ciambrone fielded questions about her personal life, typically whether she was dating, married, or had children, and whether she thought it was difficult dating or meeting people; less frequently, she was asked about her sexual or drug-use history. She was open and honest, in part because she is comfortable talking about these topics and in part because she believes in disclosure and in minimizing power imbalances in the interview. She thinks her receptiveness and candor fostered rapport and ultimately helped the women to feel comfortable and be more forthcoming (Ciambrone, personal communication). Anticipating that interviewees may ask personal questions allows the interviewer to consider how to respond.

Listening to someone describe an intimate, personal story can be a moving experience. Being "real" as an interviewer can mean acknowledging that at least some of what you hear is painful, difficult, or upsetting. The demanding nature of interviewing is not often discussed. Interviewers, even those with training, might not be prepared for or trained to handle the emotional impact of the interview process. In her work on juvenile prostitution, Melrose describes how she often felt distressed by the accounts she had heard: "[A]lone in a hotel in an unfamiliar place, struggling with my own feelings of anger and despair … I was forced to contain these feelings in order to carry on with the fieldwork the next day … I also imposed the expectation on myself that I should be able to 'manage' these feelings" (2002: 325). The two full-time interviewers in a study on low-income families found the research process to be emotionally stressful as they heard about experiences of sexual abuse, rape, domestic violence, suicide attempts, and other problems; the interviewers provided each other with emotional support through daily phone calls (Gorin, Hooper, Dyson, and Cabral, 2008).

Interviewing across the Great Divides

STOP & THINK

How effective do you think you would be as an interviewer if you were interviewing someone considerably older than you are? How about if the person were of the opposite gender, of a different race, or from a different social class? Do you think the topic of the interview would make a difference in interviewing someone whose background was very different from yours?

interviewer effect, the change in a respondent's behavior or answers that is the result of being interviewed by a specific interviewer.

As we discussed in Chapter 9, researchers using interviews need to think about **interviewer effect,** the change in a participant's behavior or the answers given to questions as the result of being interviewed by a specific interviewer. We'll expand this discussion to include all the researcher's identities, such as gender, age, social class, ethnicity, race, and sexual orientation, and note that each can affect the whole research process, including data collection. As a result, an ongoing debate concerns the desirability of matching interviewers and interviewees on specific social characteristics, mostly significantly gender, race, ethnicity, age, and class.

The most common perspective is that it's advisable to match the participants because people of similar backgrounds are thought to develop better rapport. One concern is that interviewers who differ from interviewees on important characteristics tend to get different responses and less honest answers than when the participants have the same backgrounds. Sandra Enos believes that her gender was helpful in the interview process. "A man in that setting could have been problematic, especially since the women would have had some trouble understanding why a man would be interested in mothering. Also a number of the women expressed exasperation with relying on men and others were still looking for Prince Charming" (Enos, personal communication). While it might have been better to have worked with both African American and Hispanic co-researchers, she didn't have the financial resources to hire interviewers and believes she communicated effectively across race, class, and ethnic lines (Enos, personal communication).

In Box 10.4, Nicole Banton argues that interviewing within your own group makes sense because the interviewer knows the language and traditions of the group and is more likely to be trusted by interviewees.

However, even when matching interviewer and interviewee on key characteristics, there can still be differences. Hasmita Ramji (2008), a British South Asian woman who interviewed women with the same background, found that issues of "Indianness" and religion affected interview interactions as interviewees tried to figure out commonality or difference. Doing semi-structured interviews in London's Hindu Guarati community (a community of which she was a part), Ramji (2008: 105) reported interviewees asking whether she was born in the U.K., what part of Gujarat her parents were from, what caste she was from, and so on to determine "how *Indian* I was and, more importantly, if I was *their*

BOX 10.4 On My Team? Why Intragroup Interviewing Is Better
By Nicole Banton, PhD*

What groups do you belong to? What makes your group unique—language? customs? In some groups, like those based on race, ethnicity, and/or gender, the socialization on things like language, traditions, and ideas about right and wrong begin at birth. Now, imagine an outsider asking you questions about yourself as a representative of your group. What if that person is a member of a group that has historically oppress(ed) yours? I consider these issues when I plan to collect data through interviews because they can affect how the interviewee acts during an interview.

When people talk to members of their own group, they are more likely to be comfortable because they are more likely to share a common language, including slang and jargon, as well as customs and traditions. In intragroup interviews, the interviewees are not as likely to think that they have to represent their entire group because they are

talking to one of their own. The interviewee is more likely to feel a kinship, and the interviewer is more likely to understand the interviewee.

For example, when African Americans identify each other to other blacks, they don't just say that someone is black. They describe skin tone; for example, "Nicole is light-skinned." They do this because African Americans are aware that "blackness" is a rainbow of shades, while non-blacks are less sensitive to this issue and the politics surrounding it. I want people to be relaxed during an interview, to understand them, and to get the truest information possible. The best situation is when the researcher and the researched are members of the same team(s).

*Nicole Banton's work includes a research project on the politics of infant feeding among African American mothers. This essay is published with permission.

kind of Indian." As this example demonstrates, it's not possible to match interviewer and interviewees on more than a few key characteristics.

In any event, *all* interviewers have personal characteristics and social identities that need to be considered in evaluating interviewer effect. In fact, sometimes sameness can be a problem. Researcher Don Naylor, interviewing men about gender, knew that the men in his study might not be completely honest but he didn't worry about the interviewer effect because he and the interviewees would be the same gender, age, social class, and sexual orientation (Naylor, personal communication). However, in reading the transcripts, he noticed that sometimes the men said things that suggested a need to appear masculine. Reflecting on the interview process, he concludes that there is an effect even when the interviewer and the interviewee are similar. "It is a different dynamic than when the two are different, but it is still there. Similarity can affect the content of the interview, the types of questions asked, the answers given, and the way the interviewer and interviewee interact. When the men varied in what they said and how they said it (amount of openness, attempts to control, emotions displayed), I would change my style and demeanor. They were affecting me as much as I was affecting them. I began to think not so much in terms of interviewer effect, but about interviewer–interviewee effect as we were jointly constructing the interview in a process that is gendered. We were 'doing' an interview not just as two people but as two men. We were 'doing gender'" (Naylor, personal communication). Reaching the same conclusion in a study of fathers, Robb (2004: 402) notes that masculinity should be seen as something that is "worked on" in the interview process and that men being interviewed by men will be partly motivated by a desire to prove their masculinity.

It is possible for interviewers and interviewees of different backgrounds to have a good research relationship and for those who are alike to have problems. The researchers in charge of the study *Sex in America* asked people of different races in focus groups whom they would feel most comfortable talking to (Michael, Gagnon, Laumann, and Kolata, 1994: 32). They were surprised to hear that almost everyone—men, African Americans, and Hispanics—preferred middle-aged white women as interviewers. Similarly, in her study of male clients of female prostitutes, Grenz (2005) asked prospective interviewees whether they preferred to be interviewed by a man or a woman and found that none wanted to be interviewed by a man and many indicated a preference to be interviewed by a woman.

AFTER THE INTERVIEW'S OVER

STOP & THINK   *Imagine that you're almost at the end of an interview on college students' relationships with significant others. After describing how the most recent love relationship ended, the student you're interviewing looks up and says, "I'm so depressed, I feel like killing myself." What would you do?*

At its end, the interviewer typically has the desired outcome—a completed interview—yet there may be many different responses from the interviewee. Some might rush off as soon as possible, "turn the tables" and interrogate the interviewer, bring up their own agendas, or continue talking about the topic after the official end of the interview (Warren et al., 2003: 98). When covering difficult topics, researchers might need to prepare for emotional responses or requests for help. For example, researchers interviewing hospice workers (Cain, 2012) and those talking to the families of organ donors (Shaw, 2011) reported that some interviewees cried during interviews. Researchers usually include a series of "cool down" questions at the end of an interview so the interview doesn't end immediately after talking about sensitive subjects.

Some researchers leave material with participants, most typically a list of appropriate organizations. For example, in a study on partner violence, interviewers provided referrals and information about local agencies that offered useful services (Logan, Walker, Shannon, and Cole, 2008: 1232). Although some interviewers will continue to have contact with interviewees, most researchers limit interaction beyond the interview and try to make clear the difference between an interview and therapy.

Analyzing Interview Data

If interviews have been recorded, they are usually transcribed, a time-consuming task. For example, it took Enos more than four hours to transcribe a typical one hour interview (personal communication). Speech recognition software should make transcription less expensive and time consuming in the future.

Interviewing can generate a great deal of text and the process of analyzing such data is typically more inductive than deductive. That is, although the researcher might come to the data with some tentative hypotheses or ideas from another context, the most common approach is to read with an open mind, while looking for motifs.

Here, we concentrate on some of the possible approaches to interview data and save the more technical aspects of data reduction and analysis for Chapter 15. One approach is to take the topics that the participants have talked about and to use them as a sorting scheme to find patterns. Enos, for example, found common pathways to prison (the "runaway path," involvement with drugs, family involvement with crime, etc.) and used these to categorize the women in her study.

A different approach is to construct life histories, profiles, or "types" that typify the patterns or "totalities" represented in the sample. Narratives such as life stories can be read as a whole (Atkinson, 2001) or general phenomena can be described by focusing on their embodiment in specific life stories (Chase, 1995: 2). Some caution that a "battle of representation" can arise because the researchers ultimately write in their own voices, but also present long and edited narratives drawn from informants (Fine, Weiss, Wesson, and Wong, 2000). There is also a reliability concern in categorizing data. One study used open-ended interview questions with six analysts reading verbatim transcripts of three interviews and found only moderate agreement among the analysts on the issues identified by the interviewees (Moret, Reuzel, van der Wilt, and Grin, 2007).

Cases can also be examined one at a time to see if there is a hypothesis that fits, and then the "deviant cases"—those that don't fit the hypothesis—can be examined more closely (Runcie, 1980: 186). Enos originally interviewed to hear people's stories, but as she heard them, she saw variables emerge. She was then able to see patterns in relationships among variables and then look for the exceptions (personal communication). By suspending judgment and focusing on the conditions under which the patterns did not hold, she was able to look for other explanations and consider additional independent variables.

If a description fits a series of respondents, the investigator can propose a more general statement as a theory. This is similar to what Glaser and Strauss (1967) call theoretical saturation, where the researcher determines that he or she is getting an accurate picture when successive interviewees repeat similar things.

The qualitative interview is an important and useful tool for researchers interested in understanding the world as others see it. The less structured interviews can be especially useful for exploratory and descriptive work. They allow the researcher to develop insights into other people's worlds and lend themselves to working inductively toward theoretical understandings.

SUMMARY

The qualitative interviewer typically uses either a list of topics or a set of open-ended questions, modified and adapted for each interviewee to allow participants to "tell their stories." The flexibility of the qualitative approach lends itself to good response rates, complex topics, interviewing for more lengthy sessions, and greater rapport between the participants. However, because it is an expensive and time-consuming method, the qualitative interview is most frequently used with relatively small samples.

There are choices to be made when using qualitative interviews. Among the most important are the amount of structure, whether to add visual materials, selecting a one-to-one or group interview, whether to interview or moderate, whether to "match" interviewer and interviewee, and the extent to which interviewers can share opinions and information with interviewees.

After the interviewing is completed, the creative process of looking for patterns and themes in the transcribed accounts begins. Data analysis can be a rewarding task, but is often time consuming. Qualitative interviews can't provide comparable information about each member of a large sample as easily as more structured methods can, but they can be important sources of insight into people's realities and lives.

EXERCISE 10.1

Doing an Unstructured Interview

1. Pick an occupation about which you know something, but not a great deal (police officer, waiter, nurse, veterinarian, letter carrier, high school principal, or any other occupation). Find someone who is currently employed in that occupation and is willing to be interviewed about her or his work.

2. Following the ethical guidelines discussed in Chapter 3, conduct an unstructured interview of about 30 minutes in length, finding out how the person trained or prepared for her or his work, the kinds of activities the person does on the job, approximately how much time is spent on each activity, which activities are enjoyed and which are not,

what the person's satisfactions and dissatisfactions with the job are, and whether he or she would like to continue doing this work for the next 10 years.

3. Transcribe at least 15 minutes of the interview, including your questions and comments and the interviewee's replies. Include the transcription with your exercise.

4. Consider this a case study and write a brief account of the person and his or her occupation based on your interview. If you can, draw a tentative conclusion or construct a hypothesis that could be tested if you were to do additional interviewing. Include a paragraph describing your reactions to using this method of data collection.

EXERCISE 10.2

Constructing Questions

1. Construct a semi-structured interview guide for a research project on work in an occupation of your choice.
2. Following the ethical guidelines discussed in Chapter 3, write an introductory statement that describes the research project and write at least 12 open-ended interview questions. Include both basic and follow-up questions. Focus on the same kinds of information called for in Exercise 10.1, including how the person trained or prepared for her or his work, the kinds of activities the person does on the job, the amount of time spent on each activity, which activities are enjoyed and which are not, what the person's satisfactions and dissatisfactions with the job are, and so on. Do a pilot test of your questions and then make modifications as appropriate. Turn in the introductory statement and the final version of the interview questions.

EXERCISE 10.3

Evaluating Methods

In her research with incarcerated mothers, Sandra Enos used an in-person semi-structured interview. Two other self-report methods are the in-person structured interview and the mailed questionnaire (both discussed in Chapter 9). Compare these other two methods to the method Enos used for her study.

1. Discuss the advantages and the disadvantages for Enos's study had she chosen to use a *mailed questionnaire* given her topic and the population she was interested in.
2. Discuss the advantages and the disadvantages for Enos's study had she chosen to use an *in-person structured interview* given her topic and the population she was interested in.

EXERCISE 10.4

Writing Part of Your Life Story

One kind of unstructured interview that we've discussed in this chapter is the life history interview. The Life Story Commons at the University of Southern Maine archives life stories of people of all ages and backgrounds. To get a sense of life stories, read a few at the archives on the website at http://usm.maine.edu/olli/national/lifestorycenter/.

Then be both interviewer and interviewee to write a part of your life story by answering these questions, which we've adapted from *The Life Story Interview* by Robert Atkins.

1. What was going on in your family, your community, and the world at the time of your birth?
2. Are there any family stories told about you as a baby?
3. Are there any stories of family members or ancestors who immigrated to this country?
4. Was there a noticeable cultural flavor to the home you grew up in?
5. What was growing up in your house or neighborhood like?
6. What family or cultural celebrations, traditions, or rituals were important in your life?
7. Was your family different from other families in your neighborhood?
8. What cultural values were passed on to you, and by whom?

EXERCISE 10.5

Photo-Interviewing

Find someone who has at least one pet and who has one or more photographs of the pet(s). Following the ethical guidelines discussed in Chapter 3, get permission to interview this person for about 15 minutes about his or her life with the pet(s). Ask the interviewee to bring one or more photographs to the interview and use the photograph(s) as part of an unstructured interview as you ask the interviewee to tell you about his or her pet(s). Focus on the family and social life with the pet(s), how the person handles the responsibilities of having a pet, how children, spouses, and/or neighbors relate to the pet, the quality of the emotional interaction the interviewee has with the pet, and any other issues you would like to cover. Record the interview in some way and transcribe it to attach to this exercise. Consider this a case study and write a brief account of the person and his or her relationship with his or her pet. Discuss whether using photograph(s) was helpful in conducting the interview.

Observational Techniques

CHAPTER 11

© Emily Stier Adler

CHAPTER OBJECTIVES

This chapter is designed to help students

- **Distinguish observational methods from other sorts of data collection**
- **Appreciate the reasons for doing observations**
- **Learn about the advantages and disadvantages of different observer roles**
- **Prepare for doing an observational study**
- **Do a small-scale observational study**
- **Analyze data from an observational study**
- **Get a feel for the advantages and disadvantages of observational methods**

CHAPTER OUTLINE

Are you a fan of any sports team? Have you ever seen someone you just knew was a fan because of the clothing she or he was wearing? Did you ever consider that fans of a particular sports team seem to be religious in their devotion to that team? The author of this chapter's focal research piece, Marci Cottingham, wasn't actually a fan herself of the Pittsburgh Steelers when she began working on her Master's thesis in sociology at Indiana University of Pennsylvania. But she'd grown up in western Pennsylvania and noticed that grocery stores sold Steelers paraphernalia and that an acquaintance wore a Steelers tattoo. She'd majored in philosophy and religion as an undergraduate and it seemed to her that there was something like religious fervor associated with these behaviors, so she decided to study the phenomenon. She didn't have a very good idea what she'd find through this research: it was a good example of exploratory research. This was also true for Chiu (2009) when he examined the reasons why some skateboarders in New York City persisted in skating on the streets even after the city had provided skate parks and attempted to discourage, through regulation, street skating; and for Katherine Frank (2005) when she wanted to find out about the motivations of men who go to strip clubs. All these authors practiced a brand of research that employ what we broadly call **observational techniques**, and under which rubric we include methods that are sometimes called participant and nonparticipant observation. **Participant observation** is performed by observers who take part in the activities of the people they are studying; **nonparticipant observation** is conducted by those who remain as aloof as possible. Cottingham engaged in a kind of participant observation, as she observed what Steelers fans did at games and other venues.

observational techniques, methods of collecting data by observing people, most typically in their natural settings.

participant observation, observation performed by observers who take part in the activities they observe.

nonparticipant observation, observation made by an observer who remains as aloof as possible from those observed.

FOCAL RESEARCH

The Terrible Towel and Fair-Weather Fans: Steelers Nation as a Quasi-Religion

By Marci D. Cottingham*

Introduction

The sociological study of religion stretches back to the discipline's foundation in the work of Karl Marx, Max Weber, and Emile Durkheim (Borowik 2011). Durkheim's (1915) broad conception of collective consciousness and the increasing number of those who do not identify with organized religion in America (Pew Research Center 2008) suggest a re-conceptualization of religion beyond formal, institutional definitions. In expanding definitions of religion, leading sociologists of religion have called for increased consideration of the role of quasi-religious phenomena in order to better understand social meaning and group behavior (Ammerman 2007; Luckmann 1967; McGuire 2008). The lack of attention to quasi-religions is particularly problematic given religion's proposed importance for sociology, as "religion is inscribed into the entirety of the reflection on the essence of society" (Borowik 2011:177). Understanding religion, as a

---

*Article prepared for this book

crystallization of that which a group finds deeply meaningful in all of its forms, including quasi-religions, is fundamental for understanding social life.

To advance our understanding of contemporary quasi-religious phenomena, the current study explores the role of the sacred, ritual, belief, and group meaning among fans of the Pittsburgh Steelers, a U.S. team within the National Football League (NFL). Participant observation data are used to assess the types of sacred symbols, rituals, beliefs, and overall group meaning implicated in the fan identity. Rituals and their emotional outcomes, necessary components of a quasi-religion, are framed using the microsociological approach of Interaction Ritual (IR) theory (Collins 2004; Cottingham 2012). I also examine the beliefs, emotions, and meanings that Steelers fans hold in an effort to better understand sports fandom as a increasingly popular facet of society. Doing so, this study advances our understanding of religion outside of narrow, "official" definitions and illuminates the contours of a specific fan base.

Previous Research

Religion vs. Quasi-Religion

Quasi-religion sits at the margin of formalized religion, as it embodies the functional qualities of formalized religion irrespective of the substantive elements of deities, formal theology, and historically religious organizations and institutions. Demarcating the conceptual parameters of quasi-religion is as challenging and controversial as defining religion itself, with scholars noting the political role that definitions play (Ammerman 2007; Herbrechtsmeier 1993; McGuire 2008). Scholars of alternative and new religious movements often contextualize these phenomena as extensions of particular formal religions such as Christianity, Judaism, and Buddhism (Gallagher and Aschcraft 2006), with little attention to alternative or new movements that originate outside of formal religions.

My conceptualization of quasi-religion is largely in keeping with Greil and Robbins' (1994) definition of para-religion: "apparently secular phenomena which nonetheless appear to have certain religious features" (1). Clear distinctions between the sacred and profane are foundational to categorizing phenomena as religious (Bailey 1998). The sacred may be seen as "special, even unique; in value it is important, even all-demanding; in consciousness, it is fundamental" (Bailey 1998). In contrast, the profane is that which is common, mundane, and ordinary. This dichotomous conceptualization of religious phenomena has been critiqued by scholars (Ammerman 2007; McGuire 2008), but it is particularly problematic for making sense of contemporary quasi-religions—many of which are under development and in process rather than possessing firm, unimpeachable boundaries. Increasingly, the sacred and secular (or profane) are blurred and difficult to distinguish (Greil and Robbins 1994). Given the difficulty of this distinction, one goal of this paper is to explore the notion of "sacralization" as evidenced among fans, a process by which profane objects, behaviors, and beliefs take on increased uniqueness, importance, and meaningfulness to group members.

Counter to a substantive approach to religion (Stark and Finke 2000), scholars taking a functional approach (Durkheim 1915; Luckmann 1967) have defined religion based on three key elements: sacred objects, ritual behaviors, and beliefs. Durkheim provides the classic voice on these three elements, saying that "sacred things are those which the interdictions protect and isolate," "religious beliefs are representations which express the nature of the sacred things and the

(continues)

FOCAL RESEARCH *(continued)*

relations which they sustain" and rituals are "the rules of conduct which pre-scribe how a man should comport himself in the presence of these sacred objects" (1915:56)

Sacred objects may be treated with special care and seriousness in contrast to profane, everyday use. Behaviors of group members may become collectively coordinated and ritualized, while group beliefs and cognitions take on an ideolog-ical or dogmatic status as the undergo sacralization. Through observation of fan behaviors, including treatment of objects and conversations about beliefs and convictions, we can assess the level of sacralization that these key elements have undergone for the group.

Sports Fandom as Quasi-Religious

Approaching sports fandom as quasi-religious is not new. Scholars from a variety of disciplines have considered the many ways that fandom mirrors religious involvement, as fans congregate together, engage in coordinated rituals, and experience intense communal emotions (Birrell 1981; Burstyn 2005; Frey & Eitzen 1991; Hunt 2003; Price 1984). Dunning, Murphy, and Williams (1986) note that fandom "has come to form one of the principal media of collective identifi-cation in modern society and one of the principal sources of meaning in life for many people" (p. 222). Burstyn notes how the physical struggle of competing foes resonates with fans as "these struggles provoke an intense and meaningful set of associations for their initiates: memories, fantasies, and identification" (2005:11). It seems clear from prior research that sport can lead to particularly significant identities and meanings for fans of varying types of sport.

What is missing from this literature is an understanding of the forms and interrelationships among sacred objects, rituals, and beliefs for a particular subcul-ture of fans. This deficiency in our knowledge has hampered the ability of scholars of religion to meaningfully incorporate sports fandom and quasi-religion into a broader understanding of religion. Turning to the specific case of Pittsburgh Stee-lers fans allows for a detailed look at the forms that sacred objects, rituals, and beliefs take as well as how these elements operate to create meaning for fans.

Research Questions

Based on gaps in prior work on sports fans and quasi-religions, the current study is guided by the following research questions:

1. What are important objects, behaviors, and beliefs expressed and used by Steelers fans?
2. How are objects made sacred and what beliefs do fans hold about these objects?
3. Are behaviors ritualized and what beliefs do fans hold concerning these behaviors?

Given the emphasis that these questions place on the use and expression of objects, behaviors, and beliefs by fans and the interrelated processes among these elements, naturalistic observation of fans was an appropriate methodological choice. This method allows for an understanding of fans as they typically behave in a natural setting while also incorporating informal conversations with fans.

Methodology

Following a Goffmanian approach that advances social research by attending to marginal cases (Goffman, 1961; 1967; see also Drew & Wooten, 1988), I chose fans of the Pittsburgh Steelers in an effort to generate analytically rich data on an exceptional fan base. Participant observation of a particular case is a common method of searchers seeking an in-depth understanding of how a phenomenon operates in the social world (Yin 2003). The Steelers are an NFL team based in Pittsburgh, Pennsylvania, with a following of devoted fans. The season the study took place, ESPN, an American sports news network, ranked Steelers fans as number one in the NFL based on their loyalty, creativity, vocal support, travel to away games and the number of consecutive, sold-out games (Mosley 2008).

The extent to which findings are generalizable to other fan bases is not clear. It is possible that the rituals described here characterize fans of other teams and even fans of other sports, but a case study's findings are always of questionable generalizability. Comparisons with other fans are beyond the scope of this analysis. Given the study's exploratory nature and the lack of prior theory on the development and maintenance of quasi-religions, the analysis extends prior work on the sacred, profane, rituals, and beliefs by analyzing emergent relationships among these concepts.

Pittsburgh Steelers fans may gather in a variety of public settings, but I limited systematic observation and participation to Heinz Field (the Steelers home stadium), parking lots adjacent to the stadium, and two sports bars in the city on two game days during the 2008 to 2009 season (approximately 20 hours). Tailgating occurs before and occasionally after a game as fans congregate in the surrounding stadium parking lots to prepare food, eat, drink, converse, listen to music, sing, and play games. Despite growing up in the area, I myself am not a sports fan. My informant, a long-time Steelers fan and Pittsburgh resident, introduced me to a group of 8 to 10 regular tailgaters and traveled with me to local bars as I observed fans while they watched televised games. As the researcher, I operate as the instrument of data collection, including data from my observations, conversations, and my own emotional experiences. I took notes using a note-taking application on my cell phone and transferred these notes to my laptop for analysis. I analyzed the data with attention to the presence and role of sacred objects, ritualistic behavior, cognitions that signified important beliefs, and my own and others' emotional experiences and expressions.

Results

Participant observation while tailgating, watching a game at the stadium, and at two local sports bars during games revealed a broad range of important objects, fan behaviors, and cognitions and emotions. In discussing these elements of the quasi-religion of Steelers fandom, I frame them within a process of sacralization, where the profane and sacred are not dichotomously opposed but operate along a dynamic continuum.

Sacred Objects

A variety of objects were relevant to the subculture, as team colors (black and gold) and the Steelers logo of the Steelmark, a circle encompassing three four-pointed stars with the word "Steelers," demarcated ordinary objects as set apart and special rather than ordinary and mundane. Practical items that were thus marked included hats, scarves, gloves, sweatshirts, jackets, suspenders,

(continues)

FOCAL RESEARCH *(continued)*

tablecloths, plates, umbrellas, and beer cozies (also referred to as "beer sleeves"). Novelty items were also common. These included Steelers flags and a variety of novelty headgear, including Steelers-themed Santa hats, hard hats, Viking horns, hats with fake black and gold hair, pharaoh headdresses, and vintage-style leather football helmets. Santa Claus and Ms. Claus also made an appearance, wearing a gold suit and dress, respectively, with black fur trim.

Decorating the body with Steelers merchandise could also extend to the skin, as several fans wore painted logos on their faces and local vendors offered the service. A friend of my informant showed off his Steelers tattoo, further exemplifying the length to which some fans go to mark the body with their devotion to the team. Several items conveyed the fans' working-class sensibilities as one man wore a traditional steel workers outfit, with gold overalls a black turtleneck, and black boots. Other manifestations of these sensibilities included inexpensive beer, liquor, and foods.

A particularly prominent and important object of Steelers fandom is the Terrible Towel. The towel is yellow with black writing, approximately 17 x 24 inches. It was originally created by Myron Cope, long-time Pittsburgh radio sports broadcaster, as a towel for wiping off wet stadium seats. Since its original use it has become closely identified with Cope's legacy and the charities he supported with proceeds of its sale. The towel has several uses for fans, including as a decoration in homes and at work as well as for waving throughout games. Following Cope's death, a ceremony was held in his at honor at Heinz Field, where the service finished with a minute of silent "towel-waving" (Balingit 2008).

Players, too, have become sacred objects for Steelers fans. Ben Roethlisberger, the Steelers quarterback, for instance, is revered and copies of his jerseys, sacred objects themselves, are fan favorites.

Rituals

While detailing the types of potentially sacred objects is an important task for understanding the quasi-religious nature of Steelers fans, it is through the use of collective and individual rituals that the sacralization process moves ordinary objects toward sacred status. Echoing the theoretical work of Collins (2004) on interaction ritual chains, participant observation of Steelers fans illustrated ritual ingredients and outcomes. Orchestrated rituals of chanting, towel waving, clapping, screaming, and gesturing were much more pronounced during the game in the stadium than while tailgating or watching the game at a sports bar. Entering Heinz Field was analogous to entering a religious sanctuary. Once fans entered, their rituals are more pronounced, perhaps because of their physical assembly, perhaps because the barriers (in the form of a ticket) to the stadium are greater than the ones to parking lots or sports bars, making the field seem more hallowed. The game itself dictated fan behavior, as fans booed, shook their heads, cursed, and looked away from the field during plays in which the team performed poorly and clapped, cheered, smiled, and high-fived when the team performed well.

From my field notes, a specific example demonstrated a fan's level of emotional energy, his level of personal excitement, confidence, and initiative, that results from successful rituals:

> Late in the final quarter of the game the Steelers made a spectacular play, which resulted in much cheering, screaming, clapping, and waving. A man with a stern face in front of me began to remove layers of clothing. Finally, he pulled his final shirt up

over his head and stood cheering and screaming without a shirt in temperatures around 15 degrees Fahrenheit. After a second stellar play by the Steelers, the man beside him also took off his many layers of coats and shirts and the two clutched hands and screamed in earnest.

Successful rituals lead to not only individual emotional energy as described above, but also collective effervescence: "a momentary state," but one that has lasting impact when it is "embodied in sentiments of group solidarity, symbols or sacred objects, and individual emotional energy" (Collins 2004:36). Thus, rituals play a critical role in the solidarity that group members feel and continuation of group membership and meaning outside of the stadium.

The act of tailgating itself is a ritual, one that fans enact throughout the season. One of the oldest tailgaters (probably in his late sixties) claimed he'd be tailgating for almost 18 years. On a cold December morning, I observed a group setting up in what struck me as a particularly ritualized fashion:

> One man was in charge of starting the grill; another set up the table (complete with a Steelers-themed table cloth); and others set out food and beer [...] the group beside us had an elaborate buffet-style arrangement with about 5-6 burners placed under aluminum trays, a heat blower for warmth, and a keg.

While this process was less synchronized than the cheering and clapping at games, it entailed a silent, agreed-upon division of labor that echoes certain kinds of rituals that might occur inside religious sanctuaries.

Beliefs and Emotions

While religious beliefs are considered a vital element of religious phenomena to Durkheim, observations of Steelers fans suggest that members of this quasi-religion place less emphasis on explicit, formal beliefs and more emphasis on rituals and appropriate use of sacred objects. However, cognitions were not completely absent. Fans often relied on knowledge of trivia and the details of the play-offs to demonstrate their devotion and status as "hardcore" fans. In one example, another fan confronts my informant for his response to the Steelers quarterback's poor performance:

> My informant, after watching Ben Roethlisberger throw the ball away (into the crowd, rather than to a player), shouted an obscenity and expressed frustration at the quarterback's poor play. A middle-aged man in front of us turned around and asked him if he realized how hard it was to throw a football in freezing weather and under those circumstances.

Defending the team's sacred object, Ben Roethlisberger, the interaction captures the sense of morality that fans develop in their devotion to the team.

My informant and others in the tailgating group were animated with moral disdain for those they dubbed "fair-weather" fans, noting that such fans were not as devoted or knowledgeable about the team and the playoff process. However, there was not a sophisticated set of doctrines to explain the importance of sacred objects and rituals. Fan affiliation and loyalty is worn on one's sleeve (literally), painted on one's face, expressed through a novel head-piece or an indefatigable wave of the Terrible Towel rather than through the expression of "beliefs, religious ideas, moral values (i.e., cognitive aspects of individual religion)" (McGuire 2008:16-17).

(continues)

FOCAL RESEARCH *(continued)*

Discussion and Conclusion

Results of the study provide a detailed look at the objects, behaviors, and beliefs of the community of Steelers fans. The analysis further suggests a process of sacralization through which elements of fan culture undergo transformation into their sacred counterparts: sacred objects, rituals, and profound beliefs and emotions. Through ritual chains (Collins 2004), ordinary objects become sacred objects, no longer held to the same standards or practical usage. In the case of the Terrible Towel, an item that was originally given to fans so that they might wipe off their wet stadium seats, its profane use to clean is seen as blasphemous. This was particularly evident in the case of a player from an opposing team who taunted the Steelers and their fans by wiping his feet with a Terrible Towel after scoring a touchdown ("Steelers Fans Fuming" 2008), an act that particularly angered fans. Such a heightened response suggests that the towel, and the team it represents, is anything but ordinary.

Furthermore the study reveals an important feature of this quasi-religion: its emphasis on experience and emotion over cognition and formalized beliefs. Fans do not adhere to an official set of beliefs about the team or the game, but rather hold an informal set of tacit beliefs about appropriate behavior and the knowledge base that signal a truly devoted fan. Their emotional response to profane treatment of the Terrible Towel and Ben Roethlisberger suggests a moral order among fans that is neither codified nor explicitly stated. Lacking a set of formal beliefs, however, should not disqualify the group from consideration as a quasi-religion. To do so would be to ignore the profound meaning that fandom affords members and the important role of sport in contemporary American culture.

References

Ammerman, Nancy T. 2007. *Everyday Religion: Observing Modern Religious Lives.* NY: Oxford University Press.

Bailey, Edward I. 1998. "Sacred." In *Encyclopedia of Religion and Society,* edited by W. H. Swatos, Jr. Accessed November 2012 from http://hirr.hartsem.edu/ency/sacred.htm

Balingit, M. 2008. Cope Fans Say 'Bye, Now' in Towel-Twirling Tribute. *Pittsburgh Post-Gazette,* Retrieved March 2009 from http://www.post-gazette.com/pg/08061/861772-66stm.

Birrell, S. 1981. "Sport as ritual: Interpretations from Durkheim to Goffman." *Social Forces, 60,* 354–376.

Borowik, Irena. 2011. "The Changing Meanings of Religion: Sociological Theories of Religion in the Perspective of the Last 100 Years." *International Review of Sociology* 21(1):175–189.

Burstyn, V. 2005. Sport as secular sacrament. In D. S. Eitzen (Ed.), *Sport in contemporary society: An anthology* (pp. 11–20). Boulder, CO: Paradigm Publishers.

Collins, Randall. 2004. *Interaction ritual chains.* Princeton, NJ: Princeton University Press.

Cottingham, Marci. 2012. Interaction ritual theory and sports fans: Emotion, symbols, and solidarity." *Sociology of Sport Journal, 29,* 168–185.

Drew, P., & Wooten, A. (Eds). 1988. *Erving Goffman: Explaining the interaction order.* Cambridge: Polity Press.

Dunning, E., Murphy, P., & Williams, J. 1986. Spectator violence at football matches: Towards a sociological explanation. In N. Elias & E. Dunning (Eds.), *Quest for excitement: Sport and leisure in the civilizing process* (pp. 245–266). New York, NY: Basil Blackwell.

Durkheim, Emile. 1912/1965. *The elementary forms of religious life.* New York, NY: Free Press.

Stark, Rodney and Roger Finke. 2000. *Acts of Faith: Explaining the Human Side of Religion*. Berkeley, CA: University of California Press.

Gallagher, Eugene V. and William M. Ashcraft (eds.). 2006. *Introduction to New and Alternative Religions in America*. Greenwood Press.

Gantz, W., & Wenner, L.A. 1995. Fanship and the television sports viewing experience. *Sociology of Sport Journal, 12,* 56–74.

Goffman, Erving. 1961. *Asylums: Essays on the social situation of mental patients and other inmates*. New York, NY: Archer Books.

Goffman, Erving. 1967. *Interaction ritual*. New York, NY: Doubleday.

Greil, Arthur L. and Thomas Robbins. 1994. "Introduction: Exploring the Boundaries of the Sacred." Pp. 1–23 in *Religion and the Social Order*, edited by A. L. Greil and T. Robbins. Greenwich, CT: JAI Press.

Herbrechtsmeier, William. 1993. "Buddhism and the Definition of Religion: One More Time." *Journal for the Scientific Study of Religion* 32(1):1–18.

Luckmann, Thomas. 1967. *The Invisible Religion: The Problem of Religion in Modern Society*. NY: MacMillan.

McGuire, Meredith B. 2008. *Lived Religion: Faith and Practice in Everyday Life*. NY: Oxford University Press.

Mosley, M. 2008. "NFL's best fans? We gotta hand it to the Steelers (barely)." *ESPN*. Retrieved November 2012, from http://sports.espn.go.com/nfl/preview08/columns/story?id=3530077

Pew Research Center. 2008. *U.S. Religious Landscape Survey*. Accessed December 2012, from http://religions.pewforum.org/pdf/report-religious-landscape-study-full.pdf

Price, Joseph L. 1984. "The Super Bowl as Religious Festival." *Christian Century*, Accessed November 2012, from http://www.religion-online.org/showarticle.asp?title=1375

Yin, R.K. 2003. *Case Study Research: Design and Methods*. Thousand Oaks, CA: Sage.

 THINKING ABOUT ETHICS *Cottingham consulted the Institutional Review Board (IRB) at her university before going into the field. The IRB agreed that, since she planned to observe public behaviors, she didn't need to get written consent from those she observed. The IRB did place two limitations on her proposed research, however: 1) she wasn't permitted to take pictures of those she observed, presumably to protect their identities, and 2) she wasn't permitted to go to post-season games because the IRB wanted her to select all of her dates of observation up front and post-season activity was unpredictable. Cottingham accepted these limitations without appeal and, eventually, was happy she did. They enabled her to get on with her project and limited the amount of data she had to analyze. She reminds us that "[t]he depth and ambiguity of qualitative data can sometimes leave us researchers feeling overwhelmed—'drowning in the data,' so to speak" (personal communication). Because of the IRB's decisions, her data analysis was simpler than it otherwise might have been.*

OBSERVATIONAL TECHNIQUES DEFINED

"Observational techniques" are sometimes referred to as "qualitative methods" and "field research." But both "qualitative methods" and "field research" connote additional methods (e.g., qualitative interviewing and using available data) that we cover elsewhere. Here we focus on "observations" alone. Cottingham refers to available data, in the form, for instance, of a newspaper account ("Steelers' Fans Fuming"), and information from interviews (like her informant's and others' assertions about "fair-weather fans"), but she mainly reports on observations (such as that of two male fans taking their shirts off after particularly exciting plays). In this chapter, we focus on field observation methods although we note that using more than one method is often advisable.

Of course, the data from questionnaires, interviews, content analysis, experiments, and so on *are* all based on observations of one kind or another. For questionnaires and interviews, the observations are about responses to questions. For content analysis, which we'll cover in Chapter 13, the observations are about communication of one sort or another. In experiments, however, observations are frequently of the sort that closely resemble those used by participant and nonparticipant observers. In most cases, experimental observations are of a more controlled sort than those used in the field. **Controlled (or systematic) observations** involve clear decisions about just what kinds of things are to be observed. In their experiment on the effects of exposure to violent media, Bushman and Anderson (2009) observed whether or not students who thought they'd heard someone injured in a fight would offer help and the amount of time it took to offer assistance. But such controlled observations are not the primary focus of this chapter.

Cottingham's immersion in the natural setting of her subjects distinguishes her work from these other approaches. Rather than merely reading about or asking about the behavior of National Football League fans, she observed those behaviors in person. Moreover, she believed that her observations didn't disturb what was going on (personal communication), that she was, in fact, getting an unfiltered view of fan behavior. We here confine our attention to observational techniques, then, that are used to observe actors who are, at least in the perception of the observer, acting naturally in natural settings.

controlled (or systematic) observations, observations that involve clear decisions about what is to be observed.

● **STOP & THINK**   *If Cottingham had used questionnaires instead of observing fans, what kinds of insights might she have missed?*

The origins of modern observational techniques are the subject of some current debate. Conventional histories attribute their founding to anthropologists like Bronislaw Malinowski, with his *Argonauts of the Western Pacific* (1922), and sociologists like Robert Park, who, around the same time, began to inspire a generation of students to study city life as it occurred naturally (see, for example, Wax, 1971, and Hughes, 1960). However, Shulamit Reinharz (1992) has argued that Harriet Alice Fletcher, studying the Omaha Indians in the 1880s, and Harriet Martineau, publishing *Society in America* in the 1830s, promoted essentially the same techniques that Malinowski and Park received fame for in the 1920s, thus suggesting the earliest "social science" observers might have been American and British women.

REASONS FOR DOING OBSERVATIONS

Perhaps the most important reason for using observational techniques is that they offer a relatively unfiltered view of human behavior. You can, of course, ask people about what they've done or would do under given conditions, but the responses you'd get might mislead you for any number of reasons. People do, for instance, occasionally forget, withhold, or embellish. Cottingham knew that people might not be completely forthcoming in interviews or questionnaires about the ways they participated as Steelers fans; moreover, they might not have been able to conceive of their participation as quasi-religious. She believed that only through the observation of fans in relevant natural settings such as a football stadium, parking lots and sports bars could she get a more unfiltered view of behaviors. When she tells us that, in reaction to her informant's criticism of a Roethlisberger incompletion, "[a] middle-aged man in front of us turned around and asked him if he realized how hard it was to throw a football in freezing weather and under those circumstances," we can understand that the felt need to come to the defense of Roethlisberger, a sacred object for fans, is indeed similar, in some ways, to beliefs in traditional religions.

Observation can be useful when you don't know much about the subject under investigation. A classic technique in anthropology, ethnography, or the study of culture, is primarily an observational method that presumes that the culture under study is unknown to or poorly known by the observer. Margaret Mead's classic investigation, *Sex and Temperament* (1933), of gendered behavior among three tribes in New Guinea, studied societies that were not well known to Westerners. Similarly, many well-known sociological investigations have been of things less than well known beforehand. More recently, St-Amant and others (2012) observed how family care givers participated in care decisions with and about family members living with dementia, and Mears and Finlay (2005) examined how fashion models manage bodily capital. Little was known by social scientists about these topics beforehand. Cottingham justifies her investigation of Steelers fans in part because no one had studied the sacred objects, rituals and beliefs of particular fans before.

Observational techniques also make sense when one wants to understand experience from the point of view of those who are living it or from the context in which it is lived. Norman Denzin makes a distinction between **thin description** and **thick description**, the former merely reporting on an act, the latter providing a sense of "intentions, motives, meanings, context, situations, and circumstances of action" (1989: 39). Cottingham gives us examples of both thin and thick description. When she tells us of the fans who take off their shirts, she is providing thin description. She, and we, can make inferences about their enthusiasm, but, because she doesn't ask them what they intended, we can't know fully what they meant by these actions. When she reports on her informant's disdain for "fair-weather fans," she is able, because she's talked with the informant, to get more directly at meanings. This is what other observers have done as well when they permit themselves to ask questions of others. In an example of how visual sociology can generate observations and thick description of those observations, Radley, Hodgetts, and Cullen (2005) studied how homeless people in London understand their lives in shelters and on the streets. They gave homeless

thin description, bare-bones description of acts.

thick description, reports about behavior that provide a sense of things like the intentions, motives, and meanings behind the behavior.

participants cameras to photograph typical scenes from their days. Afterward, Radley et al. interviewed the homeless about the photographs they'd taken. Different participants used the opportunity to show how they'd adapted to their homelessness in different ways. For instance, one woman photographed the public places she slept in and stressed how she could make a "normal" life despite being a "rough sleeper," while another woman photographed friends and other homeless people, using them to show how she used her relationships with others to adapt to her life without permanent shelter (Radley et al., 2005: 280).

Observational techniques are also useful when you want to study quickly changing social situations. Kai Erikson largely based his account in *Everything in Its Path* (1976) of the trauma visited on an Appalachian community by a disastrous flood in 1972 on descriptions gleaned from legal transcripts and personal interviews. He had been asked by a law firm to assess the nature and extent of the damage to the Buffalo Creek community by a flood that had been caused, in part, by the neglect of a coal company. His book presents the legal case he made then, using content analyses and unstructured interviews. When he presented the effects of the favorable legal judgment on the community he studied, he described a gathering "in a local school auditorium to hear a few announcements and to mark the end of the long ordeal." He interprets the gathering as "a kind of graduation … that ended a period of uncertainty, vindicated a decision to enter litigation, and furnished people with sufficient funds to realize whatever plans they were ready to make. But it was also a graduation in the sense that it propelled people into the future at the very moment it was placing a final seal on a portion of the past" (1976: 248–249). Erikson then used direct observation to mark this momentous occasion: the moment (two years after the flood) when Buffalo Creek began, as a community, to heal itself after its calamitous ordeal. If he hadn't been there when he was, this moment of significant change would necessarily have gone unnoticed by himself and his readers.

There are other reasons why one might prefer observational techniques. Questionnaire surveys can be very expensive, but participant or nonparticipant observation, although time consuming, can cost as little as the price of paper and a pencil. Cottingham tells us that the only monetary costs of her data collection were the price of gas to get her to and from the tailgate parties and the sports bars—the tickets to the Steelers game were a gift (personal communication). Ethically, studying homeless people in their natural settings, as Liebow (1993) did, raises fewer questions than would studying, say, the effects of homelessness by creating an experimental condition of homelessness for a group of subjects. As you can imagine, the reasons for doing observation can have complex interactions with one another, with the salience of some undermining the feasibility of others. Rebekah Nathan (2005), a college professor, became a college student for a year to gain a deep, unfiltered understanding of what it means to be a college student today, partly because she didn't think she'd get the same kind of high-quality data from simply asking her students. (Rebekah Nathan is a pseudonym that the anthropologist adopted for her study.) Nathan didn't want to accept grants from funding agencies to pay for her tuition, books, housing, and so on because she feared that would give the agency too much influence over what she could report. So she paid for these things out-of-pocket, making the research for *My Freshman Year: What a Professor Learned by Becoming a Student* the most expensive one she'd ever done.

OBSERVER ROLES

complete participant role, being, or pretending to be, a genuine participant in a situation one observes.

participant-as-observer role, being primarily a participant, while admitting an observer status.

observer-as-participant role, being primarily a self-professed observer, while occasionally participating in the situation.

complete observer role, being an observer of a situation without becoming part of it.

We suggested earlier that one of the defining characteristics of observational techniques is their relative unobtrusiveness. But this unobtrusiveness varies with the role played by the observers and with the degree to which they are open about their research purposes. Raymond Gold's (1958) typology of researcher roles suggests a continuum of possible roles: the **complete participant**, the **participant-as-observer**, the **observer-as-participant**, and the **complete observer**. It also suggests a continuum in the degree to which the researcher is open about his or her research purpose. As a complete participant, the researcher might become a genuine participant in the setting under study or might simply pretend to be a genuine participant. Mueller, Dirks, and Houts Picca (2011) were interested in learning about college students who dressed as people of other races for Halloween. Since they were professors and wouldn't have fit in with students during Halloween, they asked their own students to observe other students as a way of getting extra credit in their sociology courses. These student observers were obliged to remain complete participants as they made their observations about other students' behavior in relation to Halloween. They were explicitly instructed to "use unobtrusive research techniques so that the person(s) [they wrote] about in [their] journal[s] will not be aware that they are being studied" (from the instructions given to students, personal communication). They were also instructed NOT to interview the people they observed and to "interact with people as [they] usually would." One ethical dilemma associated with complete participant observation is informed consent. Obviously, telling the people whom the students observed in the act of cross-racial costuming that they were being observed for a possible research report might have changed their behavior, and perhaps especially what they would have said about the ways they dressed or proposed to dress. The American Sociological Association (ASA) Code of Ethics provides guidance for dealing with the issue of informed consent while conducting observational research (see Box 11.1). In general, researchers who would observe people in "private" places owe observed individuals informed consent unless "no more than minimal harm" is likely to befall such individuals and the research becomes impractical under the conditions of informed consent and an IRB confirms that both of these conditions obtain. Mueller, Dirks, and Houts Picca did receive IRB agreement that no more than minimal harm could befall participants and that the research would have been impracticable if the observed students knew they were being observed.

The complete participant, in Gold's typology, is defined by withholding the fact that the observer is also a researcher. Thus, all Mueller, Dirks, and Houts

BOX 11.1   Excerpts from the ASA Code of Ethics on Informed Consent in Observational Research

12.01 Scope of Informed Consent

(a) Sociologists conducting research obtain informed consent from research participants or their legally authorized representatives ... when behavior of research participants occurs in a private context where an individual can reasonably expect that no observation or reporting is taking place.

(b) Despite the paramount importance of consent, sociologists may seek waivers of this standard when (1) the research involves no more than minimal risk for research participants, and (2) the research could not be practicably be carried out were informed consent to be required. Sociologists recognize that waivers of consent require approval from institutional review boards or, in the absence of such boards, from another authoritative body with expertise on the ethics of research...

(c) Sociologists may conduct research in public places ... without obtaining consent. If, under such circumstances, sociologists have any doubt whatsoever about the need for informed consent, they consult institutional review boards...

Source: ASA Code of Ethics, p. 14, www.asanet.org/galleries/default-file/CodeofEthics.pdf

Picca's student observers collected data, while participating in dormitory life, school work, and Halloween-costuming preparations, telling no one that they were doing research.

STOP & THINK   *Do you agree that not informing observed students in Mueller, Dirks, and Houts Picca's study was ethical?*

The opposite of the complete participant role is that of the complete observer: one who observes a situation without becoming a part of it. In principle, complete observers might ask for permission to observe but they usually don't. Harrell and his colleagues (cited by Bakalar, 2005) made more than 400 anonymous observations of children and their parents in 14 supermarkets to see if more attractive children were treated better by their parents than others, coding not only the attractiveness of the children but also whether the parents belted their youngsters into grocery cart seats, the frequency with which parents lost track of their children, and the number of times children were permitted to do dangerous things like standing up in the cart. Harrell and others remained socially invisible to those whom they observed. They didn't tell those they watched that they were observing them. A rather more expensive version of Harrell's basic strategy was employed by observers trained at the National Opinion Research Center (NORC) in preparation for, among others, the work of Robert Sampson and Stephen Raudenbush (1999) on disorder in urban neighborhoods. These observers were driven in a sport utility vehicle at a rate of five miles per hour down every street in 196 Chicago census tracts, videotaping from both sides of the vehicle while speaking into the videotape audio. Martyn Hammersley and Paul Atkinson (1995: 96) note the paradoxical fact that complete observation shares many of the disadvantages and advantages of complete participation. Neither is terribly obtrusive: Harrell, the NORC researchers, and Mueller, Dirks, and Houts Picca's students were not likely to affect the behavior of those they watched. On the other hand, neither complete participation nor complete observation permits the kind of probing questions that one can ask if one admits to doing research. In addition to the issue of (not being able to ask scientifically interesting) questions, the lack of openness (about one's research purposes) implicit in pure participation or pure observation has raised, for some, the ethical issue of whether deceit is ever an appropriate means to a scientific end—an issue, you will recall, that also has come up in our discussion of other methods. Like complete participants, complete observers can find guidance in the ASA Code of Ethics (see Box 11.1 again), which suggests that, most of the time, observations in public places do not require the informed consent of those observed. The same applies in more private places, as long as a) no more than minimal harm is likely to befall those observed, b) the research becomes undoable when informed consent is requested, and c) an IRB agrees that these conditions exist.

Both the participant-as-observer and the observer-as-participant make it clear that they are doing research; the difference is the degree to which they emphasize participation or observation. The participant-as-observer is a participant in the setting who admits he or she is also an observer. The observer-as-participant is a self-professed observer who occasionally participates in the situation. Cottingham's study of Steelers fans is primarily an example of research conducted by a participant-as-observer. She sat at the Steelers game without announcing her observer status to most fans at the game, but she clearly announced her observer status to her informant and others at other times. Otherwise, she wouldn't have been able to come to confident conclusions, for instance, about sacred beliefs. At one point, she says, for instance, "Fans often relied on knowledge of trivia and the details of the play-offs to demonstrate their devotion and status as 'hardcore' fans," an insight that came through conversation with others. In personal communications, she confides that speaking casually to other fans wasn't always as useful as talking to her informant, another sociology graduate student, who really "got" what she was interested in. She valued that "[her] informant was a hard-core fan who could also step back and look at the situation as an outsider. I made one assessment to him that tailgating seemed a lot like 'urban camping' and the reclamation of urban/public space and he readily agreed. Being able to bounce my observations off him for confirmation or clarification was critical to my analysis of the data" (personal communication).

Elliot Liebow, in his study of homeless women, is more explicit about his alternation between being primarily participant and primarily observer in a women's shelter called the Refuge and defines the time when he was primarily observing as his research time:

> I remained a volunteer at the Refuge [throughout the study], and was careful to make a distinction between the one night [a week] I was an official volunteer and the other nights when I was Doing Research (that is, hanging around). (1993: x)

Thus, Liebow, as researcher, would fall into Gold's observer-as-participant role.

Liebow clearly announced his interest and even records the announcement of his purpose, and his request for permission:

> "Listen," I said at the dinner table one evening, after getting permission to do a study from the shelter director. "I want your permission to take notes. I want to go home at night and write down what I can remember about the things you say and do. Maybe I'll write a book about homeless women." (1993: ix)

And the book he did write, *Tell Them Who I Am*, has become a classic study of women's homelessness, and especially of the desperation and hope with which the women led their lives (Liebow, 1993: 1). He watched, for instance, as Elsie, one of the women who lived at the Refuge, fasted for three days, because, as she reported, her pastor had told his parishioners to do so if they "wanted a special favor from God." And Elsie, she told Liebow, wanted a job "sooo bad" (Liebow, 1993: 73).

One thing Liebow's, and even Cottingham's, study makes clear is that within any given research project, researchers can play several different observational roles. There were times when they were primarily participants and times when they were primarily observers.

One danger of both the participant-as-observer and the observer-as-participant roles is that they are more obtrusive than those of the pure participant or the pure observer and that those being studied might shift their focus to the research project from the activity being studied, much as they might in responding to a questionnaire or an experimental situation (see Box 11.2). Those being studied might have been aware of being such projects, but the verisimilitude and believability of what Cottingham and Liebow report tends to dispel one's doubts about the overall validity of their reports.

The danger of obtrusiveness and the value of thick description aside, there are times when announcing one's research intention offers the only real chance of access to particular sites. The only other way participant observers Becker, Geer, Hughes, and Strauss could have gained access to the student culture in medical school for their classic study *Boys in White* (1961) would have been to apply and gain admission to medical school themselves. Think of all the science courses they would have had to take!

Different research purposes require different roles. There are no hard and fast rules about what observer role to adopt in a given situation. As in all research, you should be guided by methodological, practical, and ethical considerations.

BOX 11.2   **The Continuum of Observer Roles: The Chances of Being Obtrusive and Asking Probing Questions Vary by Role**

Complete Participant	Participant-As-Observer and Observer-As-Participant	Complete Observer

− ⎯⎯⎯⎯⎯⎯⎯→ + Chances of Being Obtrusive + ←⎯⎯⎯⎯⎯ −

− ⎯⎯⎯⎯⎯⎯⎯→ + Chances of Asking Probing + ←⎯⎯⎯⎯⎯ −
Questions

GETTING READY FOR OBSERVATION

We've seen how researchers using other techniques (e.g., experiments or surveys) can spend much time and energy preparing for data gathering, in formulating research hypotheses, and developing measurement and sampling strategies, for example. Such lengthy preparations are usually not as vital for studies involving observational techniques, where design elements can frequently be worked out on an ad hoc basis. The exception to this rule, however, is controlled, or systematic, observations, which are defined by their use of explicit plans for selecting, recording, and coding data (McCall, 1984). In their controlled observations, Harrell and his colleagues (cited in Bakalar, 2005), as you know, studied parents and children in supermarkets. Because this team began with a more or less well-defined hypothesis (parents will take better care of pretty children than they will of ugly ones), one can imagine them filling out a code sheet like the one we've constructed for you in Figure 11.1.

Far more often, however, observers begin their studies with less clearly defined research questions and considerably more flexible research plans. Adler and Adler (1994b) suggest that an early order of business for all observational studies is selecting a setting, a selection that can be guided by opportunistic or theoretical criteria. The selection is made opportunistically if, as the term suggests, the opportunity for observation arises from the researcher's life's circumstances (Riemer, 1977). Cottingham reports that, initially, she planned only to

Instruction: Observe parent and young (preferably under 3 years of age) for approximately 5 minutes.

Apparent gender of child:

Assessment of child's attractiveness

| 1 | 2 | 3 | 4 | 5 | 6 | 7 | 8 | 9 | 10 |

Unattractive   Attractive

Was child belted into the grocery cart seat?

How many times did parents' attention to child lapse during the period of observation:

Describe these lapses:

Number of times the child was allowed to engage in potentially dangerous activities:

Describe the activities:

FIGURE **11.1** A Potential Codesheet for Recording Observations in the "Care of Pretty/Ugly Children Study" by Harrell and others (cited in Bakalar, 2005)

observe Steelers fans at sports bars and tailgating parties, but that she decided to take in a game when "[m]y informant and I were approached at a downtown sports bar during the beginning of a game with a pair of free tickets (this is a point where I deviated slightly from my IRB protocol—can you blame me?)" (personal communication). This is a clear case of a setting being selected opportunistically. Similarly, Liebow studied shelters for homeless women because, after he'd been diagnosed with cancer and quit his regular job in 1984, he decided to work in soup kitchens and shelters, only to find their clients extraordinarily interesting. Jeffrey Riemer (1977: 474) advocates opportunistic settings for ease of entry and of developing rapport with one's subjects and for the relatively accurate levels of interpretation they can afford.

On the other hand, Cottingham did most of her observations at sports bars and tailgate parties, settings she'd picked because she thought they would be good places to observe Steeler fans' use of objects, practice of rituals, and talk about beliefs. The choice of these settings was theoretically motivated. Similarly, Mueller, Dirks and Houts Picca asked students to observe others as they cross-racially dressed for Halloween primarily for theoretical reasons rather than because it was easy for the students to do so. Mueller reports. All three authors are "heavily influenced by 'macro' structural theories of privilege/ oppression," but they wanted to see how racism was enacted in the everyday lives of students. They felt it was appropriate to ask students to report on racialized performances surrounding Halloween (personal communication).

STOP & THINK *Do you remember Laud Humphreys's (1975) study of brief, impersonal, homosexual acts in public restrooms (mentioned in Chapter 3)? Do you imagine that Humphreys's decision to study public restrooms was informed primarily opportunistically or purposively? What makes you think so? (Humphreys's answer appears in the next "Stop & Think" paragraph.)*

The suggestion that setting selection can be made on either opportunistic or theoretical grounds might be slightly simplistic, however, insofar as it implies that one chooses only one site, once and for all, and that it is primarily opportunity or theory that informs the choice. Actually, Adler and Adler (1994a), in the study of adult-organized afterschool activities for children, were not very unusual in their decision, initially, to immerse themselves in settings that were available (their own children's afterschool activities), but then to seek out interviews that spoke to ideas they developed in the process:

> [W]e derived our ideas about what to write about from our observations of and casual interaction with kids, ... but to get quotes from kids themselves and to see how they felt or if their experiences jibed or differed from our perspectives, we conducted interviews.... These interviews were selected by theoretical [or purposive] sampling to get a range of different types of kids in different activities, age ranges, and social groups (personal communication).

Especially in seeking their interviews, then, Adler and Adler were guided by the effort to test their own theoretical notions against those of other participants, an approach they recommend elsewhere (Adler and Adler, 1987). By

theoretical saturation, the point where new interviewees or settings look a lot like interviewees or settings one has observed before.

emphasizing their efforts to "get a range of different types of kids," in fact, they suggest they attempted to approximate the ideal of **theoretical saturation** (Adler and Adler, 1994b; Glaser and Strauss, 1967)—that is, the point where new interviewees or settings looked a lot like interviewees or settings they'd observed before. Opportunity, then, might have inspired their early choices; some mix of opportunity and theory influenced their later choices.

 STOP & THINK

Humphreys says he embarked on his study of sex in public restrooms after he'd completed a research paper on homosexuality. He states that his instructor asked about where the average guy went for homosexual action and that he (Humphreys) articulated a "hunch" that it might be public restrooms. His further comment, "We decided that this area of covert deviance, tangential to the subculture, was one that needed study" (1975: 16), suggests that his decision to study public restrooms was purposively, rather than opportunistically, motivated.

purposive sampling, a nonprobability sampling procedure that involves selecting elements based on the researcher's judgment about which elements will facilitate his or her investigation.

Cottingham's study suggests what seems to be generally true of sampling in studies using observational techniques: Observations are most often done in a nonprobabilistic fashion. She practiced, as most participant observers do, **purposive** (or, as they call it, theoretical) **sampling**—she observed situations that would give her insight into her major questions: places where she could expect to see Steelers fans (e.g., sports bars, tailgate parties, a football game). Devising a sampling frame of the kinds of things studied by participant and nonparticipant observers (e.g., afterschool activities, occupations involving emotion work, homeless women, bathrooms used for homosexual encounters) is often impossible, though again, this might be less true of the (relatively infrequent) studies involving systematic observations. The researchers from NORC did, in fact, select a stratified probability sample of Chicago census tracts "to maximize variation by race/ethnicity and SES" (Sampson and Raudenbush, 1999) for its sports utility van–driven study of public spaces.

Another decision to make before entering the field is how much to tell about yourself and your research. *Disclosure* of your interests (personal and research) can help develop trust in others, but it can also be a distraction from, even a hindrance to the unfolding of, events in the field. What you're willing to disclose about your interests is obviously related to the research role you will play. Researchers who feel that disclosure will seriously endanger the validity of the data they collect will almost surely choose complete participant or complete observer roles. Mueller, Dirks, and Houts Picca's students, complete participants, weren't permitted to disclose the purposes of their observations for fear that other students' behaviors would be affected in a misleading way. Harrell and his colleagues, the ones who observed children and their care in supermarkets, were complete observers and also wanted a genuinely unfiltered view of human behavior. But researchers like Cottingham, a participant-as-observer in the Pittsburgh Steelers sports scene, and Liebow (1993), an observer-as-participant in the Refuge's shelter for homeless women, clearly felt that disclosure of their research purposes would help build the trust necessary for participants to give valid accounts of their behavior and not affect significantly the behavior they observed.

We would like to make two more general recommendations about preparing yourself for the field. First, we think potential observers (and perhaps especially novice observers) should review as much literature in advance of their observations as possible. Such reviews can sensitize the researchers to the kinds of things they might want to look for in the field and suggest new settings for study. And it was such a review that enabled Cottingham to realize that, while some scholars had considered sports fandom to be like a quasi-religion, no one had yet studied a particular fan base (like Steelers fans) in those terms. You should know, however, that our advice on this matter is not universally accepted. In fact, the advice of Jack Douglas (1976) is that researchers immerse themselves in their settings first and avoid looking at whatever literature is available until one's own impressions are, well, less impressionable. Moreover, Adler and Adler suggest that researchers should begin to "compare their typologizations to those in the literature" only once they begin to "grasp the field as members do" (personal communication).

Second, whatever decision you make about when to review the literature on your own topic, we strongly recommend spending some time leafing through earlier examples of participant or nonparticipant observation, just to see what others have done. Particularly rich sources of such examples are the *Journal of Contemporary Ethnography* (which was, in earlier avatars, *Urban Life* and *Urban Life and Culture*) and *Qualitative Sociology*.

Gaining Access and Ethical Concerns

Although choosing topics and selecting sites are important intellectual and personal tasks, gaining access to selected sites is, as Lofland, Snow, Anderson, and Lofland (2005) point out, an important *social* task. At this stage, a researcher must frequently use all the social skills or resources and ethical sensibilities he or she has at his or her command.

Lofland et al. (2005) suggest that the balance between the need for social skills or resources and ethical sensibilities, however, can depend, as much as anything, on whether or not the observer plans to announce his or her presence as observer. If the observer plans not to reveal the intention to observe, the major issues entailed in gaining (or using) access are often ethical ones (see Chapter 3), especially because of the deceit involved in all covert research. This can be a particularly knotty problem when one seeks "deep cover" in a site that is neither public nor well known, as was the case for Patricia Adler (1985) in her study of drug dealers and smugglers, but it exists nonetheless even when one plans to study public places (as Humphreys did while watching impersonal sex in public restrooms) or to study familiar places, as Mueller, Dirks, and Houts Picca's students did when they observed other students engaged in Halloween activities. Seeking the advice and consent of an IRB in such situations is particularly important.

Gaining access to, and then getting along in, a site becomes much more a matter of social skill or resources when the observer intends to announce his or her observer role. The degree to which such skills or resources are taxed, however, is generally a function of one's current participation in the setting. Furthermore, for Adler and Adler (1994a), who were already participants in many afterschool settings, at least early in their study, their pre-study participation not only made access easy, but also made it relatively easy to decide whom they needed to tell about their research, whom they needed to ask permission from, and whom they needed to consult for clarifications about meanings, intentions, contexts, and so on. Ease of access, then, is certainly one major reason why Adler and Adler advocate "starting where you are." Cottingham's informant, another graduate student in sociology, introduced her to two sports bars and to the tailgate scene, and particularly introduced her to 8 to 10 tailgaters with whom she talked.

STOP & THINK   *Suppose you wanted to study a musical rock group (The Easy Aces) on your campus, to which you did not belong. How might you gain access to the group for the explicit purpose of studying the things that keep them together?*

Access particularly taxes social skills or resources when you want to do announced research in a setting outside your usual range of accustomed settings. In such cases, Lofland et al. (2005: 41–47) emphasize the importance of four social skills or resources: connections, knowledge, accounts, and courtesy. If, for instance, you want to study a rock group (The Easy Aces) comprising students on your campus, you might first discuss your project with several of your friends and acquaintances, in the hopes that one of them knew one of the Aces well enough to introduce you. In doing so, you'd be employing a tried and true convention for making contacts—using "who you know" (*connections*) to make contact. In general, it's a good idea to play the role of "learner" when doing observations, but such role-playing can be overdone on first contact, when it's often useful to indicate some *knowledge* of your subject. Consequently, before meeting The Easy Aces, it would be a good idea to learn what you can about them—either from common acquaintances or from available public-access information. (If, for instance, the Aces have a website, you might want to listen to their music and refer to their songs, perhaps even admiringly, at your first meeting.) You'd also want to have a ready, plausible, and appealing explanation (an **account**) of your interest in the group for your first encounter. The account needn't be terribly clever, long-winded, or highbrow (something reasonably direct like "I'd like to get an idea of what keeps a successful rock group together" might do), but it might require some thought in advance (blurting out something like "I'm doing a research project and you folks seem like good subjects" might not do). Finally, and perhaps most common-sensically, a bit of common *courtesy* (such as the aforementioned expressions of appreciation for the Aces' music) can go a long way toward easing access to a setting.

All the aforementioned skills and resources—connections, knowledge, accounts, and courtesy—can be more difficult to muster effectively when there are gender, class, language, or ethnic differences between the researcher and other actors. Convincing **gatekeepers**, or persons who can let you into a setting, can be hard work, as Michelle Wolkomir discovered when she tried to

account, a plausible and appealing explanation of the research that the researcher gives to prospective participants.

gatekeepers, someone who can get a researcher into a setting.

gain access, as a heterosexual Jewish woman, to an ex-gay ministry in a southern, conservative Christian church. She spent five months meeting sporadically
with the group leader and the church board to convince them that she would
not undermine the security of the men using the ministry to "cure" their homosexuality and that she would not inhibit group talk (2001: 309–310). On the
other hand, Cottingham's informant was happy to help her and was pleased
to accompany her to the two sports bars she visited to help her understand
what she observed, and would probably have done so even if she hadn't "compensated him with a free beer" (personal communication).

Still, the eventual success of Wolkomir's and Cottingham's projects, and
that of other female observers in mixed-gender settings, as well as that of
white male observers, like Liebow, who, for *Tally's Corner* (1967), studied
poor, urban, African American males, suggests that even gender, race, class,
and sexual-orientation divides are not necessarily insurmountable.

Participant and nonparticipant observations, perhaps more than any other
techniques we've dealt with in this book, lend themselves poorly to "cookbook" preparations. We can't tell you precisely what to do when you come up
against bureaucratic, legal, or political obstacles to your observations, but we
can tell you that others have encountered them and that authors like Lofland
et al. (2005) provide some general strategies for surmounting them.

Gathering the Data

Perhaps the hardest thing about any kind of observation is recording the
data. How does one do it?

STOP & THINK   *Quick. We're about to list three ways of recording observations. Can you guess
what they are?*

Three conventional techniques for recording observations are writing them
down, recording them (perhaps with a digital camera, a phone, or an iPad),
and recording them in one's memory, to be written down later (only shortly
later, one hopes). Given the notorious untrustworthiness of memory, you
might be surprised to learn that the third technique, trusting to memory for
short periods, has probably been the most common. Despite gaining the
explicit permission of his subjects, Liebow (1993) waited until he got home at
night to write down what he'd observed about homeless women. Liebow, in
particular, was cognizant that even his relatively well-trained powers of observation and memory were subject to later distortion (whose aren't?) and so was
at pains to record his observations as soon after he'd made them as possible.
His goal, the goal of all observers, was to record observations, including quotations, as precisely as possible, as soon as possible, before the inevitably dis-
torting effects of time and distance were amplified unnecessarily.

Why not record them on the spot? Joy Browne, in her reflections on her
study of used car dealers, gives one view when she advises prospective
observers,

> [D]o not take notes; listen instead. A panicky thought, perhaps, but unless you
> are in a classroom where everyone is taking notes, note taking can be as intrusive

as a tape recorder and can have the same result: paranoia or influencing the action. (1976: 77)

The intrusiveness of note-taking, recording, and picture-taking, then, make them all potentially troublesome for recording data.

On the other hand, shorthand notes and recordings are certainly appropriate when they can be taken without substantially interrupting the natural flow. Cottingham's use of a note-taking application on her cell phone strikes us as particularly ingenious, inasmuch as emerging social etiquettes dictate the permissibility of typing on one's phone in public, and often in semi-private, spaces. Frances Maher and Mary Kay Tetreault (1994) taped professors' classes to discern varieties of feminist teaching styles. They later transferred their notes to a computer for analysis. A problem of recording, however, is that one can gather daunting quantities of data: It can take a skilled transcriber three hours to type one hour of recorded conversation (Singleton, Straits, Straits, and McAllister, 1988: 315). And even if you're able to take notes on paper or record, it is still wise to transcribe and elaborate those recordings as soon as possible, while various unrecorded impressions remain fresh enough that you can add them. One technology that already almost circumvents the need to transcribe is the microphone function available on, among other things, iPads. This function permits you (and respondents) to speak into a microphone imbedded in a device that will transcribe the materials automatically. At this writing, these transcriptions can be imperfect, however, and require immediate attention while the researcher's memory can make appropriate corrections.

Visual Sociology

visual sociology, an approach to studying society and culture that employs images as a data source.

Even more data become available, however, when one uses cameras or other recording equipment. For much of the history of sociology, pictures and images were used, when used at all, to illustrate points in an argument (e.g., Grady, 2007b). The field of **visual sociology**, which uses images to study society and culture, really didn't begin, however, until the publication of Howard Becker's "Photography and Sociology" (1974). A visual sociologist doesn't just use images for illustrative purposes, but as a source of data. Perhaps you recall that in Chapter 6 we asked you to analyze the front cover of this book for an idea of what we, Emily and Roger, were trying to tell you about research methods through the pictures. You were then practicing visual analysis, a technique employed by visual sociologists, by using the pictures to decipher what we were trying to convey. You were also clearly practicing an observational technique in doing so. But there are actually several distinct ways in which the images from cameras and video equipment can be used in observational studies.

Three techniques that visual sociologists use are:

1. analyzing visual documents
2. subject image-making
3. photo and video ethnography (see, for instance, Grady 2007c)

In analyzing visual documents, one takes pictures or video of a place one is studying and supplements field observations in this way. Chiu (2009), for instance, provides pictures of the four sites in which he did his fieldwork for the skateboarding study in New York City. An image of one park, for instance, shows an elevated plaza and wide stairs that, Chiu credibly asserts, make the park attractive to "[b]eginners [who] can practice elementary tricks on these low steps and smooth ground" (2009: 28).

In the practice of subject image-making, researchers typically provide participants with cameras to record images that are of interest to them and/or to the researcher. When combined with the process of photo-elicitation (see Chapter 10), in which participants are then interviewed about the images they (or others) have made (or found), subject image-making can make for a particularly rich source of observational data and thick description. Radley et al.'s (2005) study of homeless people, mentioned earlier, is one example. So is Glover, Stewart, and Gladdys's (2008) attempt to discover democratic ways of soliciting the views of various stakeholders in the ways land use might change in a community. Glover et al. gave all participants disposable cameras and asked them to photograph places that were "important within their daily lives" (389). They then asked each person to talk about the photos they'd taken and to expand on what was important about the place they'd photographed. One woman, for instance, took a photo of a farmer's market because, as she discovered through the process of photography, it represented to her "a great social time" (391). Then all participants were asked to bring their photos and their "discoveries" to a civic forum where many of them shared their newly articulated land use values. Glover et al. believe that they've uncovered a democratic way, via subject image-making and photo-elicitation, to help stakeholders articulate and share the ways in which they want land in their communities used in the future.

Visual sociologists also practice photo and video ethnography. Ethnography, of course, is the study of human culture. It's been the defining approach of cultural anthropology since its inception. In a relatively new development for sociologists, however, some researchers have begun to use images not only to augment their note-taking and memory, but to capture emotion and 'feel' for situations. **Video ethnography**, a cutting-edge approach to studying social life, is the video recording of participants and the reviewing of the resulting footage for insights into social life. Tutt (2008) recorded three living-room situations in which people interacted with relatively new (to them) media to get a feel for the emotional content of such interactions.

video ethnography, the video recording of participants and the reviewing of the resulting footage for insights into social life.

Participant and nonparticipant observers commonly supplement their observations with interviews and available data. Cottingham refers to an ESPN report, ranking Steelers fans as first among fans of NFL teams for their enthusiasm, and a newspaper account of a memorial service for a beloved broadcaster who had introduced the ritual of waving the Terrible Towel. Interviewing other participants, known as **informants**, frequently provides the kind of in-depth understanding of what's going on that motivates researchers to use observation techniques in the first place. Cottingham made extensive use of her main informant, both for introductions to people and places and for information about what she was seeing. In a personal communication, she says, "He [her informant] was knowledgeable about a lot of the

informants, participants in a study situation who are interviewed for an in-depth understanding of the situation.

team history and league details, so that helped me contextualize the behavior I observed." One concern raised by feminist approaches to observation, in fact, is how to retain as much of the personal perspective of informants as possible (e.g., Reinharz, 1992: 52). How, in effect, to give "voice" to those informants? Some observers (e.g., Maher and Tetreault, 1994) have even asked permission of their respondents to report their names and have thus abandoned the usual practice of offering confidentiality or anonymity. We recommend, however, that confidentiality or anonymity be offered to all interviewees whether in traditional interviews or in those conducted in the field. Informants who later see their words used to substantiate points they themselves would not have made, for instance, might regret having given permission to divulge their identities in print, no matter how generous the intention of the researcher.

ANALYZING THE DATA

The distinction between data collection and data analysis is generally not as great for participant and nonparticipant observation as it is for other methods (e.g., questionnaires, experiments) that we've discussed. (See Chapter 15, where we discuss qualitative data analysis in greater depth.) For, although these other techniques tend to focus on theory verification, observational techniques are more often concerned with theory discovery or generation. Thus, Cottingham generates a theory of Steelers fandom as a quasi-religion, one with sacralized objects (e.g., Terrible Towels) and rituals (e.g., waving Terrible Towels), and tacit, if not formalized, beliefs (e.g., Terrible Towels should not be misused by fans of other teams). In fact, a major theoretical insight of her study is that quasi-religions may sometimes lack the formal beliefs of more formal religions. Cottingham suggests, in fact, that "[l]acking a set of formal beliefs … should not disqualify the group from consideration as a quasi-religion. To do so would be to ignore the profound meaning that fandom affords members and the important role of sport in contemporary American culture." This is not to say that observational studies can't be used for theory verification or falsification. Perhaps the primary theoretical upshot of Harrell and his colleagues' (cited in Bakalar, 2005) study of parent–child interactions in supermarkets is to falsify the notion that the attractiveness of the children has no effect.

Nevertheless, more studies based on observation techniques are like Cottingham's than like that of Harrell and his colleagues—more concerned, that is, with theory generation or discovery than with theory verification or falsification. Important questions are these: When does the theory building begin and how does it proceed? The answer to the first question—when does theory building begin?—seems to be soon after you record your first observation. We know, from personal testimony, that Cottingham had begun theorizing about the religious nature of Steelers fans' devotion after observations made *before* she formally collected data for the current project. She was particularly struck, for instance, by a Steelers tattoo worn by an acquaintance. The very act of focusing on this observation, rather than whatever else she might have homed in on, implies that the observer attaches some importance or relevance to the observation. This choice is likely to be based on some preconceived notion of what's (theoretically) salient, a notion that the sensitive researcher will, sometime during or after the initial observation, try to articulate.

Once articulated, these notions become concepts or hypotheses, the building blocks of theory. At this point, the researcher can begin to look for behaviors that differ from or are similar to the ones observed before. The pursuit of similarities can lead to the kind of generalizations on which **grounded theory** (Glaser, 2005; Glaser and Strauss, 1967), or theory derived from the data, is based. Hence, the observation, repeated more than once, of Terrible Towel waving, plus conversation with informants about the significance of that waving, led Cottingham to the conviction that this behavior was, in fact, an instance of a sacralized ritual for Steelers fans.

grounded theory, theory derived from data in the course of a particular study.

The process of making comparisons, of finding similarities and differences among one's observations, is the essential process in analyzing data collected during observations. It is also the essential ingredient for making theoretical sense of those observations. Through it, Humphreys (1975) came up with his typology of those who used public restrooms for impersonal sex and Erikson (1976) came up with the concept of "collective trauma."

ADVANTAGES AND DISADVANTAGES OF OBSERVATIONAL TECHNIQUES

The advantages of observational techniques mirror the reasons for using them: studying behavior, getting a handle on the relatively unknown, obtaining an understanding of how others experience life, studying quickly changing situations, or saving money (see the section labeled "Reasons for Doing Observations" earlier in this chapter).

Adler and Adler (1994b) suggest that one of the chief criticisms of observational research has to do with its validity. They claim that observations of natural settings can be susceptible "to bias from [the researcher's] subjective interpretation of situations" (Adler and Adler, 1994b: 381). They are surely right. But compared with measurements taken with other methods (e.g., surveys and interviews), it strikes us that measures used by observers are especially likely to measure what they are supposed to be measuring and, therefore, be particularly valid. We can imagine few combinations of questions on a questionnaire that would elicit as clear a picture of the ways in which some fans express enthusiasm for football teams than Cottingham's description of the two men who stripped off layers of clothing and then screamed their approval of good plays.

A final advantage of observation is its relative flexibility, at least compared with other methods of gathering social data. Observation is unusually flexible for at least two reasons: one, because a researcher can, and often does, shift the focus of research as events unfold and, two, because the primary use of observation does not preclude, and is often assisted by, the use of other methods of data gathering (e.g., surveys, available data, interviews). Thus, once Cottingham's focus shifted from how fans wave their Terrible Towels to why they waved them, her primary method of data acquisition changed from observation to interview and newspaper research (especially about Myron Cope's creation of the towel and its use at his memorial service).

Generalizability is a perpetual question for participant and nonparticipant observation alike, however. This question arises for two main reasons. The first has to do with the reliability, or dependability, of observations and measurements. One example will stand for the rest. In Erikson's study of the effects on a flood on one community, he concludes that the community experienced "collective trauma" because all the individual presentations given by survivors were "so bleakly alike" in their accounts of "anxiety, depression, apathy, insomnia, phobic reactions, and a pervasive feeling of depletion and loneliness" (1976: 198). But it remains a question whether another observer of the same accounts would have detected the bleak sameness that Erikson did, much less come to the conclusion that sameness was an indicator of "collective trauma," rather than of many individual traumas. One way to check whether one's observations are reliable, of course, is to ask other participants whether they seem right to them. This is what Cottingham did in relation to her observations at the game, the tailgate parties, and the sports bars. But sometimes such a strategy can't work, as, for instance, when one doesn't disclose his or her research purpose to others (as, of course, Mueller, Dirks, and Houts Picca's students couldn't do) or when one thinks one might elicit inaccurate interpretations of the situation by asking other participants about it (as, for instance, Liebow [1993: 321] did in his study of homeless women).

The second reason that questions of generalizability arise is a direct result of the nonrandom sampling procedures used by most observers. In the focal research, we see examples of fan rituals that are persuasive. But whether they frequently occur among some larger population of football fans in the United States depends on how generalizable Cottingham's sample is. And because that sample was not randomly selected (how could it have been?), the question of its generalizability is one that, of necessity, must wait another day (and for replication [by one of you?] or, perhaps, the use of another method).

STOP & THINK   *Can you imagine how the generalizability of males' stripping of layers of clothing as an expression of fandom might be studied with another method? Would you be inclined to use a survey? A cross-sectional or longitudinal design? A probability or nonprobability sample?*

---

demand characteristics, characteristics that the observed take on simply as a result of being observed.

A third problem associated with observational methods, especially when they are announced, is the bias caused by **demand characteristics,** or the distortion that can occur when people know (or think) they are being observed. Cottingham plausibly asserts that her introduction to 8 to 10 tailgaters before the game probably had no real effect on their revelries (personal communication). But one wonderful example of demand characteristics in action comes from D. L. Rosenhan's classic study entitled "On Being Sane in Insane Places" (1971). Rosenhan was concerned about the accuracy of psychiatric diagnoses and so, with seven accomplices, he gained admission as a pseudo-patient to eight psychiatric hospitals. In this first instance, all pseudo-patients took the role of complete participants and, although all were as sane as you and us, none was diagnosed as sane once he or she was admitted as a patient. In a second instance, however, Rosenhan told the staff in one hospital that, during a three-month period, one or more pseudo-patients would be admitted. Of the 193 patients who were admitted in the next three months, 41 were alleged by at least one member of the staff to be pseudo-patients, 23 of those by attending psychiatrists. Because no pseudo-patients, to Rosenhan's knowledge, had attempted to be admitted, the results of this second "experiment" suggest that even the threatened presence of observers can distort what goes on in a setting. Thus, despite the effort of most observers to create a situation in which action goes on as if they weren't there, there's little reason to believe that these efforts can be totally successful—especially when their presence is announced.

Three other problems of observational methods are worth emphasizing. First, in comparison with virtually every other technique we present in this book, observation can be extremely time-consuming. Adler and Adler (personal communication) typically spent from 18 months to two years on most studies and eight years on the observations for the book, *Peer Power*; Liebow saw about nine years pass between the conception and publication of his book on homeless women. Still, Cottingham's work (which entailed about 20 hours of observation and about eight months all together) suggests that research based on observations can be reasonably time-efficient. Second, and clearly related to the issue of time consumed, is the problem of waiting around for events "of interest" to occur. In almost every other form of research (administering a questionnaire, an interview, an experiment, or collecting available data), the researcher exercises a lot of control over the flow of data. In the case of field observations, control is handed over to those being observed—and they might be in no hurry to help events unfold in an intellectually interesting way for the

BOX 11.3 Advantages and Disadvantages of Observation

Advantages

1. Provides relatively unfiltered view of human behavior.
2. Can provide an in-depth understanding of a natural situation or social context.
3. Permits the study of quickly changing situations.
4. Can provide insight into relatively poorly understood aspects of human life.
5. Can be a relatively flexible mode of data collection.

Disadvantages

1. Generalizability is a perpetual question, thanks partly to the possible unreliability of observa-tions and partly to the typical use of nonprobability samples.
2. Those observed may distort or change their behavior if they know they are being observed.
3. Ethical issues may arise about informed consent, especially if the researcher feels that providing adequate information will distort the behavior under observation.
4. Observation studies are typically time-consuming.
5. One may have to wait for "things to happen."
6. Each stage of an observational study—choosing appropriate settings, gaining access to those settings, analyzing the data—can be fraught with unexpected difficulties.

observer. Third, although observation can be fun and enlightening, it can also be terribly demanding and frustrating. Every stage—from choosing appropriate settings, to gaining access to settings, to staying in settings, to leaving settings, to analyzing the data—can be fraught with unexpected (and expected) difficulties. Observation is not for the unenergetic or the easily frustrated (see Box 11.3).

SUMMARY

Although observation is often used in experiments, the observational techniques we've focused on here—those used by participant and nonparticipant observers—are those used to observe actors who are, in the perception of the observer, acting naturally in settings that are natural for them. Cottingham, for instance, observed Pittsburgh Steelers fans at a Steelers game, at tailgate parties, and at sports bars. Observational techniques are frequently used for several purposes: to gain relatively unfiltered views of behavior, to get a handle on relatively unknown social, to obtain a relatively deep understanding of others' experience, to study quickly changing situations, to study behavior, and to save money.

A sense of the variety of roles a participant or nonparticipant observer can take is suggested by Gold's (1958) classic typology. A researcher can become a complete participant or a complete observer, both of which imply that the researcher withholds his or her research purposes from those being observed. Although both of these roles entail the advantage of relative unobtrusiveness, they also entail the disadvantages of incomplete frankness: for example, of not being able to ask probing questions and of deceiving one's subjects. The ethical issues associated with observation almost always merit the attention of an IRB, except, in most cases, when the observation occurs in public places. Both the participant-as-observer and the observer-as-participant do reveal their research purposes, purposes that can prove a distraction for other participants. On the other hand, by divulging those purposes, they earn the advantage, among others, of being able to ask other participants detailed questions about their participation.

Observers can choose their research settings for their theoretical salience or their natural accessibility or for some combination of theory and opportunity. Generally, participant and nonparticipant observation begin without well-defined research questions, measurement strategies, or sampling procedures, though systematic observations can involve some combination of these elements. In all cases, however, researchers will want to decide how much of themselves and their research they will want to disclose to others in the research setting. At some point they may also need to employ social skills to gain access to the setting from gatekeepers and to learn what they need to learn from other actors.

Once in a setting, researchers can record observations by writing them down, by using some mechanical device (e.g., a tape recorder, a camera, or a cell phone), or by trusting to memory for short periods. Most often observers have preferred trusting to memory because alternative techniques involve some element of intrusiveness. In all cases, the effort to record or transcribe "notes" should be made soon after observations are made, improving the chances for precise description and for recording less precise impressions. Increasingly, visual sociology has pressed the case for collecting or using images for making observations and studying society. The analysis of observational data involves comparing observations and searching for important differences and similarities.

The advantages of participant and nonparticipant observation mirror the purposes for using them and include the likelihood of valid measurement, and of considerable flexibility. Both types of observation permit relatively unfiltered views of behavior and enable in-depth understandings of particular settings or situations. On the other hand, disadvantages of observational techniques include almost unavoidable suspicions that they might yield unreliable and nongeneralizable results. Observation can also affect the behavior of those observed, especially when the observation is announced, and contains an element of deceit, especially when the observation is not announced. Observation can also be a particularly time-consuming and sometimes frustrating path to social scientific insight.

EXERCISE 11.1

Observing Gestures in a Film or Television Program

For this exercise, which we've adapted from ones suggested by Tan and Ko (2004: 117–118), you'll need to find a film or television program that you can watch to observe mannerisms (gestures) and facial expressions. We want you to prepare a two-page essay in which you discuss the expression of emotions through gestures and facial expressions in one scene of the film or text. Be sure your essay mentions the title of the film or television program and gives a one- or two-sentence synopsis of the characters and action in the scene. Then outline the emotions you observed in the various characters involved.

What are the actions/expressions that lead you to your conclusions about these emotions? You need to clearly distinguish the mannerisms and expressions from what they mean. (For instance, you might point out that a character raised his/her voice or motioned with an index finger [expressions or mannerisms you observe] and you might say the person looked angry [your conclusion about the emotional meaning of the expressions].) This means you need to distinguish the action you observe from how you interpret what the particular action means. **Please do not write about the storyline beyond your brief synopsis of the action.**

EXERCISE 11.2

Practicing Subject Image-Making and Photo-Elicitation

For this exercise, you'll need a digital camera and the capacity to print two pictures. You'll be replicating the subject image-making and photo-elicitation that Glover, Stewart, and Gladdys (2008) asked of their participants (see section on visual sociology).

1. Take some pictures of places that are "important within your daily life." You choose the places; you take the pictures. Then, print two of the images. (You will now

have completed the subject image-making phase of the exercise.)
2. Now practice photo-elicitation on yourself.
   a. Explain, in your own words, what the places are that you've taken pictures of.
   b. Ask yourself why the places you've depicted are so important to you and write down your responses.
   c. Now ask yourself what details from the pictures you've taken might help your reader understand why the places are so important and write down your responses. (You will now have completed the photo-elicitation phase of the exercise.)
3. Turn in both the images and your responses.

EXERCISE 11.3

Participant Observation

(Can be done during class time, if class is about 90 minutes or more. If class time is less than 90 minutes, students should work on "Planning the Data Collection," mentioned later, before class.)

In this exercise, you will pick **either** the variable *helpfulness* or the variable *friendliness* to study. Create ONE research question that you could answer about the variable you picked that can be answered on campus using an observation method. If you don't want to create your own research question, you can use one of these:

- Are women or men on campus *friendlier*?
- Are students *more helpful* than faculty or staff?
- Are people in groups *less helpful* than people who are alone?
- Do people act *more friendly* to people with whom they share one or more background characteristics (like gender or race)?

You will have 30 minutes to collect data to answer your research question. Working in teams of two or three, you will be *participant observers* and will interact with strangers on campus. At least one person in the group should take the role of **complete participant** and play the role of a new visitor to the campus rather than being a current student who already knows his or her way around campus.

Planning the Data Collection

1. Decide on *the concept* you want to study (either helpfulness or friendliness) and write down a specific *research question* you intend to answer on the page you will turn in.
2. Create a conceptual definition for the concept that is in your research question (either "helpfulness" or "friendliness").
3. Decide about the population(s) you will interact with on campus during your data collection. The population can be students, faculty, staff, or some of each.
4. Create an **operational definition** of helpfulness or friendliness using **indicators that are observations.** As your indicators, you can focus on the words, the body language, the voice tone, and/or the behavior of the sample members that you select to interact with.
5. Decide what you will do to see how people will interact with you. See some examples of things you can say to begin an interaction with the members of the sample below.
6. Based on your research question, decide what variables you will need data on and how you will collect data for EACH of the variables. [Note: If your research question includes two variables, you need to collect data on BOTH variables in each interaction.]

For each variable you can collect data with **field notes and/or by using code sheets.** For example, you could take field notes describing each interaction that include the information about each variable OR you could come up with a code sheet that you'll use that also includes both variables. If you are using code sheets, create one and make copies for the group now.

YOU CAN WRITE DOWN THE ANSWERS TO SECTIONS 1 AND 2 ON THE REPORT NOW IF YOU WANT TO. THEN COLLECT DATA AND WRITE DOWN THE REST OF YOUR REPORT.

Collecting Data during 30 Minutes in the Field

Each group will initiate and collect data on between 8 and 10 interactions on campus during the 30-minute data collection period. To start the interaction, one person in the group should *approach someone they don't know and initiate interaction with that person.* The other person or people in the group should observe the interaction (and be close enough to hear what is said). *[Be aware that there are likely to be other class members nearby also doing this exercise so don't approach the same people they do.]*

1. In each interaction, approach a person and initiate interaction to see how the person responds. Some possible things you can say are to ask the following:
   - "I'm parked illegally, I think. Do they really ticket here? Where should I park that's not too far away?"
   - "Where is [name of some building on campus]?"
   - "I'm thinking of transferring to [name of college] next year. Do you like it here?"
   (The first two questions work for students, faculty, or staff, but the last one seems most appropriate to ask of students.)

   Feel free to create other appropriate questions to use when playing the role of complete participant.

2. While one person in the group approaches someone, the other one or two people in the group should be observers (without being too obvious about it). (Take turns being the person who asks the questions). IN ADDITION to observing the interaction and focusing on the indicators of friendliness or helpfulness, if you need the information to answer your research question, you should record at least one or two pieces of infor-

mation about the *backgrounds* of the people you observe and of the *setting* to answer your research question. Some background variables you could collect data on are the person's gender, approximate age, race, or status (student, faculty, or staff). Some things you could note about the setting could be whether the person was alone or not and the location on campus.

3. After the interaction is over, record your field notes or fill in the code sheet you created with information about the degree of "friendliness" or "helpfulness" AND the other data that are necessary to answer your research question (such as information about the person's characteristics and information about the setting).

Writing the Research Report

Return to the classroom and finish writing down your report using the exercise format outlined below. Attach your data to the page (field notes, code sheets, etc.).

Writing down Your Research Report

Your names:

1. **Introduction:** State your research question. Make sure that one of the variables in your question is helpfulness or friendliness.
2. **Methodology:**
   A. State your conceptual definition by writing a sentence that says: We defined friendliness as _____ or We defined helpfulness as _____.
   B. Describe your operational definition by listing the indicators you decided to use to measure this concept.
   C. How many people did you approach? How many observations did you make?
3. **Findings:** Give your answer to your research question. Give support for your answer by summarizing the data you collected AND giving some examples from your field notes or code sheets to illustrate your findings.

Using Available Data

CHAPTER **12**

CHAPTER OBJECTIVES

This chapter is designed to help students

- **Understand what available data are**
- **Distinguish among secondary data, existing statistics, and other kinds of available data**
- **Identify major sources of secondary data and existing statistics**
- **Practice using secondary data and existing statistics**
- **Appreciate the advantages and disadvantages of using available data**

CHAPTER OUTLINE

Available Data

Kinds of Available Data

Advantages and Disadvantages of Using Available Data

Summary

Exercises

AVAILABLE DATA

You've probably heard immigration blamed for a lot of what ails us as a nation: the competition for skilled and unskilled jobs, demands on educational and other social services, and violent criminality, to name but three. Suppose you wanted to test one of these assertions. How might you do it? Let's say you wanted to focus on the connection between immigration and criminal violence. You might try participant observation (see Chapter 11), but the chances of seeing anything of interest in any particular immigrant (or non-immigrant) community probably aren't that great. You might use a questionnaire (see Chapter 9) or even qualitative interviews (see Chapter 10) to determine whether immigrants are more likely than non-immigrants to admit to having committed violent crimes, but the chances of receiving reliable responses (see Chapter 6) are pretty small (for all groups). Who's going to volunteer, after all, that they've committed a violent crime?

A concern with the crime/migrant connection moved Ramiro Martinez and Matthew Lee, the authors of this chapter's focal research. The two concerns, about crime and migration, didn't develop at the same time, though. Martinez asserts that he first became interested in crime, then in migration: "When in graduate school, about ten years ago, I asked somebody a question about Latino (Hispanic) homicides.... At the time we knew that homicide was the leading cause of death among young African American males but little attention was paid to Latinos. Only once I began to collect Latino data did the question of immigration status occur to me" (personal communication).

available data, data that are easily accessible to the researcher.

Martinez and Lee's paper illustrates the use of **available data,** or data that are available to researcher, in the social sciences. Available data are sometimes actually statistics, or summaries of data, that have been computed from data collected by a large organization—for example, governments (as when they publish census or crime statistics) or businesses (as when they prepare for annual reports). Such statistics, when they have been made available in statistical documents, are called **existing statistics.** If you're sitting at your computer, for instance, you can try the website www.fedstats.gov/, "the gateway to statistics from over 100 U.S. Federal agencies," and see how quickly you can find the population of the country or of your state.

existing statistics, summaries of data collected by large organizations.

secondary data, research data that have been collected by someone else.

primary data, data that the same researcher collects and uses.

Available data are sometimes actual *data*, as opposed to statistics, collected by others, data that the researcher analyzes in a new way. Data that have been collected by someone other than the researcher are called **secondary data.** Secondary data are frequently contrasted with **primary data,** or data that are collected and used by the same researcher. Most of the focal research that we've featured uses primary data, such as the data collected by Cottingham for the focal piece on Pittsburgh Steelers fans (in Chapter 11) and the data collected by Adler and Clark on retirement (in Chapter 7). Survey data that have been made available for others to reanalyze are perhaps the best-known kind of secondary data. The General Social Survey (GSS), conducted now every other year by the National Opinion Research Center (NORC) to collect data on a large number of social science variables, is probably the most popularly used set of survey data by social researchers, but it's just one of hundreds of sets of survey data that have been subjected to secondary analysis. Yet secondary data don't need to be survey data; they just need to have been collected by someone other than the current researcher. Thus, one of Martinez and Lee's main sources are the "logs" Miami police created to help them solve homicide cases. These became secondary data as soon as Martinez and Lee adapted them for their research purposes. We've chosen to focus on Martinez and Lee's type of secondary data analysis, not because it's especially typical of research based upon available data, but because it's a particularly interesting version of it.

The increasing popularity of available data is revealed in a study by Roger and a student, Katelyn Grandchamp. Clark and Grandchamp (2012) found only 19.7 percent of sociology articles in major journals published in 1958 used available data, while 59.3 percent did so in 2008. Of this 59.3 percent, about 11/12ths (about 54 percent) used secondary data.

A particularly accessible form of available data (for research purposes) is that which is presented in published reports of government agencies, businesses, and organizations. Many studies of crime and violence have relied on national statistics gathered and forwarded to the Federal Bureau of Investigation (FBI)'s *Uniform Crime Reports* (UCR). One such study, reported by Martinez and Lee, was the U.S. Commission on Immigration Reform (1994) investigation that, among other things, compared the city of El Paso's overall crime and homicide rates to those of similar-sized cities in the United States. El Paso is a town along Mexico's border and therefore has a high percentage of migrants within its population. The Commission's finding that El Paso's overall crime and homicide rates were much lower than those in cities of comparable size supports the belief that migrants don't commit more crimes, and particularly more homicides, than nonmigrants. The Commission's evidence, however suggestive, is nonetheless subject to a famous criticism: that it commits the **ecological fallacy**. A fallacy is an unsound bit of reasoning. The ecological fallacy is the fallacy of making inferences about certain types of individuals from information about groups that might not be exclusively composed of those individuals. The Commission wanted to see whether cities with higher proportions of migrants had higher crime rates than cities with lower proportions of migrants, and so using cities as a unit of analysis made a great deal of sense. So far so good. But does its finding that a city with higher proportions of migrants had a lower crime rate than comparable cities with lower proportions of migrants mean that you should conclude that migrants are less likely to commit crimes than nonmigrants? Not legitimately. Such a conclusion risks the ecological fallacy because it might be that the migrants are committing most of the crimes in the cities with lower proportions of migrants. The problem is that we have looked at cities as our units of analysis and that, whatever the proportions of migrants, all cities are composed of both migrants and nonmigrants. Another example of the ecological fallacy would occur if we inferred that individuals who are unemployed are more likely than those who are employed to suffer from mental illness from a finding that a city with a particularly high unemployment rate also has a high rate of mental illness. (It could be, for instance, that it is the employed who are mentally ill.)

ecological fallacy, the fallacy of making inferences about certain types of individuals from information about groups that might not be exclusively composed of those individuals.

STOP & THINK   *You might remember that Émile Durkheim (in his classic sociological study, Suicide) found that Protestant countries and regions had higher suicide rates than Catholic countries and regions. He inferred from this the conclusion that Protestants are more likely to commit suicides than Catholics. Is this a case of the ecological fallacy? Explain.*

Martinez and Lee, you will see, have devised an especially clever way of avoiding the ecological fallacy. To do so, they had to dig a little deeper than many researchers do using available data: in this case, into the "homicide logs" of the Miami Police Department. In using these logs, however, they show us many of the virtues of using available data.

FOCAL RESEARCH

Comparing the Context of Immigrant Homicides in Miami: Haitians, Jamaicans, and Mariels

By Ramiro Martinez, Jr., and Matthew T. Lee*

Are new immigrants more crime prone than native-born Americans? This issue has been debated since the early 1900s and continues to fuel controversy on the "criminal alien" problem in contemporary society (Hagan and Palloni, 1998; Short, 1997). Some scholars speculate that immigrants engage in crime more than other population groups by using the growing presence of the foreign born in the criminal justice system as evidence of heightened criminal involvement (Scalia, 1996). Other writers encourage legislation designed to rapidly deport immigrant criminals, such as California's Proposition 187 and the U.S. Congress 1996 Immigration Reform bill, despite any systematic evidence of an "immigrant crime problem" (Beck, 1996; Brimelow, 1995). Still others maintain that large labor pools of young immigrant males willing to work for lower wages than native-born Americans, in particular inner city residents, exacerbate ethnic conflict and contribute to urban disorders (Brimelow, 1995). The consequence of immigration, according to this view, is a host of social problems, especially violent crime (Beck, 1996).

This study probes the immigration and crime link by focusing on where immigrants live and how immigration might increase or decrease violent crime, at least as measured by homicide in a city that serves as a primary destination point for newcomers. Immigrants are undoubtedly influencing urban crime, but they are also potential buffers in stabilizing extremely impoverished areas (see Hagan and Palloni, 1998). Our primary argument is that while immigration shapes local crime, it might have a lower than expected effect on violent crime rates. The result is areas with high proportions of immigrants but low levels of violence, relative to overall city rates. The key question is whether immigrants have high rates of violence or if immigration lessens the structural impact of economic deprivation and thus violent crime.

*This article is adapted with permission from Ramiro Martinez and Matthew T. Lee, "Comparing the Context of Immigrant Homicides in Miami: Haitians, Jamaicans, and Mariels," *International Migration Review*, 2000, Vol. 34, 794–812.

It is, of course, preferable to study the impact of immigration and social institutions on violence across time and in diverse settings. While we do not directly test these variables with a multivariate analysis, we provide a starting point for this research agenda by conducting one of the first contemporary studies of immigrant Afro-Caribbean (Haitians, Jamaicans, and Mariel Cubans) homicide in a major U.S. city.

Previous Immigration and Crime Studies

Evidence of high immigrant crime rates is mixed (Shaw and McKay, 1931; see Short, 1997, for review of the literature). Lane (1997) notes that white newcomers (e.g., Irish, Italian) were often disproportionately involved in violent crime in the turn of the century Philadelphia and New York City. Crime rates declined, however, as settlers from abroad assimilated into mainstream society and integrated into the economy (Short, 1997). This finding is particularly instructive, since most immigrant victims and offenders were young males without regular jobs and presumably fewer ties to conventional lifestyles, suggesting that, even a hundred years ago, gender and economic opportunities played a major role in violent crime (Lane, 1997).

More contemporary studies on immigrant violence, however, have been scarce (Short, 1997). Many crime and violence studies rely on national statistics gathered from police agencies and forwarded to the FBI's Uniform Crime Reports (UCR). The UCR typically ignores the presence of ethnic groups such as Latinos or Afro-Caribbeans and directs attention to "blacks" or "whites" in compiling crime data. As a result, comparisons beyond these two groups are difficult to examine in contemporary violence research (Hawkins, 1999).

Furthermore, the handful of recent studies that have explored the connection between immigrants and crime have focused on property crimes or criminal justice decision-making processes (*cf.* Pennell, Curtis, and Tayman, 1989). And, despite studies (e.g., Hagan and Palloni, 1998) that suggest metropolitan areas with large immigrant populations have lower levels, or at least no higher levels, of violent and property crime than other metropolitan areas, the impression remains that immigrants are a major cause of crime (Hagan and Palloni, 1998: 378).

Research Setting

The city of Miami, Florida, provides a unique opportunity to examine the issue of whether or not immigration influences violent crime. Miami is one of the nation's most ethnically diverse cities and prominently known for its hot weather, natural disasters, waves of refugees, and tourist killings. It is also well known for its record high rates of violence. At various points throughout the 1980s, the violent crime rate was over three times that of similarly sized U.S. cities, and the city was widely characterized as a "dangerous" city in the popular media (Nijman, 1997).

This perception is linked, in part, to the presence of its large immigrant population (Nijman, 1997). After several distinct waves of immigration from Latin America since 1960, Latinos, mostly Cubans, comprised almost two thirds of the entire population in 1990 and had grown to almost one million in surrounding Dade County (Dunn, 1997). In 1980 alone, the number of Cubans increased dramatically as the Mariel boatlift brought 100,000 persons to the United States, and most of these settled in the greater Miami area (Portes and Stepick, 1993). The arrival of the Mariel refugees generated a moral panic in which the newcomers

(continues)

FOCAL RESEARCH (continued)

were labeled as deviants allegedly released from Cuban prisons and mental hospitals (Aguirre, Saenz, and James, 1997: 490–494).

Unlike the Mariels, who were welcomed by government officials (at least initially), Haitian immigrants were alternately detained or released by the Immigration and Naturalization Service (INS) in the early 1980s while fleeing political terror and economic despair in their home country (Portes & Stepick, 1993). Haitians, like the Mariels, however, were also stigmatized in the popular press, not only as uneducated and unskilled boat people, but also as tuberculosis and AIDS carriers (Martinez and Lee, 1998).

Miami was also a leading destination point for another Afro-Caribbean group—Jamaicans. Thousands of Jamaicans emigrated throughout the 1970s and 1980s, after open access to England was closed. The Jamaican population in Miami-Dade County tripled between 1980 and 1990, becoming the second largest Caribbean black group in the area (Dunn, 1997: 336).

As the newcomers converged in Miami, speculation arose that crime, especially drug-related types surrounding the informal economy, was growing within immigrant and ethnic minority communities (Nijman, 1997; see also Inciardi, 1992). *The Miami Herald*, for instance, reported that throughout the early 1980s (Buchanan, 1985; Schenker, 1984) Jamaican gangs operated freely in south Florida, controlling a large share of the local crack-cocaine trade and engaging in widespread violence (see also Maingot, 1992).

In sum, it stands to reason that immigration should increase absolute crime rates in Miami if it brings in a large number of people, without a corresponding emigration to surrounding areas (see Hagan and Palloni, 1998). Yet, despite the important empirical and theoretical reasons to anticipate that immigration should influence urban violence, this link has been largely ignored in the social science literature. In this article, we extend previous studies on the social mechanisms of immigration and crime by considering the violence patterns of three distinct immigrant groups in the city of Miami.

Data Collection

Data on homicides from 1980 through 1990 have been gathered directly from files in the Homicide Investigations Unit of the City of Miami Police (MPD). These "homicide logs" contain information on each homicide event, including a narrative of the circumstances surrounding each killing not readily available from other data sources. Most importantly, extensive data on victim and, when known, offender or suspect characteristics such as name, age, gender, race/ethnicity, and victim/offender relationship are also noted by the investigators on the logs.

Starting in 1980, homicide detectives began to note whether the victim or offender was white, black, or Latino. As the year progressed, and in response to media inquiries, MPD personnel would also note whether the victim or offender was a Mariel Cuban refugee, Haitian, or Jamaican. These designations were based on a variety of sources. Some Mariels were identified through INS documentation such as a "Parole ID" discovered at the scene of the crime. A temporary tourist visa or INS green card was also frequently found on the victim or at his/her residence. Still other ethnic clues were gathered from witnesses, family member, friends, or neighbors including language (e.g., speaking Creole or English with West Indian accents), providing country of birth on death certificate, or names less than common to African Americans ("Michel Pierre" or "Augustine Seaga"). All of these clues taken together helped provide a more precise definition of victim and offender ethnicity.

Each homicide type is examined, including the level of gun use and circumstances surrounding each killing. The nature of the prior contact and the circumstances of offense type are often distinct and vary by ethnicity (Hawkins, 1999). Types of homicide are therefore placed into five categories. First is "other felony" or a type of homicide committed during the course of a felony other than robbery. The second category consists of "robbery" homicides. Third is "primary non-intimate" or killings among acquaintances, neighbors, friends, or co-workers. Fourth is "family intimate homicides" or killings between spouses, lovers, estranged partners, and immediate family members. Finally, an "unknown category" is included to account for homicides not cleared with an arrest.

Since our focus is on comparing and contrasting immigrant homicides, we concentrate our account on immigrant Caribbean killings and their characteristics across two time points, 1980 and 1990. Specifically, the average number of homicide suspects (offenders) per 100,000 population from 1980 through 1984 is used for the first, and the average over 1985 to 1990 for the second. Total city homicides are also included as a baseline comparison. We begin by directing attention to suspect rates, but, given the current limitation of missing data, victim and offender relationships will be more fully explored.

We also use the 1980 and 1990 decennial censuses to provide estimates of the population in each city. Reliable population figures for our three immigrant groups are, of course, notoriously difficult to provide. We did, however, employ several strategies. First, Portes and colleagues (1985) estimated about 75 percent of the Mariel population (roughly 125,000) settled in the greater Miami area after the 1980 census was taken across the United States. Of that percentage, most (61.1 percent) moved into the city of Miami, at least as we can best determine (Portes, Clarke, and Manning, 1985: 40). Thus, we estimated roughly 57,281 Mariels were in the city limits between 1980 and 1990.

The number of Haitians and Jamaicans was equally difficult to estimate. Unfortunately, neither the 1980 nor the 1990 census provides specific information on Jamaicans and Haitians at the city level, forcing us to use county estimates by Marvin Dunn (1997: 336) on Caribbean-born blacks in 1980 and 1990. The following group-specific rates are therefore based on the number of Caribbean blacks in Miami-Dade County born in Jamaica and Haiti. While these data are useful due to their detail regarding geography and ethnicity, we acknowledge that an undercount problem exists, especially among immigrants. It is therefore likely that the level of immigrant homicides is even lower because the denominator in the calculation of the rates is likely to be larger. With these cautions in mind, we believe that the following results are reliable guides on immigrant homicides.

Results

Between 1980 and 1990, over 1,900 homicides were recorded by the Miami Police Department; 28 percent of these victims were Caribbean immigrants and 15 percent of all suspected killers were identified as Haitian, Jamaican, or Mariel. While the percentage of Mariel victims was higher than Haitians and Jamaicans (78 percent : 22 percent), the number of Mariels settling in the city of Miami

(continues)

FOCAL RESEARCH *(continued)*

(57,281) in the early 1980s was almost twice the number of Caribbean blacks in the entire greater metropolitan Miami area.

To further illustrate the distinct experiences of each group, we examine the homicide offending rate in Table 12.1. These results are especially important because much of the current anti-immigrant sentiment focuses on newcomers as killers rather than as victims. But, with a lone exception, the immigrants were not high-rate killers. In Miami during the early 1980s, the total city homicide offender rate was 69.9. This rate was higher, and in one case substantially higher, than the group specific rates for Mariels (55.5) and Haitians (26.4). One prominent contrast is the Jamaican rate of offending (102.2)—a level one and a half times the total city rate.

The offender data cannot be understood outside of their historical context, as the early 1980s proved to be a time of both social strife and social change in Miami. Social and political frustration among the city's African American population, for instance, erupted into violent riots touched off by questionable police killings in 1980, 1982, and 1989 (Dunn, 1997). In general, however, as the local turmoil subsided throughout the 1980s, the Caribbean violator rate decreased dramatically and quickly, relative to the city as a whole. The Jamaican killer rate displayed the largest decrease of all three groups (84 percent), but the Mariel and Haitian rates also declined sharply by 1990. These decreases are even more noteworthy after considering the substantial Haitian and Jamaican population growth throughout metropolitan Miami. In 1980, less than 13,000 Haitians lived in the country, but ten years later the number quadrupled to over 45,000. Similarly, the Jamaican population tripled over the same time span to more than 27,000 (Dunn, 1997: 336–337).

TABLE **12.1**

Immigrant Homicide Suspect Rates: 1980 and 1990

	1980	1990
Haitian	26.4	12.8
Jamaican	102.2	15.9
Mariel	55.5	10.5
City Total	69.9	49.9

Sources: City population estimates from 1980 and 1990 Bureau of the Census. Haitian and Jamaican population estimates cited in Dunn, 1997: 336. Mariel estimates described in Martinez, 1997: 120 and 121.

If the stereotype of Miami in the popular media as a high crime, drug-infested area, overrun by dangerous immigrants, was correct, then immigrant Caribbeans should have rates consistently higher than the city average. Apparently, immigrant murderers, at least in most cases, were engaged in killing at a much lower level than the total city rate, and the lone exception was confined to a single time point.

Table 12.2 describes two variables linked to type of homicide characteristic—gun use and victim–offender relationship. The gun distribution of victims varied in some cases. Gun inflicted homicides for Caribbeans were always higher than for the rest of Miami at both time points, except for the Mariels in 1990, although percentages for Haitians and Jamaicans dropped that same year.

The circumstances surrounding the homicide victim levels also varied. Table 12.2 reports that most Haitian victims were killed during the course of a robbery or by a non-intimate, regardless of the time frame. Jamaicans were usually murdered during the course of a felony, robbery, or by a non-intimate, but some variation occurred. For example, felony and robbery-related homicide increased in 1990, similar to the rest of Miami, while non-intimate killings dropped by half. Mariels, typically, were fatally assaulted by non-intimate acquaintances, friends, or neighbors.

One finding worthy of special attention is that of domestic-related homicides between family members and intimates. A popular myth exists that immigrant Latinos and Afro-Caribbeans are more likely to engage in domestic violence, in part because of "traditional values" or "machismo" that encourage greater control at home, including physical responses to interpersonal problems (see Buchanan, 1985). Contrary to this image, we discover lower family/intimate proportions for immigrants than for the city as a whole.

TABLE **12.2**
Characteristics of Immigrant Homicide Offense Type: 1980–1990

Offense	Victim			
	Haitian	**Jamaican**	**Mariel**	**Miami Total**
1980				
Gun	95.2	85.7	86.0	78.3
Felony	8.7	17.9	10.9	11.2
Robbery	39.1	25.0	9.2	17.0
Non-intimate	34.8	50.0	60.9	48.7
Family/Intimate	17.4	3.6	15.5	18.7
Stranger/Unknown	0.0	3.6	3.4	4.4
1990				
Gun	81.6	85.7	74.0	74.1
Felony	22.2	33.3	18.2	22.3
Robbery	31.5	33.3	15.6	18.8
Non-intimate	37.0	26.7	55.8	46.7
Family/Intimate	7.4	6.7	6.5	10.3
Stranger/Unknown	1.9	0.0	3.9	2.0

Note: All numbers in percentages

(continues)

FOCAL RESEARCH (continued)

Discussion and Summary

A popular perception exists that immigrants are responsible for crime and violence in U.S. cities with large foreign-born populations. It is important to bring empirical evidence to bear on the policy debates surrounding this concern. To address this issue, we examined Afro-Caribbean homicides in the city of Miami. The picture that emerged undercuts claims that immigrants significantly influence violent crime.

We found that, almost without exception, Afro-Caribbeans had a lower homicide rate than the city of Miami total population. Comparing the early 1980s to the late 1980s, we also found a strong pattern of declining violence, especially for Jamaicans and Mariel Cubans, while Haitians continuously maintain an overall low rate of violent crime. As immigrant groups became more established, grew in size, and were less dominated by young males, the most violence-prone group, the homicide rate rapidly dissipates.

To illustrate this process, recall the high 1980 Jamaican offending level—the only immigrant rate to exceed the city average. It is possible that a high proportion of unattached and young Jamaican males migrated to south Florida and were employed in the informal economy. As the decade progressed, they left or were removed from the drug industry, and Jamaicans with more conventional ties to society followed the earlier entrants in search of legitimate work. Although our explanation is speculative and data are not available to directly examine this proposition at the city level, we believe this is a plausible account for our findings.

The results of this study have relevance for policies in vogue, such as targeted sweeps by the INS or legislation designed to deport criminal aliens. Many of these policies are motivated by the notion that immigrants threaten the order and safety of communities in which they are concentrated (Beck, 1996; Brimelow, 1995). We discovered, however, that, overall, Afro-Caribbeans are involved in violence less than the city of Miami population. Thus, the current INS strategy to remove "criminal aliens" in an effort to combat violent crime is questionable and is a misguided basis for limiting and restricting immigration.

The question of comparability to other areas naturally arises. We stress the caveat that our findings are, perhaps, unique to Miami for a host of demographic and historical reasons. It is possible that crime differences might arise relative to other cities with large immigrant populations (e.g., New York City, Chicago, Los Angeles, San Diego). Thus, local conditions, social class in the home country, and manner of reception might be important determinants of immigrant well-being. We invite further inquiries along these lines.

Our findings suggest that some of the key propositions of social disorganization theory must be revised to account for contemporary relationships among the immigration and homicide patterns found in cities such as Miami. Rapid immigration might not create disorganized communities but instead stabilize neighborhoods through the creation of new social and economic institutions. For example, Portes and Stepick (1993) describe how immigrant-owned small business revitalized Miami by providing job opportunities for co-ethnics. The low rate of involvement of the three immigrant groups we studied demonstrates the value of exploring intra-ethnic distinctions in future research on neighborhood stability, economic conditions, and violent crime. Increased attention to the issues we have raised will likely lead scholars to challenge misleading and deleterious stereotypes and result in more informed theory, research, and policy on urban immigrant violence.

References

Aguirre, B. E., R. Saenz, and B. S. James. 1997. Marielitos ten years later: The Scarface legacy, *Social Science Quarterly*, 78: 487–507.

Beck, R. 1996. *The case against immigration*. New York: Norton.

Brimelow, P. J. 1995. *Alien nation*. New York: Random House.

Buchanan, E. 1985. Dade's murder rate was on the rise again in 1984, *Miami Herald*, January 2, B1.

Dunn, M. 1997. *Black Miami in the twentieth century*. Gainesville: University Press of Florida.

Hagan, J., and A. Palloni. 1998. Immigration and crime in the United States. In *The immigration debate*. Ed. J. P. Smith and B. Edmonston. Washington, DC: National Academy Press. pp. 367–387.

Hawkins, D. F. 1999. African Americans and homicide. In *Issues in the study and prevention of homicide*. Ed. M. D. Smith and M. Zahn. Thousand Oaks, CA: Sage.

Inciardi, J. A. 1992. *The war on drugs II: The continuing epic of heroin, cocaine, crack, crime, AIDS, and public policy*. Mountain View, CA: Mayfield.

Lane, R. 1997. *Murder in America: A history*. Columbus: Ohio State University Press.

Maingot, A. P. 1992. Immigration from the Caribbean Basin. In *Miami Now!* Ed. G. J. Grenier and A. Stepick. Gainesville: University of Florida.

Martinez, R., and M. T. Lee. 1998. Immigration and the ethnic distribution of homicide in Miami, 1985–1995, *Homicide Studies*, 2: 291–304.

Nijman, J. 1997. Globalization to a Latin beat: The Miami growth machine, *Annals of AAPSS*, 551: 164–177. May.

Pennell, S., C. Curtis, and J. Tayman. 1989. *The impact of illegal immigration on the criminal justice system*. San Diego: San Diego Association of Governments.

Portes, A., J. M. Clarke, and R. Manning. 1985. "After Mariel: A survey of the resettlement experiences of 1980 Cuban refugees in Miami," *Cuban Studies*, 15: 37–59.

Portes, A., and A. Stepick. 1993. *City on the edge: The transformation of Miami*. Berkeley: University of California Press.

Scalia, J. 1996. *Noncitizens in the federal justice system, 1984–1994*. Washington, DC: U.S. Department of Justice, Bureau of Justice Studies.

Schenker, J. L. 1984. Police: Rastafarian gangs setting up shop in Florida, *Miami Herald*, March 20, B8.

Shaw, C. R., and H. D. McKay. 1931. *Social factors in juvenile delinquency. Volume II of report on the causes of crime*. National Commission on Law Observance and Enforcement, Report No. 13. Washington, DC: U.S. Government Printing Office.

Short, J. F. 1997. *Poverty, ethnicity, and violent crime*. Boulder, CO: Westview Press.

 THINKING ABOUT ETHICS *The data about individuals that Martinez and Lee obtained were collected by others and were used without any personal identification. Although the American Sociological Association (ASA) Code of Ethics states that confidentiality is not required in the case of information available from public records, the researchers acted appropriately by collecting the data about victims and offenders from the records without names.*

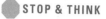**STOP & THINK**   *A good question to ask after reading a piece of research is "What was the unit of analysis, or type of subject, studied by the researchers?" What was Martinez and Lee's unit of analysis?*

STOP & THINK AGAIN   *Other good questions are "What were their major independent and dependent variables?" "How did they measure those variables?" "How did they expect those variables to be related?" "How were they related?" Try to answer as many of these questions about Martinez and Lee's article as you can.*

KINDS OF AVAILABLE DATA

Martinez and Lee take advantage of at least four sources of available data in their study: "homicide logs" from the Miami Police Department (MPD) for information about offenders and victims and their relationships; decennial census reports for information about the population of Miami; Portes, Clarke, and Manning's (1985) estimates of the number of Mariels in the city of Miami; and Marvin Dunn's (1997) estimates of the number of Caribbean-born blacks born in Jamaica and Haiti. The imaginative combination of a variety of data sources is quite common in analyses involving available data. It can make these analyses a great deal of fun, as well as illuminating. Useful in the art of such analyses is understanding the kinds of available data. Three kinds we'll introduce you to are secondary data, existing statistics, and the ever-popular "others."

Secondary Data

Secondary data are data that have been collected by others and have been made available to the current researcher. The "homicide logs," created by the Miami Police while they tried to solve murder cases, but then used by Martinez and Lee to identify the ethnicity of offenders and victims, are an unusual instance of secondary data. Martinez and Lee's access to actual police crime logs is an interesting story, indicating that access to available data isn't always just a matter of finding the right report or downloading the right computer tape. Their accounts of this story are slightly different, but both point to possible stumbling blocks in accessing "available" data. They both agree that, in the beginning, Martinez was a new assistant professor at the University of Delaware with few research contacts. He wanted to study Latino homicides but had no real access to information about Latinos. "There are few Latinos in Delaware," he reports (personal communication), "but I had a colleague with research projects in Miami. He introduced me to somebody on his research staff who knew the MPD Chief. Introductions were made and I eventually received permission to enter the police department" and access its homicide logs. His co-author, Lee, suggests that Martinez's account is overly modest. Lee points out that "police are understandably reluctant to give 'outsiders' unrestricted access to study their files" and that "only by logging countless hours accompanying the detectives on calls, and demonstrating that he was trustworthy, could Ramiro [Martinez] gain the kind of unrestricted access that I [Lee] witnessed when I visited the police department with him: the detectives greeted him enthusiastically and treated him as an insider" (personal communication).

STOP & THINK   *While the information that Martinez and Lee collected from homicide logs certainly counts as secondary data, it is not typical of the kind of secondary data most used by social scientists to produce research articles. What kind of secondary data do you imagine provides the basis for over 50 percent of all articles published in sociology?*

By far the greatest amount of social science research based on the analysis of secondary data, however, employs survey data collected by others. Our "Research in the News" box refers to another immigration study that used data provided by the U.S. government, including the Current Population Survey, to inform the then current debate about whether immigration reform was necessary. You can find hundreds of such data sets at the Inter-University Consortium for Political and Social Research (ICPSR) at the University of Michigan (http://www.icpsr.umich.edu/) and at the Roper Center for Public Opinion Research (http://www.ropercenter.uconn.edu/), both clearinghouses for survey data, mainly, but not exclusively, from the United States. Again, the point is that a large and increasing percentage of the research articles and books published in the social sciences are based on secondary data—data collected by others—and that secondary data sets are increasingly accessible to those of us interested in using them.

RESEARCH IN THE NEWS

The Cable News Network (CNN) reported the findings of a study conducted by two researchers, Pia Arrenius and Madeline Zavodny, which used economic and census data provided by the U.S. government. The study finds that highly skilled immigrants add more to our economy than they take away from it. Immigrants working in technical fields tend to complement native-born workers by moving into fields where the latter can't keep up. Far from taking jobs from American-born labor, they, as entrepreneurs, create jobs. The researchers conclude that the U.S. needs to reform its immigration system to boost legal immigration of highly educated workers.

Source: http://www.cnn.com/2011/US/03/09/texas.fed.immigration/index.html

Both the ICPSR and the Roper Center place some restrictions on access to secondary data, through either fees or the requirement of institutional membership. If you are planning a major research project, you might want to consult your professor about what data sets your college or university already has or whether there might be another convenient way for you to gain access to secondary data you've seen referred to in the literature. One of the most delightful sources of easy-access secondary data, however, is the Social Documentation and Analysis (SDA) Archive at the University of California,

BOX **12.1**	**Some Data Sets Available at SDA Archive**

(http://sda.berkeley.edu/archive.htm)

GSS Cumulative Datafile (1972–2012)
National Election Study (1948–2012)
Multi-Investigator Survey 1994, 1998–1999
Census Microdata: 2000–2008 American
    Community Surveys (over 20 million cases)

1990 and 2000 U.S. 1 percent PUMS
National Race and Politics Survey 1991
National Health Interview Survey 1991
Health Studies from Brazil

Berkeley (http://sda.berkeley.edu/archive.htm). Here you can browse a variety of data sets (see Box 12.1) and analyze the data online.

Suppose you have a research project in social psychology in which you're studying the connection between gender and love. You might want to find out whether, for instance, males or females are more likely to say they would rather suffer themselves than let their loved one suffer.

STOP & THINK   *Quick. Take a guess. Who do you think is more likely to say they'd rather suffer themselves than let loved ones suffer: males or females?*

You can use the most recent GSS, which just happens to ask respondents whether they'd rather suffer themselves than let the one they love suffer, as well as what their gender is. One way to analyze this association would be to access the SDA Archive by going to http://sda.berkeley.edu/ archive.htm.

STOP & DO   *If you have easy access to a computer, why don't you try this? Then click on the GSS Cumulative Datafile 1972–2012 or whatever GSS file contains the 2008 survey data. Then place "sex" in the column box and "agape1" in the row box, before clicking on "run the box." What do you find? Are males or females more likely to say they strongly agree with the statement "I would rather suffer myself than let the one I love suffer?"*

We were surprised to find that males are substantially more likely than females (69.3 percent to 58.0 percent) to strongly agree with the statement "I would rather suffer myself than let the one I love suffer." Think how impressive it would be to add such a finding to your social psychology paper!

The *unit of analysis* for studies based on secondary analysis of survey data is almost always the individual. We've learned something about people from our foray into GSS data on love: that men are more likely than women to say they strongly agree with the statement about suffering for the one they love. Robinson and Martin (2008) used GSS data to discover something else about individual people: that happier people differ from unhappier people in their daily activities, specifically that they report being more engaged in social activities, like religion and newspaper reading, and being less engaged in television watching. People respond to the surveys and people, therefore, tend to be the units of analysis of the research based on those surveys. The units of analysis in Martinez and Lee's study are also individuals. They are focusing on individual people, for instance, when they use the homicide logs to determine how many murders were committed by Mariel Cubans, Haitians, and Jamaicans. And when they use the various estimates of the Mariel, Haitian, and Jamaican populations in Miami, they are simply trying to supply the denominators necessary to calculate homicide rates, much as we needed to use the total number of males and females in the GSS survey to calculate the percentage of each that strongly agreed with the statement about love.

STOP & THINK   *Emile Durkheim's classic study,* Suicide *([1897] 1964), is based, in part, on comparisons of suicide rates in more or less Catholic provinces in Europe. He finds the provinces with high percentages of Catholic residents, for instance, have lower suicide rates than the provinces with low percentages of Catholic residents. Can you tell what the unit of analysis of this part of Durkheim's study was?*

Existing Statistics

Existing statistics are statistics provided by large organizations. Martinez and Lee use one of the most frequently employed sources of existing statistics: the U.S. Bureau of the Census to gather information about the overall population of Miami. Census products come in printed reports, on computer tapes, on microfiche and CD-ROM laser disks, and now via websites. But they found that they needed to combine statistics from one source with data or statistics from another (in their case, data from the homicide logs). Most users of existing statistics find the need to make similar combinations. Maynard, Shircliff and Restivo (2012) for instance, used statistics about 172 countries from the World Health Organization and the World Bank to show that being in debt to the International Monetary Fund makes government less likely to be able to fight the spread of tuberculosis successfully.

In general, and in contrast to studies using secondary data, the *unit of analysis* in studies based on existing statistics is not the individual. For Durkheim's study of suicide rates by province, the unit of analysis was the province. In Maynard et al.'s study, it was nations. Statistics, by their nature, summarize information about aggregates of people (countries, regions, states, cities, counties, organizations, etc.), and when statistics are the data or information we have to work with, we're almost surely focused on comparing aggregates of people.

Once you have a research question, it is useful to find relevant sources of data about your unit of analysis. Roger, along with a student, Christine Arthur (Arthur and Clark, 2009), wanted to know what contributes to domestic violence in a society. In our readings, we'd come across population-based indicators of the percent of women physically assaulted by an intimate partner as reported by Heise, Ellsberg, and Gottemoeller (1999). We found other indicators by following up leads in literature we read. But some we found by Googling things like "date women's vote." In this way we discovered the Inter-Parliamentary Union's (2004) *Women's Suffrage* webpage (http://www.ipu.org/wmn-e/suffrage.htm), on which we found the years in which women got the vote in different countries, and the United Nations Statistics Division (2000) *The World's Women 2000: Trends and Statistics* (http://unstats.un.org/unsd/demographic/ww2000/), where we found the level of women's labor force participation. We did, in short, what is very common in research based on existing statistics: We tried to find sources of statistics that provided information about our units of analysis (in our case, nations) that could serve as measures of our concepts of interest.

Roger and Christine actually located other sources that yielded statistics enabling them to measure variables of interest about most countries of the world. And what is true of their work is true of much work involving existing statistics: one searches for information about one's units of analysis (whether they be countries, regions, states, organizations, etc.), often from different sources, and combines that information into a single data set about those units (see Box 12.2). Creating the combination often involves listing units of analysis (as Arthur and Clark do their countries in the left-hand column of Box 12.2) and then listing values of the variables of interest (as they do in subsequent columns).

STOP & THINK *Can you use Box 12.2 to determine the year when women in Australia gained the vote? What was the women's labor force participation rate in Bangladesh in 1999?*

Once you've collected statistics from various sources, you can begin analyzing them using statistical techniques like the ones outlined in Chapter 15's section on quantitative data analysis.

The sources of existing statistics are innumerable. If you have an idea of the kind of topic you'd like to work on, and the unit of analysis, you can

| BOX **12.2** | A Portion of Arthur and Clark's Data Set on Countries: Finding Data Sources That Complement One Another |

Countries	From Heise et al. (1999) Percent of Women Physically Assaulted by an Intimate Partner	From Inter-Parliamentary Union (2004) Year Women Got Vote	From United Nations Statistics Division (2000) Women's Labor Force Participation Rate (1999)
Australia	8	1902	55
Bangladesh	47	1972	56
Barbados	30	1950	62
Canada	27	1917	61

find usable data in either of two ways: do a literature review to see what others have used or do a web search. You could, of course, do both. If you choose to search the web, finding key words is crucial. Are you interested in homicide rates in American cities? Try Googling "homicide rates in American cities." You'll get there. If you remember that the FBI warehouses information about crime in the United States in its UCR, you might Google "Uniform Crime Reports." You get right to the UCR website (http://www.fbi.gov/ucr/ucr.htm), and you'll soon be looking at interesting information.

⬢ STOP & DO | *If you're sitting at your computer, why don't you see if you can find the website for the U.S. Census Bureau?*

If you wanted the U.S. Census Bureau, you could Google "U.S. Census Bureau" and find its web address http://www.census.gov/.

We won't try to anticipate your needs, but we will provide you with some web addresses most frequently used for sources of existing statistics (Box 12.3) and the assurance that, even as recently as the last decade, finding existing statistics (which used to mean a trip to the right library and a hunt for the right volume) has never been so easy.

Of course, the use of existing statistics constitutes one of the richest research traditions in the social sciences. It provided grist, as we've already pointed out, for Émile Durkheim's classic study, *Suicide* ([1897] 1964), and for Karl Marx's argument for economic determinism in his *Das Kapital* ([1867] 1967). It is hard to imagine what these early social scientists would make of the ever increasing and increasingly accessible amounts of statistics available today—and, frankly, there's no way we could do justice to them here—but it's almost a certainty that Marx and Durkheim would have done a lot with them.

Other Kinds of Available Data

You don't need to depend on outside agencies to make your "available data" available. In their now classic list of alternative data sources, Webb, Campbell, Schwartz, and Sechrest (2000) mention, for instance, **physical traces** and personal records. Physical traces include all kinds of physical evidence left by humans in the course of their everyday lives and can generally be divided into what Webb and his associates call **erosion measures** and **accretion measures**. An erosion measure is one created by the selective wear of a population on its physical environment. Webb and associates cite Frederick Mosteller's (1955) study of the wear and tear on separate sections of the *International Encyclopedia of the Social Sciences* in various libraries as a quintessential example of the use of erosion measures, in this case to study which sections of those tomes received the most frequent examination. An accretion measure is one created by a population's remnants of its past behavior. The "Garbage Project" at the University of Arizona has studied garbage since 1973. By sorting and coding the contents of randomly selected samples of household garbage, William Rathje and his colleagues were able to learn a great deal, including what percentage of newspapers, cans, bottles, and other items

physical traces, physical evidence left by humans in the course of their everyday lives.

erosion measures, indicators of a population's activities created by its selective wear on its physical environment.

accretion measures, indicators of a population's activities created by its deposits of materials.

BOX **12.3** Some Good Sources of Existing Statistics

Units of Analysis: American states, cities, counties, and so on

- *The Bureau of the Census Home Page* (http://www. census.gov/) allows users to search census data on population. In particular, the bureau's *Statistical Abstract* (http://www. census.gov/statab/www/) provides useful information on crime, sex and gender, social stratification, and numerous other topics.
- *FBI UCR Page* (www.fbi.gov/ucr/ucr.htm) offers access to recent crime data.
- *Bureau of Labor Statistics Data Home Page* (www.bls.gov/data/) has, among other things, data on men's and women's labor force participation.
- *Department of Education Home Page* (http://www.ed.gov/index.jhtml) leads to much information about educational achievement, and so on.

- *The National Center for Health Statistics Home Page* (http://www.cdc.gov/nchs/) is a wonderful source of information about health and illness.

Units of Analysis: Countries

- *The United Nations Yearbook Statistical Division* (http://unstats.un.org/unsd/default.htm) provides basic statistical information, including data on population (in its *Demographic Yearbook*), trade, health, education, and many other topics.
- *The International Labor Office* (http://www.ilo.org/) is a great source of information about work worldwide.
- *CIA World Factbook* (http://www.cia.gov/cia/publications/factbook/) provides a wide variety of information on most countries in the world.
- *The World Bank* (http://www.worldbank.org) offers basic information and much financial information about countries of the world.

---

unobtrusive measures, indicators of interactions, events, or behaviors whose creation does not affect the actual data collected.

aren't recycled and how much edible food is thrown away (Rathje and Murphy, 2001). Because people frequently misreport their dietary habits, the proponents of "garbology" argue that this type of available data can sometimes be more useful than asking people questions about their behavior. One great advantage of using physical traces, like garbage, is that their "collection" is **unobtrusive** (or unlikely to affect the interactions, events, or behavior under consideration).

 STOP & THINK

Suppose that, like Kinsey and associates (1953), you wanted to study the difference between men's and women's restrooms in the incidence of erotic inscriptions, either writings or drawings. Would you be taking erosion or accretion measures?

---

personal records, records of private lives, such as biographies, letters, diaries, and essays.

Personal records include autobiographies, letters, e-mails, diaries, Twitter feeds, and essays. With the exception of autobiographies, which are often published and available in libraries, these records tend to be more difficult to gain access to than the public records we've already discussed. Frequently, however, personal records can be used to augment other forms of data collection. Roger has used autobiographies, for instance, to augment information from biographies and questionnaire surveys to examine the family backgrounds of famous scientists, writers, athletes, and other exceptional achievers (Clark, 1982; Clark and Ramsbey, 1986; Clark and Rice, 1982). Private diaries, when they are accessible, hold unusually rich possibilities for providing insight into personal experience. Historians like Laurel Thatcher Ulrich have used them to create classics, such as her *A Midwife's Tale* (1990), to piece together accounts of times gone by. Letters can be an especially useful source of information about the communication between individuals. Karen Lystra's *Searching the Heart: Women, Men and Romantic Love in Nineteenth-Century America* (1989) uses letters almost exclusively to build an argument that nineteenth-century conventions of romantic love required people to cultivate an individual, unique, interesting self that they could then present to the beloved other—that is, that romance is an historical and psychological root of individualism.

Lystra's works points to one great advantage of available data: the potential for permitting insight into historical moments that are inaccessible through more conventional social sciences techniques (such as interviews, questionnaires, and observation). Other sources of such historical data are newspapers and magazines, and organizational and governmental documents. Beverly Clark (2003) has scrutinized nineteenth- and early twentieth-century newspapers and magazines for insight into the popular reception of (children's) books written by, among others, Louisa May Alcott. Clippings from the *Cleveland Leader* and the *Albany Daily Press and Knickerbocker* indicate that the Cleveland public library required 325 copies of *Little Women* in 1912 to satisfy constant demand; the New York City branch libraries required more than a thousand copies.

Texts, like the diary studied by Ulrich (1990), the letters studied by Lystra (1989), and the newspapers and magazines studied by Clark (2003), are amenable to both quantitative and qualitative data analysis. In the next chapter you'll see how others have done these kinds of analyses.

Sociologists have frequently used archives of unofficial and official records as sources of data, but they've tended to neglect visual materials (Grady, 2007b: 66). Increasingly, though, visual sociologists have taken advantage of such archives and repositories, just as the number of easily accessible archives and repositories has grown. These repositories can constitute an excellent source of primary data for social research. Grady (2007a), as mentioned in Chapter 6, studied his own archive of all advertisements containing a black figure in *Life* magazine from 1936 to 2000 and found in them evidence of increasing "white commitment to racial integration," despite also finding areas of social life where whites are still wary of blacks. Anden-Papadopoulos (2009) investigated homemade videos uploaded to YouTube by coalition soldiers in Afghanistan and Iraq and found considerable evidence that their firsthand experiences contradicted the official narrative of the war offered by the U.S. government. Excellent sources of still and video images are the "image" and "video" functions of Google.

STOP & DO

Check out Google Video. You might type in something like "Iraq war YouTubes by soldiers" and watch two or three. Do you agree with Anden-Papadopoulos (2009) that these YouTube videos created by soldiers tend to be critical of the government's conduct of the war?

Grady (2007b: 66) suggests you might get fascinating social insights from studying family albums, yearbooks, postcards, greeting cards, magazine advertisements, and newspaper advertisements. You could also look at television advertisements, television programs, and movies. Students in Roger's classes have done interesting research on the changing portrayal of female characters in animated films by Disney, in motion pictures winning the Academy award for best actress, and in television situation comedies. The possibilities are endless.

The larger point here is, however, this: You really don't want to be guided in your search for available data by any of the "usual suspects" we've suggested. A much sounder approach is to immerse yourself in a topic by reading the available literature. In the course of your immersion, look at the sources other researchers have used to study the topic. If you discover that they've typically used the FBI UCR, you might want to use that too. If they've used the United Nations' *Demographic Yearbook*, you also might want to look into that. Very often, using earlier research as a resource in this way is all you'll need to do to find the data you'll want to use.

ADVANTAGES AND DISADVANTAGES OF USING AVAILABLE DATA

Available data offer many distinct advantages over data collected through questionnaires, observation, or experiments. Most striking among these are the savings in time, money, and effort they afford a researcher. The analysis of secondary data or existing statistics might cost little more than the price of your Internet connection (though depending on your institutional affiliation, there can be a charge of several hundred dollars—still relatively little—for access to and the right to use such data) and the time it takes to figure out how you want to analyze them. Countries might spend billions on their national censuses, but comparing 200 nations based on the results of those collections might cost you only the price of a library card or a few keystrokes on your computer.

Martinez and Lee's work with police logs, however, provides a useful caveat to the general rule that available data are cheap and time-saving. In fact, Martinez estimates that he spent roughly five years, off and on, collecting data on 2,500 homicide cases that occurred from 1980 to 1995 and probably spent about $200,000, using grants, salary, and sabbatical supplements, to do so. Part of this expense was because he started his work while working in Delaware. He says, "Flying to Miami from Delaware sounded great at first but it was not cheap. You needed airfare, hotel and food monies, as well as money for supplies. You'd manually pull out homicide records from a file room, copy them, and ship them back to your home base. Once you were at home, you needed to store, read and code the data, then have them entered and readied for statistical analysis [usually by a hired student]" (personal communication). Martinez speaks of "maxing out two credit cards" while getting the project going and hoping that a "grant proposal would hit." Eventually, he did "hit" on grants, but, when he didn't, even later on, "it was depressing. At times I felt like quitting. But then I remembered my credit cards."

Matthew Lee suggests that we should think of available data as being on a "continuum of availability" that reinforces the point that, although much available data is, in fact, easily available, some can come at considerable cost to the researcher (personal communication). He points out, however, that sometimes the less available data are the only means to true conceptual breakthroughs. He points out that he got interested in the migrant/crime question because he'd been so impressed by Cliff Shaw and Henry McKay's "idea that social processes like immigration and residential turnover (rather than 'bad' individuals) weaken community levels of social control and thereby facilitate crime" (1931). Yet Lee and Martinez found little support for that notion. In fact, immigrants appear to be stabilizing urban communities and suppressing levels of violent crime. Lee believes that one reason many people see newcomers as a disorganizing influence on "institutions and sensibilities" is that the most easily available data, such as the FBI's UCR, cannot be used to address the validity of conventional wisdom on immigration and crime. And it's hard work to get alternative data. But, Lee points out, "Our experiences suggest that taking a gamble on collecting secondary data that are not easily available can lead to a bigger theoretical payoff" (personal communication).

A related advantage of using available data is the possibility of studying social change over time and social differences over space. Robert Putnam, in his national bestseller, *Bowling Alone: The Collapse and Revival of American Community* (2000), analyzes available data from many sources (election data, survey data, organizational data) to substantiate his argument that political, civic, religious, and work-related participation have substantially declined in America during the past quarter century, as have informal social connections and volunteer and philanthropic activities. And using U.S. Census data collected since 1790, gerontologists have been able to discern that the proportion of the American population that is 65 years or older has grown from 2 percent in 1790 to 4 percent in 1890 to 13 percent in 2010. Most individual researchers don't live long enough to observe such trends directly, but because of available data and the imagination to use them, they don't have to.

STOP & THINK   *Imagine you want to find out how much the size of the "average" American family household has changed over the last two centuries. Would you be more inclined to use a survey, participant observation, an experiment, or U.S. Census reports as your data source? Why?*

Available data also mean that shy people who don't like the prospect of persuading others to answer a questionnaire or to participate in a study can make substantial research contributions. Moreover, the researcher can work pretty much where he or she wants and at whatever speed is convenient. Roger once coded data from a source he'd borrowed from the library while waiting (along with 150 other people who could do no more than knit or play cards or read) in a courthouse to be impaneled on a jury. (At other times, when unable to sleep, he's collected data in the wee hours of the morning—a time when researchers committed to other forms of data collection pretty much have to shut down.) Again, this advantage doesn't always exist. Martinez found it useful, for instance, to cultivate many MPD officers, if only to provide details about cases that were left out of the logs he studied (personal communication).

A final, and most important, advantage of using available data is that the collection itself is unobtrusive and extremely unlikely to affect the nature of the data collected. Unlike researchers who use questionnaires, interviews, observations, or experiments, those of us who use available data rarely have to worry that the process of data collection will affect our findings.

On the other hand, the unobtrusiveness of using available data is bought at the cost of knowing less about the limitations of one's data than if one collects the data firsthand. One major disadvantage of available data is that one is never quite sure about its reliability. Martinez and Lee, for instance, were evidently concerned that the "homicide logs" they investigated provided adequate definitions of victim and offender ethnicity. Martinez observes, "If a homicide record written in 1980 was not clear on certain things, you were stuck with it." Eventually, though, he had access to an atypical resource for researchers relying on available data. "With time," he reports, "I became acquainted with detectives, especially the 'old-timers,' and asked them about certain cases.... If I had not had direct access, it would have been very difficult. I was very fortunate" (personal communication).

Most researchers, using available data, are not so fortunate. Cross-national researchers (like Arthur and Clark, 2009) using data on female labor force participation, for instance, have been plagued by their awareness that, in many countries, especially in those where women's work outside the home is frowned upon, there is considerable undercounting of that work.

STOP & THINK   *What kinds of nations, do you think, would be most likely to undercount women's work outside the home? Why?*

Sometimes the best you can do is to be explicit about your suspicion that some of your data are not reliable. You can at least try to clarify the kinds of biases the sponsors of the original data collection might have had, as we did when we used available data on female labor force participation. This kind of explicitness provides readers with a chance to evaluate the research critically. At other times, checks on reliability are possible, even without the capacity to perform the kinds of formal tests of reliability (e.g., inter-rater tests—see Chapter 6) that you could use while collecting primary data, or access, such as Martinez's, to the collectors of the original data. Steven Messner (1999) found that cities where people watched lots of violent television programs were likely to have lower violent crime rates than other cities were. Messner was concerned, however, that the "underreporting" of violent crimes in cities where many people were desensitized to crime by the viewing of violent television programming might have been responsible for his unexpected finding that such viewing was negatively associated with the presence of such crimes. But he knew that one kind of violent crime, homicide, was very unlikely to be underreported under any circumstances. So, when he found that the unexpected negative association between viewing violent programs and violent crime rates persisted even for homicide rates, he concluded that it was less plausible that his other unexpected findings were only the result of systematic underreporting.

STOP & THINK *Why is homicide less likely than, say, shoplifting to be underreported, whatever the television viewing habits of inhabitants?*

A second disadvantage of using available data is that you might not be able to find what you want. In some cases, there just might not have been anything collected that's remotely appropriate for your purposes. Modern censuses are designed to enumerate the entire population of a nation, so you can reasonably expect to be able to use their data to calculate things such as sex ratios (the number of males per 100 females, used, for instance, by Messner as a control variable in his analysis). But ancient Roman censuses focused on the population of adult male citizens (the only group, after all, owning property and eligible for military service and therefore worth knowing about for a government interested primarily in taxation and military recruitment [Willcox, 1930, cited in Starr, 1987]), and so the data necessary for the computation of, say, sex ratios, not to mention population density or the percentage of the population that's of a certain age, might simply not be available. Also, available data might not present data the way you want. Martinez notes, for instance, that crime data from the FBI are presented in racial categories ("black" and "white") and not ethnic ones (e.g., "Mariel," "Haitian," and "Jamaican").

Sometimes, however, you can locate data that very nearly (if still imperfectly) meet your needs, so the issue of validity arises. Numerous cross-national studies (e.g., Arthur and Clark in their study of domestic violence mentioned earlier) have used the women's labor force participation rate as a rough measure of women's economic status in a nation. A women's labor force participation rate refers to the proportion of women in an area (in this case, a country) who are employed in the formal workforce. The rationale behind its use as a measure of women's economic status is the notion that, if women don't work for pay, they won't have the power and prestige within the family and in the community that access to an independent income can buy. But what if women and men don't always (as we know they don't) receive equal monetary returns for their work? Wouldn't the economic status of women then depend on, among other things, how closely their average income approximated that of men's—something that their labor force participation rate doesn't begin to measure? We're sure you can think of other problems with labor force participation as a valid measure of economic status, but you probably get the idea: Sometimes researchers who employ available data are stuck using imperfectly valid indicators of the concepts they want to measure because more perfect indicators are simply not available.

STOP & THINK *Can you think of other problems with using women's labor force participation, or the proportion of women over 16 years old who are in the labor force, as an indicator of women's economic status in an area? (Hint: What other things, besides working at all, contribute to our sense that a person has a high economic status?)*

One peculiar danger of using available data, especially existing statistics, is the possibility of becoming confused about your unit of analysis (or subjects) and committing the ecological fallacy. Much of the research on the migrant-crime connection (e.g., Hagan and Palloni, 1998; U.S. Commission on Immigration Reform, 1994), before Martinez and Lee's study, had looked at the relationship between the proportion of migrants in a city and its crime rates. This research, however plausible, was nonetheless subject to the ecological fallacy because its units of analyses (cities) were not the same as units of analyses (people) researchers wanted to make inferences about. Martinez and Lee avoid the ecological fallacy by obtaining information that enables them to directly compare the homicide rates of various immigrant groups and of immigrants and non-immigrants.

SUMMARY

In contrast with primary data, or data that the researcher collects himself or herself, available data exist before you begin your research. Secondary data, or data that have been collected by someone else, have become increasingly popular in social science research. More and more social scientists are using survey data that have been collected by other researchers and are made accessible for secondary analysis. Two major repositories for such data are the ICPSR at the University of Michigan and the Roper Center at the University of Connecticut.

A major advantage of using available data is that the data are usually cheap and convenient. Martinez and Lee's experience indicates, however, that there is a "continuum of availability" and that sometimes the least available of available data are well worth exploring. Available data also allow the study of social change over long periods and of differences in many societies. Finally, their collection is unobtrusive (or unlikely to affect the interactions, events, or behavior under consideration) even if their original collection might not have been.

Common sources of existing statistics are governmental agencies, like the U.S. Bureau of the Census, or supra-governmental agencies, like the United Nations. These statistics tend to be summaries at some aggregate level of information that has been collected at the individual level. These data have become increasingly available on the Internet.

The main disadvantage of using available data is that, because someone else has collected the data, their reliability might be unknown or unknowable by their user. Moreover, one frequently has to settle for indicators that are known to be imperfectly valid. Finally, especially when using existing statistics, the researcher has to be especially cognizant of the temptation to commit the ecological fallacy—of making inferences about individuals from data about aggregates.

EXERCISE 12.1

Urbanization and Violent Crime: A Cross-Sectional Analysis

This exercise gives you the opportunity to examine one assumption that seems to lie behind Martinez and Lee's analysis: Cities, or more particularly standard metropolitan statistical areas (SMSAs), are especially good places to study violent crime. We'd like you to see whether, in fact, population aggregates that are more highly urbanized (i.e., have a greater proportion of their population in urban areas) have higher violent crime rates than those that are less highly urbanized.

Let's examine a population aggregate that can be more or less urbanized: states in the United States. Find a recent resource, perhaps in your library, that has data on two variable characteristics of states: their levels of urbanization and their overall violent crime rate. Collect data on these variables for the first 10 states, listed alphabetically. (Hint: *The Statistical Abstract of the United States* [http://www.census.gov/compendia/statab/] is a good source of the data you'll want. Most college libraries will also have this in their reference section.)

1. Prepare 10 code sheets like this (actually, you can arrange all data on one page):
   State: Total Violent Crime Rate: Percent Urban:

2. Go to your library (or the website) and find the data on crime and urbanization. (If you use the website, you can just type in "crime rates" and "percent urban" in the "Search the Abstract" box. Using the code sheets you've prepared, collect data on 10 states of your choice.)

3. Once you've collected the data, analyze them in the following ways:
   a. Mark the five states with the lowest percent urban as **LU** (low urban) states. Mark the five states with the highest percent urban as **HU** (high urban) states.
   b. Mark the five states with the lowest total violent crime rates as **LC** (low crime) states. Mark the five states with the highest total violent crime rates as **HC** (high crime) states.
   c. Calculate the percentage of **LU** states that you have also labeled **HC**. Calculate the percentage of **HU** states that you have also labeled **HC**.

4. Compare your percentages. Which kinds of states tend to have the highest violent crime rates: highly urbanized or not-so-highly urbanized states? Interpret your comparison relative to the research question that sparked this exercise.

EXERCISE 12.2

Life Expectancy and Literacy: A Cross-National Investigation

This exercise asks you to investigate the connection between education and health by examining the research question: Do more educated populations have the longest life expectancies? Investigate this hypothesis using data from 10 countries (your choice) about which you obtain data on the percentage of the population that's literate and life expectancy. [You can find the data in most world almanacs and in a volume of Human Development Reports, any year, online. You can search for them by using key words "Human Development Reports" and then "Life Expectancy."]

1. Prepare 10 code sheets like this (again, you can arrange all data on one page):
   Country: Literacy: Life Expectancy:

2. Go to your library (or the website) and find the appropriate data. Using the code sheets you've prepared, collect data on 10 countries of your choice.

3. Once you've collected the data, analyze them using the technique introduced in Exercise 12.1.
   a. Mark the five countries with the lowest literacy rate as **LL** (low literacy) nations. Mark the five countries with the highest literacy rates as **HL** (high literacy) nations.
   b. Mark the five countries with the lowest life expectancy as **LLE** (low life expectancy) nations. Mark the five countries with the highest life expectancy as **HLE** (high life expectancy) nations.
   c. Calculate the percentage of **LL** nations that you have also labeled **HLE**. Calculate the percentage of **HL** nations that you have also labeled **HLE**.

4. Compare your percentages. Which kinds of nations tend to have the highest life expectancy: those with high literacy rates or those with not-so-high literacy rates? Interpret your comparison relative to the research question that sparked this exercise.

5. What is the ecological fallacy? Do you think you succumbed to this fallacy in your interpretation of your analysis in this exercise? Why or why not?

EXERCISE 12.3

For Baseball Fanatics

This exercise gives you the chance to examine an age-old social science dispute: whether hitting or pitching wins ball games. Help shed light on this dispute by, again, turning to a recent almanac and the Major League Baseball website (http://mlb.mlb.com/index.jsp), this time to collect winning percentages, team batting averages, and team earned-run averages for all the teams in either the American or the National League for one year. Use data collection techniques similar to those you used for the last two exercises, this time collecting data on three variables, rather than two. Moreover, you should use the data analysis techniques of each exercise, this time examining the relationship between winning percentage and team batting average, on the one hand, and winning percentage and team earned-run average, on the other. Are both batting average and earned-run average associated with winning percentage? Which, would you say, is more strongly associated with it? Why?

Write a brief essay in which you state how your analysis bears on this research question.

EXERCISE 12.4

A Photo Finish

This exercise gives you a chance to exercise your skills as a visual sociologist through a comparison of two photos used to announce Barack Obama's win of the 2008 Presidential election. One of these will be the one that appeared on the first page of the *New York Times* the day after the election; the other will be from the first page of the *Philippine Daily Inquirer* that day. Your job will be to compare the messages of these photos and to try to understand how the editors of these two newspapers were similar and how they differed in their take on the election.

1. Find Google Images and type in "Obama wins election New York Times 2008." Find the largest picture you can of the picture the

New York Times put on its front page and study the image. What, would you say, are three or four main messages that the editors of the *New York Times* may have tried to convey by using this image?

2. Using Google Images again, type "Obama wins election Philippines Daily Inquirer 2008." Find the largest picture you can of the picture the *Philippine Daily Inquirer* put on its front page and study the image. What, would you say, are three of four main messages that the editors of the *Philippine Daily Inquirer* may have tried to convey by using this image?

3. What are clearly some of the messages that both the *New York Times* and the *Philippine Daily Inquirer* editors tried to convey through their images?

4. What is one message conveyed by the *New York Times* image that is not conveyed by the *Philippines Daily Inquirer* image?

5. What is one message conveyed by the *Philippines Daily Inquirer* image that is not conveyed by the *New York Times* image?

6. How might you interpret the similarities and differences you've discovered?

CHAPTER 13

Content Analysis

CHAPTER OBJECTIVES

This chapter is designed to help students

- Learn what content analysis is
- Get a sense of the range of materials that are appropriate for content analysis
- Prepare for doing quantitative and qualitative content analysis
- Appreciate visual analysis
- Do a small-scale content analysis
- Understand the advantages and disadvantages of content analysis

CHAPTER OUTLINE

Appropriate Topics for Content Analysis

Quantitative Content Analysis

Sampling

Creating Meaningful Variables

Qualitative Content Analysis

Visual Analysis

Advantages and Disadvantages of Content Analysis

Summary

Exercises

© Jonathan Littlejohn/Alamy

337

In 1972, four sociologists, Lenore Weitzman and three colleagues, noticed that female characters were essentially absent from award-winning children's picture books published in the late 1960s. They argued that this omission may have conveyed to little girls to whom these books were read the idea that they weren't as important as little boys. Does this kind of reasoning seem persuasive to you? It did to Roger and two students, Julie Kessler and Allyssa Coon, so they decided to see, first, whether the omission of female characters in children's picture books persisted into the twenty-first century and, second, whether the omission, if still present, differed by the racial, ethnic, sexual sympathies and gender of the authors and illustrators. Their piece illustrates the technique of **content analysis**, a technique that is particularly useful for doing historical investigations. Content analysis is a method of data collection in which some form of communication (speeches, television programs, newspaper articles, films, advertisements, blogs, tweets, even children's picture books) is studied systematically. Content analysis is one form of available data analysis.

content analysis, a method of data collection in which some form of communication is studied systematically.

FOCAL RESEARCH

Who Envisions the Distaff Side?: Female Visibility in Children's Picture Books over Time and by Authors' Race, Ethnicity, Sexual Sympathies and Gender*

By Roger Clark, Julie Kessler and Allyssa Coon

It is forty years since Lenore Weitzman and colleagues published their classic (1972) piece, "Sex-Role Socialization in Picture Books for Preschool Children." They showed that children's picture books of the late 1960s rendered female characters essentially invisible. McCabe et al. (2011) showed that the level of female invisibility varied throughout the twentieth century but that, throughout, it threatened the "symbolic annihilation" of women and girls, an annihilation that could have negative consequences for girls' self-esteem. Our goal here is not only to continue the focus on the visibility of female characters in well-regarded picture books into the twenty-first century, but also to expand the investigation by examining the significance of illustrators' race, ethnicity, sexual sympathies and gender.

McCabe et al. (2011) found that female visibility in children's picture books, though decreasing after the first women's movement achieved its goal of the vote for women in 1920, increased after the 1960s, probably the result of the second women's movement. Multicultural and lesbian feminists (e.g., hooks, 1987; Collins, 1990; Anzaldua, 1987; Lorber, 2001) consistently emphasize the importance of women and girls for helping their communities "get over" forms of oppression and make them central to their theorizing about social progress.

*This piece is redacted from Clark, Kessler and Coon's "Women and Girls Last?: Female Visibility in Children's Picture Books by White, Black and Gay-Sympathetic Authors," forthcoming in the Spring, 2013, edition of the *International Review of Modern Sociology*. Published with permission.

The women's movement generally has argued for the need for female access to places previously occupied by males.

Our understanding of this movement and our reading of McCabe et al. (2011) and of feminist theorists leads us to three hypotheses: 1) 21st-century picture books make female characters more visible than books of the 1960s; 2) books by African American, Latino/a and Gay-Sympathetic authors and illustrators today make female characters more visible than books by white authors and illustrators today; 3) books by female authors make female characters more visible than books by male authors.

Methodology

Caldecott Medalists and runners-up from 2007 to 2011 are our source for White-authored and illustrated books. The Caldecott is the most prestigious American award given for children's picture books but it is not given just to White illustrators, so we have deleted books by known Black illustrators in certain analyses. (There were no Latino/a authors among Caldecott illustrators.) Whereas the Caldecott award is given only to illustrators of "outstanding picture" books, the Coretta Scott King and Pura Belpré awards are given to books that meet more specific goals. The King award, for instance, is given to African American illustrators whose works "promote understanding and appreciation of the culture of all peoples and their contribution to the realization of the American dream of a pluralistic society" (Coretta Scott King Award home page, 2011). And the Pure Belpré award honors the Latino/Latina illustrator whose "work best portrays, affirms, and celebrates the Latino cultural experience in an outstanding work of literature for children and youth" (Pura Belpré Award Home page, 2011). We used the Pura Belpré award winners and runners-up from 2007 to 2011 for our sample of Latino/Latina-illustrated books. To obtain comparable numbers of main characters in the King books, books that from 2007 to 2011 tended to focus on generalized subjects, like illustrations of poems by Langston Hughes, rather than stories with clear-cut characters, we examined King award winners and runners-up from 2002 to 2011 for our sample of books by Black illustrators. We define Gay-Sympathetic authors as ones whose works evince sympathetic interest in gay or lesbian characters and their family members. Our Gay-Sympathetic books are from Patricia Sarles (2011) webpage entitled "Gay-Themed Picture Books for Children." Sarles' goal is to acknowledge books that deal with the "experience of being a child in a gay family, or having a gay friend or family member" (Gay-Themed Picture Books Home page, 2011). We examined only books that had already been adopted by 100 or more libraries by the time of our study (Sarles supplies data on library adoptions) and examined all such books that had been published between 2003 and 2010. Our sample in this study includes 70 books: 18 Caldecotts, 26 Kings, 15 Pura Belprés and 11 Gay-Sympathetic books. We also present, for comparative purposes, data on 18 books analyzed by Clark, Lennon and Morris in 1993, the 18 Caldecott winners and honor books from 1967 to 1971.

We have followed Clark et al. (1993) in their measurement strategies and employed three indicators of female visibility: the proportion of human single-gender illustrations depicting only female characters; the proportion of books with no female character at all; and the proportion of books with a central female character. We agree with Williams et al. (1987) and Clark et al. (1993) that it isn't

(continues)

FOCAL RESEARCH *(continued)*

always possible to definitively determine the gender of characters, even by studying illustrations and parallel texts. In *Red Sings from the Treetops*, Joyce Sidman (2009) creates a remarkably vivacious character and an apparent sibling, and neither character's gender identity is perfectly clear. The egg in Seeger's (2007) *First the Egg* is even less clearly gendered. Two recent Gay-Sympathetic books, Newman and Thompson's (2009) *Daddy, Papa, and Me* and *Mommy, Mama, and Me*, feature infants (the titular "me's") of unidentifiable gender. With these exceptions, however, we made reasonably confident gender assignments. Our counts of single-gender human pictures and single-gender animal pictures were, for the three of us, correlated at the .93 level or more. We took the average of our three counts to be our final score for a book. We agreed on whether books had any female characters and whether they had a central female character over 90% of the time. When we disagreed, the identification made by two out of the three of us dictated how we coded the variable.

Findings

Weitzman et al. (1972) claimed that females were essentially invisible in children's picture books (Caldecotts) of the late 1960s. Hypothesis 1 suggests that books by White, Black, Latino/a and Gay-Sympathetic authors today make females more visible than books by Caldecott authors in the late 1960s did. Table 13.1 supplies a good deal of evidence to support this hypothesis. Following Clark et al. (1993), we have defined human single-gender illustrations as ones depicting only male or female human characters. Clark et al. (1993) found that the percentage of such illustrations that depicted females was 11.7 percent of all single-gender illustrations for the Caldecotts of the late 1960s. By the late 2000s, we find that this percentage was 28.7 for Caldecotts, substantially and significantly greater than it was in the late 1960s. The Kings, Pura Belprés and the Gay-Sympathetic books also made female characters substantially and significantly more visible by this measure than the 1960s Caldecotts.

Our other two indicators of female visibility, the percentage of books with no female character, and the percentage of books with a central female character, tell much the same story. Among 1960s Caldecotts, 33.3% of the books had no female character, higher than in any of the categories of 2000s books: 22.2% of Caldecotts, 15.4% of Kings, 0% of Pura Belprés, and 9.1% of Gay-Sympathetic Books. All four of the categories of 2000s books were substantially and significantly more likely than the 1960s Caldecotts to have central female characters. In fact, while only 11% of the 1960s books had female central characters, around half of all categories of the 2000s books had them: 44% of Caldecotts did; 58% of Kings; 60% of Pura Belprés, and 64% of Gay-Sympathetic books.

Our second hypothesis, that books by Black, Latino and Gay-Sympathetic authors and illustrators are more likely to make females more visible than White authors and illustrators, is also supported by Table 13.1. Thus, the percentages of human single-gender illustrations that are of females in the King (41.9%), Pura Belpré (34.2%) and Gay-Sympathetic books (34.2%) are all greater than in recent Caldecotts (28.7%). None of these differences is statistically significant, however, unless one eliminates from the Kings Kadir Nelson's (2008) *We Are the Ship*, a book that is a history of the Negro baseball league and therefore presents 44 pictures only of men. Then, the percentage of human single-gender

TABLE **13.1**
Comparisons of Female Visibility

	Caldecotts 1967-1971	Caldecotts Late 2000s	Kings 2000s	Pura Belprés 2000s	Gay- Sympathetics 2000s	Male Illustrators 2000s	Female Illustrators 2000s
Human Single-Gender Illustrations							
Total Number	186	356	245	111	76	416	394
Percentage Female	11.7	28.7	41.9	34.2	34.2	20.0	51.3
Books							
Total Number	18	18	26	15	11	31.1	43.2
Percentage with No Female Character	33.3	22.2	15.0	0	9.1	29.1	9.1
Percentage with Central Female Character	11.1	44.4	57.7	60	63.6	32.2	53.4

illustrations that are of females jumps to 51.2% among the Kings, almost twice as great as the percentage in the Caldecotts and significantly different, at the .01 level, from that percentage.

The two indicators based on books, rather than illustrations, as units of analysis tell the same story: that females are more visible in the current Kings, Pura Belprés and Gay-Sympathetic books than the Caldecotts. Thus, while 22.2% of the current Caldecotts have no female character, only 15.4% of the Kings, 0% of the Pura Belprés and 9.1% of the Gay-Sympathetic books do so. (Only the difference between the Pura Belprés and the Caldecotts is statistically significant, however.) And, while 44.4% of the Caldecotts have a central female character, fully 57.7% of the Kings, 60% of the Pura Belprés and 63.6% of the Gay-Sympathetic books do so.

To our knowledge, no previous analysis of the gender of characters in children's picture books has examined whether the gender of the author and illustrator is related to the visibility of female characters. We find that it is, consistent with our third hypothesis. In fact, the last two columns of Table 13.1 clearly show that female authors and illustrators are substantially and significantly more likely

(continues)

FOCAL RESEARCH *(continued)*

than male authors and illustrators to feature female characters. Thus, for instance, while the percentage of human single-gender illustrations in 2000s books by male authors and illustrators is 20.0 per cent, it is 53.3 per cent in books by female authors and illustrators. Moreover, while 29.1 per cent of the books by males lacked any female character at all, only 9.1 per cent of the books by females did so. And, while 32.2 per cent of books by males had a central female character, fully 51.3 per cent of books by females did. Incidentally, through analyses not presented here, we discovered that this finding, that females were more likely to create female characters than males, held for every category of author we examined: White, Black, Latino and Gay-Sympathetic.

Conclusion

Consistent with McCabe et al.'s (2011) study, we find that there's been a substantial increase in the visibility of female characters in children's picture books since the 1960s. This increase holds for award-winning books by White, Black, Latino/a and Gay-Sympathetic authors and illustrators, at least in comparison with the primarily White authored and illustrated Caldecotts of the 1960s. The "symbolic annihilation" of girls and women, an annihilation that concerned McCabe et al. (2011), is no longer an issue in children's picture books, as fully half of the 2000s books we examined featured central female characters. We believe, as McCabe et al. (2011) did, that this achievement reflects the consciousness raising entailed by the second-wave women's movement, whose attention Weitzman et al. (1972) directed to children's books. Weitzman et al.'s study shone the light of feminist criticism on an industry that, since the end of the first women's movement, had gone unlit.

This paper also expands on previous findings by showing that: a) books by Black, Latino/a and Gay-Sympathetic illustrators are substantially more likely to feature female characters than books by White illustrators, and b) books by women authors, whether White, Black, Latina or Gay-Sympathetic, are substantially more likely to make female characters central than their male counterparts. We had anticipated the first findings through our reading of multicultural feminist scholars, each of whom had suggested a special place for women in helping their communities "get over" oppression based upon their communities' marginal status in American society. As a rule, the characters created by the Black, Latino/a and Gay-Sympathetic authors and illustrators in our sample are intended to help young readers see what it takes to successfully deal with oppression. At times, and depending on their point of view, these creators provide historical models of what it once took; at times, they focus on making what may seem unusual in their readers' lives seem more typical. That this involves the creation of more visible female characters, especially when the authors and illustrators are themselves female, may not be a great surprise.

Nonetheless, one of the major findings of the current research is the degree to which female authors are more likely to create female characters than male authors, whatever their race, ethnicity or sexual sympathies. We anticipated this finding in terms of women's greater likelihood of being attuned to various modes of feminist thought, though we didn't anticipate the strength of the relationship between gender of author and the presence of female characters. It is not clear whether this relationship has always existed in the production of children's picture books, but its strength in our sample of 21st-century books suggests that gender of author needs to be included in any subsequent studies of the presentation of gender in children's literature.

References

Anzaldua, Gloria E. 1987. *Borderlands/La Frontera: The New Mestiza*. San Franciso: Aunt Lute Books.

Clark, Roger, Rachel Lennon & Leanna Morris. 1993. Of Caldecotts and Kings: Gendered Images in Recent American Children's Books by Black and non-Black Illustrators. *Gender & Society* 5: 227–245.

Collins, Patricia Hill. 1991. *Black Feminist Thought: Knowledge, Consciousness, and the Politics of Empowerment*. Boston: Unwin Hyman.

Coretta Scott King Award Home Page. 2011. American Library Association. Retrieved 3/11/12. http://www.ala.org/emiert/cskbookawards

hooks, bell. 1984. *Feminist Theory: From Margin to Center*. South End Press: Boston.

Lorber, Judith. 2001. *Gender Inequality: Feminist Theories and Politics*. Los Angeles: Roxbury.

McCabe, Janice, Emily Fairchild, Liz Grauerholz, Bernice Pescosolido and Daniel Tope. 2011. Gender in Twentieth-Century Children's Books: Patterns of Disparity in Titles and Central Characters. *Gender & Society* 25(2): 197–226.

Newman, Leslea and Carol Thompson (ill.). 2009. *Daddy, Papa, and Me*. Berkeley: Tricycle Press.

Newman, Leslea and Carol Thompson (ill.). 2009. *Mommy, Mama, and Me*. Berkeley, CA: Tricycle Press.

Pura Belpré Award Home Page. 2011. American Library Association. Retrieved 3/11/12. http://www.ala.org/alsc/awardsgrants/bookmedia/belpremedal

Sarles, Patricia. 2011. Gay-Themed Picture Books for Children. Retrieved 3/11/12. http://booksforkidsingayfamilies.blogspot.com/2011_07_01_archive.html

Seeger, Laura Vaccaro. 2007. *First the Egg*. New Milford, CT: Roaring Book Press.

Sidman, Joyce and Pamela Zagarenski (ill). 2009. *Red Sings From Treetops*. Houghton Mifflin Books for Children.

Weitzman, Lenore, Deborah Eifler, E. Hokado & C. Ross. 1972. Sex-role Socialization in Picture Books for Preschool Children. *American Journal of Sociology* 77: 11-25-1150.

 THINKING ABOUT ETHICS *In this research, Clark, Kessler, and Coon are studying published materials rather than human subjects, so their main ethical responsibility is to report their findings honestly.*

APPROPRIATE TOPICS FOR CONTENT ANALYSIS

Although Clark, Kessler, and Coon applied content analysis to children's picture books, content analytic methods can be applied to any form of communication[1] including movies, television shows, speeches, letters, obituaries,

---

[1]In fact, although content analysis is most often applied to communications, it may, in principle, be used to analyze any content. Thus, for instance, Rathje and Murphy's (1992) "garbology" project, mentioned in Chapter 12, employed content analysis to study the contents of garbage cans.

editorials, YouTube clips, online chats, tweets, blogs, and song lyrics. Gilbert Shapiro and Philip Dawson (1998) looked at the notebooks of revolutionaries to gain insight into the causes of the French Revolution; Cottingham (2012) analyzed newspaper stories to see how loyalty to the Pittsburgh Steelers, a football team, provided a way of structuring weddings and funerals in the Pittsburgh area; Zurbriggen and Allen (2010) examined editorial cartoons for racist and sexist messages about three major-party candidates for the Presidency in 2008, Barack Obama, Hillary Clinton and John McCain; and Zhao, Grasmuck, and Martin (2008) studied Facebook accounts to examine how users create digital identities. Classic content analyses on racism include one by the freed slave Ida Wells (1892, cited in Reinharz [1992]) and G. J. Speed (1893, cited in Krippendorff [2003]), who analyzed newspaper articles to show, respectively, the extent to which Black men were being lynched in the South and how gossipy newspapers in New York were becoming during the 1880s. More recently, Hazell and Clarke (2008) examined recent advertisements in two Black-oriented magazines (*Essence* and *Jet*), finding that ideologies of racism persist in them.

Materials Appropriate for Content Analysis

Studies have focused on suicide notes (Bourgoin, 1995, to amend Durkheimian suicide theory), letters (Kenschaft, to examine nineteenth-century efforts to reinvent marriage, 2005), magazines (Friedan, to define the Feminine Mystique, 1963, and Taylor, to see how they construct the sexuality of young men, 2005), tweets (Goh and Lee, to discover the nature of reactions to Michael Jackson's death, 2011), blogs (Attwood, 2009), chat rooms (Donelle and Hoffman-Goetz, 2008), wills (Finch and Wallis, 1993), radio programs (Albig, 1938), personal advertisements (Jaggar, 2005), speeches (Rutherford, 2004), verbal exchanges (Bales, 1950), diaries (Ulrich, 1990), parenting-advice books (Krafchick, Zimmerman, Haddock, and Banning, 2005), films—both conventional (Leslie, 2005) and pornographic (Cowan and Campbell, 1994)—and websites (Holster, 2004). The list is endless. Systematic content analysis seems to date back to the late 1600s, when Swedish authorities and dissidents counted the words in religious hymns and sermons to prove and disprove heresy (Dovring, 1973; Krippendorff, 2003; Rosengren, 1981). More recently, Roger and a student, Katie Grandchamp, used content analysis to show that, in recent years, over 10 percent of articles in major sociology journals relied on content analysis (Clark and Grandchamp, 2012).

Questions Asked by Content Analyzers

Generally the questions asked by content analyzers are the same as those asked in all communications research: "Who says what, to whom, why, how, and with what effect?" (Lasswell, 1965: 12). Clark, Kessler, and Coon hypothesized that today's children's picture books would make female characters more visible than books did forty years ago, that Black, Latino/a and gay-sympathetic authors of such books would make female characters more visible than White authors and that female authors would make female characters more visible than male authors. They found support for all three hypotheses. Evaluating communications against hypotheses, or some other standards, and describing trends in communication content are two general purposes of content analysis (see Holsti, 1969: 43).

 STOP & THINK   *Do you think local, regional, or national newspapers are more likely to "head-line" international stories on the top of the front page or at the top of their websites? What is your hypothesis? Why? (You might test this hypothesis in Exercise 13.1.)*

QUANTITATIVE CONTENT ANALYSIS

quantitative content analysis, analysis focused on the variable characteristics of communication.

When researchers want to test hypotheses or to describe trends in communication, they are apt to engage in quantitative content analysis, in the way Clark, Kessler, and Coon did. **Quantitative content analysis,** like all other quantitative research, focuses on variable characteristics (in the case of the focal research, the degree of minority visibility, the year of publication, etc.) of units of analysis (in this case, of children's picture books).

Units of Analysis

units of analysis, the units about which information is collected.

Sampling for quantitative content analysis involves many of the same issues that sampling for any type of social research involves. First, you identify the **units of analysis** (or elements), the kinds of subjects you want to focus on. In the focal research on the children's picture books, this task was simplified because the authors were replicating earlier studies (Weitzman et al., 1972 Clark et al., 1993) whose units of analysis they wanted to use again, if they could. And when it came to questions of female visibility, such as how many books had female main characters, or even how many human single-gender illustrations were of females, determining units of units was a snap: They were clearly books, in the first instance, and human single-gender illustrations, in the second. These were the units the Weitzman and earlier Clark studies had used.

 STOP & THINK   *Suppose you wanted to examine the question of whether recent heroines in Disney animated films, like Pocahontas and Mulan, were more or less stereo-typically feminine than heroines of earlier Disney films, like Sleeping Beauty and Snow White. What would your unit of analysis be?*

units of observation, the units from which information is collected.

Deciding on the appropriate unit of analysis is not always as simple as it was in the picture-book study. One of Roger's students, Katie Grandchamp (2009), for instance, examined the question mentioned in the last "Stop & Think": whether more recent heroines in Disney animated films were portrayed more or less stereotypically than heroines in earlier Disney films (she found that they were portrayed less stereotypically, by the way). In this case, the unit of analysis wasn't the film, but the heroine within the film. To use a distinction made by Earl Babbie (2012), the unit of analysis—the unit about which information is collected (heroines)—was imbedded within the **units of observation**—the units from which information is collected (films). The distinction between units of analysis and observation is one without a difference when the unit of analysis and observation are the same. The units of observation in the focal research were always children's books (that is, the authors *always* got information from the books) and in some parts of the analysis (when the authors were determining the gender of the main characters of the *book*), the units of analysis were also books. But when the units of analysis and observation are different, it is important to remember that the members of one's sample are units of analysis, not the units of observation. Thus, when Clark, Kessler, and Coon shifted their focus (their units of analysis) to illustrations, they needed to remember that they were then interested in variable characteristics of illustrations (within books) and not the books themselves. Thus, while their unit of observation was always children's books, their units of analysis were sometimes pictures and sometimes the books themselves.

If you were interested in the sexual content of television shows, as the Kaiser Foundation (2005) was, television programs would be both your unit of analysis and your unit of observation. In Box 13.1, some of the highlights of this interesting study are presented.

SAMPLING

In content analysis, units of analysis can be words, phrases, sentences, paragraphs, themes, photographs, illustrations, chapters, books, characters, authors, audiences, or almost anything you want to study. Once you've chosen the units of analysis, you can sample them with any conventional sampling technique (see the various probability and nonprobability sampling techniques of Chapter 5) to save time or effort. In their analysis of human single-gender illustrations, for instance, Clark, Kessler, and Coon might have numbered each one and systematically sampled 25 percent of them by random sampling the first one and picking every fourth illustration after that. As things turned out, however, they decided that, because Weitzman and colleagues had examined the whole population of human single-gender illustrations between 1967 and 1971, they would do the same thing in their books. In fact, nonprobability sampling and the study of whole populations are typical of quantitative content analysis, though probability sampling does occur. For example, the Kaiser Foundation (2005), for its study of sex on television reported in Box 13.1, random-sampled television programs from October 2003 to March 2004.

STOP & THINK   *How might Clark, Kessler, and Coon have collected a stratified sample of picture books? (Hint: Can you think of any characteristics of authors they might have used to divide their sample?)*

CREATING MEANINGFUL VARIABLES

Quantitative content analysis is a technique that depends on the researcher's capacity to create and record meaningful variables for classifying units of analysis. Thus, for instance, in classifying books, Clark, Kessler, and Coon were interested in knowing whether they had female characters or not and whether they had female central characters or not, and in classifying human single-gender illustrations in terms of whether they depicted a female or not. In each of these cases, variables were of nominal scale, but that wasn't necessary. And, in fact, the authors did ask, for analyses not appearing in the focal research piece, about such things as the total number of adult roles and the number of these that were female. Both of these were interval-scale variables.

Actually, the portion of their code sheet used for this paper, depicted in Figure 13.1, makes it look as if Clark, Kessler, and Coon have already collected interval-scale information about single-gender illustrations. Question 3 in Figure 13.1 asks each coder, for instance, about the number of illustrations that depict characters of clear and singular gender (compared with, for example, those that show characters of both genders or characters of ambiguous gender). But the appearance of interval categories is a sleight-of-hand, made possible because the unit of observation (picture books), in most cases, contains several units of analysis (single-gender illustrations), and because all the

Book: Author: Year of Publication:

Publisher: Award or Type:

1. Does the book have any female character?
2. Does the book have a central female character?
3. Number of human illustrations that depict a single and clearly determinate gender:
4. Number that depict females:

FIGURE **13.1** Sample Coding Sheet from "Who Envisions the Distaff Side" Study researchers were really interested in, after all, was the proportion of all single-gender illustrations in a given sample of books that depicted females.

STOP & THINK   *How do you think Clark, Kessler, and Coon used the data from their coding sheets to calculate the proportion of all single-gender illustrations that showed females?*

Clark, Kessler, and Coon calculated the proportion of all single-gender illustrations that showed females by dividing the response to item 3 (number of human illustrations that depict a single and clearly determinate gender) into the response to item 4 (number of these that depict females). The unit of analysis here was the human single-gender illustration and the variable was gender (a nominal, not an interval, variable).

As in any measurement operation, the scheme left room for unreliable observations, or the possibility, in this case, that parallel observations of the same phenomenon would yield inconsistent data. Thus, the three coders (in this case, the three authors) disagreed about several books even on the number of unambiguous, single-gender illustrations. The authors handled these disagreements by taking the arithmetic average of their three estimates of the number of single-gender illustrations as their final, single estimate for each book.

STOP & THINK   *If Clark, Kessler, and Coon hadn't trusted the averaging method as "good enough," what could they have done to arrive at a final estimate of the number of single-gender illustrations in a book?*

If the authors didn't feel satisfied with averaging their individual estimates, they might have discussed those estimates and, by referring to the books themselves, come to an agreement about the number of single-gender illustrations.

QUALITATIVE CONTENT ANALYSIS

We have discussed the piece by Clark, Kessler, and Coon to give you some appreciation for the potential of a relatively quantitative kind of content analysis, a kind that was designed with the statistical analysis of variables in mind. Quantitative content analysis is the most common kind in the social sciences. But we wouldn't want to leave you with the impression that content analysis must be quantitative to be effective (see also Reinharz, 1992). Actually, we think content analysis is well employed when qualitative social scientists aim for primarily verbal, rather than statistical, analysis of various kinds of communication. Roger did—with two students, Joel McCoy and Pamela J. Keller—a **qualitative content analysis** of 20 recent award-winning books for young adults, 10 Newbery medalists or runners-up written by White women about White female protagonists, and 10 Coretta Scott King award winners or runners-up written by Black women about Black female protagonists (Clark, McCoy, and Keller, 2008). We came to these books with few preconceived ideas but "listened" to them as if we were eavesdropping on White

qualitative content analysis, content analysis designed for verbal analysis.

and Black mothers talking to their adolescent daughters about life's struggles. We did so not to try to find ways to "quantify the data," as the previous section suggests we might have. Instead, we found ourselves "listening" to themes of struggle and noting patterns.

Compared with the relatively well-accepted techniques (e.g., for quantifying and ensuring the reliability of data) for doing quantitative content analysis, however, guidelines for doing qualitative content analysis are few and far between. (For those interested in such guidelines, however, we recommend Reinharz, 1992; Strauss, 1987: 28ff.) Roger, Joel, and Pamela tried to discern theme patterns by a process of comparison and contrast that would be congenial to the advocates of grounded theory (see also Glaser, 2005; Glaser and Strauss, 1967). Thus, we noticed Black literary mothers were much more likely than White literary mothers to bring up the topic of oppression and much more likely, when "talking" about oppression, to locate it in the present, rather than in the past.

Generally speaking, then, quantitative content analyses tend to be of the deductive sort described in Chapter 2—analyses that begin, perhaps, with explicit or implicit hypotheses that the researcher wants to test with data. Explicit in the focal research of this chapter, for instance, is a hypothesis that female authors and illustrators are more likely than male authors and illustrators to feature female characters. Qualitative content analyses, on the other hand, tend to be of the inductive sort, analyses that might begin with research questions but are then likely to involve observations about individual texts (or portions of texts) and build to empirical generalizations about texts (or portions of texts) in general. Clark, McCoy, and Keller, for instance, began with the general research question "What differences are there in the messages about struggle presented by white and black authors?" and then analyzed the kinds of presentations they discovered as they read, and viewed, such books.

VISUAL ANALYSIS

visual analysis, a set of techniques used to analyze images.

A relatively new set of techniques, closely related to those of content analysis, is **visual analysis**. Visual analysis, a branch of the general field of visual sociology, we would argue, is actually a kind of content analysis. Visual analysts have studied photographs, video images, paintings, drawings, maps, and other kinds of images using a variety of analytic approaches. The earliest sociological research based on visual analysis was probably Erving Goffman's *Gender Advertisements* (1979). Here Goffman argues that advertisers, to create a kind of comfort with their product, will often depict them in social situations that we all might try to emulate in our day-to-day lives:

> Thus, just as a Coca-Cola ad might feature a well-dressed, happy looking family at a posh beach resort, so a real family of modest means and plain dress might step up their level of spending during ten days of summer vacation, indeed, confirming that a self-realizing display is involved by making sure to photograph themselves onstage as a well-dressed family at a posh summer resort. (1979: 27)

Goffman examined magazine advertisements to see what they implied about gender relations in society. A famous example is his assertion that relative size, especially height, in pictures is one way to mark relative power or authority and his observation that, typically, in images associated with products, men are depicted in ways that exaggerate their height advantage over women, much as powerful men are depicted in ways that make them look taller than other men (Goffman, 1979: 28).

This kind of close analysis of images in advertisements also occurs in John Grady's (2007a) study of advertisements depicting Blacks in *Life* magazine, from 1936 to 2000. Following procedures typical of qualitative analysis, largely developed after Goffman's work, Grady coded the images, identifying variables as he noted that some photos were easily grouped together and others weren't. He found, for instance, that some advertisements caricatured Blacks stereotypically and some did not. And then he discovered that virtually all of the caricaturing advertisements occurred in editions of *Life* published before 1968 and that virtually none of them occurred afterward. This kind of finding led him to the conclusion that White commitment to racial integration, implicit in the advertisements approved by *Life*'s editors and used by companies that advertised in *Life*, grew in the period from 1936 to 2000.

STOP & DO   *Try using Google Images to look at the first few advertisements that pop up when you type in "Aunt Jemima pancakes"? Would you be inclined to code these advertisements, using Grady's scheme, as ones that caricature Blacks or ones that don't? Why or why not?*

Some visual analysts advocate the kind of qualitative approach to images employed by Goffman and, at least in the analysis described in the previous paragraph, by Grady, one that explicitly involves observing the "overtones and subtleties" of images (e.g., Collier, 2001: 39). Some (e.g., Bell, 2001) advocate the value of a kind of quantitative content analysis of pictures. Grady did this kind of analysis when he looked at all the advertisements that appeared in *Life* between 1936 and 2000 and calculated the percentage of all racially identifiable advertisements that contained Blacks. (He found that this percentage grew from a very low level, never more than about 2 percent before 1961, to about 12 percent between 1998 and 2000.)

STOP & DO   *Try using Google Images again, but this time see what pops up when you type in "Sports Illustrated ads"? Examine the first page and identify the advertisements that were indeed from Sports Illustrated. Then count the number of these advertisements that include women or girls and calculate the percentage of all the advertisements on the page that include women or girls. Does this percentage surprise you at all? Why or why not?*

Van Leeuwen and Jewitt's *Handbook of Visual Analysis* (2001) is perhaps the best guide to the variety of approaches taken by visual analysts; *Visual Studies* is the journal in which one can find the best examples of its uses by sociologists.

ADVANTAGES AND DISADVANTAGES
OF CONTENT ANALYSIS

One great advantage of content analysis is that it can be done with relatively little expenditure of time, money, or person power. Clark, Kessler, and Coon worked as a team of three on their examination of children's picture books, all of which they borrowed from local libraries. They worked as a team partly for the fun of working with one another and partly for the enhanced data reliability that comes from cross-checking one another's observations in various ways. However, their project could have been done by one of them, with no special equipment other than a library card.

Another related plus is that content analysis doesn't require that you get everything right the first time. Frankly, the partial coding sheet shown in Figure 13.1 is part of a larger one that represented a compilation (and distillation) of several separate efforts. Indeed, the authors had made a first pass at kinds of children's books (notably the 2000s Caldecotts) before realizing they wanted to look at books by other kinds of authors and illustrators (e.g., Black, Latino/a, and gay-sympathetic). They then returned to their first subjects (Caldecotts) to record newly relevant information (e.g., the award type and, as it turns out, the percentage of adult characters that were female). This kind of recovery from early oversights is much less feasible in survey or experimental research.

STOP & THINK *Compare this to having surveyed 500 people and then deciding you had another question you wanted to ask them.*

Yet another advantage is that content analysis is unobtrusive; our content analysis itself can hardly be accused of having affected the content of the books we looked at. Nor can the Kaiser Foundation's (2005) study of "Sex on TV" be criticized for having affected the sexual content of the television programming it examined. This is not to say that content analysis might not affect the content of subsequent communications. In fact, we believe that the findings of Weitzman and colleagues (1972) did contribute to the kinds of attitude shifts (in award givers, in particular, and in writers) that can account for some of the changes Clark, Kessler, and Coon document. And, when the Kaiser Foundation introduces its study by referring to the fact that "a third (34 percent) of young women still become pregnant at least once before they reach the age of 20—about 820,000 a year" (2005: 2), it seems to be encouraging a diminution of the number of sexual messages on television.

STOP & THINK *Can you imagine whose attitudes would need to be affected so that the "Sex on TV" study would have its desired effect?*

Like using other kinds of available data, content analysis permits the examination of times (and peoples) past—something surveys, experiments, and interviews don't do so easily. Clark, Kessler, and Coon's analysis of books from the 1960s and the 2000s is a good example, as are Shapiro and

Dawson's (1998) examination of French revolutionaries' notebooks and Sheldon Olson's (1976) use of the Code of Hammurabi, a list of laws and punishments from ancient Babylonia.

STOP & THINK

The Code of Hammurabi consists of 282 paragraphs, each specifying an offense and its punishment. One could code each of the paragraphs by the severity of the punishment associated with an offense, for instance, to gain a sense of what kinds of behavior the Babylonians most abhorred. Based on the punishments associated with stealing from a temple and striking another man, mentioned in the following paragraphs, which of these offenses was held in greater contempt?

Paragraph 6. If a man steal the property of a god (temple) or palace, that man shall be put to death.

Paragraph 204. If a freeman strike a freeman, he shall pay ten shekels of silver.

One of Emily's future projects is an analysis of life expectancy, family structure, and gender roles in previous centuries by doing content analysis of the inscriptions on headstones in New England cemeteries.

Yet another advantage of content analysis is that, because it typically permits one to avoid interaction with human subjects, it also permits the avoidance of the usual ethical dilemmas associated with research involving human subjects and with the time involved in submitting a proposal to an Institutional Review Board, unless one plans to use private documents. Somewhat more questionable, ethically, are types of communication, say Facebook postings, for which the researcher needs to get permission, say through "friending," to access. It is probably worth consulting an Institutional Review Board about using these less public types of communication.

Some disadvantages of content analysis might have occurred to you by now. Because content analysis is usually applied only to recorded communication, it can't very well be used to study communities that don't leave (or haven't left) records. Thus, we can study the legal codes of ancient Babylonia because we've discovered the Code of Hammurabi, but we can't study the codes of societies for which no such discoveries have been made. And Clark, Kessler, and Coon can content-analyze picture books with some sense that what children read, or heard, might have predictable consequences for their attitudes. But to tap those attitudes directly, it would be more valid to analyze literature produced *by*, not *for*, children. Peirce and Edwards (1988) actually did content-analyze stories written by 11-year-olds, but this would be pretty hard to do for pre-literate, and certainly pre-verbal, children, the not-infrequent audience for picture books. Content analysis can also involve social class biases because communications are more likely to be formulated by educated persons.

There are also questions about the validity of content-analytic measures. For example, is the invisibility of female characters in award-winning and other children's books of the late 1960s really a valid indicator of sexist attitudes? Perhaps, but maybe not. And even if one were pretty sure that the invisibility was a valid measure, whose attitudes are being measured: those of the authors, the publishers, the award givers, or the adult purchasers and readers?

STOP & THINK

Emily and a colleague thought they'd found evidence of significant social change when they content-analyzed 20 years of the "Confidential Chat" column (which published reader contributions) in the Boston Globe and found the column increasingly dealing with women's work outside the home between 1950 and 1970. They were discouraged when a reviewer of their research report for a scholarly journal suggested that their results might reflect changes in the editorial climate at the newspaper, rather than changes in the attitudes of the readership. When they interviewed the editor of the column, they found, indeed, that he and the former editor had published only about 10 percent of the letters he received and that changes in content were likely to reflect his sense of what his readership (mostly women) wanted to read. Emily and her co-author never resubmitted their paper because they now felt it failed to say anything significant about social change. What do you think of this decision?

Finally, content analysis doesn't always encourage the sensitivity to context that, say, literary criticism often does. Weitzman et al. (1972), McCabe et al. (2011) and Clark, Kessler, and Coon all assume that the presence of a central female character should enhance the self-esteem of female readers. But what if the female character, for instance, is portrayed as so dependent on others as to make girls reading the story feel impotent themselves, thus diminishing, not enhancing, their self-esteem? Questions of validity plague those who do content analysis just as much as they plague those who employ other methods of data collection.

STOP & THINK

Suppose your main concern was whether authors were creating empowered (and empowering) female characters, rather than simply visible ones. What questions might you ask of your books to code for "empowered"? Must quantitative content analysis be as insensitive to context as the previous paragraph suggests?

Validity concerns are less likely to be legitimate in the case of qualitative content and visual analyses, where the researchers focus on things like context and author (artist) intention and the verbal analysis is more likely to take such things into account. Thus, a project that focused on the research question, "What kinds of struggle does the author emphasize?" is more likely to identify such struggles validly than one that begins with the hypothesis "Books about male characters are more likely to be about personal struggle than books about female characters."

SUMMARY

Content analysis (as well as its offspring, visual analysis) is an unobtrusive, inexpensive, relatively easy, and flexible method of data collection that involves the systematic and—typically, but not necessarily—quantitative study of one kind of available data—some form of communication. It, like other forms of available data analysis, provides a way to study the past. Content analysis can be used to test hypotheses about communication, to compare the content of such communication with some standard, or to describe trends in communication, among other things. Clark, Kessler, and Coon simultaneously looked at trends in the visibility of female characters in children's picture books and tested hypotheses about levels of female visibility in books by different groups of authors and illustrators.

An early step in quantitative content analysis is to identify a unit of analysis. This is the unit about which information is collected and can be words, phrases, or paragraphs; stories or chapters; themes; photographs; illustrations; authors; audiences; or almost anything associated with communication. In Clark, Kessler, and Coon's study, the unit of analysis was sometimes whole books and sometimes illustrations.

The quality of a content analysis depends on the creation and meaningful usage of variables for classifying units of analysis. Clark, Kessler, and Coon used the presence of female characters, the presence of a central female character, and a high proportion of human single-gender illustrations being of females as their measures of female visibility. Explicit and reliable coding schemes for recording variation are most important. Checks on or enhancements of measurement reliability can be pretty simple but are always appropriate. Clark, Kessler, and Coon took averages of their individual classification efforts to enhance the reliability of their final measures.

On the down side, content analysis cannot tell us much about communities that have left no recorded communications. Moreover, content analysis is frequently criticized on validity grounds, mainly because those criticizing aren't sure whether the content analyzers are measuring what they think they are. Validity concerns are less well founded, however, in the case of qualitative content and qualitative visual analyses.

EXERCISE 13.1

How Newsworthy Is the News?

This exercise offers a chance to observe, and speculate about, differences in the treatment of a given news story in daily newspapers. Your research question in this exercise is this: Does the treatment of a given news story vary from one newspaper to another and, if so, what might account for this variation?

1. Select a current national or international news story. What story did you choose?
2. Select four newspapers (from either a newsstand or your library) or four newspaper websites. Get copies of these papers, or examine their websites, on the same day. List the names of the newspapers and the date.

Date:
Newspaper 1:
Newspaper 2:
Newspaper 3:
Newspaper 4:

3. Decide how you will measure the importance given a story in a newspaper. (Hint: You might consider one of the following options or combinations thereof: What page does the story appear on? Where does the story appear on the page? Or, in the case of newspaper websites, is this the first, second, third, etc. story on the opening webpage? On some subsequent webpage? How big is the typeface used in the headline? How many

words or paragraphs are devoted to the story? Is it illustrated? Does the newspaper devote an editorial to the story?) Describe the measure you will use to determine the relative importance given your story in each of the papers.

4. Report your findings. Which paper gave the story the greatest play, by your measures? Which paper, the least? Interpret your findings by trying to account for differences in terms of characteristics of the papers you've examined. Please submit copies of the stories with the assignment.

EXERCISE 13.2

Analyzing Website Commentaries on School Shootings

This exercise adapts an idea from Professor Molly Dragiewicz of George Mason University to give you the chance to analyze website commentaries on school shootings in the United States.

1. Do a web search using a search engine or a subject guide such as Google or Yahoo, using the terms "school shootings" and "commentary." Find four commentaries about school shootings that have taken place in the past decade and print them out.

2. Your research questions are the following:
   a. Do website commentaries tend to focus on the perpetrators or the victims of school shootings?
   b. Do these commentaries tend to focus most on gender, race, or class in their interpretations of shootings in American suburban schools?

3. Prepare four code sheets like this:
   Commentary: Author: Publication date:

   a. Would you say that the commentary focuses more on the perpetrators or the victims of school shootings?
   b. If the commentary focuses on the perpetrators, in your view, does it focus on the race, gender, or class of these perpetrators? (Yes or no)
   c. If so, which of these characteristics seems most important to the author?
   d. If commentary focuses on the victims, in your view, does it focus on the race, gender, or class of these victims? (Yes or no)
   e. If so, which of these characteristics seems most important to the author?

4. Write a paragraph summarizing your findings. (You might summarize, for instance, whether the commentaries seem more concerned with the perpetrators or the victims of the shootings. Then you might summarize whether race, gender, or class seems to be the "lens" through which the commentaries tend to be made.)

5. Include the copies of the commentaries as an appendix to the exercise.

EXERCISE 13.3

The Utopianism of Television Commercials

It has been said that commercials for children's television programs are utopian—that they tap into utopian desires for something different, something better; that they express utopian values of energy, abundance, and community

(see, for example, Ellen Seiter, 1993: 133)—rather than extolling the virtues of things as they are. But are commercials for children's programs really all that different in this regard from those for adult programs? This is your research question for this exercise.

1. Select a children's television show and an adults' television show on commercial stations. (Hint: Saturday mornings are particularly fruitful times to look for children's commercial television programs.) Pick shows of at least 30 minutes in duration (preferably one hour).

2. Indicate the shows you've chosen and the times you watched them.

    Children's show:
    Time watched:
    Adults' show:
    Time watched:

3. Describe the measures you will use to determine whether a commercial appealed to utopian sentiments or values. (For example, you might ask whether the commercial makes an explicit appeal to the viewer's desire for a better life [for instance, by showing an initially depressed-looking character fulfilled by the discovery of a soft drink or an initially bored breakfaster enlivened by the acquisition of a new cereal] or if it makes implicit appeals to, say, the viewer's wish for energy [is there dancing or singing associated with a product?], abundance [is anyone shown to be constrained by considerations of price?], or community [do folks seem to get along better after acquiring brand X?].) Show an example of the code sheet you will use, including a space for the product advertised in the commercial and a place for you to check if you observe the presence of what it is you're using to measure an appeal to utopian sentiments or values.

4. Use your coding scheme to analyze the first five commercials you see in both the children's and the adults' show.

5. Report your findings. In which kind of programming (for children or adults) did you find a higher percentage of commercials appealing to utopian sentiments or values? Interpret your findings.

EXERCISE 13.4

Gender among the Newberys: An Internet Exercise

Knowing Your Responsibilities

The American Library Association (ALA) awards the Newbery medal for adolescent fiction. For this exercise, you'll want to use the ALA website (www.ala.org) to answer questions about the gender of characters in Newbery winners and honor books over the last decade. (Hint: One way of finding such a list, after getting on the ALA site, is to select the "search website" option, type "Newbery" at the request line, request information about one of the recent winners, then request the list of other winners and honor books at the bottom of that page.)

1. Find a list of Newbery award winners and honor books for the last 10 years.

2. Begin your content analysis of the titles of these books by listing the titles that suggest to you that there is a female character in the book.

3. Now list those books whose titles suggest to you that there is a male character in the book. (The two lists might overlap. For instance, the title Lizzie Bright and the Buckminster Boy, an earlier title, suggests to us that there are at least two characters in the book, one male and one female. It would, therefore, make both your lists, if it fell into the period of your study.)

4. Does your brief content analysis of recent Newbery award and honor books suggest that more attention has been paid to males or females?

EXERCISE 13.5

Gender Advertisements

This exercise gives you an opportunity to practice visual analysis à la Erving Goffman's *Gender Advertisements* (1979). Your job will be to find four magazine advertisements that seem to tell a story using human models and to identify ways in which the advertisement displays "ideal" male and/or female behavior or traits.

1. Find four advertisements that seem to tell stories or portray relationships in magazines using human models that you can get easily from magazines or from *Google Images* by using the keywords "magazine ads." Cut out the advertisements or print them out and display them on a sheet or sheets of paper that you'll turn in with the exercise. Label the four advertisements with the numbers 1, 2, 3, and 4.

2. For each advertisement, briefly (in a sentence or two) tell the story of the advertisement as you see it.

3. Now, describe ways in which the model or models in each advertisement are posed to create an image of the "ideal" male and/or female.

Applied Social Research

CHAPTER 14

© ENRIQUE CASTRO-MENDIVIL/Reuters/Landov

CHAPTER OBJECTIVES

This chapter is designed to help students

- Distinguish between basic and applied research
- Learn about the history of and reasons for applied research
- Understand evaluation research as evidence-based research
- Appreciate the kinds of evaluation research, including outcome evaluation, cost-benefit analysis, and needs assessments
- See participatory action research as a new research paradigm
- Learn how to design traditional evaluation research and participatory action research
- Understand the nature of "real world" research
- Recognize the ability to use sociology to help solve social problems

CHAPTER OUTLINE

BASIC AND APPLIED RESEARCH

applied research,
research intended to be useful in the immediate future and to suggest action or increase effectiveness in some area.

By now you've read quite a bit of social research. We hope you found that at least some of it focused on topics of interest to you personally or professionally. Perhaps one or two of the studies will be helpful in a current or future work setting. In many work settings, the kind of research we'll be considering in this chapter—**applied research**—is useful.

We've explored a variety of methodological concerns: creating research questions and hypotheses, considering ethical and practical issues, selecting populations, sampling strategies and research designs, and choosing one or more methods of data collection. Before covering data analysis, we want to expand on our discussion of applied social science research. While all social research is designed to increase our understanding of human behavior and can be useful to individuals, groups, or the whole society, some work is more immediately useful than other research. As we noted in Chapter 1, **basic research** seeks to create new knowledge for the sake of that knowledge, whereas applied research is designed to help solve immediate social problems and answer broad research questions about settings in complex social and political environments. While basic research can have practical applications, it is designed primarily to provide greater understanding of our social world, to describe individuals or groups, to see relationships between variables, to develop or test theories, and the like. Applied work has a practical focus and is designed to provide organizations such as schools, legislatures, communities, social service agencies, health care institutions, and other organizations with information that can be used in the immediate future. Purity of theory is much less important in applied research; utility of theory is what counts as researchers look for concepts, variables, and theories that will help solve a problem (Bickman and Rog, 2009: xi). As many research projects fall somewhere between the poles of basic and applied research, identifying a study's goal or emphasis is helpful in determining where a study fits on the continuum.

basic research, research designed to add to our fundamental understanding and knowledge of the social world regardless of practical or immediate implications.

Social Problems and Social Solutions

Awareness of social problems and programs directed toward solving them became common in the twentieth century. The efforts before World War I focused on literacy, occupational training, and controlling infectious diseases, with projects designed to determine whether funding initiatives in these areas was worthwhile (Rossi and Freeman, 1993: 9). Through World War II and into the 1960s, federal human service projects in health, education, housing, income maintenance, and criminal justice became widespread, as did the efforts to demonstrate their effectiveness.

By the end of the 1970s, many interested parties started questioning the continued expansion of government programs, not only because of changes in ideology, but also because of disenchantment with the outcomes and implementation of many of the programs advocated by public officials, planners, and politicians (Rossi and Freeman, 1993: 24). Despite disappointment with earlier efforts, programs continued to be created and implemented to solve both old and newly identified social problems. With many millions of program dollars spent on local, state, and federal levels, policy makers wanted answers: Are the programs effective? Should they be continued? Modified or eliminated entirely? Applied research is a way to answer these questions.

What do you think are the most serious social problems we face today? Unemployment? Violence in our schools and communities? Access to health care? Poverty? Racism? Obesity? Teenage pregnancy? Homelessness? Drug and alcohol abuse? Are you aware of a program nationally or in your state or local community designed to help solve one of these problems? If so, do you think it is effective? Are those served by this program involved in evaluating it?

Evidence-based social programs and policies, those that present persuasive evidence of successful achievement based on careful monitoring of outcomes.

You may believe that all the problems listed above warrant attention. In fact, there have been programs addressing each of them. Sometimes we read or hear about efforts to improve things, but many such efforts are ignored by the media. We rarely learn whether programs and policies designed to address social problems actually work; even more rarely do we hear of the concerns of those affected by the programs and policies. With pressure for effective and accountable use of public dollars, recent years have brought a growing interest in **evidence-based programs and policies** (Haskins and Baron, 2011; Schorr, 2012), which are those that present persuasive evidence of successful achievement based on careful monitoring of outcomes.

EVALUATION RESEARCH

evaluation research, research specifically designed to assess the impacts of programs, policies, or legal changes.

There are several kinds of applied research. One of the best known is **evaluation research**—research specifically designed to assess the impacts of programs, policies, or legal changes. Most often the focus of an evaluation is whether a program, policy, or law has succeeded in effecting intentional or planned change. Rather than being a unique research method, evaluation research has a specific purpose. Rossi and Freeman call evaluation research "a political and managerial activity, an input into the complex mosaic from which emerge policy decisions and allocations for the planning, design, implementation, and continuance of programs to better the human condition" (1993: 15).

Outcome Evaluation

outcome evaluation, research that is designed to "sum up" the effects of a program, policy, or law in accomplishing the goal or intent of the program, policy, or law.

The most common evaluation research is **outcome evaluation**, which is also called impact or summative analysis. This research seeks to estimate the effects of a treatment, program, law, or policy and thereby determine its utility. Outcome evaluation typically begins with the question "Does the program accomplish its goals?" or the hypothesis "The program or intervention (independent variable) has a positive effect on the program's objective or planned outcome (dependent variable)." Key to outcome evaluation is identification of a program's goal. Evaluations most useful to practitioners and policy makers will answer questions about a program's success and suggest possible improvements.

A wide variety of programs, policies, or laws can be evaluated—ones that serve many or few people; last for days, weeks, or years; have a large or limited geographic target area; and have broad or narrow goals. Examples of the many hundreds of research questions that have guided outcome evaluations in the past decade include:

1. Does the Experience Corps program, which brings older adults into public elementary schools to tutor and mentor at-risk children, actually improve student academic outcomes? (Lee et al., 2012)
2. Are the health and development of first-time mothers and their infants and their future life prospects improved by prenatal and infancy home visits by nurses? (Olds, 2012)
3. Does a peer-led suicide prevention program increase high school students' referrals of suicidal friends to adults? (Wyman et al., 2010)

The answer to all three questions was "yes"—that is, the program accomplished its goals. However, it is also important to know what *doesn't*

BOX **14.1**   Outcome Evaluation of a School-Based Intervention Program

The Chicago public schools administer Child-Parent Centers (CPCs), a center-based early intervention program. Since 1967, this program has provided educational services to preschool children and their parents who live in targeted, low income areas. There have been many evaluations of the program to measure its success. One evaluation followed a cohort of 1,539 low-income minority children born in 1980, who were in either the intervention program or an alternative program, to determine whether CPC program participation was associated with lower rates of reported child maltreatment (Reynolds and Robertson, 2003) while others focused on adult well-being. The intervention group began with 989 children in a CPC program (preschool, kindergarten, and child support services) in one of the 20 CPCs in the city; the matched comparison group was composed of 550 children who participated in a full-day kindergarten program for low-income families in areas of the city that were not served by a CPC program (Reynolds and Robertson, 2003). The evaluation researchers found that at the start, the program and comparison group members were similar on many child and family characteristics. In one evaluation, using child protective services records and court records as data, the groups differed in the outcome variable of levels of reported child

maltreatment through age 17. After controlling for the influences of preprogram maltreatment, family risk, and other explanatory factors, the researchers found CPC participants had "a 5.5% lower rate of child maltreatment than the comparison group (5% versus 10.5%).... This translates into a 52% reduction in the rate of child maltreatment associated with [program] participation" (Reynolds and Robertson, 2003: 13–14). Evaluators also found program effects on the participants at age 24, including fewer felony arrests and greater occupational prestige (Jacobson, 2007; Reynolds and Ou, 2011).

The following diagram of the CPC evaluation presents some of the research data and the study's alternate-treatment, quasi-experimental design, an experimental design described in Chapter 8. Due to the researchers' not being able to control who entered the intervention program and who attended the alternative program, there was no true control group. However, the evaluation used a matched comparison group, following the two groups for decades. The results give support to the explanatory hypotheses that the Chicago CPC program decreases child maltreatment and produces better outcomes in educational attainment, occupational achievement, and criminal behavior.

work. Reynolds, Mathieson, and Topitzes (2009) looked at 15 studies of 14 programs designed to reduce rates of child abuse and neglect to determine the extent to which early childhood interventions actually prevented child maltreatment. In each of the evaluations, the program was the independent variable and child abuse or maltreatment was the dependent variable. Most of the evaluation studies used an experimental design with both control and experimental groups and used official reports of maltreatment from child protective services or the courts as indicators of the dependent variable. Review of the outcome evaluations led Reynolds et al. (2009: 200) to conclude that program success may not be typical. "Of the 12 intervention models investigated, only 4 or one third reported that program participants had significantly lower rates of maltreatment than comparison groups.... Although 5 studies reported reductions in either substantiated or parent reported maltreatment, only for 3 programs was there consistent evidence of enduring effects." Some of the outcome evaluations of an especially successful program are discussed in Box 14.1.

STUDY DESIGN used in evaluating the Child–Parent Center Programs

Groups	Time 1 (1985–1986)	Introduction of independent variable	Time 2 (14 years later) Measure dependent variables	Time 3 (21 years later) Measure dependent variables
Experimental group N = 989 children	Measured family and social characteristics	Children attended a CPC program for preschool and kindergarten that includes family support services	5% parental maltreatment rate	71% high school completion 15% four-year college attendance 16% felony conviction 39% employed full-time
Comparison group N = 550 children	Measured family and social characteristics	Children attended other programs	10.5% parental maltreatment rate	64% high school completion 10% four-year college attendance 20% felony conviction 37% employed full-time

Cost-Benefit Analysis

cost-benefit analysis, research that compares a program's costs to its benefits.

cost-effectiveness analysis, comparisons of program costs in delivering desired benefits based on the assumption that the outcome is desirable.

Even if a program has accomplished its goals, there is always the issue of cost and whether it might be better to expend the resources consumed by the program in other ways. A **cost-benefit analysis** is a study designed to weigh all expenses of a program (its costs) against the monetary estimates of the program's benefits (even putting dollar values on intangible benefits); a **cost-effectiveness analysis** estimates the approach that will deliver a desired benefit most effectively (at the lowest cost) without considering the outcome in economic terms. Comparing program costs and benefits can be helpful to policy makers, funding agencies, and program administrators in allocating resources. With large public expenditures on health care, correctional facilities, social welfare programs, and education in the United States, saving even small percentages of costs adds up.

Typical questions asked in cost-benefit and cost-effectiveness analyses are: How effective is the program? How expensive is it? Is it worth doing? How does this program compare with alternative programs? Outcomes for program participants are compared to those not in programs or with those in alternative programs.

Evaluations of the previously described Chicago Child-Parent Center (CPC) programs have estimated the costs of the Center programs versus expenditures for those not in the program, including remedial education, criminal justice expenses, and economic benefits for program participants. The evaluators judge it to be a very worthwhile investment: the preschool program alone had a societal return of $10.83 for every program dollar invested (Reynolds et al., 2011).

Another illustrative cost-benefit analysis is of Minnesota's Challenge Incarceration Program (CIP), an intensive, structured, and disciplined program for offenders (commonly called "boot camp"). To evaluate whether the program lowered recidivism (effectiveness) and saved money (cost), the researchers operationalized the outcome or dependent variable as a rearrest, a felony reconviction, and a return to prison for a new criminal offense or a technical violation. Although the rearrest, felony reconviction, and reimprisonment rates were lower for CIP offenders compared to those in the control group, the two groups did return to prison for all reasons (including technical violations) at about the same rate (Duwe and Kerschner, 2008: 627). However, the boot camp program was judged to be cost effective because when the program's offenders returned to prison, they stayed 40 fewer days than control group offenders since they were less likely to return for a new crime. Duwe and Kerschner (2008) estimated that over the life of the program, given the per diem costs of prison, it saved Minnesota at least $6.2 million by providing early release to program graduates and reduced the future prison time of the program participants.

STOP & THINK *Once a program is in place, it can be evaluated. But what about social problems for which there are no programs? What might be done to identify a problem and suggest solutions?*

Needs Assessments and Other Kinds of Evaluations

stakeholders, people or groups that participate in or are affected by a program or its evaluation, such as funding agencies, policy makers, sponsors, program staff, and program participants.

needs assessment, an analysis of whether a problem exists, its severity, and an estimate of essential services.

The evaluator's tasks are determined by the stage of the program (whether it is a new, innovative program or an ongoing one) and the needs and interests of the **stakeholders** (those involved in or affected by the program in some way). It's critical to determine the needs of a geographic area or a specific population *before* a program or project is designed. This can be accomplished with a **needs assessment**—an analysis of the existence and severity of a problem and the number of people affected. Such assessment may also include an appraisal of existing resources and suggestions for new services or programs to ameliorate the problem. Needs assessments can focus on a wide variety of concerns—educational, health, housing, and the like—and a variety of populations—students, the elderly, the incarcerated, and the like. Assessments can be designed by those within a community or by those who are outside of it.

Hurricane Katrina caused many to leave the Gulf Coast area and move to distant towns and cities. One needs assessment was done by researchers working with disaster relief and service providers focusing on the needs of evacuees who had relocated to Kansas City. While the emphasis was on mental health needs, service providers also described the displacement-related challenges faced by the evacuees, such as material loss, unemployment, and other stressful life events as well as provider issues including the lack of knowledge about the cultural backgrounds of the evacuees (Legerski, Vernberg, and Noland, 2012).

Two researchers, Jill Harrison and Maureen Norton-Hawk, conducted a needs assessment far from home. In the focal research piece that follows, you'll read about their study of incarcerated women in Ecuador and the many inmate needs that their research documented. Jill Harrison had become familiar with Ecuador's prisons when she visited Ecuador and was unfairly arrested and imprisoned as an illegal alien. She remembers having no money in her pocket to buy water or food and having to read by candlelight even during the day. Her experience was one of the reasons she was interested in studying the current conditions in the women's prisons (Harrison, personal communication).

FOCAL RESEARCH

A Needs Assessment of Incarcerated Women in Ecuador

By Jill Harrison and Maureen Norton-Hawk*

The relationships between abuse, poverty, drugs, incarceration, and the inequitable impact on women have been well documented. We conducted a needs assessment of the inmates in Ecuador's largest women's prison. This exploratory research indicates that gendered injustice extends well beyond American borders.

(continues)

*Jill Harrison is an Associate Professor of Sociology at Rhode Island College. Maureen Norton-Hawk is an Associate Professor of Sociology at Suffolk University. This article is published with permission.

FOCAL RESEARCH (continued)

Incarceration of women in Ecuador has increased dramatically since this small South American country adopted U.S. foreign policy whereby favorable trade relations and military support increases as Ecuador cracks down on the distribution and possession of drugs (Nuñez, 2006; Merolla, 2008). Most favored nation trading status with the United States is an incentive for a partnership in the War on Drugs, and such laws do not differentiate between large scale traffickers and low level carriers or "mules," who are mostly women (Preston & Roots, 2004; Merolla, 2008). In 2001, 2,310 people were incarcerated for drug violations in Ecuador, and women's incarceration doubled for drug offenses from 376 in 2001 to 840 in 2004 (Boletin Estadistico, 2001–2004). The women are relatively easy to convict because they are generally caught "red-handed" and have no money to mount a legal defense. Whether the carrier has two ounces of marijuana or two kilos of unprocessed coca, the minimum mandatory sentence is between 6 and 8 years (Associated Press International, 2008).

Method

El Centro de Rehabilitación Social, colloquially called el Inca, the largest women's prison in Ecuador, was selected as the primary research site. Maureen Norton-Hawk had visited incarcerated women at el Inca frequently over the past 5 years and both researchers visited several times in 2007 and 2008. Formal IRB approval was secured from Suffolk University in Boston, Massachusetts. As there is no formal review process for researchers at el Inca, informal approval was granted by the Comite de las Mujeres, an inmate group that negotiates donations, and the prison director.

The Comite de las Mujeres "advertised" our research the day before we arrived for data collection. Inmates are generally free to move about the prison between 7:00 AM and 10:00 PM and were invited to come to a classroom in the morning to be part of the study. We talked about the purpose of the research and explained the concept of informed consent. As this was a normal visiting day, those without visitors were more likely to hear about the study. Each woman was provided with a small gift: a plastic bag filled with a roll of toilet paper, tampons, shampoo, conditioner, a bar of soap, and 10 hard candies. They were free to leave at that point, although none did. We went over informed consent with each woman individually and offered the option of not signing the form in order to protect confidentiality if papers were seized by guards, committee members or the director. Most of the women chose the option of signing with an X rather than a signature while a researcher served as the witness.

A convenience sample of 50 female inmates, or approximately 15% of the total prison population in December 2008, completed a needs assessment questionnaire anonymously. Ten women had the questions read to them by a researcher who also recorded their answers because they could not read, had poor eyesight or did not have corrective eye glasses. All the women surveyed were incarcerated on drug charges and were serving an average sentence of 6.5 years. Most were under age 30, mothers of small children, with limited access to education and were unemployed when arrested.

The women were asked to assess their housing, educational needs, medical treatment, access to legal counsel, visitation rights, and nutritional needs during incarceration. The needs assessment survey also asked them about their children and how they were being cared for either inside or outside the prison.

Results

Our findings indicate that the women experience hardships during their incarceration that would be unconstitutional in the United States. As Table 14.1 indicates, on almost every measure, from visitation to violence, medical care, legal assistance to the quantity and quality of food, the vast majority of women reported that their basic needs were not being met.

TABLE **14.1**
Distribution of Selected Questions from Needs Assessment

Survey (n = 50)

Received some legal assistance	64%
Thought assistance was helpful	50%
No lawyer	40%
Family Visitation: Yes	69%
No family visitation	31%
Average frequency of visits	Weekly
Procedures for visitation satisfactory	40%
Quantity of prison food satisfactory	50%
Quality of prison food satisfactory	16%
Job training in prison satisfactory	48%
Health care received in prison	
Medical	20%
Obstetric/gynecological	12%
Other	10%
Dental	10%
No medical assistance reported	22%
No response	26%
Reporting fear for personal safety in prison	76%

Many women complained that the one meal per day supplied by the prison was simply not enough to eat. Most women worked inside the prison and/or relied on family members or charitable organizations to get supplemental food. Supplementing prison rations is expected and encouraged by prison administration, so typically bags of groceries arrive with visitors. It is not uncommon for illegal drugs or other contraband, such as cell phones, to be smuggled in with the food. Some women prostituted themselves during visiting days while others sold their crafts, such as dolls, cards, and jewelry.

(continues)

FOCAL RESEARCH *(continued)*

The women reported that they provide their own clothes, and if family members or other inmates cannot assist them, the poorest of women wear the clothes they had on their backs at the time of their arrest. A hierarchy develops among the inmates based on quantity and quality of supplies each woman is able to obtain from the outside. Power is a function of those who have enough or more than enough to get by.

Medical care is the number one service the women want to see improved. Just over 50% of the women said they received any prison medical services at all, and, for those that reported receiving medical services, the most common was for either obstetric or gynecological care. No preventative care is available and mental health services do not exist. Even inmates with severe health problems, such as HIV-AIDS and advanced diabetes, receive no treatment.

Seventy-six percent of women often felt afraid for their safety while in prison as inmates and visitors have free access within the prison and prison guards do not monitor inmate interactions or patrol cell blocks. Drugs are reported to be widely available inside the prison and are used by visitors and inmates alike. Over 50% of the women reported that they wished the guards had more training in order to deal with the violence they are confronted with periodically from guests or other inmates.

We find the conditions in the prison problematic for the overall custody and care of inmates that typically spans 6 to 8 years. Every woman we encountered at el Inca recounted very difficult experiences before and during incarceration, including poverty, isolation, abuse, sexual assault, rape, and in some cases, torture. Both long standing trauma and chronic anxiety fuel women's inability to resist similar mistreatment by others and may be a dominant factor in recidivism.

Conclusion

The U.S. "War on Drugs" has an impact on Ecuador. Curbing illegal employment in the drug trade is a hard sell where poverty, corruption, and deep political, economic, and social unrest leave many women struggling for survival (Sherret, 2005). Patriarchal structures within the culture exacerbate this problem in which dependent, naïve, poor, young women often feel as though they have little choice but participate and know that they must endure or suffer the consequences (Alvirez, Bean, & Williams, 1981). At the same time, their participation is vilified by the police and they easily become targets for mistreatment, neglect, and violence at all levels of the drug trade (Epele, 2003). The women feel they are easy prey for the police who use the arrest data to argue that they are doing their part to stop the flow of drugs into the United States.

Our assessment points to the need for large scale policy change as well as for more immediate programs for the incarcerated women of el Inca. We will be sharing our assessment with the prison director and Marcha Blanca, a leading Ecuadorian quasi-governmental organization that works with national security issues that include a focus on incarcerated women and their children. We will seek external funding to develop cottage industries for incarcerated and paroled women, as well as to facilitate other programs related to health care and employment. For example, a group of volunteers has organized a women's cooperative of recently released inmates to make reusable grocery bags. We will also help those in Ecuador to evaluate the long term impact of the U.S. War on

Drugs, especially as it affects the women and the children of the incarcerated. Our efforts along with the work of local organizations may begin the process of long term social change.

References

Associated Press International, July 7, 2008.

Alvirez, D., Bean, F., & Williams, D. 1981. The Mexican American Family. In C. Mindel and R. Habenstein, eds., *Ethnic Families in America.* New York: Elsevier.

Boletin Estadistico El Sistem Penitenciario Ecuatoriano. 2001–2004. Direccion Nacional de Rehabilitacion Sociale (DNRS).

Epele, Maria E. 2003. Changing Cocaine-Use Practices. *Substance Use and Misuse.* 38:9, 1189–2126.

Merolla, D. 2008. The War on Drugs and the Gender Gap in Arrests: A Critical Perspective. *Critical Sociology,* 34:2, 255–270.

Nuñez, Jorge. 2006. Caceria de Brujos Drogas Ilegales y Sistema de Carceles en Ecuador Quito Ecuador: Digitale Abya-Yala.

Preston, F. W., & Roots, R. I. 2004. Law & its Unintended Consequences. *American Behavioral Scientist,* 47:11, 1371–1376.

Sherret, Laurel. 2005. Futility in Action: Coca Fumigation in Colombia. *Journal of Drug Issues,* 35:1.

THINKING ABOUT ETHICS

While prisons in the United States have their own internal review boards to protect inmates against ethical violations, no such review boards exist in Ecuador. For this reason, the researchers sought and received Institutional Review Board (IRB) approval from Suffolk University, Norton-Hawk's home institution. Of special concern to the IRB was that the gift the researchers intended to give each inmate could be construed as coercive in nature if the gift was too large to refuse and would make the inmate feel obligated to participate. The researchers satisfied the IRB concern by giving a small bag of necessities valued at under $3.00 as a token of appreciation for participation.

After collecting and analyzing the data, Harrison and Norton-Hawk wrote their report describing the situation and proposing some changes in the area of health care and employment. In June 2009, Jill Harrison met with the director of the prison and also with the director of Marcha Blanca, the quasi-governmental organization that works with national issues including the concerns of incarcerated women. Both were very pleased to have the data identifying the more immediate and long-term needs of the inmates (Harrison, personal communication). Identifying medical care as a primary issue led Harrison to contact Brown University's Alpert School of Medicine. As a result of her efforts, Brown University held a conference on HIV prevention in Ecuador and planned a program for their medical students to do clerkships in el Inca and other public health facilities (Harrison, personal communication). This needs assessment is likely to lead to some very successful outcomes in the future.

BOX **14.2**   Practicing Sociology as a Vocation

Kristy Maher believes in practicing sociology in a way that uses her knowledge and skills to effect a positive change in her local community of Greenville, North Carolina. Her first project was an assessment of the health care needs of the local Hispanic population and her second was an outcome evaluation of a prenatal outreach program for women with high-risk pregnancies. The first resulted in the hospital system's hiring full-time translators for the Hispanic community and the second was helpful in securing continuing funding for the successful program. Since then she has worked on a variety of projects, including serving as the medical sociologist in a community health care assessment and as an evaluator for a program aimed at providing quality of health care access for children in low-income schools. Maher (2005) says that she believes in practicing sociology as a vocation, which means using her expertise as a medical sociologist, a researcher, and a statistician to better her community.

While sociologists can work far away, they also do important work closer to home. In Box 14.2 is Kristy Maher's view of practicing applied sociology in her own town.

If a new program is funded and there is time in the early stages of design or implementation to make improvements, program staff and developers can benefit from a **formative analysis**. This analysis is when an evaluator provides program staff with an evaluation of the beginning stages of a program that makes suggestions for improvement in program design and service delivery. Suppose, for example, a high school principal seeks assistance with a new anti-bullying program. A formative analysis would carefully review the program's goals and its current instructional materials (including comparisons of the program with programs and materials used in other school systems) and collect data about the ongoing program. Administrators, teachers, and students might be interviewed, teacher training and program sessions might be observed, and the first students involved in the program might be asked to complete a questionnaire or to participate in focus groups. After analyzing the data, evaluators should be able to offer advice, including suggestions for staff development, for modifying instructional materials or techniques, and for making adjustments in program management.

A related kind of evaluation is **process evaluation**, which is research to determine whether a program is being implemented as designed and if modifications, including alternative procedures, would be useful. If there are variations in program delivery in different sites or for different recipients, then the program and its outcome can vary. It's especially important for the program staff responsible for delivering subsequent versions of a program to know exactly how the initial version was implemented and what parts are working. A process evaluation of a community-based volunteer-led health awareness program that had peer educators focused on the implementation and ways to make improvements (Karwalajtys et al., 2009). In this program, peer volunteers in two Canadian communities were recruited, trained, and supported by a local organization. The peer volunteers helped the attendees in a cardiovascular health program by giving information, explaining automated blood pressure instruments, recording

formative analysis, evaluation research focused on the design or early implementation stages of a program or policy.

process evaluation, research that monitors a program or policy to determine if it is being implemented as designed.

readings, and the like. The process evaluation focused on the experiences of the volunteers and found that, while there were some differences between the experiences at the two program sites, the majority reported positive experiences and rewarding aspects of their involvement as well as a desire for additional, ongoing training. The process evaluation helped the program planners strengthen support for and expand the roles of peer educators (Karwalajtys et al., 2009).

Designing Evaluation Research

As in all kinds of research, the evaluation researcher needs to select appropriate research strategies. Making good choices for study design, measurement, data analysis, and the like depends on the requirements of the evaluation, including intended audience, resources available, ethical concerns, and the project's time frame.

Typically, many parties will have an interest in an evaluation, including program administrators, legislators, client groups, interest groups, and the general public. Different constituencies or stakeholders may want different information and have different expectations about the initiative. Scientists are most often concerned with conducting high-quality research and obtaining knowledge; the public and legislators may focus on using tax dollars wisely and improving the well-being of individuals and society; program administrators want to manage resources effectively and efficiently (Trochim, Marcus, Mâsse, Moser, and Weld, 2008).

As the objective of most evaluation research is to determine if a program, policy, or law accomplishes what it sets out to do, one concept or variable must be the goal or desired outcome of the program, policy, or law. A typical evaluation research question asks whether the program accomplished its goal(s). If there is a hypothesis, typically the program, policy, or law will be the independent variable and the goal(s) or outcome(s) will be the dependent variable(s). In the evaluation of the CPC program discussed previously, the independent variable was the program itself, which offered comprehensive education to preschool and kindergarten children and services for their families for between four and six years. A variety of outcomes were the dependent variables. One hypothesis, which was supported by evaluation research, was that CPC program participation was associated with lower rates of reported parental maltreatment of children. Another was that CPC program participation was associated with better educational outcomes for the participants through their teens and twenties.

A critical challenge for evaluators is to articulate appropriate dependent variables and select valid measurement techniques. If a program has broad or amorphous goals, it can be especially difficult to select dependent variables even when program personnel help define the specific goal(s). Some issues evaluation researchers must confront are specified in Box 14.3.

STOP & THINK   *Evaluation researchers almost always have an explanatory purpose and typically test a causal hypothesis about the effect of an independent variable such as a program, law, or policy on a desired outcome. Can you remember some of the study designs discussed in Chapters 7 and 8? If you have a study with an explanatory purpose, what is a particularly useful study design?*

When evaluating a program, policy, or law, the following must be decided:

- What is the desired outcome?
- Is the outcome to be short or long term?
- Are attitudinal changes sufficient, or is it also essential to study behavioral change?

- How should change be determined?
- Should all aspects of the program be studied or only certain parts?
- Should all the targets of an intervention be studied or only some of them?

causal hypothesis, a testable expectation about an independent variable's effect on a dependent variable.

internal validity, agreement between a study's conclusions about causal connections and what is actually true.

In Chapter 8, we discussed experiments to test **causal hypotheses.** Built into the experimental design are ways to decrease systematic differences between the experimental and control groups before introducing the program or policy. The pretest–posttest experimental design can handle many of the possible challenges to **internal validity** because there are two measurements (before and after) and the two groups are exposed to the same conditions except for the program, policy, or law. Preexisting group differences before the program can be determined and the impact of the independent variable (program, policy, or law) on the dependent variable (goal of the program, policy, or law) can be determined.

 STOP & THINK   *The traditional experimental design is possible only when participants can be assigned to experimental and control groups. Can you think of evaluation research situations where a researcher either couldn't or wouldn't want to put people into a control group?*

Not all situations lend themselves to using the true experimental model. Outside a laboratory, the evaluator might not be able to control all aspects of the design. A well-known field study of the impact of police intervention on wife abuse in Minneapolis (Berk, Smyth, and Sherman, 1988) was supposed to be a true experiment. Police were to randomly assign domestic violence calls to one of three treatments (mediating the dispute, ordering the offender to leave the premises for several hours, or arresting the abuser). Despite the intended design, however, in about one-third of the calls, officers used individual discretion, not random assignment, to determine treatment. Without a pretest measurement or random assignment, there was no way to estimate prior differences between the experimental and control groups.

Sometimes practical concerns of time or money limit design choices. Evaluation research can take five or more years from design to the final report—often too much time for administrators and policy makers (Rossi and Wright, 1984). In other situations, as in the Chicago CPC evaluation, researchers know from the beginning that assigning participants to experimental or control groups is not possible or desirable and therefore select designs such as the quasi-experimental design. Similarly, studying the impact of "no fault" divorce laws on husbands' and wives' post-divorce economic statuses, Lenore Weitzman (1985) could not regulate where a couple would live when filing for divorce nor when couples would want to end their marriages. And clearly, in situations where there is an ethical responsibility to offer services to the most needy—such as AIDS treatment studies and child abuse programs—using random selection as the criterion for assignment to treatment or no-treatment groups is not acceptable. Instead, it is ethical to give treatment and services to the whole sample and have no control group.

If there is no possible control or comparison group, a useful design for evaluation is the panel study—a design that uses before and after comparisons of a group that receives program services or is affected by policy changes. If, for example, a school intended to change a discipline code by increasing the use of in-school suspension and decreasing out-of-school suspension, student behavior for the months preceding the change could be compared with behavior afterward. If student behavior improved after the policy was introduced, school personnel might decide to keep the new policy, even if they weren't sure about the extent to which other factors (time of year, a new principal, a new course scheduling pattern, and so on) contributed to the differences in behavior. Because of the problems in identifying causal relationships with a panel study, it will be better for documenting changes in the group than for identifying the causes of any changes.

Cross-sectional studies are sometimes used in evaluating programs or policies, especially if other designs are problematic. In a cross-sectional design, data about the independent and dependent variables are collected at the same time. Say, for example, that a city has been using a series of public service announcements about AIDS prevention in all buses for the past six months. Officials might commission a survey of city residents to determine the effectiveness of the advertisements. Interviewers could interview a random sample of respondents asking if they had seen the advertisements and soliciting their opinions of the advertisements, and then include a series of questions about AIDS and AIDS prevention. If the survey's data show that a higher percentage of those who remember seeing and liking the advertisements are more knowledgeable about AIDS and its prevention than those who either don't remember or didn't like the advertisements, the city officials could conclude that the advertisement campaign was successful.

◉ STOP & THINK *If those were our results, why would we need to be cautious in concluding that the advertisement campaign was successful?*

In our hypothetical city with our imagined advertisement campaign, with this relatively inexpensive and easy-to-do study, we could determine that a correlation exists between two variables. Although the connection between noticing advertisements and knowledge is suggestive of causation, we don't have evidence of the time order. For example, it's possible that those already most interested in and best informed about AIDS prevention would be those who paid the most attention to the advertisements. And those least informed about AIDS to begin with could be the ones who didn't read the public service announcements on the topic. In other words, the proposed independent variable could actually be the dependent variable!

Evaluation research can be judged by many standards: usefulness for intended users, degree of practicality, meeting ethical responsibilities, balancing client needs against responsibilities for public welfare, and issues of validity and reliability. For a statement of program evaluation standards, see the American Evaluation Association's Guiding Principles for Evaluators, posted on their website at http://www.eval.org/Publications/GuidingPrinciples.asp.

In most evaluation projects, like the one by Jill Harrison and Maureen Norton-Hawk on the Ecuadorian prison, the researchers are typically "outsiders" to the organization under study. Although Harrison and Norton-Hawk were originally viewed with suspicion by the local prison authorities, the researcher from outside the setting is, in many situations, seen as a credible professional, detached and objective. Such researchers can use what William Foote Whyte called the "professional expert model," where the researcher is "called upon by a client organization (or gets himself or herself invited) to evaluate existing policies and to suggest policy changes to meet new objectives" (1997: 110). When outside evaluators are hired, they are expected to give priority to the perspectives and needs of their clients or the funding agency. One of the limitations of traditional evaluation research is that the research participants and other stakeholders are given very limited opportunities to get involved and, as a result, they might not be invested in the outcome.

Some researchers want more equity for research participants. For example, psychologist Anne Brodsky (2001) and a team of students were invited to evaluate job training and education programs for low-income women. During the two-year project, the research team learned a great deal about research and the program. They provided feedback to the program staff but Brodsky (2001: 332) was troubled by the difficulty balancing the needs of the staff and those of the researchers, noting that the research questions were theoretical in nature and probably more important in the long run to her career and her students' education than to the program's effectiveness. Moreover, the research team was aware of power differences among its own members, based on things like position (faculty/student) and prior experiences and was cognizant of the need to make it clear to the program participants that, as researchers, they were outsiders and distinct from program staff (Brodsky et al., 2004). While priorities, insider/outsider status, power differences, access, and the like affect all kinds of applied research, they may be less of a concern in more activist and participatory kinds of applied research to which we now turn.

PARTICIPATORY ACTION RESEARCH

participatory action research, or PAR, research done by community members and researchers working as co-participants, most typically within a social justice framework, to empower people and improve their lives.

In the past few decades, a growing body of activist research from different disciplines, historical traditions, and geopolitical environments has emerged with an explicit value system and view of stakeholders. The related kinds of activist research include empowerment evaluation, liberation sociology, participatory research, collaborative research, action research, critical action research, and participatory action research (PAR).

Explicit Goals

In our discussion, we'll concentrate on PAR, although much of the commentary applies to the other forms as well. **Participatory action research, or PAR,** differs from more conventional sociological research in purpose and roles and views of study participants, including the researcher. PAR is defined as research in which "some of the people in the organization or community under study participate actively with the professional researcher throughout the research process from the initial design to the final presentation of results and discussion of action implications.... The social purpose underlying PAR is to empower low-status people in the organization or community to make decisions and take actions that were previously foreclosed to them" (Whyte, 1997: 111–112).[1] Such research may involve collaborations among professional researchers and local stakeholders or may be completed entirely by those who share the lived experience of the problem (Boser, 2006: 11).

An essential aspect of this kind of applied research is that those who would normally be the subjects of a study are also actively engaged in all stages of the process as researchers. The other essential aspect is that action comes from the research—action that can help the participating community.

Some argue that PAR can have conservative goals, as when corporations, foundations, and government bureaucracies use participation in research to get people at the grassroots level to comply with the organization's goals, such as profit-making (Feagin and Vera, 2001: 165). However, PAR usually aims to have an emancipatory cast. Feagin and Vera (2001) argue, in fact, that the roots of PAR are in the very beginnings of U.S. sociology—in the efforts of female sociologists, such as Jane Addams, working out of and in the Hull House settlement in Chicago in the 1890s. Addams sought the advice of women at all class levels, brought them onto the stage at gatherings, and provided them with research data to help them understand community patterns in their decision making (Feagin and Vera, 2001: 66).[2]

The practical significance of research and its potential impact are critical to this activist tradition. PAR rejects the view of research as a value-free endeavor. Instead, it is an unapologetic approach to knowledge creation through and for action, including as a goal the challenging of existing power relationships so as to make positive changes in the lives of individuals and their communities (Brydon-Miller and Maguire, 2009). Focusing on social justice, participatory work centers pragmatically on what works. Early advocate Peter Park (1993: 2) described PAR as seeking to "bring about a more just society in which no groups or classes of people suffer from the deprivation of life's essentials, such as food, clothing, shelter, and health, and in which all enjoy basic human freedoms and dignity.... Its aim is to help the downtrodden be self-reliant, self-assertive, and self-determinative, as well as self-sufficient." Unlike mainstream social science research, PAR has traditionally been a methodology of the margins, working with subordinate or oppressed groups to better their circumstances within society, rejecting research objectivity, and acknowledging that research is inherently political (Jordan, 2003).

---

[1]Whyte, W. F. 1997. Creative problem solving in the field: Reflections on a career. Walnut Creek, CA: AltaMira Press.

[2]Feagin and Vera indicate that Addams very clearly saw herself as a sociologist and that her "demotion" to mere reformer and social worker, by subsequent generations of sociologists, is more a reflection of the loss of the original emancipatory vision of sociology than of Addams's lack of concern with empirical research.

Participation and the Researcher's Role

Doing participatory research means working in partnership with those in the community to obtain and use the knowledge that is generated to empower the community. John Gaventa, a practitioner of PAR, says these researchers consider the questions of "Knowledge for whom?" and "Knowledge for what?" but also ask "Knowledge by whom?" as they understand that knowledge is a form of power (O'Neill, 2008).

In more traditional work, researchers may be seen by community organizations and social service agencies as coming in to judge how well they have been performing their jobs. Often in nonparticipatory evaluation research, the researcher treats the members of organizations and communities being studied as passive subjects, and the relationship between researcher and researched is hierarchical, with the researcher above others. But in participatory action work, the relationship is more equal, with researchers acting collaboratively with community and organization members throughout the process.

In PAR, the researcher must be involved intimately in the life of a community and its problems and be in dialogue with other researcher collaborators. The research groups are usually composed of professionals and ordinary people, with all viewed as authoritative sources of knowledge and experience. Project successes depend on researchers' abilities to communicate as well as their technical skills.

Researchers in activist traditions can select from all methods of data collection, including questionnaires, interviews, focus groups, observation, archival material, and visual methods. The more naturalistic methods of inquiry, such as participant observation and informal interviewing (perhaps in a more reciprocal way), are more typically selected because the research focus is usually on the participants and their interpretations of the social context. Kemmis and McTaggart (2003: 375) argue that in participatory action and related kinds of research, the researchers may make sacrifices in methodological and technical rigor in exchange for better immediate face validity so that the evidence makes sense within its context. They argue that this tradeoff is acceptable because a goal of the research is transformative social action, where participants live with the consequences of the transformations. Others such as Tina Cook (2006) worry that, by using participatory methods to both develop and then evaluate data, critical perspective can be lost.

The essence of PAR is that members of the community are involved in making the methodological decisions, developing or adapting the research instruments, and carrying out the research and analysis. Every aspect of the research is seen as an opportunity for interaction and discussion. In this perspective, the process is not doing research *on* subjects but is instead doing research *with* community participants (Cook, 2012).

One PAR project illustrates the approach's goals, methodology, range of participants, and the way authority is distributed in the research process. Youth in Focus (YIF) is a nonprofit organization in northern California that fosters youth development and organizations by supporting youth-led research, using interviews, focus groups, surveys, observation, photography, and video as methods of data collection in efforts to investigate inequities and to advocate

---

BOX **14.4**   **Planning PAR**

- Find one or more community agencies, organizations, and/or individuals to be your partner(s).
- With your community partners, make joint decisions on a research topic or question and the time frame for the study.
- Find out about the requirements for approval of your project from the IRB at your college or university and apply for approval in a timely manner.

- Think about your stake in the research, your expertise, and what you want to contribute.
- With your community partners, estimate the time and the costs of doing the project, locate possible resources, and put together an action plan before proceeding to do the research.

---

for justice in fields such as health, juvenile justice, education, and community planning (Sanchez, Lomeli-Loibl, and Nelson, 2009). One YIF Health Justice Initiative project was conducted in partnership with the Sacramento Gay and Lesbian Center and adult allies in local gay–straight alliances. The broad research question tackled was "What are the needs of lesbian, gay, bisexual, transgender, and questioning (LGBTQ) youth in Sacramento?" Over the course of the project, the research team included 25 young people ages 13 to 22 and 3 professionals. The initial effort was a needs assessment survey with an emphasis on mental health using a sample of students and counselors at local high schools. Based on the results, the research team made a series of recommendations that included increases in affordable and accessible mental health services for LGBTQ youth, greater visibility of and training for providers who work with this population, and strong outreach to community centers and schools. The team also filmed a documentary focused on the narratives and needs of LGBTQ youth throughout Sacramento. The PAR project has been well received by local agencies and has attracted attention from local service providers. "With an overwhelming sense of both need and optimism, the youth research team developed its strategy to be not only an informative process but also an exercise in advocacy for the many who are silenced by homophobia" (Sanchez et al., 2009: 7–8).

The steps in doing PAR are described in Box 14.4. If you'd like to read about some interesting work in the field, check out the website of The Public Science Project, an organization committed to democratizing the systematic production of knowledge and collaborating with communities to design research examining the impact of policy and structural injustice, at http://www.publicscienceproject.org/.

FINAL CONSIDERATIONS

A Middle Ground

Some argue that while "flattening" the research hierarchy may be an ideal in community research, in most settings the roles of researcher and participant are not equal. These researchers advocate acknowledging that "power and resource differences that typically exist and, thereby, accept additional responsibilities as researchers that are not necessarily balanced by parallel or complementary responsibilities on the part of participants" (Bond, 1990: 183). Although at first glance it appears that researchers must choose between the more conventional and the more activist kinds of applied work, there is a middle ground in the continuum of community involvement and subject participation. Many organizations have barely sufficient staff and volunteers to offer their programs and services and have no additional resources to commit to evaluation or action research. As a result, groups might be willing to work in collaboration with researchers to obtain information about the effectiveness of programs for their own internal use or to use in seeking recognition or funding for future projects. In addition, PAR and its approach are gaining acceptance and are increasingly given credence by researchers and international organizations such as the World Bank and government agencies (Jordan, 2003).

PAR is not likely to replace mainstream or expert-directed work but it is possible and sensible for the practitioners from both traditions to collaborate. For example, A. Javier Trevino (2012: 11), president of a social science organization, called for "service sociology" where research is a problem-solving endeavor that relates to individuals as real people and works toward helping them meet their pressing social needs.

Social scientists can play a major role in initiating identification of and response to oppressive conditions, contributing macro-level policy analyses, and providing cross-national analyses (Krimerman, 2001). Social research should be of help to "those who have experienced patterns of exclusion and subordination, as well as people and communities who are victims of natural disasters and man-made catastrophes" (Trevino, 2012: 12). Using a combined approach, described by some as engaged evaluation or community-based research, the researcher and the organization or group work together to share responsibilities for assessing a program and charting future directions. Evaluation is typically different when done for internal rather than external purposes. In internal evaluations, questions are asked about "our program," whereas external evaluation asks about "your program" (Fetterman and Eiler, 2000). The researcher working with the community can bridge the gap between these approaches.

The researcher and community group can equitably divide the work. Examination of the experiences of 20 teams of evaluators and project directors concluded that that most had shared many of the roles (Cartland, Ruch-Ross, Mason, and Donohue, 2008). While in some projects there was tension and confusion in role sharing, in most, the problems were resolved quickly with participants willingly adopting collaborative efforts. Few evaluators avoided the participation of others in project development and almost all invited stakeholders to make some decisions about the evaluation itself (Cartland et al., 2008). This cooperative and role-sharing model of evaluation may become more prevalent in the future.

An organization that balances conventional evaluation research perspectives with more participatory viewpoints has been the Center for AIDS

Prevention Studies (CAPS), which has worked to stimulate collaboration among academic researchers, public health professionals, and community-based organizations. Established in 1986, the Center for AIDS Prevention Studies (CAPS) at UC San Francisco is the world's largest research center dedicated to social, behavioral, and policy science approaches to HIV. CAPS works with communities to disseminate information dissemination, provides technical assistance to community-based organizations, and does collaborative research with communities (Fernández-Peña et al., 2008).

One CAPS project examined 11 partnerships between community-based organizations and university-based researchers and found the following: the community-based organizations took the lead in developing the research question, delivering the program, and collecting the data; the academic researchers took the lead in developing the instrument, consent procedures, and data analysis; and together the groups trained the staff, interpreted the data, and disseminated the findings (Goldstein et al., 2000). CAPS supported many collaborative research and evaluation projects with their efforts. A useful resource is "Working Together" (CAPS, 2001), a manual for doing collaborative research with community organizations.

STOP & THINK   *Applied research is usually field research. What do you think are some of the consequencs of real-world settings on research outcomes?*

Politics and Applied Research

Regardless of approach, all research is political in nature and is affected, to some degree, by social, political, and economic climates. Knowledge is socially constructed, so the choice of research topics and questions and all methodological decisions are related to their context. In participatory action research, the politics of the setting are sometimes problematic, especially as researchers work as non-privileged members of a team despite their professional expertise. Communities make initial judgments about the research and researchers and decide whether or not to invite them into the setting. Because of shared responsibility, the working group will determine the form and direction of the research project.

In evaluation research, the research project is affected not only by societal values and the priorities of funding agencies, but also by the perspectives of various constituencies and program stakeholders. Evaluations are typically conducted in field settings, within ongoing organizations such as school systems, police departments, health care facilities, and social service agencies. The politics or power within the organizations and the world beyond them must be considered. There is always ideology to consider, "turf" to protect, loyalties and long-standing personal relationships to look after, and careers and economic survival to think about. Organizational participants and their allies can either help or hinder evaluation research efforts depending on their assessment of the politics of the research and its social context.

Sara McClelland and Michelle Fine (2008) demonstrated the way that politics and evaluation interacted in their fascinating discussion of the federally sponsored abstinence-only-until-marriage (AOUM) programs. There was increased spending on these programs between 2001 and 2006, during the presidency of George W. Bush, with a total of almost $800 million spent on them by the end of 2006. As result, there was great interest in evaluating these programs. One way that politics affected evaluation was by selecting *program outcome*. One group of evaluations that assessed various AOUM programs focused on *attitudes*, not behavior. The evaluation studies asked young people about their attitudes toward pre-marital sex and their intentions to abstain from sex until marriage and found evidence of pre- and post-attitudinal shifts among preteens, but most did not measure behavior as an outcome variable (McClelland and Fine, 2008). It wasn't until 2007 that Mathematica Policy Research, Inc., a highly regarded, nonpartisan social science research institute, released the final results of its quasi-experimental, longitudinal evaluation of abstinence-only education. After nine years of evaluation, the reports were released and presented convincing evidence on *behavior*: The federally funded abstinence-only education programs do not delay sexual activity, do not increase abstinence rates, and do not change rates of condom use (McClelland and Fine, 2008).

The Obama administration increased emphasis on "evidence-based" policy. In 2012, the Acting Director of the Office of Management and Budget advised agencies that funding approval for the 2014 budget would favor requests demonstrating commitment to developing and using evidence and invited agencies to propose new evaluations, including those of existing programs (Zients, 2012). Of course the issue is whether values and ideology will continue to affect funding or whether evaluation results will become most important. David Bornstein (2012), a social innovation analyst, is optimistic as "the administration has brought this focus to home visitation, teen pregnancy prevention, educational innovation, social innovation, career training and workforce innovation—programs that account for close to $4 billion of spending. What's different is not just that these programs have to be evaluated using scientific methods, but that funding is being tied to evidence-based models in the legislation. In the case of the administration's home visitation initiative, 75 percent of the funding must go to support programs that have been shown to produce results."

Beyond Our Control

Organizations are not "neutral territory," and in most cases, the researcher is an outsider, working in someone else's sphere. The results of any assessment will have the potential to affect the organization and the individuals under study. As a result, a special challenge in evaluation research can be obtaining the cooperation of program staff for access to data. Sometimes this is especially challenging. When one of us, Emily, was hired by a special education professor and researcher to help with the evaluation of special-needs educational policies in several communities, feasibility became a concern. The research plan was for Emily to observe school meetings at which individual education plans for special-needs students were designed by school personnel in consultation with parents. There was no difficulty in getting cooperation at three of the schools selected, but, at the fourth school, the principal kept "forgetting" to communicate when meetings were canceled or had been rescheduled because of "unavoidable problems and crises." Several times, on arriving for a meeting at this particular school, Emily was told that the meeting had already been held—the day or even the hour before. The message was clear: The principal did not want an observer. Because the schools selected were a purposive, rather than random, sample, the meetings at another school were substituted and observed.

Harrison and Norton-Hawk also had problems obtaining data in their needs assessment of the prison in Ecuador. Even with letters from their home institutions, access to inmates for the research was difficult. The researchers were not initially viewed as credible or as having any authority (Harrison, personal communication). The Comite de las Mujeres, the inmate group they had worked with before, was helpful in gaining access to the prison director as there is no formal review process for researchers. The director was concerned that they would exploit the inmates by using the information in inappropriate ways. After they explained the IRB approval process at Suffolk University and talked about their experience studying female incarceration in prisons in the United States, he allowed the research. However, they were not able to use a random selection process and were told that they were lucky to be permitted access at all (Harrison, personal communication).

The sentiment that evaluators aren't welcome isn't unusual. Evaluators and their work may be distrusted by program personnel who often feel that the time and money spent on evaluation would be better spent on programs. Data collection for an evaluation can make it harder to run to the program; evaluators must be sensitive to the fact that program staffs have jobs to do that might be made more difficult by ongoing evaluations.

In contrast to evaluation researchers, participant action researchers have less control over and responsibility for the research. In PAR, the professional researchers must be willing to contribute expertise and work within a team rather than in a traditional hierarchy. They may find that while they can be invited in, they can also be asked to leave.

Having an Impact

As education and action are explicit parts of their work, participatory action researchers can usually see the impact of the work. For example, at the end of the Youth in Action needs assessment project, a documentary film was produced and recommendations for services were made to a variety of local organizations (Sanchez et al., 2009). Working with local groups and agencies, participatory action researchers know that their work will have an impact because a group of knowledgeable and committed community participants remain when the research is completed.

In conventional evaluation, a major challenge is implementing change after the research is completed. This is not a new story. In 1845, for example, the Board of Education of the city of Boston initiated the use of printed tests to assess student achievement. The resulting evaluation of test scores showed them to be low, but rather than analyze the causes of the poor performance (Traver, cited in Chambers, Wedel, and Rodwell, 1992: 2), the board decided to discontinue the test! Unfortunately, this example, more than a century and a half old, still has relevance.

Evaluation work can influence policy and programs in several ways. It can be used to end or modify programs or can justify continued support and expansion of programs. However, even if evaluators are trusted and allowed to collect data, and do a credible job of evaluating a program or policy, their conclusions might have little impact on the decision to keep a program, modify it, or eliminate it. Ideological and political interests can sometimes have more influence on decisions than evaluative feedback. Even if shown to be ineffective, a program might be kept if it fits with prevailing values, satisfies voters, or pays off political debts.

Outside evaluators usually have an advisory role and furnish information about effectiveness of programs. They can provide reports with significance for policy making, discuss the larger context of their work, argue forcefully for their positions, and work toward disseminating their findings widely, but they rarely have the power to save, close, or institute changes in programs or policies. The following Research in the News describes a small a piece of the political battle over one educational program, long after the evaluation research was completed.

RESEARCH IN THE NEWS

Jeanette Rundquist (2012) and Rebecca Forand (2013), education reporters respectively for the *New Jersey Star Ledger* and the *South Jersey Times*, both wrote about DARE, a drug abuse program used in about 750 New Jersey schools. A new DARE program, Keepin' it REAL, had been created by the national organization. However, the New Jersey Association of School Administrators did not support the new program and New Jersey DARE officials were concerned that the new national program had not been proven to work.

Forand reported that the New Jersey group had decided to break away from the national organization and teach a different curriculum. She quotes Nick Demauro, chairman and CEO of DARE NJ, as saying "Basically the issue is our state officials are insisting, justifiably so, that we only offer evidence-based curricula. This [program] has been evaluated, peer reviewed and submitted to a recognizable entity to demonstrate the program shows effectiveness." Rundquist noted that the original DARE program has long been plagued by outside evaluation findings that DARE programs are ineffective or counterproductive and described the legal battle over whether DARE America can revoke the charter of its New Jersey chapter over its use of a different curriculum.

For more information, see the complete articles at:
http://www.nj.com/news/index.ssf/2012/12/widely_used_anti-drug_program.html
http://www.nj.com/gloucester-county/index.ssf/2013/02/evolution_of_njs_dare_program.html

Drug Abuse Resistance Education (DARE) provides an interesting case study of how programs that receive negative evaluations can continue to flourish. DARE originated on a small scale in the early 1980s and over the next decade spread around the country. In 2001, estimates of its costs varied from about $200 million to around $1 billion annually, with the DARE organization estimating the annual costs of the officer services to be about $215 million (Shepard, 2001).

Over the years, many evaluations indicated that DARE programs were unsuccessful (Ennett, Tobler, Ringwalt, and Flewelling, 1994; Perry, Komro, Veblen-Mortenson, and Bosma, 2000; West and O'Neal, 2004; Wysong, Anis-kiewicz, and Wright, 1994). The research found that positive effects on students' knowledge, attitudes, and behavior (often observed right after the program) were temporary so that by late adolescence students exposed and not exposed to the program were indistinguishable (Birkeland, Murphy-Graham, and Weiss, 2005). The negative findings received a great deal of attention in the media and spread the message that DARE doesn't work (Weiss et al., 2005). However, some stakeholders—in this case, organizations with direct and indirect involve-ment—accumulated sufficient resources and legitimacy to protect the program.

School districts have had mixed responses to the negative evaluations. Some ignored the evidence and continued to offer DARE for a variety of reasons, including the view that some unmeasured long-term effect is at work such as positive relationships between police officers who teach the programs and stu-dents who take them (Birkeland et al., 2005). Other systems were persuaded by the evaluations or dropped the program for other reasons (Weiss et al., 2005). An analysis of how decisions were made in 16 communities in four states that had DARE programs found that there was no single way that communities made the decision to keep or cut the program (Weiss et al., 2005).

Even when evaluations are very good and the results are indisputable, programs can be cut due to economic and political factors. Box 14.5 tells about the demise of the very successful educational program (CPC) for par-ents and children that we described earlier in this chapter.

BOX **14.5**   Successful Programs and Political and Economic Realities

A blog entry on the website of New American Foundation, a nonprofit, nonpartisan public policy institute, tells a depressing story.

Consider an education program so effective that its impact can be measured 19 years later, so well-studied that it can be backed up with dec-ades of scientific evidence on children's improved skills in math and reading, and so impressive to policy makers that it continues to be championed around the country 40 years after its launch. These are the superlatives that come with Chicago's Child Parent Centers. So you might figure they're flourishing as part of the Chicago Public Schools' early childhood pro-grams, right? Not so. Their numbers are dwin-dling. In the mid-1980s, there were at least 25 CPCs serving more than 1,500 children. By 2006, there were 13. Today, 11 are still open, according to the Promising Practices Network. Enrollment in 2009, as reported by the Chicago Public Schools, is down to 670, less than half of what it once was. It now represents just 2 percent of the system's total preschool enrollment.... Shifts in budget priorities within the school system and across the state played a big role, along with changing neighborhood demographics that have complicated the CPC's mission of helping low-income children.... CPCs offer some of the strongest lessons of all, and their closures send a warning about how difficult it can be to sustain the programs that have been shown to do the most good (Guernsey, L. 2009. What's been cut: The story of the child parent centers. Retrieved on February 26, 2009, from http://www.newamerica.net/blog/early-ed-watch/2009/whats-been-cut-story-child-parent-centers-10341).

SUMMARY

Applied research is research with a practical purpose. It can be undertaken by a researcher with the participation of one or more stakeholders. In evaluation research, the researcher or a specific stakeholder develops a research question or hypothesis, most typically a causal hypothesis with a program or its absence as the independent variable and the goal of the program as the dependent variable. Evaluation research can provide reasonably reliable and valid information about the merits and results of particular programs that operate in particular circumstances. Necessary compromises in programs and research methodologies can mean that the users of the information will be less than fully certain of the validity of the findings. Because it is better to be approximately accurate than have no information at all, researchers might have to settle for practical rather than ideal evaluations.

Participatory action researchers and others in the activist tradition work with specific goals, most often toward achieving social justice. Professionals outside of the setting work collaboratively with participants in the setting. Together they decide on the research questions, the study's methodologies, and ways to apply the findings.

Applied work is done in real-world settings with political and social contexts. In each project, there will be multiple stakeholders. At the end of each study, the analyses and their practical implications can be valuable to program staff, clients, other stakeholders, other researchers, and the general public. If used wisely, applied research can help to create, modify, and implement programs and activities that make a difference in people's lives.

EXERCISE 14.1

Designing Evaluation Research

Find a description of a social program in a daily newspaper or use the following article about a hypothetical program. (If you select a newspaper article, attach it to the exercise.)

"Pets Are Welcome Guests"

Residents of the Pondview Nursing and Rehabilitation Center have a series of unusual guests once a week. The VIPs (Volunteers Interested in Pondview) have organized a "Meet the Pets Day" at the local facility. Each week, one or more volunteers bring a friendly pet for short one-to-one visits with residents. On a typical day, a dozen or so owners will bring dogs, cats, and bunnies to Pondview, but sometimes companions include hamsters and gerbils.

Last week, Buffy, a spirited golden retriever with a wildly wagging tail, made her debut at

Pondview. In a 15-minute visit with Mrs. Rita Williams, an 85-year-old widow recovering from a broken hip, Buffy managed to bestow at least several "kisses" on the woman's face and hands. Mrs. Williams said she has seen more sedate pets, but wasn't at all displeased with today's visit.

Margaret Collins, facility administrator, said that the program had been adapted from one she had read about in a nearby city. She was glad that the VIPs had organized the new program. "It gives the residents something to look forward to," she said. "I think it makes them more alert and attentive. If it really does aid the residents' recoveries and results in their improved mental health, we'll expand it to several days a week next year."

Design a research project to evaluate either "Meet the Pets Day" or the program described in your local newspaper by answering the questions that follow.

1. What is an appropriate research question that your evaluation should seek to answer or a hypothesis that you would test?
2. Describe the social program that is being offered.
3. Who are the program's participants?
4. In addition to the program's participants, who are the other stakeholders?
5. What is the goal or intended outcome of the program?
6. Describe how you would decide if the program was successful in meeting its goal by designing an outcome evaluation study. Be sure to describe the study design you would use, whom your sample would include, and how you would measure the dependent variable.
7. Comment on the internal validity of the study you designed.
8. What are the ethical considerations you would need to consider if you were interested in doing this study?
9. What are the practical issues (time, money, and access) that you would need to consider if you were interested in doing this study?

EXERCISE 14.2

Participatory Action Research

Imagine that you would like to do a piece of applied research that focuses on a group in your community, such as a shelter for homeless people, a literacy program, a neighborhood preservation association, a local boys or girls club, a volunteer services for animals organization, or any group of interest.

Visit one such group or organization in your community. See if you can meet either formally or informally with members of the staff and with clients. Ask them what they feel their most pressing needs are and what problems the organization or group has had in meeting these needs.

Based on what you find out, write a short paper describing a possible PAR project, which includes the following:

1. A description of the organization, agency, or group.
2. A listing and description of the major stakeholders.
3. A possible research question that could be answered if you were to work collaboratively with the major stakeholders.
4. The benefits of doing the work you're proposing.
5. Any practical or ethical concerns you think you'd run into in doing the project.

EXERCISE 14.3

Evaluating Your Course

The course you are taking this semester is an educational program designed to help students understand and use social research methods. As such, it can be evaluated like any other program. For this exercise, select one aspect of your course (such as the textbook, the instructional style of the teacher, the frequency of class meetings per week, the length of each class session, the lectures, or the exercises) as an independent variable. Design an evaluation research project that could be conducted to test the effectiveness of the aspect of the course you've selected on the dependent variable of student learning.

Answer the following questions.

1. What aspect of the course are you focusing on as your independent variable?
2. Who are the significant stakeholders in this setting?
3. How would you define and measure the dependent variable "student learning of social research methods?"
4. What would you do to determine the effectiveness of the aspect of the course that you are interested in?
5. Using your imagination, what are some results that you think you'd get from your evaluation?
6. Comment on the practical and ethical considerations you would need to consider if you wanted to do this study.

EXERCISE 14.4

Comparing Evaluation Research Projects

The website of the Office of Juvenile Justice and Delinquency Prevention has a large database of programs archived by topic, such as delinquency prevention. The archive is available at http://www.ojjdp.gov/programs/ProgSearch.asp. Think about your interests and issues your community is facing and find two programs described in the database that have addressed this concern. Read about any evaluations of the programs. For this assignment, describe and analyze the programs and evaluations using the following format:

1. Describe and compare the programs in terms of cost, populations served, approach, and so on, based on the information you can find online about them.
2. Discuss the evaluations of the programs. Do they have different outcome variables? How are the variables measured? Does one program seem to be more effective than the other? Do you think one or more of the programs should be modified? Expanded? Eliminated?
3. Print out the listings of the two programs from the database and attach them to your exercise.

Quantitative and Qualitative Data Analysis

CHAPTER 15

© Ocean/Corbis

CHAPTER OBJECTIVES

This chapter is designed to help students

- **Distinguish between quantitative and qualitative data analysis**
- **Do elementary quantitative analyses, including**
  a. **Calculating means, medians, modes, ranges, and standard deviations of individual variables**
  b. **Constructing crosstabulations and appreciating other kinds of bivariate analyses**
  c. **Understanding elaborations of crosstabulations**
- **Appreciate the difference between social anthropological and interpretivist approaches to qualitative data analysis**
- **Get a feel for the steps involved in qualitative data analysis, including**
  a. **Transcribing data**
  b. **Coding data**
  c. **Data displaying**
  d. **Drawing conclusions**

CHAPTER OUTLINE

Perhaps you've asked yourself, as we've discussed all the ways of collecting data, "Once I've collected all the data, how do I organize all these facts so that I can share what I've found?"

We'd now like to introduce you to the most rewarding part of the research process: analyzing the data. After all the work you've done collecting your information, there's nothing quite like putting it together in preparation for sharing with others—a little like finally cooking the bouillabaisse after shopping for all the individual ingredients. But like the excitement of cooking bouillabaisse, the excitement of data analysis needs to be balanced by having a few guidelines (or recipes) and principles to follow. In this chapter, we'll introduce you to quantitative and qualitative data analyses. Rather like an elementary cookbook, this chapter offers rudimentary data-analytic recipes, and some insights into the basic delights and principles of data analysis.

QUANTITATIVE OR QUALITATIVE?

What is the difference between quantitative and qualitative data? One answer is that quantitative data results from quantitative research and qualitative data results from qualitative research. This sounds like question begging, and of course it is, but it carries an important implication: The distinction drawn between quantitative and qualitative data isn't as important (for data-analytic purposes) as the distinction between the strategies driving their collections. Sure, you can observe, as James Dabbs (1982: 32) and Bruce Berg (1989: 2) have, that the notion of qualitative refers to the essential nature of things and the notion of quantitative refers to their amounts. But, as we've hinted before (e.g., in chapters on qualitative interviewing and observation techniques), there's really nothing about qualitative research that precludes quantitative representations (see Morrow, 1994: 207). Enos (see Chapter 10), of course, did count the number of inmates' children who had been placed in foster care. Similarly, research that might seem primarily quantitative never totally ignores essences. Martinez and Lee (in Chapter 12) used qualitative judgments about which murders were committed, for instance, by non-intimates before they used quantitative techniques to decide what percent of murders were committed by non-intimates. Indeed, although the mix of quantitative and qualitative approaches does vary in social science research, almost all published work contains elements of each. Some research involves about equal amounts of quantitative and qualitative research. Thus, in their work on college women's sexual enjoyment, Armstrong, England, and Fogarty (2012) use secondary data from a survey of over 13,000 heterosexual women (a quantitative analysis) to confirm their hypothesis that college women find sex within relationships more satisfying than sex during hookups. But they use 89 in-depth interviews (a qualitative analysis) of both women and men to come up with their explanation of why this association exists: "[B]oth men and women question women's (but not men's) entitlement to pleasure in hookups but believe strongly in women's (as well as men's) entitlement to pleasure in relationships"

(Armstrong et al., 2012: 435). Consequently, the distinction we draw (and have used to organize this text) is not one of mutually exclusive kinds of analysis but, rather, of kinds that, in the real world, stand side by side and, in an ideal world, would always be used to complement one another.

An Overview of Quantitative Data Analysis

quantitative data analysis, analysis based on the statistical summary of data.

However artificial the distinction between quantitative and qualitative research and data might be, there's no denying that the motives driving quantitative and qualitative researchers are distinguishable. Although there are exceptions, **quantitative** researchers normally focus on the relationships between or among variables, with a natural science–like view of social science in the backs of their minds. Thus, in the fashion of the physicist who studies the relationship between the gravitational attraction and distance between two bodies, Ramiro Martinez and Matthew Lee (in Chapter 12) looked at whether migrant communities were more likely to have high crime rates than non-migrant communities. Projects like these engaged in other natural science–like activities as well: for example, looking at aggregates of units (whether these were characters in picture books or communities), more or less representative samples in some studies (see Chapter 5 on sampling) of these units, and the measurement of key variables (see Chapter 6 on measurement). For data-analysis purposes, the distinguishing characteristic of this type of study is attention to whether there are associations among the variables of concern: Whether, in general, as one of the variables (say, whether one is an immigrant or not) changes, the other one (whether one is likely to commit homicide or not) also changes. To demonstrate such an association, social science researchers generally employ *statistical* analyses. Thus, we will aim toward a basic understanding of statistical analyses in the part of the chapter on quantitative data analysis.

🛑 **STOP & THINK**   *Recall Adler and Clark's study (in Chapter 7) of retirement. Did people who said they expected to work after retirement actually end up being more likely to work after retirement than people who didn't expect to work after retirement? Do you see that the focus of this article, like that of others using quantitative data, is on variables (in this case whether the expectation to work after retirement is associated with actually working after retirement) and not really on people, groups, or social organizations?*

An Overview of Qualitative Analyses

qualitative data analysis, analysis that results in the interpretation of action or representations of meanings in the researcher's own words.

The strategic concerns of qualitative researchers differ from those of quantitative researchers and so do their characteristic forms of data analysis. Generally, **qualitative** researchers tend to be concerned with the interpretation of action and the representation of meanings (Morrow, 1994: 206). Rather than pursuing natural science–like hypotheses (as Durkheim advocated in his *The Rules of Sociological Method* [1938]), qualitative researchers are moved by the pursuit of empathic understanding, or Max Weber's *Verstehen* (as described in his *The Theory of Social and Economic Organization* [1947]), or an in-depth description (e.g., Gertz's [1983] *thick description*). Sandra

Enos (in the focal research in Chapter 10) aspired to an empathic understanding of the meaning of motherhood for women inmates and of the differences in those meanings for different inmates. Cottingham (in Chapter 11) provides a thick description of Steelers nation as a quasi-religion. Qualitative researchers pursue single cases or a limited number of cases, with much less concern than quantitative researchers have about how well those cases "represent" some larger universe of cases. Qualitative researchers look for interpretations that can be captured in words rather than in variables and statistical language. Later in the chapter, we'll seek a basic understanding of how qualitative researchers reach their linguistic interpretations of data.

QUANTITATIVE DATA ANALYSIS

Quantitative data analysis presumes one has collected data about a reasonably large, and sometimes representative, group of subjects, whether these subjects (or units of analysis) are individuals, groups, organizations, social artifacts, and so on. Ironically, the data themselves don't always come in numerical form (as one might expect from the term *quantitative* data).

A brief word to clarify this irony: You'll recall that the focus of quantitative data analysts isn't really the individual subjects being studied (e.g., college students in sociology classes or communities) but, rather, the variable characteristics of those subjects (e.g., whether they are exposed to film clips designed to illustrate key concepts or whether they have high crime rates). Remember that variables come in essentially three levels of measurement: nominal, ordinal, and interval-ratio. The most common level of measurement is the nominal level (all variables are at least nominal), because all variables (even marital status) have categories with names (like "married" and "single"). Thus, an individual adult might be married or single, and the information you'd obtain about that person's marital status might, for instance, be "single." But, "single" is obviously not a number. Nonetheless, it is information about marital status that, when taken together with data about marital status from other subjects, can be subjected to numerical or quantitative manipulation, as when we count the number of single people in our sample.

This is just a long way of saying that the descriptor "quantitative" in "quantitative data analysis" should really be thought of as an adjective describing the kind of analysis one plans to do, rather than as an adverb modifying the word "data." When you do a quantitative analysis, the data might or might not be in the form of numbers.

Sources of Data for Quantitative Analysis

coding, assigning observations to categories.

Data appropriate for quantitative analysis can come from many sources, including surveys conducted by researchers themselves. When the data are collected by the researcher, an important first step often is **coding**. In quantitative analyses, this frequently means making data computer usable. In qualitative analyses, this means associating words or labels with passages and then collecting similarly labeled passages into files. Coding is the process by which raw data are given a standardized form. In most cases, quantitative coding involves assigning a number to each observation or datum. Thus, when coding gender, you might decide to assign a value of "1" for each "female" and "2" for each "male." The assignment of a number is often pretty arbitrary (there is no reason, for instance, that you couldn't assign "1" for females and "0" for males), but it should be consistent within your data set.

As we suggested in Chapter 12, more and more published research in the social sciences is now produced by researchers who have analyzed secondary survey data—data collected by large research organizations and made available to other researchers, sometimes for a fee. According to one study (Clark and Grandchamp, 2012), about 56 percent of articles published in major sociological journals in 2008 and 2009 were based on analyses of secondary survey data. You might recall that the General Social Survey (GSS) is the most popular of these sets of survey data. One of the great advantages of secondary survey data is that they usually come pre-coded and ready for analysis via one of the software packages social scientists use for statistical analyses. The GSS comes, in an abridged version, for instance, along with abridged versions of a number of other data sets, in *The SPSS Guide to Data Analysis* (Norusis, 2011). This guide introduces students to statistical analysis using the software package, the Statistical Package of the Social Sciences (SPSS). GSS data are so widely available that you can even visit a website, the Social Documentation and Analysis (SDA) Archive (http://sda.berkeley.edu/archive.htm), and do your own analysis of GSS data from 1972 to 2010, and other data sets. In some of the embedded "Stop & Do" sections that follow, we'll be asking you to do so, so you can see what can be done with GSS data online.

In any case, we will be using GSS data to illustrate some of the kinds of quantitative analysis most frequently used by social scientists in our discussion of elementary quantitative analyses.

Elementary Quantitative Analyses

Perhaps the defining characteristic of quantitative analyses is their effort to summarize data by using statistics. Statistics, being a branch of mathematics, is an area of research methods that some social science students would just as soon avoid. But the advantages of a little bit of statistical knowledge are enormous for those who would practice or read social science research, and the costs of acquiring that knowledge, we think you'll agree, aren't terribly formidable, after all. Our primary goal in the section on quantitative analysis is to provide you with the kind of overview that will give students who have never had a course in statistics, and those with more experience, a feeling for how statistics advance the practice of social research.

To organize and limit our discussion of social statistics, however, we'd like to begin with two basic sets of distinctions: one distinction, between **descriptive** and **inferential** statistics, relates to the generalizability of one's results; the other distinction among **univariate**, **bivariate**, and **multivariate** statistics relates to how many variables you focus on at a time. Descriptive statistics are used to describe and interpret data from a sample. We find that about 62 percent of the people sampled in the GSS from 1972 to 2010 were married, so that the modal (the most common) marital status of the GSS

descriptive statistics, statistics used to describe and interpret sample data.

inferential statistics, statistics used to make inferences about the population from which the sample was drawn.

univariate analyses, analyses that tell us something about one variable.

bivariate analyses, data analyses that focus on the association between two variables.

multivariate analyses, analyses that permit researchers to examine the relationship between variables while investigating the role of other variables.

sample is "married" (data and analysis done using GSS data from the SDA Archive, 2012). The mode is a descriptive statistic because it describes only the data one has in hand (in this case, data on 55,121 respondents). Inferential statistics, on the other hand, are used to make estimates about characteristics of a larger body of data (population data—see Chapter 5 on sampling). We find, using the GSS data again, men significantly more likely than women to have been employed full time (SDA Archive, 2012). The word "significantly" in this context means that there's a good chance that, in the larger population from which the GSS sample has been drawn (basically adult Americans), men are more likely to be employed full time than women. We used an inferential statistic to draw this conclusion. In the rest of this chapter, we'll focus on descriptive statistics, but we'll briefly introduce inferential statistics, too.

Much powerful quantitative research can be done with an understanding of relatively simple univariate (one variable) or bivariate (two variable) statistics. Univariate statistics, like the mode (the most frequently occurring category), tell us about one characteristic. The fact that the modal marital status is "married" in the GSS sample tells us about the typical marital status of that sample. Bivariate statistics, on the other hand, tell us something about the association between two variables. When we wanted to find out whether men or women have been more likely to be employed full time—that is, that there is a relationship between "gender" and "work status"—we used bivariate statistics. Multivariate statistics permit us to examine associations while we investigate the role of additional variables. We hypothesized that the association between people's genders and their likelihood of being employed full time might have been accounted for in terms of their level of education. We used GSS data (from the SDA Archive, 2012) and found that the association between gender and employment status could not be accounted for in terms of education by using multivariate statistical methods.

UNIVARIATE ANALYSES

Measures of Central Tendency

mode, the measure of central tendency designed for nominal level variables. The value that occurs most frequently.

median, the measure of central tendency designed for ordinal level variables. The middle value when all values are arranged in order.

mean, the measure of central tendency designed for interval level variables. The sum of all values divided by the number of values.

Perhaps the most common statistics that we use and encounter in everyday life are ones that tell us the *average* occurrence of something we're interested in. In 2012 Miguel Cabrera, the Detroit Tigers' third baseman, the American League Most Valuable Player and its Triple Crown winner (with the highest batting average, most home runs, and most runs batted in) sometimes got a hit and sometimes didn't, but his batting average for the year was .330, meaning that he got a hit 33 out of every official 100 times he came to bat. In most distributions, values hover around an average, or central tendency. The most common of these are the **mode**, the **median**, and the **mean**.

In Chapter 6, we observed that the mode, median, and mean are designed for nominal, ordinal, and interval level variables, respectively. The mode, designed for nominal level variables, is that value or category that occurs most frequently. Have a look at the following data about five students' gender, age, and height. Because more students are female than male, the modal gender for the sample is "female."

Student	A	B	C	D	E
Gender	Male	Female	Female	Female	Female
Age	18	18	17	19	20
Height	Tall	Tall	Short	Short	Medium

STOP & THINK   *Look at those data again to see if you can decide what the mode is for "age" in this sample.*

Although the mode has been designed for nominal level variables, like gender, it can be calculated for ordinal and interval level variables (like age) because ordinal and interval level variables, whatever else they are, are also nominal. As a result, the mode is an extremely versatile measure of central tendency: It can be computed for any variable.

STOP & THINK   *We say a variable is unimodal when, like "gender" and "age" in the sample of students mentioned earlier, it has only one category that occurs most frequently. We say it is bimodal when it has two categories that occur most frequently. Check out the data for the variable "height." Is it unimodal or bimodal?*

STOP & DO   *The mode is easy enough to calculate when, as in our example of students, there are five cases. But let's see if you can apply the principles to a much larger sample? Let's look at data from the GSS on the SDA Archive. Access the Archive by going to http://sda.berkeley.edu/archive.htm. Now hit on the most recent GSS Cumulative Data File, and see if you can figure out what the modal highest educational degree has been for respondents to this survey. (Hint: You'll want to browse the codebook for "education" and then look at the information for "Rs highest degree," a variable called DEGREE. View DEGREE and identify the mode.)*

The median, the measure of central tendency designed for ordinal level variables, is the "middle" case in a rank-ordered set of cases. The variable "height," when measured as it is for our students mentioned earlier in categories like "short," "medium," and "tall," is an ordinal level variable. The height for the Students A through E is reported to be "Tall," "Tall," "Short," "Short," and "Medium," respectively. If you arrange these data from shortest to tallest, they become "Short," "Short," "Medium," "Tall," and "Tall." The third, or middle, case in this series is "Medium," so the median is "Medium." Though it is designed for ordinal level variables, the median can also be used with interval variables, because interval variables, whatever else they are, are also ordinal variables.

STOP & THINK   *Can you calculate the median age for the sample mentioned earlier?*

The five students mentioned earlier have ages of 18, 18, 17, 19, and 20. Arranging these in order, from lowest to highest, they become 17, 18, 18, 19, and 20. The third, or middle, case in this series is 18, so the median is 18 years of age.

The median is pretty easy to calculate (especially for a computer) when you have an odd number of cases (provided, of course, you have at least ordinal level information). What you do with an even number of cases is locate the median halfway between the two middle cases. If, for instance, the "number of years of education" for eight cases were 9, 9, 2, 9, 9, 7, 9, and 12, you'd want to arrange them in order so that they would become 2, 7, 9, 9, 9, 9, 9, and 12. Then, you'd take the two middle cases (9 and 9) and find that the median for these cases is $(9 + 9)/2 = 9$.

The mean, the measure of central tendency designed for interval level variables, is the sum of all values divided by the number of values. The mean years of education for the first seven cases mentioned in the previous paragraph is

$$\frac{9 + 9 + 2 + 9 + 9 + 7 + 9}{7} = 7.7 \text{ years.}$$

How does a researcher know which measure of central tendency (mode, median, or mean) to use to describe a given variable? Beyond advising you not to use a measure that is inappropriate for a given level of measurement (such as a mean or a median for a nominal level variable like gender), we can't give you hard and fast rules. In general, though, when you are dealing with interval level variables (such as age, years in school, or number of confidantes), variables that *could* be described by all three averages, the relative familiarity of reading audiences with the mean makes it a pretty sound choice. Thus, although you could report that the modal number of years of education among the first seven cases listed above is 9 or that their median is 9, we'd be inclined to report that the mean is 7.7 years, other things being equal. See Box 15.1 for an example of a time when things are not equal.

Variation

Measures of central tendency can be very helpful at summarizing information, but as Stephen Jay Gould (1997) eloquently argued, they can also be misleading. Gould gave his own bout with a rare form of cancer, abdominal mesothelioma, as an example. He'd been diagnosed with this disease as a 40-year-old and learned, almost immediately, that mesothelioma was incurable and that people who had it had a median life expectancy of eight months. Gould reported that his understanding of statistics provided him with almost immediate consolation (talk about knowing about statistics being personally useful!) because he knew that if all the journals were reporting life expectancy in terms of the median, then it was very likely that life expectancy for individuals was strongly skewed. Otherwise, the journals would be reporting a mean life expectancy. Moreover, he reckoned, the extreme values of life expectancy after diagnosis were almost surely on the high end, perhaps even many, many years after diagnosis. (One or two life expectancies of zero, after all, couldn't skew a distribution with a median of eight months that much.) Gould also realized that his own personal life expectancy was almost surely going to be on the high end of the scale. He was, after all, young, eager to fight the disease, living in a city (Boston) that offered excellent medical treatment, gifted with a supportive family, and "lucky that my disease had been discovered relatively early in its course" (Gould, 1997: 50).

Gould's larger point is a crucial one: Human beings—indeed, anything we're likely to study as social scientists—are not measures of central tendency, or means, or medians, or modes. All things worth studying entail variation. In his case, he needed to be able to place himself in a region of variation based on particulars of his own case. Similarly, if we're interested in the educational level, income, or wealth of a group of people, averages will only get us so far: We need to know something about the variation of these variables.

frequency distribution, a way of showing the number of times each category of a variable occurs in a sample.

skewed variable, an interval level variable that has one or a few cases that fall into extreme categories.

We can discover a lot of important information about variation by describing the distribution of a sample over various categories of a variable. One of the most commonly used techniques for such descriptions is the **frequency distribution,** which shows the frequency (or number) of cases in each category of a variable. To display such a distribution, the categories of the variable are listed, and then the number of times each category occurs is counted and recorded. Box 15.1, with its display of hypothetical data about the income of 11 people, provides an example of a frequency distribution.

BOX **15.1**   **Another Factor in Choosing an Average: Skewness**

Although our suggestion that you use the mean as your measure of central tendency for describing interval level variables works some of the time, it can yield pretty misleading "averages" for those many real-world interval level variables that are skewed. A **skewed variable** is an interval level variable that has one or a few cases that fall into extreme categories. Although we made up the following distribution of incomes for 11 people, it is not completely unlike the real distribution of incomes (and wealth) in the United States and the world, inasmuch as it depicts few very high-income people and many low-income people.

This income variable is skewed because it contains at least one extreme case (the one falling into the $20 million category. (The one falling in the $1 million category is "out there" too.) Because of this case, the mean of $1,926,363.60 provides a sense of average that somehow seems to misrepresent the central tendency of the distribution. In this case, as in the case of most skewed variables, we would advise reporting an average, like the median, that is less sensitive to extreme cases.

Do you agree that the median provides a better sense of this distribution's central tendency than the mean?

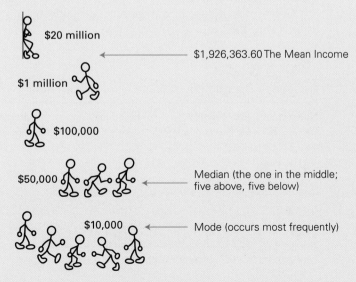

$20 million

$1,926,363.60 The Mean Income

$1 million

$100,000

$50,000

Median (the one in the middle; five above, five below)

$10,000

Mode (occurs most frequently)

TABLE **15.1**

Frequency Distribution of Gender of Respondent

Gender	Frequency (f)	Percentages
Female	17	85
Male	3	15
Total	N = 20	100

Let's imagine how we might create a frequency distribution for data about the gender of a hypothetical sample of respondents. Suppose we had 20 people in our sample and we found that there were 17 females and 3 males. The frequency distribution for this variable, then, is displayed in Table 15.1.

One thing this frequency distribution demonstrates is that the overwhelming majority of our respondents were female. The table has three rows. The first two rows display categories of the variable (gender). The third row displays the total number of cases appearing in the table.

The middle column shows the number of cases in each category (17 and 3). The number is called the *frequency* and is often referred to by the letter *f*. The total of all frequencies, often referred to by the letter *N*, is equal to the total number of cases in the sample examined (in this case, 20). The third column shows the percentage of the total number of cases that appears in each category of the variable. Because 17 of the 20 cases are female, the percentage of cases in this category is 85 percent.

The variable in this example, gender, is a nominal level variable, but frequency distributions can be created for variables of ordinal and interval level as well. Let's confirm this by working with an interval level variable: the number of close friends our respondents have. Suppose our 20 respondents told us that they had 1, 0, 0, 0, 2, 1, 1, 1, 3, 0, 2, 2, 2, 1, 1, 1, 0, 0, 1, and 1 close friends, respectively.

Create the outline of an appropriate table on a separate piece of paper, using Table 15.2 as a guide.

TABLE **15.2**

Frequency Distribution of Number of Close Friends

Number of Close Friends	Frequency (f)	Percentages
0		
1		
2		
3		
Total	N = 20	100

Now complete the distribution. First, count the number of respondents who report having "0" close friends and put that number (frequency) in the appropriate spot in the table. Do the same for categories "1," "2," and "3." Do the frequencies in each of these categories add up to 20, as they should? If not, try again. Once they do, you can calculate the percentage of cases that fall into each category. (Calculate percentages by dividing each f by the N and multiplying by 100. We count six respondents with 0 close friends, so the percentage of respondents with 0 close friends is $6/20 \times 100 = 30$.) Into which category does the greatest number of respondents fall? (We think it's category "1," with nine respondents.) In which category does the smallest number of respondents fall? What percentage of the respondents said they had three close friends?

Frequency distributions, simple as they are, can provide tremendous amounts of useful information about variation. One of our students, Angela Lang (2001), did her senior thesis on Americans' attitudes toward income inequality. Some of her sources had suggested that the American public is basically apathetic toward income inequality in the country and that a majority, in fact, felt that income inequality provided an "opportunity … for those who are willing to work hard" (Ladd and Bowman, 1998: 36). Lang was therefore surprised to find, upon examining GSS data from 1996, that, when asked whether they agreed or disagreed with the statement "Differences in income in America are too large," Americans gave responses that are summarized in the frequency distribution depicted in Table 15.3.

Most striking to Lang was the percentage of the 1,468 respondents who answered the question and either strongly agreed or agreed with the statement "Differences in income in America are too large:" 32.3 percent and 32.5 percent, respectively. You can spot these percentages in the column labeled "percent." Using the cumulative percentage column, which presents the percentage of the total sample that falls into each category and every

TABLE **15.3**

Frequency Distribution of Responses to the Statement "Differences in Income in America Are Too Large"

Response	Frequency	Percent	Cumulative Percent
Strongly agree	464	32.3	32.3
Agree	467	32.5	64.8
Neither	178	12.4	77.2
Disagree	169	11.8	89.0
Strongly disagree	117	8.1	97.1
Can't choose	41	2.9	100.0
Total	1,468	100.0	

Source: General Social Survey, 1996 (see Davis and Smith, 1998).

category above it in the table, Lang was also able to discern that 64.8 percent of this sample either strongly agreed or agreed with the statement. This was a much higher level of agreement than she expected based on her reading. Notice that the median and the mode for this variable were both "agree." The point, though, is that reporting that the median response was "agree" doesn't convey nearly as much information about the level of agreement as saying "About 65 percent of respondents either agreed or strongly agreed with the statement, whereas only about 20 percent disagreed or strongly disagreed."

STOP & THINK   *Can you see where the figure of 20 percent comes from?*

STOP & DO   *Over the years the GSS has asked about 26,885 respondents whether, if their party nominated a woman for president of the United States and she were qualified for the job, they would vote for her. What percentage, would you guess, has said, "Yes"? What percentage has said, "No"? Check your answers against the actual percentage by going to http://sda.berkeley.edu /archive.htm and finding the frequency distribution for the variable FEPRES.*

Measures of Dispersion or Variation for Interval Scale Variables

Examining frequency distributions, and their associated percentage distributions, is a pretty good way of getting a feel for dispersion or variation in nominal or ordinal level variables. If, for instance, you're looking at gender and discern that 100 percent of your sample is female and 0 percent is male, you know that there is no variation in gender in your sample. If, on the other hand, you find that 50 percent of your sample is female and 50 percent is male, you know that there is as much variation as there could be over two categories. Statisticians have actually given us more elegant ways of describing variation for interval level variables. Again, though, we start with the observation that the mean, perhaps the most frequently used measure of central tendency for interval level variables, can hide a great deal about the variable's *spread* or *dispersion*. Thus, a mean of 3 could describe both of the following samples:

Sample A: 1, 1, 5, 5

Sample B: 3, 3, 3, 3

measures of dispersion, measures that provide a sense of how spread out cases are over categories of a variable.

Inspection shows, however, that Sample A's values (varying between 1 and 5) are more spread out or dispersed than those of Sample B (*all* of which are 3). To alleviate this problem, researchers sometimes report **measures of dispersion** for individual variables. The simplest of these is the **range**: the difference between the highest and the lowest value. The range of Sample A would be 5 − 1 = 4, while that of Sample B is 3 − 3 = 0. The two ranges tell us that the spread of Sample A is larger than the spread of Sample B.

range, a measure of dispersion or spread designed for interval level variables. The difference between the highest and the lowest values.

Calculate the range of Samples C and D:

Sample C: 1, 1, 5, 5; Sample D: 1, 3, 3, 5

Having calculated the range, can you think of any disadvantage of the range as a measure of spread or dispersion?

standard deviation, a measure of dispersion designed for interval level variables that accounts for every value's distance from the sample mean.

There are several measures of spread that, like the range, require interval scale variables. The most commonly used is the **standard deviation.** The major disadvantage of the range is that it is sensitive only to extreme values (the highest and the lowest). Samples C and D in the last "Stop & Think" exercise have ranges of 4, but the spreads of these two samples are obviously not the same. In Sample C, each value is two "units" away from the sample mean of 3. In Sample D, two values (1 and 5) are two "units" away, but two values (3) are zero "units" away. In other words, the average "distance" or "variation" from the mean is greater in Sample C than in Sample D. The average variation is 2 in Sample C, but less than 2 in Sample D. The standard deviation is meant to capture this difference (one that isn't caught by the range) and to assign higher measures of spread to samples like Sample C than to those like Sample D. And, in fact, it does.

It does so by employing a computational formula that, in essence, adds up the "distances" of all individual values from the mean and divides by the number of values—a little like the computation of the mean in the first place. That's the essence. In fact, the computational formula is

$$s = \sqrt{\frac{\sum (X - \overline{X})^2}{N}}$$

where s stands for the standard deviation
$\overline{X}$ stands for the sample mean
X stands for each value
N stands for the number of sample cases

Although this formula might look a bit formidable, it's not very difficult to use. We'll show you how it works for Sample D.

First notice that the computational formula requires that you compute the sample mean ($\overline{X}$). For Sample D, the mean is 3. Then subtract this mean from each of the individual values, in turn ($X - \overline{X}$). For Sample D (whose values are 1, 3, 3, and 5), these differences are -2, 0, 0, and 2. Then square each of those differences $[(X - \overline{X})^2]$. For Sample D, this results in four terms: 4, 0, 0, and 4 (-2 squared $= 4$). Then sum these terms $[\sum (X - \overline{X})^2]$. For Sample D, this sum is 8 ($4 + 0 + 0 + 4 = 8$). Then the formula asks you to divide the sum by the number of cases $[\sum (X - \overline{X})^2/N]$. For Sample D, this quotient is 2 ($8/4 = 2$). Then the formula asks you to take the square root of this quotient. For Sample D, this is the square root of 2, or about 1.4. So the standard deviation of Sample D is about 1.4.

STOP & THINK

Now try to calculate the standard deviation of Sample C. Is the standard deviation of Sample C greater than (as we expected it would be) or less than the standard deviation of Sample D?

normal distribution, a distribution that is symmetrical and bell shaped.

We hope you've found that the standard deviation of Sample C (we calculate it to be 2) is greater than the standard deviation of Sample D. In any case, we're pretty sure you'll see by now that the standard deviation does require variables of interval or ratio scale. (Otherwise, for instance, how could you calculate a mean?) You might also see why a computer is helpful when you are computing many statistics.

The standard deviation has properties that make it a very useful measure of variation, especially when a variable is normally distributed. The graph of a **normal distribution** looks like a bell, with its "hump" in the middle and cases diminishing on both sides of the hump (see Figure 15.1). A normal distribution is symmetric. If you folded it in half at its center, one half would lie perfectly on the other half. One of the nice mathematical properties of a variable that has a normal distribution is that about 68 percent of the cases would fall between one standard deviation above the mean (the center of the distribution) and one standard deviation below the mean.

The standard deviation becomes less intuitively useful when a variable's distribution does not conform to the normal distribution, however. Many socially interesting variables do not so conform. Most of these nonconformers, like income in Box 15.1, are skewed. Figure 15.2 shows another skewed distribution: the distribution of the number of children respondents said they had in the 1996 GSS. As you might expect, most of the 2,889 who answered this question said they had zero, one, or two children. But just enough reported that they had five, six, seven, and eight (or more), that

FIGURE **15.1** Normal Distribution

FIGURE **15.2** Number of Children Reported by General Social Survey Respondents. Mean = 1.8. Skewness = 1.0

Source: The General Social Survey, 1996 (see Davis and Smith, 1998). Graph was produced using SPSS.

the distribution has a "tail" ... this time to the right. Statisticians have provided another very useful statistic that we won't try to show you how to compute: skewness. We discussed the concept of skewness earlier, but skewness is so important that it also merits its own statistic. If the computed skewness statistic is zero, the variable is very likely to be nearly normally distributed. The skewness statistic for number of children, as depicted in Figure 15.2, is actually about 1, indicating that the "tail" of the distribution is to the right. Skewness values can be both positive and negative (when the "tail" is to the left) and can vary from positive to negative infinity. But when skewness gets near 1 (or −1), as it does in the case of number of children, or in the case of life expectancy at diagnosis, as it did for mesothelioma when Gould was diagnosed with the disease, one can no longer pretend that one is dealing with a normal distribution. As Gould's case indicates, however, distributions don't have to be normal to be interesting, enlightening, and even life-affirming ... especially if one has some appreciation of statistical variation.

BIVARIATE ANALYSES

Analyzing single variables can be very revealing and provocative. (Think of Gould's musing about the "life expectancy at diagnosis" variable.) But sometimes the real fun begins only when you examine the *relationship* between variables. A relationship exists between two variables when categories of one

FIGURE **15.3** Our Hypothesis About the Relationship Between Time and the Willingness to Vote for a Woman President

variable tend to "go together" with categories of another. You might have found, if you tried to do the last "Stop & Do" exercise, that about 87.1 percent of adult Americans queried by the GSS from 1972 to 2010 have said that they would vote for a woman if she were nominated by their party and if she were qualified. But do you think this percentage has gone up, down, or remained the same over time since 1972? Let's suppose you think it might have generally gone up since 1972, perhaps because the women's movement since the late 1960s has made it more acceptable to Americans to envision women in leadership positions. If you think of this hypothesis in the variable language introduced in Chapter 2 and the passage of time and people's willingness to vote for a woman for president as variables, then you're saying that you expect the passage of time and people's willingness to vote for a woman to "go together." One way of depicting this expectation is shown in Figure 15.3.

One way of showing such a relationship is to crosstabulate, or create a bivariate table for, the two variables. **Crosstabulation** is a technique designed for examining the association between two nominal level variables and therefore, in principle, applies to variables of any level of measurement (because all ordinal and interval-ratio variables can also be treated as nominal level variables).

crosstabulation, the process of making a bivariate table to examine a relationship between two variables.

Bivariate tables (sometimes called contingency tables) provide answers to questions such as "Has the percentage of adult Americans who are willing to vote for a woman president increased since 1972?" or, more generally, to questions such as "Is there a difference between sample members that fall into one category of an independent variable and their counterparts in other categories in their experience of another characteristic?"

STOP & DO

Let's see what's actually happened to the percentage of adult Americans who say they'd be willing to vote for a woman president over time by going back to http://sda.berkeley.edu/archive.htm. This time, after you've gotten into the most recent GSS Cumulative Data File, and started the "Frequencies or crosstabulation (with charts)," put FEPRES in the "row" box and YEAR in the "column" box. You'll have crosstabulated the variable "Would You Vote for a Female President?" with the variable "Year of Survey." Examine the resulting table and see whether the percentage of adult Americans who say they would vote for a woman president has gone up, down, or remained the same since 1972.

Isn't it interesting how the percentage of American adults who say they would vote for a qualified woman for president inched up after 1972? At 74.0 percent in 1972, it rose, with minor exceptions, pretty steadily until it reached 96.2 percent in 2010. In fact, if we just focused on the years 1972 and 2010, as categories of the independent variable, "Year of Survey," and on "Yes" and "No," as categories of the dependent variable, "Would Respondent Vote for a Qualified Woman for President?", the crosstabulation of these two variables could be reduced to the bivariate (or contingency) table, Table 15.4. (We've calculated a couple of other statistics, "Cramer's V" and "p," and put them in Table 15.4 as well. We'll be getting to them soon.)

This table demonstrates that, as we hypothesized, the year of the survey is related to or associated with whether respondents answered that they would vote for a qualified woman for president; it makes a difference what year it was.

Let's create a contingency table using data that Roger collected with students and colleagues (Clark et al., 2006) for a recent study of children's picture books. In this study, the researchers were interested to see, among other things, whether award-winning picture books created by Black illustrators are more likely to make female characters central than award-winning picture books created by non-Black illustrators. They collected data on, among other things, Caldecott award and honor books from 2000 to 2004 (created, as it turns out, by non-Black illustrators and authors) and Coretta Scott King award and honor books from 2000 to 2004 (created by Black illustrators and authors). The researchers hypothesized that the King books would make females more central to their books than the Caldecotts, because the King committee, unlike the Caldecott committee, is enjoined to pick "educational" books and because women have been central to the survival of the African American community in the United States.

TABLE **15.4**

Whether Respondent Would Vote for a Woman for President By Year of Survey, 1972 or 2010

Would Respondent Vote for Woman?	Yes	No	Total
1972	1,134 (74.0%)	398 (26.0%)	1,532 (100.0%)
2010	1,361 (96.2%)	53 (3.8%)	1,414 (100.0%)
	Cramer's V = .31	p < .05	

Here are data on two variables for the 36 relevant award-winning books from 2000 to 2004: the type of award (Caldecott or King) and whether there is a female central character:

Award	Central Female?	Award	Central Female?
1. King	Yes	19. Caldecott	No
2. Caldecott	Yes	20. Caldecott	No
3. King	Yes	21. King	No
4. Caldecott	No	22. Caldecott	No
5. Caldecott	Yes	23. Caldecott	No
6. King	Yes	24. King	No
7. King	Yes	25. King	Yes
8. Caldecott	No	26. Caldecott	No
9. Caldecott	No	27. King	No
10. King	Yes	28. Caldecott	No
11. Caldecott	No	29. King	No
12. Caldecott	Yes	30. Caldecott	Yes
13. Caldecott	No	31. King	No
14. King	No	32. Caldecott	No
15. King	No	33. Caldecott	Yes
16. Caldecott	No	34. King	No
17. Caldecott	No	35. Caldecott	No
18. King	Yes	36. Caldecott	No

Note that 15 of these 36 books were Coretta Scott King award or honor books and that 21 were Caldecott award or honor books. Of the 15 King books, seven (46.7 percent) have female central characters and eight (53.3 percent) do not. Of the 21 Caldecott books, five (23.8 percent) have female central characters and 16 (76.2 percent) do not. We'll put this information into a bivariate (or contingency) table, Table 15.5.

TABLE **15.5**
Do Caldecotts or Kings from 2000 to 2004 Have More Female Central Characters?

Award	Does Book Have Female Central Character?		
	No	Yes	Total
Caldecott	16 (76.2%)	5 (23.8%)	21 (100.0%)
King	8 (53.3%)	7 (46.7%)	15 (100.0%)
	Cramer's V = .24	$p > .05$	

What percentages would you compare from the table to show that King books were more likely to have female central characters than Caldecott books?

Note that the way we've formulated our hypothesis forces us to look at the relationship between type of book award and presence of central female character in a certain way. Using the variable language introduced in Chapter 2, we expected "type of book award" to affect "presence of central female character," rather than "presence of central female character" to affect "type of book award." In other words, "type of book award" is our independent variable and "presence of central female character" is our dependent variable. Given the formulation of our hypothesis and given the analysis of Table 15.5, we're in a good position to say that King books are more likely than Caldecott books to have central female characters.

Measures of Association

measures of association, measures that give a sense of the strength of a relationship between two variables.

Before we leave the topic of bivariate analyses, let's note a few more important points. Table 15.4 contains some curious new symbols that are worth mentioning, mainly because they illustrate a whole class of others. One such class is called **measures of association,** of which "Cramer's V" is a specific example. Measures of association give a sense of the strength of a relationship between two variables, of how strongly two variables "go together" in the sample. Cramer's V can vary between 0 and 1, with 0 indicating that there is absolutely no relationship between the two variables and 1 indicating that there is a perfect relationship. A perfect relationship exists between two variables when change in one variable is always associated with a predictable change in the other variable. The closer Cramer's V is to 0, the weaker the relationship is; the farther from 0 (closer to 1), the stronger the relationship.

You'll note, from Table 15.5, that Cramer's V for the relationship type of award and the presence of female main characters is .24 and, from Table 15.4, that Cramer's V for year of survey and the willingness of respondents to vote for a qualified woman for president is .31. Which relationship is stronger?

Measures of Correlation

measures of correlation, measures that provide a sense not only of the strength of the relationship between two variables but also of its direction.

Cramer's V is a measure of association that can be used when both variables are nominal level variables (and have at least two categories each). Statisticians have cooked up literally scores of measures of association, many of which can be distinguished from others by the levels of measurement for which they're designed: some for nominal, some for ordinal, and some for interval level variables. A particular favorite of social science researchers is one called Pearson's r, designed for examining relationships between interval level variables. Pearson's r falls in a special class of measures of associations: it's a **measure of correlation**. As such, it not only provides a sense of the strength of the relationship between two variables, it also provides a sense of the *direction* of the association. When variables are intervally (or ordinally) scaled, it is meaningful to say that they can go "up" and "down," as well as to say that they vary from one category to another.

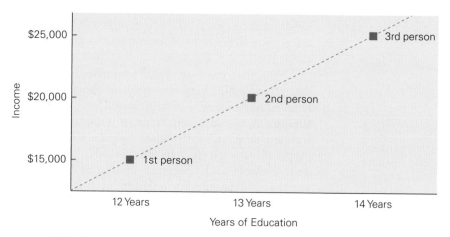

FIGURE **15.4** A Graph of the Relationship Between Education and Income for Our Hypothetical Sample

Suppose, for instance, that you had data about the education and annual income of three people. Suppose further that the first of your persons had 12 years of education and made $15,000 in income, the second had 13 years of education and made $20,000, and the third had 14 years of education and made $25,000. In this case, not only could you say that the two variables (education and income) were related to each other (that every time education changed, income changed), but you could also say that they were *positively* or *directly* related to each other—that as education rises so does income. You might even be tempted to graph such a relationship, as we have done in Figure 15.4.

One feature of our graphical representation of the relationship between education and income is particularly striking: All the points fall on a straight line. One striking feature of this line is that it rises from the bottom left of the graph to the top right. When such a situation exists (i.e., when the data points from two variables fall on a straight line that rises from bottom left to top right), the correlation between two variables is said to be perfect (because all points fall on a line) and positive (because it goes from the bottom left up to the top right), and the Pearson's r associated with their relationship is 1.

STOP & THINK *Suppose your sample included three individuals, one of whom had 14 years of education and made $15,000 in income, one of whom had 13 years of education and made $20,000, and one of whom had 12 years of education and made $25,000. What would a graphical representation of this data look like? Can you imagine what the Pearson's r associated with this relationship would be?*

If, on the other hand, the data points for two variables fell on a line that went from the "top left" of a graph to the "bottom right" (as they would for the data in the "Stop & Think" exercise), the relationship would be perfect and negative and the Pearson's r associated with their relationship

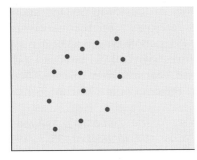
r about equal to .70

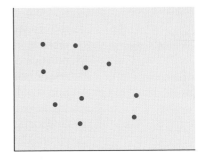
r about equal to – .70

FIGURE **15.5** Idealized Scattergrams of Relationships of r = .70 and –.70

would be – 1. Thus, Pearson's r, unlike Cramer's V but like all other measures of correlation, can take on positive *and* negative values, between 1 and –1.

In general, negative values of Pearson's r (less than 0 to –1) indicate that the two variables are *negatively* or *indirectly* related; that is, as one variable goes up in values, the other goes down. Positive values of Pearson's r (greater than 0 to 1) indicate that the two variables are directly related—as one goes up, the other goes up. In both cases, r's "distance" from 0 indicates the strength of the relationship: the farther from 0, the stronger the relationship. Thus, a Pearson's r of .70 and a Pearson's r of –.70 indicate relationships of equal strength (quite strong!) but opposite directions. The first (r = –.70) indicates that as one variable's values go up, the other's also go up (pretty consistently); the second (r = –.70) indicates that as one variable's values go up, the other's go down (again pretty consistently). Figure 15.5 provides a graphical representation (known as a *scattergram*) of what relationships (with many data points) with r's of .70 and ——.70 might look like. Notice that, although in neither case do the points perfectly conform to a straight line, in both cases they all hover around a line. r's of .30 and –.30 would, if you can imagine them, conform even less well to a straight line.

STOP & THINK   *Would you expect the association between education and income for adults in the United States to be positive or negative?*

STOP & DO   *Try doing a correlation analysis for the association between years of education and income for adults in the United States in 2006. Go back to http://sda .berkeley.edu/archive.htm, enter the most recent GSS Cumulative Data File, start the "correlation matrix" action, and enter the variable names "educ" and "rincom06" into the first two input boxes before hitting "run correlations." Check the correlation in the middle of the page. Look at the intersection of a row headed by "educ" and a column headed by "rincom06" (or a row headed by "rincom06" and a column headed by "educ") and read the correlation coefficient (the Pearson's r).*

We did the analysis in the "Stop & Do" and found what we hope you did as well, that the Pearson's r for the association between education and income in the 2006 GSS sample was .33. As it turns out, the association is a positive one (the positive "sign" of .33 indicates a direct relationship), but the relationship is nowhere near a perfect one (.33 is closer to 0 than to 1)—so one would not expect the data points to conform very well to a straight line.

STOP & THINK   *Suppose the Pearson's r for the relationship between years of education and incomes had proven to be –.67, instead of .33. How would you describe the relationship?*

We could tell you more about correlation analysis, but any standard statistics book could do so in more detail. We think you'll find correlation analysis a particularly handy form of analysis if you ever want to study the associations among interval level variables.

Inferential Statistics

We draw your attention to a final detail about Table 15.4 before we conclude our discussion of bivariate relationships. On the same row as Cramer's V in the table is a funny-looking symbol, "p < .05." This symbol, standing for "probability is less than .05," enables the reader to make an inference about whether a relationship, like the one shown in the table for the sample (of 2,976 respondents), exists in the larger population from which the sample is drawn. This "probability" is estimated using what we've called inferential statistics, in this case chi-square, perhaps the most popular inferential statistic used in crosstabulation analysis. There are almost as many inferential statistics as there are measures of association and correlation, and a discussion of any one of them is outside the scope of this chapter. Inferential statistics are properly computed when one has a probability sample (see Chapter 5). They are designed to permit inferences about the larger populations from which probability samples are drawn.

You might recall from Chapter 5, however, that, even with probability sampling, it is possible to draw samples whose statistics misrepresent (sometimes badly misrepresent) population parameters. Thus, for instance, it is possible to draw a sample in which a strong relationship exists between two variables even when no such relationship exists in the larger population. Barring any other information, you'd probably be inclined to infer that a relationship also exists in the larger population. This kind of error—the one where you infer that a relationship exists in a population when it really doesn't—is called a Type 1 error. Such an error can cause serious problems, practically and scientifically.

Social scientists are a conservative lot and want to keep the chances, or probability, of making a Type 1 error pretty small—generally lower than 5 times in 100, or "p < .05." When the chances are greater than 5 in 100, or "p > .05," social scientists generally decline to take the risk of inferring that a relationship exists in the larger population.

STOP & THINK   *When "p < .05," social scientists generally take the plunge and infer that a relationship, like the one in the sample, exists in the larger population. Would social scientists take such a plunge with the relationship between year of survey and respondents' willingness to vote for a woman for president as displayed in Table 15.4?*

The probability of making a Type 1 error, the "p," is related to the size of the sample. The larger the sample, the easier it is to achieve a p of less than .05, and therefore it is easier to feel comfortable with an inference that a relationship exists in the larger population. Consequently, it is not surprising that the association between year of survey and respondents' willingness to vote for women for president as displayed in Table 15.4 is associated with a p less than .05, even though the strength of the association (as measured by a Cramer's V of .31) is only moderately strong. The number of cases in the sample is 2,976, after all.

Measures of association, measures of correlation, and inferential statistics are not an exotic branch of statistical cookery. They are the "meat and potatoes" of quantitative data analysis. We hope we've presented enough about them to tempt you into learning more about them. We'd particularly recommend Marija Norusis's *The SPSS Guide to Data Analysis* (2011), which, in addition to offering a pretty good guide to SPSS (the statistical package we used to generate several tables and figures), offers a good guide to the kinds of statistics we've referred to here.

MULTIVARIATE ANALYSIS AND THE ELABORATION MODEL

Examination of bivariate relationships is the core activity of quantitative data analysis. It establishes the plausibility of the hypothesized relationships that inspired analysis in the first place. Suppose we expected that the happiness of a person's marriage would be positively associated with the person's overall, or general, happiness. To use the variable language introduced in earlier chapters, we'd expect that an independent variable, marital happiness, would be associated with the dependent variables, general happiness. We could then turn to GSS data to test our proposition. Table 15.6 uses such data (from 2010) and demonstrates that our guess was right: 55.6 percent of people who reported being very happy with their marriages reported being very happy generally, while only 11.7 percent of people who reported being less than very happy with their marriages reported being very happy generally. Our bivariate analysis would demonstrate what we expected: that happily married people are more likely to be happy than other people. But the examination of bivariate relationships is frequently just the beginning of quantitative data analyses.

You might ask: Why would a researcher want to examine more than two variables at a time? One answer is that we might be worried that there's some obvious challenge to the thesis we're proposing: in this case, that marital happiness causes general happiness. You may recall, from Chapter 2, that one such challenge is the possible presence of a third variable, antecedent to the

TABLE **15.6**
Relationship Between Marital and General Happiness (2010)

| | | General Happiness | | |
		Very Happy	Less than Very Happy	Total
Marital Happiness	Very Happy	55.6%	44.4%	100%
		367	292	659
	Less than Very Happy	11.7%	88.3%	100%
		45	340	385
		Cramer's V = .41	p < .05	

Source: General Social Survey, 2010 (http://sda.berkeley.edu/archive.htm).

independent and dependent variables, which is responsible for its association. Hans Zeisel, in his classic *Say It with Figures* (1968: 108), reminds us of an old Chinese statistical joke that people who visit a doctor have considerably higher mortality rates than people who don't. Obviously, in this case, we should look at the association between doctor visits and mortality separately for those who were sick and those who were not. This, essentially, is the point of multivariate (more-than-two-variable) analyses.

As the doctor/mortality example suggests, a researcher will often want to pursue questions like "Is there a third variable (say, sickness), associated with both the independent (say, doctor visiting) and the dependent variable (say, mortality), that can account for the association between the two?" or "Are there certain conditions under which the association between the independent and dependent variables is particularly strong or weak?" or "Are there any other variables, affected by the independent variable and affecting the dependent variable, that can help me understand how the independent variable affects the dependent variable?" The pursuit of any such question can lead researchers to elaborate on the original bivariate relationship by introducing additional variables. In doing so, they will employ the **elaboration** model. A complete description of Paul Lazarsfeld's (see Rosenberg, 1968) elaboration model is beyond the scope of this book. But we think you deserve at least a taste of it.

Recall our finding, from Table 15.6, that people who are happy with their marriages are more likely to be generally happy than others. We may want to figure out why. Perhaps being happy in one area of life makes people happier in general. But before we jump to this conclusion we might want to consider that marital and general happiness are *not* causally related at all. We might fear that others will argue that they are not really related and their association reflects their common association with a third variable, say gender, that forces them to "go together," as it were. The explanation might go something like this: Women tend to be happier with their marriages than men and women tend to be happier in general than men, therefore the relative grumpiness of men (both with their marriages and in general) and the relative happiness of women (with both) accounts for the apparent association

elaboration, the process of examining the relationship between two variables by introducing the control for another variable or variables.

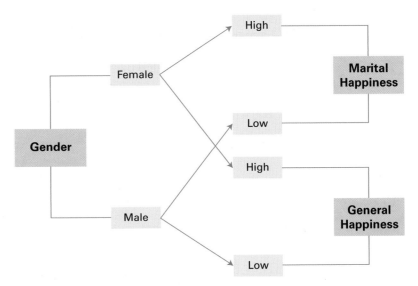

FIGURE **15.6** Gender as an Antecedent Control Variable of Marital Happiness and General Happiness

control variable, a variable that is held constant to examine the relationship between two other variables.

between marital and general happiness. In this case, gender is hypothesized to be an antecedent **control variable** because it is expected to affect both marital happiness and general happiness. (In the variable language of earlier chapters, it is a possible independent variable for both marital and general happiness.) This possibility is depicted in Figure 15.6.

Let's be clear. We've entertained the possibility that gender might prove the reason why marital and general happiness are related, but we're doing it partly because we think our critics might think it's the reason. We don't really want gender to prove to be the reason. (That would undermine our argument that marital happiness itself has a significant impact on general happiness.) To show that it wasn't the reason, we need to understand how one would go about showing that it was (or, at least might be). To do that, one would have to do three things: (1) show that gender and marital happiness are related, (2) show that gender and general happiness are related, and (3) show that when gender is *controlled*, the association between marital and general happiness "disappears." Thanks to Lazarsfeld's elaboration model, we know that all three of these conditions have to exist to conclude that gender "explains" the relationship between marital and general happiness. Actually, explanation is the term Lazarsfeld gave to that form of elaboration that shows a spurious relationship: a relationship that is explained away by an antecedent control variable.

Can you imagine what it would mean to show that when gender is *controlled*, the association between marital and general happiness "disappears"? Understanding the introduction of a control variable is really the key to understanding the elaboration model, so we'd like to show you how it's done.

In the context of crosstabulation, the concept of *controlling* a relationship (e.g., between marital and general happiness) for a third variable (e.g., gender) is reasonably straightforward. It means looking at the original association separately for cases that fall into each category of the control variable. For the control variable, gender, there are two categories: female and male. Controlling the relationship between marital and general happiness for gender means, then, examining that relationship only for females, on the one hand, and only for males, on the other.

But what does it mean to say that a relationship "disappears" when you control for a third variable? Table 15.7 illustrates with a hypothetical example.

The first segment of this table simply replicates the information about the association between marital and general happiness from Table 15.6 and shows that, for the full sample of 1,044 respondents, a substantial relationship exists. We could emphasize the strength of this relationship by comparing the percentages (e.g., 55.6 percent of very happily married people reported being generally very happy, whereas only 11.7 percent of people reporting

TABLE **15.7**

Hypothetical Example Demonstrating How the Relationship Between Marital and General Happiness Could "Disappear" with Gender Controlled

		General Happiness		
		Very Happy	Less than Very Happy	Total
Full Sample				
Marital Happiness	Very Happy	55.6%	44.4%	100%
		367	292	659
	Less than Very Happy	11.7%	88.3%	100%
		45	340	485
		Cramer's V = .41	p < .05	
Females				
Marital Happiness	Very Happy	100%	0%	100%
		367	0	367
	Less than Very Happy	100%	0%	100%
		45	0	45
		Cramer's V = .00	p > .05	
Males				
Marital Happiness	Very Happy	0%	100%	100%
		0	292	292
	Less than Very Happy	0%	100%	100%
		0	340	340
		Cramer's V = .00	p > .05	

themselves less than happily married reported being generally very happy) or by looking at the measure of association (a Cramer's V of .41 indicates a strong association).

The second and third segments of this table reflect the introduction of the control variable, gender. The second segment presents only hypothetical information about females. Here we examine what is sometimes called a **partial relationship** (or partial, for short) because we're only examining the original relationship for part of the sample—for females. Among females, the table suggests, all 367 respondents who were reported being happily married were also reported being happy generally, and all 45 respondents who were reported being less than very happily married were also reported being happy generally. In other words, there is no relationship between marital and general happiness for women—a fact that is underscored by a Cramer's V of .00. Similarly, the third segment of the table presents hypothetical information about males. Within this group too, there is no relationship between marital and general happiness.

partial relationship, the relationship between an independent and a dependent variable for that part of a sample defined by one category of a control variable.

🟤 **STOP & THINK**   *How can we tell there's no relationship in the partial involving males?*

Consequently, because there's no relationship between marital and general happiness for females and males, the relationship that existed with no control has "disappeared" in each of the controlled situations. Table 15.7, then, makes the kind of demonstration that would lead researchers to believe that gender "causes" the association between marital and general happiness. It shows a situation where the control for an antecedent variable makes the relationship between an independent and a dependent variable "disappear."

This table was based on hypothetical data. As it turns out, nothing like what the table shows occurs in reality. Table 15.8 shows what really occurs when we control the relationship between marital and general happiness for gender, using data collected in the 2010 GSS. The real data indicate that the partial relationship between marital and general happiness does not "disappear" for females and males. It is, in fact, reasonably strong for both females (Cramer's V = .50) and males (Cramer's V = .36), though it is clearly stronger for female than it is for males.

Thus, the introduction of a control for gender does not do what we feared it might: "explain away" the original relationship. As a result, we're in a stronger position to assert that the relationship might be "causal" (rather than spurious, or caused by the action of an antecedent variable that makes the variables in the relationship appear to vary together).

🟤 **STOP & THINK**   *Of course, you cannot entirely eliminate the possibility that a relationship is spurious. In Chapter 2, we suggested that such an elimination would require showing that there is no antecedent variable whose control makes the original relationship disappear. Can you see why this is not possible?*

TABLE **15.8**

What Really Happened to the Relationship Between Marital and General Happiness with Gender Controlled

		General Happiness		
		Very Happy	Less than Very Happy	Total
Females				
Marital Happiness	Very Happy	60.4%	39.6%	100%
		210	137	247
	Less than Very Happy	9.2%	90.8%	100%
		18	178	196
		Cramer's V = .50	p < .05	
Males				
Marital Happiness	Very Happy	50.3%	49.7%	100%
		157	155	312
	Less than Very Happy	14.3%	85.7%	100%
		27	162	189
		Cramer's V = .36	p < .05	

Source: General Social Survey, 2010 (http://sda.berkeley.edu/archive.htm).

To summarize and, well, elaborate, elaboration permits us to better understand the meaning of a relationship (or lack of relationship) between two variables. We've actually discussed two types of elaboration earlier, but there are others as well. Let us briefly introduce you to four kinds of elaboration: replication, explanation, specification, and interpretation.

explanation, a kind of elaboration in which the original relationship is explained away as spurious by a control for an antecedent variable.

1. **Explanation.** The original relationship can be *explained away* (or found to be *spurious*) through the introduction of the control variable. Two variables that were related in the bivariate situation might no longer be related when an *antecedent* variable is controlled. This happened in the hypothetical (but unreal) situation, shown in Table 15.7, when the relationship between marital and general happiness "disappeared" both for females and for males.

specification, a kind of elaboration that permits the researcher to specify conditions under which the original relationship is particularly strong or weak.

2. **Specification.** Introducing a control variable and finding that there are conditions under which the original relationship is particularly strong and weak. This was what we found when we controlled the relationship between marital and general happiness by gender (see Table 15.8) The partial relationship for women (Cramer's V = .50) is considerably stronger than the one for men(Cramer's V = .36). We can therefore *specify* a gender for which the original relationship is particularly strong (for females) and one for which it is particularly weak (males).

TABLE **15.9**

Relationship Between Marital and General Happiness with Whether Respondent Has Ever Had a "Reborn Again" Experience Controlled

		General Happiness		
		Very Happy	Less than Very Happy	Total
People Who Have Been Reborn				
Marital Happiness	Very Happy	59.4%	40.6%	100%
		147	101	148
	Less than Very Happy	15.6%	84.4%	100%
		20	117	137
		Cramer's V = .43	p < .05	
People Who Have Not Been Reborn Again				
Marital Happiness	Very Happy	53.1%	46.9%	100%
		213	188	401
	Less than Very Happy	10.2%	89.8%	100%
		25	218	243
		Cramer's V = .43	p < .05	

Source: General Social Survey, 2010 (http://sda.berkeley.edu/archive.htm).

replication, a kind of elaboration in which the original relationship is replicated by all the partial relationships.

3. **Replication.** The original relationship can be replicated in each of the partial relationships. Table 15.9 provides an example of replication, using data from the 2010 GSS. Here, again, the major concern is with the relationship between marital and general happiness, but now we're controlling for whether the respondent reports ever having had a "born again" religious experience. When we separate the original sample into those reporting that they've had such an experience and those who say they haven't, we find that the relationship between marital happiness and general happiness is essentially the same for both groups (where the Cramer's Vs are both .43) as in the original, undivided sample (Cramer's V = .41).

interpretation, a kind of elaboration that provides an idea of the reasons why an original relationship exist without challenging the belief that the original relationship is causal.

4. **Interpretation.** A control variable can give the researcher an idea about why the independent variable is related to the dependent variable *without challenging the belief that the one causes the other*, as it does in explanation. In this case, the control is said to *interpret* rather than *explain* the relationship. Interpretation occurs when, in the mind of the researcher, the control isn't antecedent to both the independent and the dependent variable (as it would be in explanation) but, rather, intervenes between

them. An *intervening* control variable is one that "comes between" the independent and dependent variable and, therefore, at least partially accounts for their association. In such a situation, Lazarsfeld tells us, the partials (as measured by, say, Cramer's V) will be lower than the original. We provide an example of what we believe to be interpretation in Table 15.10, which is based on 2006 GSS data. Here the major concern is with whether a person's income is associated with whether he or

TABLE **15.10**

An Example of Interpretation: The Association Between Family Income and Presidential Vote for Democrats, Independents, and Republicans

	Voted for Kerry	Voted for Bush
Full Sample		
Income less than $35,000	57.5%	42.6%
	614	453
Income $35,000 or more	43.1%	56.9%
	1,034	1,364
	Cramer's V = .10	p < .05
Democrats		
Income less than $35,000	85.7%	14.4%
	527	88
Income $35,000 or more	86.0%	14.0%
	868	141
	Cramer's V = .01	p > .05
Independents		
Income less than $35,000	50.0%	50.0%
	52	52
Income $35,000 or more	40.0%	60.0%
	87	125
	Cramer's V = .08	p > .05
Republicans		
Income less than $35,000	7.8%	92.2%
	26	307
Income $35,000 or more	6.3%	93.7%
	71	1,061
	Cramer's V = .03	p > .05

Source: General Social Survey, 2006 (http://sda.berkeley.edu/archive.htm).

she voted for John Kerry, the Democratic candidate, or George Bush, the Republican candidate, in the 2004 presidential election (people who voted for other candidates were eliminated). What the table shows is that people whose family income was less than $35,000 were more likely to have voted for Kerry than for Bush and those whose family income were $35,000 or more were more likely to have voted for Bush than for Kerry.

STOP & THINK   *What percentage of the people with family incomes of less than $35,000 voted for Kerry? What percentage of these people voted for Bush? Can you imagine why higher-income people might have voted for Bush, while lower-income people might have voted for Kerry?*

The Cramer's V for the association, .10, is modest at best, but noticeable. We then asked ourselves "What variable might intervene between people's income and their vote for president?" and considered their political party affiliation as a possibility. In particular, we thought it possible that people who had lower incomes might well affiliate themselves with the Democratic party and, then, in turn, vote for the Democratic candidate, in this case John Kerry. We also thought it possible that people who had higher incomes might affiliate with the Republican Party and, in turn, vote for the Republican candidate, in this case George Bush. We depict this expectation, that party affiliation might be an intervening variable between income and presidential vote, in Figure 15.7. Note that, in this figure, unlike the situation in Figure 15.6, the control variable is envisioned as falling between the independent and dependent variables.

We coded three categories of political party affiliation: Democrat, Independent, and Republican. Table 15.10 shows the association between income and presidential vote for each of these categories. Given our expectation that party affiliation would help "interpret" the relationship between income and vote, we expected each of the partials to be lower than the original (Cramer's V = .10). And they were. The partial for Democrats, as you can see from the table, was .01. The partial for Independents was .08, and the partial for Republicans was .03. Crucially, however, the diminution in this case has a very different meaning than it does when the control variable is antecedent

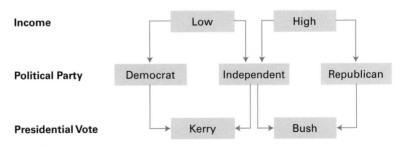

FIGURE **15.7** Political Party as an Intervening Variable Between Income and Presidential Vote

to both the original variables. Now it suggests that income affects political party affiliation, which in turn affects presidential vote. Income's causal effect on presidential vote is not questioned; it's simply interpreted as income's effect on party affiliation and party affiliation's effect on voting.

The great advantage of elaboration is that it gives us the tools to think beyond mere two-variable relations: to question, and test, whether they might not be causal (as in the case of explanation); to wonder, and see, whether they might be stronger under some conditions and weaker under others (as in the case of specification); and to think about, and assess, possible mechanisms through which the effects of an independent variable work on a dependent variable (as in the case of interpretation). The model even permits us to anticipate why others might believe a favored two-variable relationship is not causal and show that their objection does not withstand the test of empirical investigation (as in the case of replication).

QUALITATIVE DATA ANALYSIS

> Qualitative data are sexy. They are a source of well-grounded, rich descriptions and explanations of processes in identifiable local contexts.... Then, too, good qualitative data are more likely to lead to serendipitous findings and to new integrations.... Finally, the findings from qualitative studies have a quality of "undeniability." Words, especially organized into incidents or stories, have a concrete, vivid, meaningful flavor that often proves far more convincing to a reader ... than pages of summarized numbers. (Miles and Huberman, 1994: 1)

As we indicated at the beginning of this chapter and as the quotation from Matthew Miles and Michael Huberman (1994) suggests, a major distinction between qualitative and quantitative data analyses lies in their products: words, "especially organized into incidents or stories," on the one hand, and "pages of summarized numbers," on the other. Even though, as Miles and Huberman argue, the former are frequently more tasty to readers than the latter, the canons, or recipes, that guide qualitative data analysis are less well defined than those for its quantitative counterpart. As Miles and Huberman elsewhere observe, "qualitative researchers ... are in a more fluid—and a more pioneering—position" (1994: 12). There is, however, a growing body of widely accepted principles used in qualitative analysis. We turn to these in this chapter.

If the outputs of qualitative data analyses are usually words, the inputs (the data themselves) are also usually words—typically in the form of extended texts. These data (the words) are almost always derived from what the researcher has observed (e.g., through the observation techniques discussed in Chapter 11), heard in interviews (discussed in Chapter 10), or found in documents (discussed in Chapter 13). Thus, the data analyzed by Cottingham were Pittsburgh Steelers fans' words and behaviors, and the data analyzed by Enos were the words of incarcerated women. The outputs of both data analyses were the essays themselves (see Chapters 10 and 11).

SOCIAL ANTHROPOLOGICAL VERSUS INTERPRETIVIST APPROACHES

Although there might be general agreement about the nature of the inputs and outputs of qualitative analyses (words), there are nonetheless some fundamental disagreements among qualitative data analysts. One basic division involves theoretical perspectives about the nature of social life. To illustrate, and following Miles and Huberman (1994: 8ff.), look at two general approaches: a social anthropological and an interpretative approach. Social anthropologists (and others, like grounded theorists and life historians) believe that there exist behavioral regularities (e.g., rules, rituals, relationships) that affect everyday life and that it should be the goal of researchers to uncover and explain those regularities. Interpretivists (including phenomenologists and symbolic interactionists) believe that actors, including researchers themselves, are forever interpreting situations and that these, often quite unpredictable, interpretations largely affect what goes on. As a result, interpretivists see the goal of research to be their own (self-conscious) "accounts" of how others (and they themselves) have come to these interpretations (or "accounts"). The resulting concerns with law-like relationships, by social anthropologists, and with the creation of meaning, by interpretivists, can lead to distinct data-analytic approaches. Given this difference, our discussion in this chapter will generally be more in line with the beliefs and approaches of the social anthropologists (and those with similar perspectives) than it is with those of the interpretivists. Nevertheless, we believe that much of what we say applies to either approach and that their differences can be overstated. Thus, although Enos clearly sought to discover law-like patterns in the experience of inmate mothering (remember how white mothers were more likely to have their children placed in foster care than black mothers), she was nonetheless sensitive to variation in the interpretations or accounts given by all mothers.

STOP & THINK   *When Enos pursued law-like patterns in placement of children, she was doing what the social anthropologists advocate. What about when she looked to the accounts given by the mothers?*

DOES QUALITATIVE DATA ANALYSIS EMERGE FROM OR GENERATE THE DATA COLLECTED?

In qualitative field studies, analysis is conceived as an *emergent* product of a process of gradual induction. Guided by the data being gathered and the topics, questions, and evaluative criteria that provide focus, analysis is the fieldworker's *derivative ordering* of the data (Lofland and Lofland, 1995: 181).

In ethnography the analysis of data is not a distinct stage of the research. It begins in the pre-fieldwork phase, in the formulation and clarification of research problems, and continues into the process of writing up (Hammersley and Atkinson, 1983: 174).

These quotations from John Lofland and Lyn Lofland (1995) and Martyn Hammersley and Paul Atkinson (1983) seem to point to another debate concerning qualitative analysis: whether qualitative analysis *emerges from* or *generates* the data collected. At issue seems to be, to paraphrase the more familiar chicken-or-egg dilemma, the question of which comes first: data or ideas about data (e.g., theory). Lofland and Lofland stress the creative, after-the-fact nature of the endeavor by describing it as a "process of gradual induction"—of building theory from data. Hammersley and Atkinson emphasize how data can be affected by conceptions of the research topic, even before one enters the field. These positions, however different in appearance, are, in our opinion, both true. Rather than being in conflict, they are two sides of the same coin. Patricia Adler and Peter Adler (1994b), in discussing their own practice in studying afterschool activities, have told us, in personal communication, about their efforts to immerse themselves in their setting first, *before* examining whatever literature is available on their topic, to better "grasp the field as members do" without bias. But *something* compelled their interest in afterschool activities, and whatever that was surely entailed certain preconceptions. Sandra Enos (the author of the focal research report in Chapter 10) is much more explicit about her preconceptions and, in conversation, has confided that one of her expectations before observing mother child visits was that she'd see mothers attempting to create "quality time" with their children during the visits. Still, she remained open to the "serendipitous findings" and "new integrations" mentioned by Miles and Huberman. When she found mothers casually greeting their kids with "high-fives," rather than warmly embracing them, as she'd expected, she dutifully took notes. Because Enos has been explicit about various stages of her data analysis, and because she's been willing to share her field notes with us, we use her data to illustrate our discussion of qualitative data analysis in this chapter.

THE STRENGTHS AND WEAKNESSES OF QUALITATIVE DATA ANALYSIS REVISITED

Before we actually introduce you to some of the techniques that are increasingly associated with qualitative data analysis, we want, once again, to suggest one of its great strengths. As Gary Fine and Kimberly Elsbach suggest, qualitative data can produce theories that "more accurately describe real-world issues and processes than do quantitative data" (2000). Because qualitative data usually come from some combination of the observation or interview of real-world participants, they are more likely to be grounded in the immediate experiences of those participants than in the speculations of researchers. Thus, rather than simply handing out a set of closed-ended questions to people who had participated in activities surrounding Pittsburgh Steelers' games, Cottingham (see Chapter 11) observed those activities and asked participants about them as they were going on. Consequently, she didn't need to worry quite as much about whether she was asking the right questions (and supplying the right answers, as she would if she'd used closed-ended questions). She made observations as well as accounts of what the observations meant. The resulting capacity of qualitative data to yield insights into the longer-term dynamics of behavior is also a great advantage in producing theories that "describe real-world ... processes." We've mentioned earlier (see, for instance, Chapter 11) that one of the disadvantages of qualitative data is the questionable generalizability of the theories they lead to. Because Enos's study was based on a convenience sample of incarcerated mothers, for instance, the extent to which her findings may be typical is unclear. In general, though, we feel that the richness and accuracy of theories generated from qualitative data analysis are well worth the trouble, especially if their limitations are kept in mind and, perhaps, investigated further through future research.

ARE THERE PREDICTABLE STEPS IN QUALITATIVE DATA ANALYSIS?

Quantitative data analysis often follows a fairly predictable pattern: Researchers, after they have coded their own data or acquired computer-ready available data, almost always begin by doing some sort of univariate analyses, perhaps followed by bivariate and multivariate analyses. The "steps" involved in executing good qualitative analyses, however, are much less predictable, and a lot "more fluid," as Miles and Huberman suggest. Miles and Huberman prefer to see qualitative analysis as consisting of "concurrent flows of activity," rather than as "steps." They dub these flows "data reduction," "data display," and "conclusion drawing/verification" and depict the interactions among these flows as we do in Figure 15.8.

This depiction of qualitative data analysis involves several notable assumptions. First, even data collection is not immune to the "flows" of data analysis, being itself subject to the (perhaps tentative) conclusions a researcher entertains at various stages of a project (note the arrow from "conclusions" to "data collection"). Enos had expected inmate mothers to greet their children with warm embraces (even before she entered the field) and so focused attention on the gestures of greeting and, as you'll see, collected data about them. Second, Figure 15.8 suggests that data reduction and data display (both discussed later) are not only products of data collection but of one another and of conclusions, even if such conclusions are not yet fully developed. Thus, Enos's coding of her

FIGURE **15.8** Based on Components of Data Analysis: Interactive Model from Miles and Huberman: 12, 1994

Source: Based on Interactive Model in Drawing Meaning from Qualitative Data: Toward a Shared Craft by Matthew B. Miles and Michael A. Huberman.

data (a data reduction process described later) not only contributed to her final conclusions about mothering in prison, but might have been itself a product of earlier, perhaps more tentative, conclusions she'd reached. Finally, the figure emphasizes that the conclusions a researcher draws are not just products of earlier flows (data reduction and data display) but continually inform both of the other "flows," as well as the data collected in the first place.

We'd like to emphasize that one danger of the model depicted in Figure 15.8 is that it could give the impression that qualitative data analysis invariably follows a predictable pattern of data collection and transcription, followed by data reduction (particularly coding and memoing), followed by data displaying, followed by conclusion drawing and verification. It's not that simple. We think, however, that the advantages of such a model outweigh its dangers. One important advantage is that it gives new (and even experienced) practitioners a sense of order in the face of what otherwise might feel like terribly uncharted and, therefore, frightening terrain. Angela Lang (in Clark and Lang, 2001: 1) speaks of the anxiety she experienced when confronted with the demands of generating theory from qualitative data for an undergraduate project: "I soon realized I had to reinvent my creative side. I was nervous that

I spoke to a woman who had remained silent in the group. She was sitting alone at a table. I asked if I could talk with her. She has six children, two of her children will be visiting today. Her mother has always had her kids. Last year her mother passed away and now her sister is taking care of the kids. When she's released, she'll have responsibility for the whole bunch. She says it was hard to see your kids while you were in jail but that if you were DCYF[1] involved sometimes it was easier. (This has to do with termination of rights.) She remarked that it was important not to be a "stranger" to your kids while you were in jail. It made everything worse. I asked about telling your kids where you are. She said she just straight out told them and they accepted. "Mommy's in jail. That's it." She thought having such direct experience with jail that maybe her kids would "avoid" it when they got older. I asked her when she'd be released.

[1]Department of Children, Youth & families.

FIGURE **15.9** A Segment from Enos' Field Notes

I would discover that I wasn't creative at all." But, once she'd finished the project, she reflected on the model implicit in Figure 15.8 as a source of inspiration and comfort. She said that, by doing qualitative data analysis, she "had gained a tremendous amount of confidence in [her]self." And that, "If I were to do a qualitative project again, I'd probably try to organize my work in much … [the same way]. I'd probably plan to create fieldnotes, code data, memo, do data displays, review literature, write drafts of the final paper … and hope that the ideas and research questions began to float into my head, as they did this time" (Clark and Lang, 2001: 5–6).

DATA COLLECTION AND TRANSCRIPTION

We've discussed at some length the process of collecting qualitative data in earlier chapters: through qualitative interviewing (Chapter 10), observation techniques (Chapter 11), studying the content of texts (Chapter 13), and studying the results of various forms of visual analysis (Chapters 10, 11, and 13). So far, however, we haven't shown you any actual data. We'd like to present a small portion of Enos's field notes, as she composed them on the evening after an interaction with one mother in the prison parenting program. These notes were based on observation and conversation in the off-site parenting program,

In a few months, but she'd be
going to a drug treatment program.
She's been in prison many times
before and this time she's going to
do things right. She's made a
decision to change, to do something
new.
She is very happy with this program.
She likes the fact that there are
no officers around, that kids can
run and scream, that there is an
extended time out of the joint and
that she can spend more time with
her kids.
Her children arrived, a boy five
years old or so and his older
sister. They walked over to us.
The mother greeted her son, "Hey,
Dude! How you doing?" He didn't say
anything. The woman asked for the
children's coats and hung them on the
chair backs. She introduced me to
the kids, who were quiet and
beautiful, very well cared for. I
complimented her on these kids and
excused myself so she could be alone
with them.

FIGURE **15.9** Continued

where women got together with their children. The excerpt in Figure 15.9 is from nearly 150 pages of field notes Enos took before she engaged in the qualitative interviewing she reports on in the essay within Chapter 10.

STOP & THINK *One of Enos's expectations had been that women would greet their children with hugs. How did this woman greet her son? How did he greet her?*

The physical appearance of these field notes brings up the issue of computer usage in qualitative data analysis. Enos did all her written work, note collection, and analysis with an elementary word-processing program, a practice that is extremely common these days. Enos did not use one of the increasingly available software packages (e.g., AnSWER, Kwalitan, and QUALPRO) designed to facilitate the processing of qualitative data, until she entered the qualitative interview phase of her work, when she used a package called HyperResearch. If you think you'd like to try one of these packages, we strongly recommend Gibb's (2012a) online guide for helpful advice about getting started (http://onlineqda.hud.ac.uk/Introduction/index.php#thinking). If you would like to try free, online, web-based software for doing qualitative analysis, you might try The University of Massachusetts' (2012) Qualitative Data Analysis Program (QDAP) package (http://www.umass.edu/qdap/). We believe that much good qualitative analysis can be done without the use of anything more sophisticated than a word processor—and are inclined to point out that much has been achieved with nothing more sophisticated than pen and paper. (See Box 15.2 for some sense of the advantages and disadvantages of computer-assisted qualitative data analysis software [CAQDAS].)

Before we return to Enos's transcribed field notes, however, we'd like to point to one trend in CAQDAS that has shown increasing promise: the direct analysis of audio- and video-recorded data using computer software. The transcription of audio (and video) recordings, as mentioned in Chapter 10, is

---

BOX 15.2   The Debate about CAQDAS

Advocates of CAQDAS claim that it will

- Speed up and enliven coding (see "Coding")
- Offer a clear-cut structure for writing and storing memos (see "Memoing")
- Provide a "more complex way of looking at the relationships in the data" (Barry, 1998: 2.1)

Opponents see the possibility that CAQDAS will

- Distance analysts from their data
- Lead to qualitative data being used quantitatively
- Generate an unhealthy sameness in the methods used by analysts
- Continue to favor word-processed text over other forms of data such as sketches, maps, photos, video images, or recorded sound (Dohan and Sanchez-Jankowski, 1998: 483), though this disadvantage is has been much ameliorated by recent advances in software, like Transana, that analyses audio and video data

Barry (1998) gives a relatively balanced appraisal of these positions and offers specific insights into two of the most popular varieties of CAQDAS, NUDIST, and Atlas.ti. You can read it on the web at www.socresonline.org.uk/socresonline/3/3/4.html.

enormously time consuming, sometimes taking, in the case of interviews, up to four hours for an hour's worth of interview. Woods and Dempster (2011) describe how multiple researchers used Transana to do analyses of video games, both of their video and audio data, without ever having to transcribe the data at all. Transana is just one of several software packages enabling the analysis of audio and video data (see, e.g., Gibbs, 2011, http://onlineqda.hud.ac.uk/Intro _CAQDAS/what_packages_are_available/audio_video_analysis.php). Transana, for instance, enables the transcription of audiotapes, the arrangement and rear-rangement audio and video clips, the discovery of relationships among keywords applied by analysts, and the sharing of analyses among colleagues. For now, though, back to Enos's field notes.

The physical appearance of these field notes might be the first thing that strikes you. Enos uses only half a page, the left column, for her field notes in anticipation of a need to make notes about the notes later on (see "Data Reduction"). Leaving room for such notes about notes is excellent practice. Enos didn't have an interview guide at this stage of her investigation, but she evidently had certain questions in mind as she approached this woman. Her notes indicate that she asked about how the woman explained her incarceration to her children and about her expectations for release and her current child-care arrangements.

Enos's interest in the aforementioned questions indicates that she's probably engaged in a fair amount of what Miles and Huberman (1994: 16) call **anticipatory data reduction**—that is, decision making about what kinds of data she'd like to collect. Admittedly, her research strategy (more or less "hanging out" at an off-site parenting program for female prisoners) is pretty loose at this stage.

There is quite a bit of debate about the relative merits of "loose" and "tight" (those that entail explicit conceptual frameworks and standard instruments) research strategies in field research. Those advocating relatively loose strategies or "designs" (e.g., Wolcott, 1982) point to the number of questions and answers that can be generated through them. Susan Chase, for instance, in the study mentioned in Chapter 10, was primarily interested in getting female school superintendents to tell their life stories and was willing to use any question that stimulated such stories. She was willing to "hang out" during these interviews, though she does confess that she eventually found asking female superintendents to describe, generally, their work histories, was a pretty good way to induce "lively, lengthy, and engrossing stor[ies]" (Chase, 1995: 8).

But the advantages of relatively tight strategies, especially for saving time spent in the field, are worth noting. Enos developed an appreciation of tight strategies during the course of her fieldwork. She's indicated how tempting it is to pursue every idea that occurs in a setting:

> So many questions occurred to me after I left the site. I would consider what women had said to me and then try to figure out if that was related to race/ ethnicity, to class, to previous commitments to prison, to number of children, to street smarts, to sophistication about the system, and so on and so on. (Enos, personal communication)

anticipatory data reduction, decision making about what kinds of data are desirable that is done in advance of collecting data.

In the end, she realized that she couldn't pursue all those questions that she needed, in her words, "to keep on track." She claims that the most important task was to continually sharpen her focus as the fieldwork went on and that this meant "finally coming to the conclusion that you can't look at everything" (personal communication).

By the time Enos entered the interview stage of her project, she'd developed a very tight design indeed, having, among other things, composed the interview guide shown at the end of her report in Chapter 10. This guide led Enos to ask very specific questions of her respondents, about their age of first incarceration, and so on.

DATA REDUCTION

data reduction, the various ways in which a researcher orders collected and transcribed data.

Qualitative **data reduction** refers to the various ways a researcher orders collected data. Just as questioning 1,000 respondents in a closed-ended questionnaire can generate more data on most variables than the researcher can summarize without statistics, field notes or transcripts also can generate volumes of information that need reduction. Some researchers can do a reasonably good job of culling out essential materials, or at least can tell compelling stories, without resorting to systematic data reduction procedures in advance of report writing. But most can't, or don't. Especially in long-term projects with ongoing data collection, most researchers engage in frequent data reduction. In the following section, we'll focus on two common data reduction procedures: coding and memoing. In doing so, we'll be emphasizing one thing implicit in Figure 15.8: Data reduction is affected by the kinds of data you collect but is also affected by, and affects, the conclusions you draw.

CODING

Coding, in both quantitative and qualitative data analysis, refers to assigning observations, or data, to categories. In quantitative analysis, the researcher usually assigns the data to categories of variables he or she has already created, categories that he or she frequently gives numbers to in preparation for computer processing. In qualitative analysis, coding is more open-ended because both the relevant variables and their significant categories are apt to remain in question longer. In any case, coding usually refers to two distinct, but related, processes: The first is associating words or labels with passages in one's field notes or transcripts; the second is collecting similarly labeled passages into files. The goal of coding is to create categories (words or labels) that can be used to organize information about different cases (whether situations, individuals, or so on). When you don't have much of a framework for analysis, as Enos didn't at first, the codes (words or labels), at least initially, can feel a little haphazard and arbitrary. They can nonetheless be useful guides to future analysis. Thus, Enos's own coding of the initial paragraph of her notes (from Figure 15.9) looked something like what appears in Figure 15.10.

Here we see Enos, in her first informal interview, developing codes that will appear again and again in the margins of her field notes: codes that

Field Notes	Codes
I spoke to a woman who had remained silent in the group. She was sitting alone at a table. I asked if I could talk with her. She has six children, two of her children will be visiting today. Her mother has always had her kids. Last year her mother passed away and now her sister is taking care of the kids. When she's released, she'll have responsibility for the whole bunch. She says it was hard to see your kids while you were in jail but that if you were DCYF involved sometimes it was easier. (This has to do with termination of rights.) She remarked that it was important not to be a "stranger" to your kids while you were in jail. It made everything worse. I asked about telling your kids where you are. She said she just straight out told them and they accepted. "Mommy's in jail. That's it." She thought having such direct experience with jail that maybe her kids would "avoid" it when they got older.	women as birthgivers, not caretakers motherhood management permanent substitute temporary substitute

FIGURE **15.10** Enos' Coding of Field Notes

permit her to characterize how social mothering is managed while a biological mother is in prison. The codes "permanent substitute" and "temporary substitute," referring to this particular woman's mother and sister, respectively, become categories for classifying the kinds of mothering that all other children receive.

STOP & THINK   *Suppose you wanted to compare all the data that Enos coded with the words "temporary substitute." What might you do to facilitate such a comparison?*

Assigning a code to a piece of data (i.e., assigning "temporary substitute" to this woman's sister or "permanent substitute" to her mother) is just the first step in coding. The second step is physically putting the coded data together with other data you've coded in the same way. There are basically two ways of doing this: the old-fashioned way, which is to put similar data into a file folder, and the new-fashioned way, which is to do essentially the same thing using a computer program that stores the "files" in the computer itself. Old-fashioned manual filing is simple enough: you create file folders with code names on their tabs, then place a copy of every note you have coded in a certain way into the folder with the appropriate label. In these days of relatively cheap photocopying and the capacity of computers to make multiple copies, you can make as many copies of each page as there are codes on it and slip a copy of the page into each file with the corresponding code. Thus, Enos might have placed all pages on which she placed the codes "temporary substitute" into one folder and all pages on which she wrote the code "permanent substitute" into another. Later in her analysis, when she wanted to see all the data she'd coded in the same way, she'd just have to pull out the file with the appropriate label.

The logic of the new-fashioned, computer-based filing is the same as that of the old-fashioned one. The newer way requires less paper and no actual file folders (and can therefore result in a great savings of money and trees), but it does generally require special computer software (so the initial outlay can be substantial). "Code-and-retrieve programs," for instance, like AnSWER, Kwalitan, and QUALPRO, help you divide notes into smaller parts, assign codes to the parts, and find and display all similarly coded parts (Gibbs, 2012b). Computer-based filing also makes revising codes simpler. For those who are interested, we recommend Gibbs (2012b: http://onlineqda.hud.ac.uk /Intro_CAQDAS/what_packages_are_available/code_retrieve_packages.php).

Types of Coding

One purpose of coding is to keep facts straight. When you code for this purpose alone, as you might be doing if you were to assign a person's name to all paragraphs about that person, you'd be engaged in purely *descriptive coding*. Descriptive codes can emerge from the data and usually do from when "loose" research designs are employed. Initially Lang (in Clark and Lang, 2001: 11), who looked at respondents' open-ended reflections on an image of Sojourner Truth, found herself coding the reflections as "complex" or "simple," with no particular idea where she might be going with those codes.

Sometimes, and especially when research designs are of moderate "tightness," as in the qualitative interview portion of Regnerus and Uecher's (2011) study of premarital sex and marrying, the codes can be suggested by participants themselves. Responses, in this study, to questions about why their college-student participants planned to postpone marriage led naturally to seven explanations: "can't afford it," "be your own person," "it's too soon to have children," "the travel narrative," "parental resistance," "the pursuit of sexual chemistry," and "deflated confidence in the institution." Sometimes, especially when a very "tight" research design is employed, descriptive codes actually exist before the data are collected. Thus, Dixon, Roscigno, and Hodson (2004: 3–33) report that even before their coders read the 133 book-length ethnographies of workplaces in the United States and England, they'd been prepared to look for and code evidence of strikes, union organization, internal solidarity, and a legacy of workplace activism. The coders found 21 workplaces (out of the 133) that had strikes during the period of observation, 67 that were unionized, 78 that exhibited average to strong levels of solidarity, and 16 that had a legacy of frequent strike activity.

Because the basic goal of coding is to advance your analysis, you'd eventually want to start doing a more *analytic* type of coding. Enos's simultaneous development of the two codes "permanent substitute" and "temporary substitute" is an example of analytic coding. Kathy Charmaz calls this preliminary phase of analytic coding *initial coding* and defines it as the kind through which "researchers look for what they can define and discover in the data" (1983: 113). This is also a point when one's early, provisional conclusions (see Figure 15.8) can inform one's choice about codes.

Eventually, according to Charmaz, initial coding gives way to *focused coding*, during which the initial codes themselves become the subject of one's coding exercise. At this stage, the researcher engages in a selective process, discarding, perhaps, those codes that haven't been terribly productive (e.g., Enos never used the "women as birthgivers" code after its first appearance) and concentrating or elaborating on the more enlightening ones. After Enos saw all the ways in which women found temporary substitutes, for instance, she began to use variations in those ways—for example, relatives, foster care, friends—as codes, or categories, of their own.

Memoing

memos, more or less extended notes that the researcher writes to help herself or himself understand the meaning of codes.

All data analyses move beyond coding once the researcher tries to make sense of the coded data. In quantitative analyses, the researcher typically uses statistical analyses to make such sense. In qualitative analyses, other techniques, such as memoing and data displaying, are used. **Memos** are more or less extended notes that the researcher writes to help understand the meaning of coding categories. Memos can vary greatly in length (from a sentence to a few pages) and their purpose, again, is to advance the analysis of one's data (notes). An example (labeled distinctly as a theoretical note) of a memo appears about 100 pages after Enos's write-up of the encounter with the woman whose children had been (permanently) cared for by her mother until the mother died, but were now being (temporarily) cared for by her sister. It appears to be a reflection on all the material that's been coded "temporary substitute" and "permanent substitute" throughout the field notes. This memo is shown in Figure 15.11.

> Theoretical note
> Consider the work on the breakdown of the family and compare that to the efforts of these families to keep their children in their midsts. These children are being raised by persons other than their birth mothers but they are well dressed, taken care of, appear healthy, and so forth. If they are on welfare, there appears to be another source of support.

FIGURE **15.11** Illustration of Memo from Enos

STOP & THINK   *Enos tells herself to think about the literature on family breakdown, something that the mother's imprisonment can also lead to. What do you think are some of the more familiar causes of family breakdown?*

Notice that Enos makes meaning by drawing an analogy between family separations brought about by imprisonment and those brought about by divorce, separation, death, and so on. She's focusing her attention on what happens to the care of children when they lose daily access to a parent (in this case, the mother). This has clearly become a central concern for future consideration. With this memo as a springboard, it's not surprising that child care and relationships with family members are a key issue in the research she reports in Chapter 10.

Because fieldwork and coding can be very absorbing, it's easy to lose track of the forest for the trees. The practice of memoing, of regularly drawing attention to the task of analysis through writing, is one way that many qualitative researchers force themselves to "look for the big picture." Memos can be thought of as the building blocks of the larger story one eventually tells. Some memos, such as Enos's, still need some refinement (e.g., through the reading of the literature on family breakdown), but others qualify for inclusion in the final report without much alteration. Some, especially those that are memos about linkages among memos, can provide the organizing ideas for the final report. Memos are a form of writing and have the potential for creative insight and discovery that writing always brings. One can easily imagine that Hannah Frith and Celia Kitzinger's (1998) key theoretical question came from a memo about whether qualitative data are transparent reflections of reality or context-dependent constructions of participants. In their own analysis of young women in focus groups talking about how they refuse men sex, Frith and Kitzinger decided that the talk needed to be seen as generative and perhaps reflective of group constructions. The women spoke of their efforts to protect the feelings of the man they'd refused. Frith and Kitzinger saw the women's talk as "constructing" men as needy, and perhaps weak, and themselves as knowledgeable agents, in opposition to traditional stereotypes of men as sexual predators and women as vulnerable prey. The talk was not a "transparent" window on what went on during the refusals, Fritz and Kitzinger conclude. Roger (Clark, 2001) was also made aware of such group constructions when he asked faculty at a particularly successful urban middle school to talk, in focus groups, about how they'd managed to make differences in test performance between high-income and low-income students "disappear." His memos led to this awareness as he tried to interpret the focus-group interactions in which one faculty member would come up with an explanation and others would agree vigorously. Thus, for instance, one faculty member mentioned at one point that he or she maintained high standards for all of his or her students, both high- and low-income. When others voiced immediate agreement (and still others independently wrote of high standards in response to an open-ended question), Roger didn't necessarily feel he'd "seen" what went on in individual classrooms, but he did feel he had evidence of a common definition of faculty values at this school.

Methodological note
While it has been easy to use my
role as researcher and "playmate" to
gain some information about mothers
and children in prison, it has also
been frustrating from a data
collection perspective. I am
involved in some long and wideranging
conversations with the women
and trying to remember the tone, the
pacing, and the content of discussion
while engaging in these as an active
member. I can disappear for a few
minutes to an adjoining room to
scribble notes but this is difficult.
I focus on the points that are most
salient to my interests but can get
sidetracked because so much is of
interest to me. A more regimented
conversation in an atmosphere that is
more controlled than this one would
probably yield "better" notes.

FIGURE **15.12** Methodological Memo from Enos

Memos can be potent theoretical building blocks, but they can also be vital methodological building blocks. Enos tells us, for instance, that memos were a particularly useful way for her to figure out what she needed to "learn next" during her study and how to actually learn it (personal communication). One memo (presented in Figure 15.12) shows Enos deciding to adopt a more formal interview strategy than she'd used before. This memo constituted a transition point between the fieldwork and the interview stages of Enos's study.

DATA DISPLAYS

If memos, in their relative simplicity, are analogous to the relatively simple univariate examinations of quantitative analyses, data displays can be seen as a rough analogue to the more complex bivariate and multivariate analyses used there. Miles and Huberman (1994) and Williamson and Long (2005) are particularly adamant about the value of data displays for enhancing both the processes of data reduction, admittedly a means to an end, and conclusion making and verifying, the ultimate goals of data analysis.

data display, a visual image that summarizes information.

A **data display** is a visual image that summarizes a good deal of information. Its aim is to present the researcher himself or herself with a visual image of what's going on and help him or her to discern patterns. Miles and Huberman's figure (Figure 15.8) itself can be seen as an example of such a display, summarizing the interaction among various processes in qualitative analysis.

The crucial feature of data displays is that they are tentative summaries of what the researcher thinks he or she knows about his or her subject. Therefore data displays permit him or her to examine what he or she thinks by engaging in more data examination (and reduction) and, eventually, to posit conclusions. Enos, for instance, tends to use the space on the right half of her field note pages not only for coding but also for data displays, often in the form of exploratory typologies about the kinds of things she's seeing. Thus, next to notes she's taken about one mother, "Pam," and her interaction with her children, "Tom" (a baby) and "Joy" (a three-year-old), we find a display (in Figure 15.13) about how mothers seem to manage their weekly interactions with their children in the parenting program.

Enos's coding of the first paragraph (referring to Pam's attempt to normalize her interaction with Tom and Joy) with a label that she's used before ("attempt at normalization") clearly has pushed her, by the time she rereads the second paragraph, to play with a list of ways she's seen mothers interact with their children at the parenting program. (Nothing in the second paragraph, in fact, is likely to have sparked this reflection.) This typology is the kind of data display that Miles and Huberman advocate. It pictorially summarizes observations Enos has made about the varieties of mothering behavior

There were a number of infants visiting today. One of them, Tom[2], is the son of Pam who also has a three-year-old girl. Pam plays actively with her children, is with them on the carpeted area, holding the baby and roughhousing with her daughter. Joy is an active and very outgoing kid who asks to have her hair fixed by her mom who is struggling to feed the baby and pay attention to her daughter.	attempt at normalization
The baby is tiny for his age and pale. These children are being taken care of by Pam's boyfriend's mother. Pam has a problem with cocaine and got high the day before Tom was born. She says that before he was born all she was interested in was getting high and the "chase." During her pregnancy with Joy, she quit smoking and was drug free.	Setting management Mother Behavior 1) no playing with child 2) active playing 3) sporadic conversation 4) steady conversation 5) steady interaction 6) kid in other's care

[2]All names are pseudonyms.

FIGURE **15.13** Enos' Emergent Typology of Mothering Behavior in the Parenting Program she's seen, and its presence facilitates her evaluation of its adequacy for subsequent presentation.

The forms in which data displays occur are limited only by your imagination. Figure 15.8 is a flowchart; Enos's display is a more conventional-looking typology. Miles and Huberman (1994) also advocate using various kinds of matrices, something that Roger and a student co-author, Heidi Kulkin, used in preparing their findings about 16 young-adult novels about non-white, non-American, or non-heterosexual main characters. Heidi and Roger found themselves most interested in themes of oppression and resistance in such novels and needed a quick way of summarizing the variation they'd discovered. The matrix shown in Table 15.11 worked. Preparing this matrix was helpful in two ways. First, it revealed some things they definitely wanted to say in their final analysis. It gave them the idea, for instance, of organizing their written presentation into four parts: stories about contemporary American characters, historical American characters, non-American characters, and historical non-American characters. Reviewing the table also led them to valid observations of between-group differences for their sample. Thus, for instance, they observed that authors writing about contemporary American characters were less likely to emphasize the importance of formal education and extraordinary self-sacrifice than those writing about other subjects.

Preparing the matrix, then, led Heidi and Roger to some observations they directly translated into their final report (Clark and Kulkin, 1996). It

TABLE **15.11**

Content Analytic Summary Matrix Used in Preparation of Clark and Kulkin's (1996) "Toward Multicultural Feminist Perspective on Fiction for Young Adults"

Place	Contemporary	Historical
United States	No. of books: 4 Varieties of oppression: racism, classism, cultural imperialism, heterosexism Strategies of resistance: escape, resort to traditional custom, communion with similarly oppressed others	No. of books: 3 Varieties of oppression: sexism, racism Strategies of resistance: escape, cooperation, extraordinary self-sacrifice, learning, writing
Outside United States	No. of books: 4 Varieties of oppression: sexism, classism, racism, imperialism Strategies of resistance: education, communion with similarly oppressed persons, creative expression, self-reliance, self-sacrifice	No. of books: 5 Varieties of oppression: all the usual suspects Strategies of resistance: escape, communion with similarly oppressed others, writing, self-definition, education, self-sacrifice

also helped them in another, perhaps more important, way: It warned them against saying certain things in that report. Thus, for instance, although each mode of oppression and resistance listed in each "cell" of the matrix was prominent in some novel that fell within that cell, each was not equally evident in all the novels of the cell. Heidi and Roger knew, then, that their final written presentation would have to be much more finely attuned to variation within each of their "cells" than the matrix permitted.

Sometimes, after coding, a researcher will have essentially comparable and simple information about cases. Such was Lang's case after she'd coded respondents' reflections on images of Sojourner Truth. Initial coding had led her to observe that some respondents wrote relatively complex reflections. A memo, which consciously located all reflections in the early morning class in which they'd been received, led Lang to ask respondents whether they considered themselves "morning," "afternoon," or "night" people. Lang then organized the two sets of information (about complexity and whether people considered themselves "morning" people, etc.) using a kind of display that Miles and Huberman call a meta-matrix or "monster-dog." Table 15.12 is one of the

TABLE **15.12**

Example of Lang's Monster-Dog Data Displays: Elaborateness of Response Versus When Respondent Feels Most Awake

Respondent	Elaborate	Morning	Afternoon	Night
A				
B		X		
C	X		X	
D			X	
E	X	X		
F		X		
G	X			
H	X	X		
I			X	
J	X			X
K				X
L		X		
M	X	X		
N	X	X		
O				X
P	X	X		

Note: X indicates presence of a trait; no X indicates absence.

Source: Adapted from Clark and Lang, 2001.

monster-dogs that Lang created, albeit a relatively small and tame one. Lang's monster-dog, like all good examples of its breed, is a "stacking up" of "single-case displays" (Miles and Huberman, 1994: 178), one on top of the other, so that the researcher can make comparisons. As Lang reports, "I looked at several Monster Dogs, each displaying different sets of variables. I was soon able to see helpful patterns. For instance, self-described morning people looked as though they gave more elaborate responses to the photograph of Sojourner Truth than [others] did. When I crosstabulated these two variables, there was in fact clearly a relationship" (Clark and Lang, 2001: 14).

Data displays of various kinds, then, can be useful tools when you analyze qualitative data. Miles and Huberman would take this as too weak an assertion. They claim

> You know what you display. Valid analysis requires, and is driven by, displays that are focused enough to permit a viewing of a full data set in the same location, and are arranged systematically to answer the research questions at hand. (Miles and Huberman, 1994: 91–92)

Although we think this position is overstated, if only because we know of good qualitative analyses that make no explicit reference to data displays, we think you could do worse than try to create data displays of your own while you're engaged in qualitative data analysis. When you do, you might well look to Miles and Huberman (1994) for advice.

CONCLUSION DRAWING AND VERIFICATION

The most important phase of qualitative data analysis is simultaneously its most exciting and potentially frightening: the phase during which you draw and verify conclusions. Data displays are one excellent means to this end because they force you to engage in two essential activities: pattern recognition and synthesis. Coding, of course, is a small step in the ongoing effort to do these things, and memoing is a larger one. But you can hardly avoid these activities when you're engaged in creating a data display.

Enos emphasizes that one very important resource for verifying conclusions, especially for researchers who are doing announced field observations or qualitative interviews, is the people being observed or interviewed. These people, she reminds us, can quickly tell you "how off-base you are": "This was very helpful to me, as was asking questions they could theorize about, like 'It seems that white women use foster care a lot more than black women do. Is that right?' [And, if they agree, 'Why is that the case?']" (personal communication).

Elliot Liebow (1993) advocates, however, an "Ask me no questions, I'll tell you no lies" principle, especially relative to participant observation. In his study of homeless women, he consciously decided not to bring to the women the kind of theoretical questions that Enos recommends. In doing so, he claimed that he was less concerned with the "lies" he might receive in return for his questions than with "contaminating the situation with questions dragged in from the outside" (Liebow, 1993: 321).

Whatever you decide to do about the issue of "bouncing ideas" off those you're studying, we think you'll find that a vital phase of conclusion drawing and verification comes when you compose your final report. The very act of writing, we (and others [e.g., Lofland, Snow, Anderson and Lofland, 2005]) find, is itself inseparable from the analysis process. As a result, we divide the remainder of the section on drawing and verifying conclusions into a discussion of two phases: the pre–report-writing and the report-writing phase.

Pre-Report Writing

Lofland, Snow, Anderson and Lofland (2005) point out that the growing use of computers for qualitative data analysis has helped reinforce two points. The first is that qualitative analysis can be facilitated by the kinds of things computers do, like storing and retrieving information quickly. The new programs that facilitate data coding testify to these strengths. The second is that good qualitative analysis requires what computers aren't really much good at. One of these tasks, and something that really distinguishes qualitative research from other brands of social research, is deciding what problems are important. Enos began her study of women in prison with an open mind about what the real problems were. The same was true of Cottingham (see Chapter 11) as she approached the study of Pittsburgh Steelers' fans, and Clark and Kulkin (1996) as they began to read nontraditional novels for young adults. These authors all came to their settings with open eyes, open ears, and, particularly, open minds about what they'd find to be the most important problems.

● STOP & THINK   *Contrast this openness with the hypothesis-driven approach of Martinez and Lee (see Chapter 12).*

Finding out just what problem you should address requires something more than what computers are good at: the imagination, creativity, and moral insight of the human brain.

Imagination and creativity are necessary even after you've chosen your topic and are well into the data analysis. One way to stimulate such imagination and creativity is to use the kinds of data displays advocated by Miles and Huberman. But, as Lofland, Snow, Anderson, and Lofland (2005) point out, other practices can create fertile conditions for qualitative analysis. One practice is to *alter data displays*. If you've created a data display that, somehow, isn't useful, try a different form of display. We can imagine, for instance, Miles and Huberman building the flowchart that we've reproduced in Figure 15.8, indicating the various interrelationships that might exist among data collection, data reduction, data display, and conclusions, out of a simple typology of (with no indication of relationships among) activities involved in qualitative analysis: data collection, data reduction, data display, and conclusions. Note that altering data displays involves conclusion verification and dismissal, both important goals of data analysis.

Another pre-report-writing activity, advocated by Lofland, Snow, Anderson and Lofland, and before them by Glaser and Strauss (1967), is that of *constant comparison*. The key question here is this: How is this item (e.g., person, place, or thing) similar to or different from itself at a different time or comparable items in other settings? One can, again, easily imagine how such comparisons led Enos to her eventual interest (see Chapter 10) in how substitute mothering differs for women from different racial backgrounds, or Cottingham (Chapter 11) to her focus on the similarity between the rituals of Steelers fans and those of religious congregations, or to Clark and Kulkin's interest in how themes of oppression and resistance vary by geographical or historical context, or to Lang's (see Clark and Lang, 2001) belief that "morning" and other people might differ in their capacity to give complex responses early in the morning.

Lofland et al. (2005) actually recommend several other pre–report-writing techniques for developing creative approaches to your data, but we'd like to highlight just one more of these: *speaking and listening to fellow analysts*. Conversations, particularly with friendly fellow analysts (and these needn't, of course, be limited to people who are actually doing qualitative data analysis but can include other intelligent friends) can not only spark new ideas (remember the adage "two heads are better than one"?) but also provide the kind of moral support we all need when ideas aren't coming easily. Friendly conversation is famous not only for showing people what's "right in front of their noses," but also for providing totally new insights. Lang is explicit about the value of sharing her ideas with other students in a data-analysis class devoted to this purpose: "The discussion was similar to having someone proofread a paper for you. Sometimes outsiders help you find better ways to phrase sentences, etc. In this case, they helped me articulate to myself what I was most interested in. The class felt like one big peer review session" (Clark and Lang, 2001: 11–12).

Qualitative Report Writing

A qualitative report writer on report writing had this to say:

> When I began to write, I still had no order, outline, or structure in mind....
> I plunged ahead for many months without knowing where I was going, hoping and believing that some kind of structure and story line would eventually jump out. My original intention was to write a flat descriptive study and let the women speak for themselves—that is, to let the descriptive data and anecdotes drive the writing. In retrospect, that was a mistake. Ideas drive a study, not observations or unadorned facts.... I shuffled my remaining cards again and discovered that many cards wanted to be grouped under Problems in Day-to-Day Living, so I undertook to write that chunk. The first paragraph asked the question, How do the women survive the inhuman conditions that confront them? After I wrote that paragraph it occurred to me that most situations, experiences, and processes could be seen as working for or against the women's survival and their humanity—that the question of survival might serve as the opening and organizing theme of the book as a whole and give the enterprise some kind of structure. (Liebow, 1993: 325–326)

Lang's reflections on her writing of an undergraduate paper, based on qualitative data, made related, but slightly different, points:

> I finally sat down and began writing the paper. It was now that the project actually began to make sense and fit together. One of my biggest concerns going into this step was how I would incorporate the literature reviewed with the rest of my findings. I'd found literature, I knew, that was somewhat relevant, but nothing that spoke directly to my findings that morning people would write more elaborately in the morning. As I wrote, I discovered that my worries were misplaced. Things did flow. The fact that the literature I reviewed didn't speak directly to what I'd found became a virtue; I could simply say that I was addressing a hole in what the literature already said about sleep patterns. Once I began to put it all into words, the paper "practically wrote itself." That was neat. (Clark and Lang, 2001: 17)

Most authorities on qualitative data analysis attest to the difficulty of distinguishing between report writing and data analysis (see Hammersley and Atkinson, 1983; Lofland, Snow, Anderson and Lofland, 2005). Data collection, data reduction, data display, and other pre-report-writing processes all entail an enormous amount of writing, but, whatever else it is, *report writing is a form of data analysis*. Liebow's confession that he came to the report-writing phase of *Tell Them Who I Am*, his touching portrayal of the lives of homeless women, with "no order, outline, or structure in mind" could probably have honestly been made by many authors of qualitative social science—and, frankly, by many authors of quantitative social science. Lang might have had a better sense of "order" going into her short essay than Liebow did going into his book, but writing clearly helped her to refine her ideas, too. The discovery, through writing, of one vital organizing question or another is probably the rule rather than the exception for qualitative analyses.

The writing involved in pre-report phases and the report-writing phase of qualitative analysis can be distinguished, notably by their audiences. Early phases involve writing in which the researcher engages in conversations with himself or herself. We're privy to Enos's codes, her memos, and her typologies, for instance, only because she has generously permitted us to peer over her shoulder, as it were. (Lang's [in Clark and Lang, 2001] self-conscious renderings are exceedingly unusual.) Enos's intended audience was Enos and Enos alone. Formal report writing, on the other hand, is much more clearly a form of communication between the author and some other audience. This is not to say that the author himself or herself won't gain new insights from the process. We've already said that he or she will. But it does imply that at least a nodding acquaintance with certain conventions of social science writing is useful, even if you intend only to "play" with those conventions for effect. As a result, we'd like to introduce you to a few such conventions and do so in our Appendix C on report writing.

Liebow's confession suggests a few guidelines we don't mention in Appendix C. One is that, even if you're not sure just where a particular segment of a report will fall within the whole, as long as you've got something to say, write. Don't insist on seeing the forest, before you write, especially if you've got a few good trees. Liebow wrote and wrote, long before he'd gotten a sense of his main theme. He wrote with the kind of hope and belief that an experienced writer brings: that a story line will emerge. But in writing what he took to be a minor theme, the "Problem of Day-to-Day Living," he first glimpsed what would be the organizing principle of his whole book, "the question of survival." Once he discovered this principle, he found that much of what he'd written before was indeed useful.

Another principle, implicit in Liebow's confession, is that you needn't look for perfection as you write. Especially if you have access to a word processor and time, you can write relatively freely, in the certain knowledge that whatever you write can be revised. Don't assume you need to say what you want to say in the best (the most grammatical, the most concise, the most stylish) way possible the first time around. Assure yourself that you can always go back, revise, and rethink as necessary. Such assurances are terribly liberating. Just remember to keep your implicit promise to yourself and revise whenever you can.

SUMMARY

The most common goal of quantitative data analysis is to demonstrate the presence or absence of association between variables. To demonstrate such associations, and other characteristics of variables, quantitative data analysts typically compute statistics, often with the aid of computers.

Statistics can be distinguished by whether they focus on describing a single sample, in which case they are descriptive statistics, or on speaking to the generalizability of sample characteristics to a larger population, in which case they are inferential statistics. Statistics can also be distinguished by the number of variables they deal with at a time: one (as in univariate statistics), two (bivariate statistics), or more than two (multivariate statistics).

The most commonly used univariate statistical procedures in social data analysis are measures of central tendency, frequency distributions, and measures of dispersion. Measures of central tendency, or average, are the mode, the median, and the mean. Measures of dispersion include the range and standard deviation. All the univariate statistical procedures described in this chapter focus on describing a sample and are therefore descriptive, rather than inferential, in nature.

We demonstrated bivariate and multivariate statistical procedures by focusing on crosstabulation, one of many types of procedures that allow you to examine relationships between two variables. In passing, we discussed descriptive statistics, like Cramer's V and Pearson's r, used to describe sample relationships, and inferential statistics, like chi-square, used to make inferences about population relationships. We also illustrated ways of controlling the relationship between two variables through the introduction of control variables.

The canons for analyzing qualitative data are much less widely accepted than those for analyzing quantitative data. Nonetheless, the various processes involved in such analysis are broadly accepted in general terms: data collection itself, data reduction, data display, and conclusion drawing and verification. These processes are interactive in the sense that they do not necessarily constitute well-defined and sequential steps and insofar as all the individual processes can affect or be affected by, directly or indirectly, any of the others.

Word processors are very useful tools in data collection. Soon after each field or interview session, it is useful to word-process field notes or transcripts. It's often a good idea to leave plenty of room in your notes for coding, comments, memos, and displays.

Data reduction refers to various ways a researcher orders collected and transcribed data. We've talked about two sub-processes of this larger process: coding and memoing. Coding is the process of associating words or labels with passages in your field notes and then collecting all passages that are similarly coded into appropriate "files." These files can be traditional file folders or computer files. Codes can be created using predefined criteria, based on a conceptual framework the researcher brings to his or her data, or using post-defined criteria, that "emerge," as it were, as the researcher reviews his or her field notes. Codes can be merely descriptive of what the researcher's seen or analytic, in which case the codes actually advance the analysis in some way. Memos, unlike codes, are more or less extended notes that the researcher writes to help create meaning out of coding categories. Memos are the building blocks of the larger story the researcher eventually tells about his or her data.

Data displays are visual images that present information systematically and in a way that allows the researcher to see what's going on. The forms in which data can be displayed are limitless, but we've discussed four types in this chapter: charts, typologies, matrices, and "monster-dogs."

Data displays are particularly useful for drawing conclusions that are then subject to verification or presentation. We discussed a few other pre-report techniques for developing such conclusions: altering data displays, making comparisons, and talking with and listening to fellow analysts. We also described report writing itself as an important stage in the analysis process.

EXERCISE 15.1

Analyzing Data on the Web

Visit the GSS Cumulative Data file by going to http://sda.berkeley.edu/archive.htm and hitting on the most recent GSS Cumulative Data File. Do your own univariate, bivariate, and multivariate analyses of marital status (called "marital" in the codebook) and labor force status ("wrkstat"), controlling for gender ("sex").

1. Do you remember what we said the modal marital status (married, divorced, separated, or never married) is among adults in the United States? Why don't you check for yourself? Start the "Browse codebook" action, hit the "standard codebook," hit the "sequential variable list," and click on

"Personal and Family Information." Find the variable named "marital" and hit on it. Which, in fact, has been the modal marital status among adults in the United States?

2. Now, consider the variable "labor force status" ("wrkstat" in the codebook). Which labor force status would you expect to be the most common among adults in the United States: working full time, working part time, temporarily not working, retired, in school, keeping house, or something else? State your guess. Now find the variable named "wrkstat." Which, in fact, is the modal labor force status among adults in the United States? Were you right?

3. Now consider both marital status and labor force status: Which marital status would you expect to have the greatest representation in the full-time labor force? State your guess in the form of a hypothesis. Now return to the original GSS menu (where you found the "browse codebook" option before). This time, click "frequencies and crosstabulation." Where the menu asks for "row" variable, type "wrkstat." Where it asks for "column" variable, type "marital." Then "run the table." Which marital status does, in fact, have the greatest percentage of its members employed "full time"? Were you right?

4. Now consider the relationship between marital status and labor force participation, controlling for gender. Which marital status would you expect to have the greatest representation in the full-time labor force **for men?** Which marital status would you expect to have the greatest representation in the full-time labor force **for women?** Now return to the original GSS menu. This time, click "frequencies and crosstabulation." Where the menu asks for "row" variable, type "wrkstat." Where it asks for "column" variable, type "marital." Where it asks for "control" variable, type "sex." Which marital status does, in fact, have the greatest percentage of members employed "full-time" **among men? Among women?** Were you right?

EXERCISE 15.2

Interpreting a Bivariate Analysis

After Lang (2001) had analyzed Americans' GSS responses to the question of how they felt about income differences in America, she decided to look at how this feeling was related to other variable characteristics of the respondents.

Look at the following table from her paper and write a paragraph about what it shows about the association between respondents' satisfaction with their own financial situation and their sense of whether income differences in America are too large.

TABLE
Crosstabulation of Satisfaction with Present Financial Situation and Response to Whether Differences in Income in America Are Too Large

Satisfaction with Financial Situation	Differences in Income in America Are Too Large					
	Strongly Agree	Agree	Neither	Disagree	Strongly Disagree	Total
Pretty well satisfied	27.6%	29.7%	16.3%	13.7%	12.6%	100%
More or less satisfied	30.0%	37.5%	13.2%	13.0%	6.3%	100%
Not satisfied at all	44.7%	30.7%	8.5%	8.5%	7.7%	100%

Source: Lang (2001: 25).

EXERCISE 15.3

Reconstructing a Qualitative Data Analysis

Return to either Cottingham's essay (in Chapter 11) or Enos's essay (in Chapter 10). See if you can analyze one of these articles in terms of concepts developed in this chapter.

1. See if you can find two portions of transcribed field notes or interviews that made it into the report. (Hint: Your best bet here is to locate quotations that might have been taken from those notes.) Write these out here.
2. See if you can imagine two codes or categories that the author or authors may have used to organize his or her/their field notes. (There are no wrong answers here. We aren't actually told, in either article, what the codes were. You're looking here for categories that

come up in the analysis and that seem to have linked two or more segments of the field notes.) Briefly discuss why you think these codes might have been used.
3. See if you can imagine a memo that the author or authors might have written to organize his or her/their memos. (Again, there are no wrong answers here.) Briefly discuss why you think this memo might have been used.
4. See if you can imagine and draw a data display that the author or authors might have used to organize his or her/their essay. (Again, there are no wrong answers, though a very good answer might be the presentation of a typology.)

EXERCISE 15.4

Data Collection and Coding

For this exercise, you'll need to collect data through either observation or qualitative interview. If you prefer to do observations, you'll need to locate a social setting to which you can gain access twice for at least 20 minutes each time. You can pick some easily observable activity (e.g., a condiment table at a cafeteria, an elevator on campus, a table at a library where several people sit together). You can choose to tell others of your intention to do research on them or not. You can work out some research questions before you go to the setting the first time or not. You can take notes at the setting, tape-record what's going on, or choose to do all your note-taking later.

If you prefer to do two interviews, we recommend that you do something like the work suggested in Exercise 10.1: pick an occupation you know something, but not a great deal, about (such as police officer, waiter, veterinarian, letter carrier, or high school teacher) and find two people who are currently employed in that

occupation and willing to be interviewed about their work. Conduct 20-minute semi-structured interviews with each of these people, finding out how they are trained or prepared for the work, the kinds of activities they do on the job, about how much time is spent on each, which activities are enjoyed and which are not, what their satisfactions and dissatisfactions with the job are, and whether or not they would like to continue doing their work for the next 10 years.

Whatever the setting or whoever the interviewee, write down your notes afterward, preferably using a word processor and preferably leaving room for coding and memoing (i.e., use wide right margins).

After writing down your notes, spend some time coding your data by hand. Remember: This means not only labeling your data but also filing away similarly labeled data (in this case, in file folders you make or purchase).

EXERCISE 15.5

Memoing and Data Displaying

Some time shortly after you've recorded your notes (for Exercise 15.4) and finished coding them (at least for the first time), reread both the notes and your coding files.

Spend about 20 minutes or so writing memos (either word processing them or hand-writing them) at the end of your notes. You can, of course, spend more than 20 minutes at this activity. The important thing is to develop some facility with the process of memoing.

Once you've written your memos, spend some time trying to create a data display that helps you understand patterns in the observations or interviews you've done. This display can be in the form of a chart, a typology, a graph, a matrix, or a mini-monster-dog. The purpose is to help you "see" patterns in the data that you might not have already seen.

EXERCISE 15.6

Report Writing

Having completed Exercises 15.4 and 15.5, write a brief report (no more than two or so typed pages) about your project. You should consult Appendix C on report writing for guidance here. You can omit a literature review (though, of course, if you found a couple of relevant articles, they might help), but you might otherwise be guided by the list (of items to include in a report) provided in Appendix C. (In particular, consider giving your report a title; providing a brief abstract, an overview, and a sense of your methods; and highlighting your major findings and your conclusions.) Good luck!

APPENDIX A

Comparing Methods

The text covers the major methods of data collection in the social sciences: questionnaires, interviews, observations, and available data (including content analysis). We believe that no one method is inherently better than others. Instead, we believe that some methods are more or less useful for studying specific topics and in specific situations.

Researchers should consider the following issues when deciding whether to observe something, to ask questions about it, or to use some available data as the data collection method:

1. Which method(s) will allow the collection of valid information about the social world?
2. Which method(s) will allow the collection of reliable information about the social world?
3. To what extent will the results be generalizable?
4. To what extent are these data collection methods ethical?
5. To what extent are these data collection methods practical?
6. Is it appropriate and practical to use more than one method?

Each method of data collection has advantages and disadvantages. A brief overview of these is found in the following table:

TABLE
Advantages and Disadvantages of Methods of Data Collection

Method	Advantages	Disadvantages
Asking Questions		
Group-administered questionnaire	Inexpensive Time efficient Can be anonymous, which is good for "sensitive" topics Good response rate	Not all populations possible Must have literate sample Must be fairly brief Needs permission from setting or organization
Mailed questionnaire	Fairly inexpensive Completed at respondent's leisure Time efficient Can be anonymous, which is good for "sensitive" topics	Response rate may be relatively low Needs names and addresses Must have literate sample Can't explain questions
Individually administered questionnaire	Good response rate Completed at respondent's leisure	Most expensive questionnaire Needs names and addresses Must have literate sample
Web or Internet-based questionnaire	Very inexpensive and time efficient Can be completed at respondent's leisure	Not all populations possible Random samples might not be possible Low response rate
Structured phone interview	Good response rate Little missing data Can explain questions and clarify answers Can use "probes" Don't need names or addresses Good for populations with difficulty reading/writing	Expensive Time consuming Respondents might not feel anonymous Some interviewer effect Lower response rate than for in-person interview
Structured in-person interview	Good response rate Little missing data Good for populations with difficulty reading/writing Can assess validity of data Can explain questions and answers and use "probes"	Expensive Time-consuming Needs names and addresses Interviewer effect Confidential, not anonymous

Method	Advantages	Disadvantages
Semi-structured interview	Allows interviewer to develop rapport with study participants Good response rate Questions can be explained and modified for each participant Useful for discussing complex topics because it is flexible and allows follow-up questions Can be used for long interviews Useful when the topics to be discussed are known	Time-consuming Expensive Requires highly skilled interviewer Interviewer effect Data analysis is time consuming
Unstructured interview	Allows interviewer to develop rapport with study participants Good response rate Questions can be explained and modified for each participant Participant can take lead in telling his or her own story Useful for exploratory research on new topics Useful for discussing complex topics and seeing participant's perspective Can be used for long interviews	Time-consuming Expensive unless small sample Requires highly skilled interviewer Interviewer effect Data analysis is time consuming May not have comparable data for all participants
Focus group interview	Allows participants to have minimal interaction with interviewer and keep the emphasis on participants' points of view Typically generates a great deal of response to questions Often precedes or supplements other methods of data collection	Attitudes can become more extreme in discussion Possibility of "group think" outcome Requires highly skilled interviewer Data less useful for analysis of individuals

(Continued)

Method	Advantages	Disadvantages
Participant and nonparticipant observation	Permits researcher to get handle on the relatively unknown Permits researcher to obtain understanding of how others experience life Permits the study of quickly changing situations Relatively inexpensive Permits valid measurement, especially of behavior Relatively flexible, at least insofar as it permits researcher to change focus of research	Results are often not generalizable Being observed sometimes elicits characteristics (demand characteristics) in observed simply because of observation Especially in unannounced observations, requires deceit Time-consuming
Available Data and Content Analysis		
Available data	Savings in time, money, and effort Possibility of studying change and differences over space Good for shy and retiring people and for those who like to work at their own convenience Data collection is unobtrusive	Inability to know limitations of data Might not be able to find information about topic Sometimes one has to sacrifice measurement validity Possibility of confusion about units of analysis
Content analysis	Savings of time, money, and effort Don't need to get everything right the first time Good for shy and retiring people and for those who like to work at their own convenience Data collection is unobtrusive	Can't study communities that have left no records Validity issues, especially resulting from insensitivity to context of message examined

APPENDIX B

Multiple Methods

RESEARCH IN THE NEWS

Stephanie Pappas (2012), writing on the website **LiveScience,** writes about
an "east–west culture clash" in discussing Facebook CEO Mark Zucker-
berg arriving at a Wall Street meeting wearing a hoodie and the headlines
that followed, which compared Silicon Valley's laid-back culture with East
Coast formality. Her article went on to describe social research that supports the view that life is different on the coasts. In the study, researchers
Victoria Plaut and colleagues (2012) told a tale of two cities—Boston and
San Francisco—that used multiple methods of data collection to compare them.

USING MULTIPLE METHODS

Each individual method has advantages and disadvantages, so it is useful to
employ more than one method of data collection whenever practical. Using
multiple methods, or what has been called *triangulation* (Denzin, 1997),
allows the strengths of one method to compensate for the weaknesses of
another and increases the validity of the study's conclusions. For example, in
a study focusing on the connections among violence, masculinity, and football (or soccer), Ramon Spaaij (2008) collected data on football fan culture
in several countries. He was able to understand the creation of the "hooli-
gan" identity within social and historical contexts by using multiple methods,
including semi-structured interviews, participant observation, and available
data such as websites, media reports, and fan magazines.

The study comparing Boston and San Francisco, discussed above, was
described in various media outlets including the *Huffington Post* and *The
Boston Globe* because of the interesting cultural differences between the cities
and the researchers' innovative methods. To compare the cities and their residents, Plaut and her colleagues (2012) used several kinds of data from the
two areas, including answers to questions on a survey with a representative
sample of midlife adults; surveys of college students; content analyses of view-
books of universities in the two cities; an online newspaper in each city; and

venture capital and hospital websites in both cities. The data supported the study's conclusions that sense of self and well-being are in some ways local—with Boston and its residents emphasizing tradition, status, and community and San Francisco and its residents emphasizing unlimited possibility, egalitarianism, and innovation (Plaut et al., 2012).

Of course, using additional methods means increases in time and cost, the need to obtain access more than once, and the obligation to reconcile discrepant findings. These considerations must be weighed against the increases in validity or generalizability that the use of multiple methods brings.

Examples of Multiple Methods for Student Research

Research questions: What do high school students think about careers in law enforcement? Do attitudes differ by race?

Possible Methods: Interview high school guidance counselors concerning student views; do an Internet-based survey of a sample of high school students; be a complete observer at career fairs that include law enforcement opportunities; and use available data from law enforcement recruitment efforts.

Research questions: Do nurses treat terminally ill patients differently from those who are expected to recover?

Possible Methods: Being a participant observer in a hospital for at least several weeks coupled with semi-structured, in-person interviews with samples of nurses, patients, and patients' families, and an analysis of a sample of patient records.

Research questions: How do two communities in the same urban area differ? Do they receive different levels of services? Do residents in one community feel more socially isolated than those in the other? Does one community experience higher levels of crime than the other?

Possible Methods: Available data (such as the annual reports from local agencies and organizations, editorials, news stories in local newspapers, information on websites and blogs, National Crime Victim Survey data, and the Federal Bureau of Investigation (FBI)'s *Uniform Crime Reports*), along with observations in public places taking the role of a complete observer, and structured phone interviews conducted with a random sample of residents.

APPENDIX C

Writing the Research Report

The research report can be communicated in a book, a written paper, or an oral presentation. The intended audience(s) can be the teacher of a class, the supervisor at work, an agency that funded the research, an organization that gave access to the data, or the general public. Although the report might differ for each audience, each version should describe the research process and the research conclusions in an orderly way. Students wanting a detailed discussion of the "nuts and bolts" of technical writing, including coverage of footnoting, referencing, and writing style, should consult one of the many excellent guides, such as Turabian's *A Manual for Writers of Term Papers, Theses, and Dissertations* (1996), Pyrczak and Bruce's *Writing Empirical Research Reports: A Basic Guide for Students of the Social and Behavioral Sciences* (2007), or Lester and Lester, Jr.'s *Writing Research Papers* (2009).

We'd like to tell you about some organizational conventions of report writing, even though we feel a bit presumptuous in doing so. After all, what will work for book-length manuscripts probably won't work equally well for article-length reports. Moreover, generally speaking, research reports involving qualitative data are less formulaic than those involving quantitative data. Nonetheless, because report writing does involve an effort to reach out to an audience, and because social science report writing frequently involves an effort to reach out to a social science audience, it's not a bad idea to have some idea of the forms expected by such audiences.

We'd like to suggest, then, a checklist of features that often appear in article-length research reports. (Books tend to be much less predictable in their formats.) Audiences can include policy makers or practitioners, who typically will expect practical information about either policy or practice, and the general public, who are interested in facts, new ideas, or information on their particular interests (Silverman, 2004: 369). Many of you, however, will write for an academic audience, so we'll focus on what that audience will expect most of the time. Those features are the following:

1. *A title* that tells what your research is about. We think the title should give the reader some idea of what you have to say about your subject. Four good examples are from reports presented in Chapters 7, 9, 10, and 11:
   - "Moving On? Continuity and Change after Retirement" (Adler and Clark, Chapter 7)

- "Environmentalism among College Students: A Sociological Investigation" (Lang and Podeschi, Chapter 9)
- "Managing Motherhood in Prison" (Enos, Chapter 10)
- "Who Envisions the Distaff Side?" (Clark, Kessler, and Coon, Chapter 13)

2. *An abstract* or a short (usually more than 100 but less than 150 words) account of what the paper does. Such accounts should state the paper's major argument and describe its methods. We haven't included abstracts for any of our focal research articles because we wanted to keep the length of the text manageable. But we've been very self-conscious about not following this convention and think we should give you at least one good example of an abstract, this one from Clark, Kessler, and Coon's original article, "Women and Girls Last?: Female Visibility in Children's Picture Books by White, Black and Gay-Sympathetic Authors," on which "Who Envisions the Distaff Side?" was based:

> We mark the fortieth anniversary of Lenore Weitzman et al.'s (1972) study that found women and girls to be essentially invisible in children's picture books of the late 1960s by examining female visibility in books by White, Black, Latino/a and Gay-Sympathetic authors of the early 21st century. We find female characters enjoy much more attention today than they did in the 1960s in books by all four sets of authors. We find that they are more visible, however, in books by current Black, Latino/a and Gay-Sympathetic authors than in books by White authors, in books by female authors than books by male authors, and in books by Black, Latina and Gay-Sympathetic women authors than by White women authors. We interpret these findings in terms of multicultural feminist theory and, particularly, the role assigned by that theory to women in helping marginalized communities "get over" oppression. (2013, forthcoming)

We think this abstract does a good job of encapsulating key points of the article while also noting the method and the context of the research. Abstracts frequently come before the main body of a report—just after the title—but are generally (and ironically) written *after* the rest of the report's been put together, so the author knows what points to summarize. Here again, however, there are no rules: Drafting an abstract can also be a useful heuristic earlier in the process—to help the writer focus (in a process sometimes called "nutshelling"). Notice, however, that in this abstract, the authors have actually engaged in a final act of conclusion drawing—the ultimate goal, after all, of data analysis itself.

3. *Overview of the problem* or *Introduction* and *literature review*. These two features are sometimes treated distinctly but can also be combined. The basic function of these features is to justify your research, although they're also good places to give credit to those who have contributed ideas, like theories, conceptual definitions, or measurement strategies, which you're using in some way. The literature review, in particular, should justify the current research through what's already been written or not written about the topic. In Chapter 10, Enos does this with two kinds of "literatures": literature that has discussed the "imprisonment boom" in the United

States and has pointed to the special problems of women in prison and literature that implies that family forms should have certain "normative" aspects, such as mothers who engage in full-time mothering. If you've never written a research report before, we think you could do worse than study the ways in which the authors of articles in this book have used literature reviews to justify their projects. Literature reviews are an art form, and social science literature reviews are a specialized genre of the larger one. As an art form, there are few hard-and-fast rules, but there are some. One is this: Never simply list the articles and books you've read in preparation for your report, even if you include some of what they say while you're listing them. Always use your literature review to build the case that your research has been worth doing.

Literature reviews themselves offer an excellent chance to engage in analysis. Whether your review occurs before, after, or throughout your field research, it will compel you to make decisions about what your study's really about. (See Chapter 4 for further tips on preparing your literature review.)

4. *Data and methods.* This section involves the description and explanation of research strategies. Authors generally lay out the study design, population, sample, sampling technique, method of data collection, and any measurement techniques that they used. They also sometimes discuss practical and ethical considerations that influenced their choice of method. Authors also sometimes justify methodological choices by discussing things such as the validity and reliability of their measures in this section.

5. *Findings.* Being able to present findings is usually the point of your analysis. You might want to subdivide your findings section. In the findings section, researchers always incorporate data or summaries of data. If the project involved quantitative data analyses, this is where researchers display their statistical results, tables, or graphs (see, e.g., Martinez and Lee, Chapter 12). If the research involved qualitative data analyses, this is where the authors present summary statements or generalizations, usually supported by anecdotes and illustrations from their interview or field notes. Qualitative researchers usually spice up and document their accounts with well-selected quotations in this section (see, e.g., Enos, Chapter 10). In both kinds of studies, writers generally offer interpretations of their data in this section.

6. *Summary and conclusions.* This section should permit you to distill, yet again, your major findings and to say something about how those findings are related to the literature or literatures you've reviewed earlier. It's very easy to forget to draw such relationships (i.e., between your work and what's been written before), but once you get started, you might find you have a lot more to say than you anticipated that could interest other social scientists.

7. *References.* Don't forget to list the references you've referred to in the paper! Here you'll want to be guided by certain stylistic conventions of, perhaps, a formal guide (e.g., American Sociological Association, 1997). If you're preparing a paper for a course or other academic setting, you might even ask the relevant faculty member what style he or she prefers.

8. *Appendices.* It's sometimes useful to include a copy of a questionnaire, an interview schedule, or a coding scheme if one was used. Sometimes appendices are used to display elements of a sample (e.g., particular states, countries, or organizations) that were used in the research and sometimes even the raw data (especially when the data were difficult to come by and other researchers might be interested in using them for other purposes). In general, researchers tend to use appendices to display information that is germane, but not central, to the development of the paper's main argument or that would require too much space in the article.

There are few absolutes here, however. Perhaps the most important rule is to give yourself oodles of time to write. If you think it might take about 25 hours to write a report, give yourself 50. If you think it might take 50, give yourself 100. Writing is hard work, so give yourself plenty of time to do it.

Report writing is an art form. Although guidance, such as what we've presented in this section, can be useful, this is one of those human activities where experience is your best guide. Good luck as you gather such experience!

Checklist for Preparing a Research Proposal

A good research proposal should impress someone (e.g., a teacher, a faculty committee, an Institutional Review Board [IRB], or a funding agency) with the project's worthiness, feasibility, ethical treatment of participants and the community, and appropriateness of its design. The following items usually, but not always, appear in a research proposal.

I. **A Title** A title captures the theme or thesis of the proposed project in a nutshell.

II. **A Statement of the Project's Problem or Objective** In this section, you should answer questions such as
   - What exactly will you study?
   - Why is it worth studying?
   - Does the proposed study have practical significance?

III. **Literature Review** In general, a good literature review justifies the proposed research. (See Appendix C for a discussion of literature reviews in research reports. Literature reviews in research proposals should do the same things that literature reviews in research reports do.) In a literature review, one normally cites references that appear in the proposal's reference section (see later) using a style that is appropriate to one's discipline (e.g., American Sociological Association style for sociology, American Psychological Association style for psychology and education). It is often appropriate to end the literature review with a statement of a research question (or research questions) or of a hypothesis (or hypotheses) that will guide the research.

   The literature review normally accomplishes this goal (of justification) by addressing some of the following:
   - What have others said about this topic and related topics?
   - What research, if any, has been done previously on the topic?
   - Have other researchers used techniques that can be adapted for the purposes of the proposed study?
   - References used
   - Statement of research question or hypothesis

IV. **Methods** In a methods section, you should answer questions such as:
- Whom or what will you study to collect data?
- How will you select your sample?
- What, if any, ethical considerations are relevant?
- What method(s) of data collection will you use—a questionnaire, an interview, an observation, and/or an available data?

You might also, depending on the nature of the study (e.g., whether it is quantitative or qualitative), want to answer questions such as
- What are the key variables in your study?
- How will you define and measure them?
- Will you be borrowing someone else's measures or using a modified form of measures that have been used before?
- What kind of data analysis, or comparisons, do you intend to do, or make, with the data you collect?

V. **Plan for Sharing Your Findings** In this section, you will want to answer questions like these:
- Will you write up your results in the form of a paper (or book) to be shared with others?
- What kinds of reporting outlets might you use: a journal or magazine, a conference, or some other audience? (If you might prepare a book manuscript, what publisher might be interested in the material?)
- With whom will you share your findings?

VI. **Budget**
- What are the major foreseeable costs of your project?
- Sources of funding or contributions?

VII. **References** List of the references (books, articles, websites) already cited in the proposal, using some standard style format appropriate for your discipline (e.g., American Sociological Association style for sociology).

Random Number Table

3250	2881	2326	9108	3702	6844	9827	3210	4895
7961	7952	3660	1503	4644	7656	3856	1672	0695
7739	5850	5998	9821	9476	2165	7662	5742	5048
0417	0262	7442	0873	6101	8592	2862	5346	7284
5816	4578	4061	7262	6328	1355	9386	5446	0666
7843	1922	3038	8442	5367	5952	6782	4206	4630
4912	7287	8526	9602	4545	3755	4159	3288	2900
2105	4791	5097	7386	6040	3575	4143	4820	5547
3263	3102	3555	4204	9411	0088	4873	1407	0858
4619	0105	9946	3234	1265	9509	0227	6671	3992
4460	7800	4476	3613	1252	8579	0611	6919	4208
6041	6103	3043	4634	1191	7886	4579	6301	4488
6718	8181	1425	6396	5312	4700	8077	1604	0870
8253	0766	2292	0146	6302	4413	6170	9764	3377
2061	5261	7862	8368	6533	9481	4098	1313	8527
4712	2096	6000	4173	8920	6913	3092	2028	2678
3344	6985	0656	2416	8367	5673	3272	8878	8714
5965	7229	0064	8356	7767	0960	5365	4980	9568
1203	8216	0354	1005	8063	3853	6732	0012	5391

(Continued)

6619	7927	0067	5559	2929	8706	6366	1111	9997
5545	8811	0833	1259	1344	0579	2735	9419	8533
8446	2198	8304	0803	5947	6322	9623	4419	8007
6017	2981	0880	6879	0193	6721	9130	2341	4528
7582	0139	5100	9315	5571	4508	1341	3450	1307
4118	7411	9556	9907	0612	0601	9545	8084	7117
8443	2486	1121	6714	8133	9113	4613	2633	6828
0231	2253	0931	9757	3471	3080	6369	6270	6296
8393	2801	8322	2871	6504	4972	1608	4997	2211
6092	5922	5587	8472	4256	2888	1344	4681	0332
8728	4779	8851	7466	9308	3590	3115	0478	6956
8585	4266	8533	9121	1988	4965	1951	4395	3267
9593	6571	7914	1242	8474	1856	0098	2973	2863
9356	6955	4768	1561	6065	3683	6459	4200	2177
0542	9565	7330	0903	8971	5815	2763	3714	0247
6468	1257	2423	0973	3172	8380	2826	0880	4270
1201	1959	9569	0073	8362	6587	7942	8557	4586
3616	5075	8836	1704	9143	6974	5971	0154	5509
4241	9365	6648	1050	2605	0031	3136	5878	9501
1679	5768	2801	5266	9165	5101	7455	9522	4199
5731	9611	0523	7579	0136	5655	6933	7668	8344
1554	7718	9450	9736	4703	9408	4062	7203	8950
1471	3213	7164	3788	4167	4063	7881	5348	4486
9563	8925	5479	0340	6959	3054	9435	6400	7989
3722	8984	2755	2560	1753	1209	2805	2763	8189
4237	1818	5266	2276	1877	4423	7318	1483	3153
2859	9216	5303	2154	3243	7359	1393	9909	2482
8303	7527	4566	8713	0538	7340	6224	3135	8367

4150	4056	1579	9286	4353	3413	6618	9524	9284
1507	1745	7221	9009	4769	9333	6803	0198	7879
1365	3544	3753	1530	7384	9357	2231	6166	3217
7414	5143	8912	6611	3396	7969	6559	9463	9114
9658	7935	7850	4297	4908	9653	9540	8753	3920
3169	0990	6016	0631	8451	2488	8724	3625	1805
7294	6007	0887	9667	5211	0082	2522	7055	4431
0514	8862	2713	9944	7880	5101	7535	1200	2583
7016	0451	2039	5827	2820	4979	1997	0920	0988
0389	0734	3322	0822	7189	7229	1031	2444	5827
4903	9574	0313	6611	6589	4437	8070	4480	6481
3746	6799	3082	6987	3766	8230	4376	7150	6963
7695	8918	3141	8417	7200	6066	4097	8883	5254
8423	2024	8911	9430	1711	6618	6768	0251	8926
5009	0421	9412	0227	4881	2652	8863	2900	1878
3265	7124	8628	0386	7584	3412	9002	7013	2898
2263	6500	7451	5922	8289	3552	5546	7389	1959
4629	5201	3214	6479	0313	4100	0596	9274	0415
6833	0952	8529	2108	2219	8249	2184	6631	7768
1991	4762	7349	3282	6394	3786	5792	0337	7781
9628	1897	4009	1946	2765	3552	9865	0398	4612
5860	8540	7938	2799	6545	0922	9124	4352	2057
3394	0297	8151	6445	6994	4566	0323	6185	3117
6976	4713	7478	9092	7028	8140	2490	9508	9481
5359	7743	8669	8469	5126	1434	7695	8129	3184
7127	6156	8500	7593	0670	9534	3945	9718	0834
8690	6983	6359	3205	6167	4362	5340	4104	3004
5705	2941	2505	3360	5976	2070	8450	6761	7404

(Continued)

7210	4415	4744	2061	5102	7796	2714	1876	3398
0777	2055	8932	0542	1427	8487	3761	3793	0954
0270	9605	7099	9063	2584	5289	0040	2135	6934
7194	7521	3770	6017	4393	3340	8210	6468	6144
6029	0732	9672	6507	1422	3483	2387	2472	6951
6631	3518	5642	2250	3187	8553	6747	6029	1871
6958	4528	6053	9004	6039	5841	8685	9100	4097
1589	6697	9722	9343	3227	3063	6808	2900	6970
2004	8823	1038	1638	9430	7533	8538	6167	1593
2615	5006	1524	4747	4902	7975	9782	0937	5157
6487	6682	2964	0882	2028	2948	6354	3894	2360
7238	7432	4922	9697	7764	7714	4511	3252	6864
0382	2511	7244	7271	4804	9564	8572	6952	6155
8918	3452	3044	4414	1695	2297	6360	2520	5814
9331	0179	6815	5701	0711	5408	0322	8085	0080
1065	9702	9944	9575	6714	9382	8324	3008	9373
6663	5272	3139	4020	5158	7146	4100	3578	0683
7045	5075	6195	5021	1443	2881	9995	8579	7923
6483	7316	5893	9422	0550	3273	5488	2061	1532
4396	9028	4536	8606	4508	3752	9183	5669	5927
5766	7632	9132	7778	6887	8203	6678	8792	3159
2743	8858	5094	2285	2824	9996	7384	4069	5442
3330	8618	1974	2441	0457	6561	7390	4102	2311
4791	2725	4085	3190	9098	2300	6241	7903	9150
6833	9145	2459	2393	7831	2279	9658	5115	6906
9474	9070	4964	2035	7954	7001	8688	7040	6924
8576	0380	5722	6149	4086	4437	9871	0850	3248

GLOSSARY

access the ability to obtain the information needed to answer a research question.

account a plausible and appealing explanation of the research that the researcher gives to prospective participants.

accretion measures indicators of a population's activities created by its deposits of materials.

anonymity when no one, including the researcher, knows the identities of research participants.

antecedent variable a variable that comes before both an independent variable and a dependent variable.

anticipatory data reduction decision making about what kinds of data are desirable that is done in advance of collecting data.

applied research research intended to be useful in the immediate future and to suggest action or increase effectiveness in some area.

authorities socially defined sources of knowledge.

available data data that are easily accessible to the researcher.

basic research research designed to add to our fundamental understanding and knowledge of the social world regardless of practical or immediate implications.

biased samples samples that are unrepresentative of the population from which they've been drawn.

bivariate analyses data analyses that focus on the association between two variables.

case study a research strategy that focuses on one case (an individual, a group, an organization, and so on) within its social context at one point in time, even if that one time spans months or years.

causal hypothesis a testable expectation about an independent variable's effect on a dependent variable.

causal relationship a nonspurious relationship between an independent and a dependent variable with the independent variable occurring before the dependent variable.

closed-ended questions questions that include a list of predetermined answers.

cluster sampling a probability sampling procedure that involves randomly selecting clusters of elements from a population and subsequently selecting every element in each selected cluster for inclusion in the sample.

coding assigning observations to categories. In quantitative analyses, this frequently means making data computer usable. In qualitative analyses, this means associating words or labels with passages and then collecting similarly labeled passages into files.

cohort a group of people born within a given time frame or experiencing a life event, such as marriage or graduation from high school, in the same time period.

cohort study a study that follows a cohort over time.

complete observer role being an observer of a situation without becoming part of it.

complete participant role being, or pretending to be, a genuine participant in a situation one observes.

composite measures measures with more than one indicator.

concepts words or signs that refer to phenomena that share common characteristics.

conceptual definition a definition of a concept through other concepts. Also called a theoretical definition.

conceptualization the process of clarifying what we mean by a concept.

concurrent criterion validity how well a measure is associated with behaviors it should be associated with at the present time.

confidentiality when no third party knows the identities of the research participants. Also called privacy.

construct validity how well a measure of a concept is associated with a measure of another concept that some theory says the first concept should be associated with.

content analysis a method of data collection in which some form of communication is studied systematically.

content validity a test for validity that involves the judgment of experts in a field.

contingency questions questions that depend on the answers to previous questions.

control variable a variable that is held constant to examine the relationship between two other variables.

controlled (or systematic) observations observations that involve clear decisions about what is to be observed.

convenience sample a group of elements that are readily accessible to the researcher.

cost-benefit analysis research that compares a program's costs to its benefits.

cost-effectiveness analysis comparisons of program costs in delivering desired benefits based on the assumption that the outcome is desirable.

cover letter the letter accompanying a questionnaire that explains the research and invites participation.

coverage error an error that results from differences between the sampling frame and the target population.

cross-sectional study a study design in which data are collected for all the variables of interest using one sample at one time.

crosstabulation the process of making a bivariate table to examine a relationship between two variables.

data display a visual image that summarizes information.

data reduction the various ways in which a researcher orders collected and transcribed data.

deductive process a process that moves from more general to less general statements.

demand characteristics characteristics that the observed take on simply as a result of being observed.

dependent variable a variable that is seen as being affected or influenced by another variable.

descriptive statistics statistics used to describe and interpret sample data.

descriptive study research designed to describe groups, activities, situations, or events.

dimensions aspects or parts of a larger concept.

double-blind experiment an experiment in which neither the subjects nor the research staff who interact with them knows the memberships of the experimental or control groups.

ecological fallacy the fallacy of making inferences about certain types of individuals from information about groups that might not be exclusively composed of those individuals.

elaboration the process of examining the relationship between two variables by introducing the control for another variable or variables.

element a kind of thing a researcher wants to sample. Also called a sampling unit.

empirical generalizations statements that summarize a set of individual observations.

erosion measures indicators of a population's activities created by its selective wear on its physical environment.

ethical principles in research the set of values, standards, and principles used to determine appropriate and acceptable conduct at all stages of the research process.

evaluation research research specifically designed to assess the impacts of programs, policies, or legal changes.

evidence-based social programs and policies those that present persuasive evidence of successful achievement based on careful monitoring of outcomes.

exhaustive the capacity of a variable's categories to permit the classification of every unit of analysis.

existing statistics summaries of data collected by large organizations.

experimental design a study design in which the independent variable is controlled, manipulated, or introduced in some way by the researcher.

experimenter expectations when expected behaviors or outcomes are communicated to subjects by the researcher.

explanation a kind of elaboration in which the original relationship is explained away as spurious by a control for an antecedent variable.

explanatory research research designed to explain the cause of a phenomenon and that typically asks "What causes what?" or "Why is it this way?"

exploratory research groundbreaking research on a relatively unstudied topic or in a new area.

extraneous variable a variable that has an effect on the dependent variable in addition to the effect of the independent variable.

face validity a test for validity that involves the judgment of everyday people, like you and me.

feasibility whether it is practical to complete a study in terms of access, time, and money.

field experiment an experiment done in the "real world" of classrooms, offices, factories, homes, playgrounds, and the like.

focus group interview a type of group interview where participants converse with one another and have minimal interaction with a moderator.

formative analysis evaluation research focused on the design or early implementation stages of a program or policy.

frequency distribution a way of showing the number of times each category of a variable occurs in a sample.

gatekeeper someone who can get a researcher into a setting or facilitate access to participants.

generalizability the ability to apply the results of a study to groups or situations beyond those actually studied.

grounded theory theory derived from data in the course of a particular study.

group-administered questionnaire questionnaire administered to respondents in a group setting.

group interview a data collection method with one interviewer and two or more interviewees.

history the effects of general historical events on study participants.

honest reporting the ethical responsibility to produce and report accurate data.

hypothesis a testable statement about how two or more variables are expected to relate to one another.

in-person interview an interview conducted face-to-face.

independent variable a variable that is seen as affecting or influencing another variable.

index a composite measure that is constructed by adding scores from several indicators.

indicators observations that we think reflect the presence or absence of the phenomenon to which a concept refers.

individually administered questionnaires questionnaires that are hand delivered to a respondent and picked up after completion.

inductive reasoning reasoning that moves from less general to more general statements.

inferential statistics statistics used to make inferences about the population from which the sample was drawn.

informants participants in a study situation who are interviewed for an in-depth understanding of the situation.

informed consent the principle that potential participants are given adequate and accurate information about a study before they are asked to agree to participate.

informed consent form a statement that describes the study and the researcher and formally requests participation.

institutional review board (IRB) the committee at a college, university, or research center responsible for evaluating the ethics of proposed research.

internal consistency method a method that relies on making more than one measure of a phenomenon at essentially the same time.

internal validity agreement between a study's conclusions about causal connections and what is actually true.

interobserver (or interrater) reliability method a way of checking the reliability of a measurement strategy by comparing results obtained by one observer with results obtained by another using exactly the same method.

interpretation a kind of elaboration that provides an idea of the reasons why an original relationship exists without challenging the belief that the original relationship is causal.

intersubjectivity agreements about reality that result from comparing the observations of more than one observer.

interval level variables variables whose categories have names, whose categories can be rank-ordered in some sensible way, and whose adjacent categories are a standard distance from one another.

intervening variable a variable that comes between an independent and a dependent variable.

interview a data collection method in which respondents answer questions asked by an interviewer.

interview guide the list of topics to cover and the order in which to cover them that can be used to guide less structured interviews.

interview schedule the list of questions and answer categories read to a respondent in a structured or semi-structured interview.

interviewer effect the change in a respondent's behavior or answers that is the result of being interviewed by a specific interviewer.

keywords the terms used to search for sources in a literature review.

laboratory research research done in settings that allows the researcher control over the conditions, such as in a university or medical setting.

life story interview a short account of an interviewee's life created in collaboration with an interviewer.

literature review the process of searching for, reading, summarizing, and synthesizing existing work on a topic or the resulting written summary of a search.

longitudinal research a research design in which data are collected at least two different times, such as a panel, trend, or cohort study.

mailed questionnaire questionnaire mailed to the respondent's residence or workplace.

margin of error a suggestion of how far away the actual population parameter is likely to be from the statistic.

matching assigning sample members to groups by matching the sample members on one or more characteristics and then separating the pairs into two groups with one group randomly selected to become the experimental group.

maturation the biological and psychological processes that cause people to change over time.

mean the measure of central tendency designed for interval level variables. The sum of all values divided by the number of values.

measure a specific way of sorting units of analysis into categories.

measurement the process of devising strategies for classifying subjects by categories to represent variable concepts.

measurement error the kind of error that occurs when the measurement we obtain is not an accurate portrayal of what we tried to measure.

measures of association measures that give a sense of the strength of a relationship between two variables.

measures of correlation measures that provide a sense not only of the strength of the relationship between two variables but also of its direction.

measures of dispersion measures that provide a sense of how spread out cases are over categories of a variable.

median the measure of central tendency designed for ordinal level variables. The middle value when all values are arranged in order.

memos more or less extended notes that the researcher writes to help herself or himself understand the meaning of codes.

mode the measure of central tendency designed for nominal level variables. The value that occurs most frequently.

multidimensionality the degree to which a concept has more than one discernible aspect.

multistage sampling a probability sampling procedure that involves several stages, such as randomly selecting clusters from a population, then randomly selecting elements from each of the clusters.

multivariate analyses analyses that permit researchers to examine the relationship between variables while investigating the role of other variables.

mutually exclusive the capacity of a variable's categories to permit the classification of each unit of analysis into one and only one category.

natural experiment a study using real-world phenomena that approximates an experimental design even though the independent variable is not controlled, manipulated, or introduced by the researcher.

needs assessment an analysis of whether a problem exists, its severity, and an estimate of essential services.

nominal level variables variables whose categories have names.

nonparticipant observation an observation made by an observer who remains as aloof as possible from those observed.

nonprobability samples samples that have been drawn in a way that doesn't give every member of the population a known chance of being selected.

nonresponse error an error that results from differences between nonresponders and responders to a survey.

normal distribution a distribution that is symmetrical and bell-shaped.

objectivity the ability to see the world as it really is.

observational techniques methods of collecting data by observing people, most typically in their natural settings.

observer-as-participant role being primarily a self-professed observer, while occasionally participating in the situation.

open-ended questions questions that allow respondents to answer in their own words.

operationalization the process of specifying what particular indicator(s) one will use for a variable.

ordinal level variables variables whose categories have names and whose categories can be rank-ordered in some sensible way.

outcome evaluation research that is designed to "sum up" the effects of a program, policy, or law in accomplishing the goal or intent of the program, policy, or law.

panel attrition the loss of subjects from a study because of disinterest, death, illness, or inability to locate them.

panel conditioning the effect of repeatedly measuring variables on members of a panel study.

panel study a study design in which data are collected about one sample at least two times where the independent variable is not controlled by the researcher.

participatory action research (PAR) research done by community members and researchers working as co-participants, most typically within a social justice framework, to empower people and improve their lives.

parameter a summary of a variable characteristic in a population.

partial relationship the relationship between an independent and a dependent variable for that part of a sample defined by one category of a control variable.

participant-as-observer role being primarily a participant, while admitting an observer status.

participant observation observation performed by observers who take part in the activities they observe.

passive consent when no response is considered an affirmative consent to participate in research; also called "opt out informed consent," this is sometimes used for parental consent for children's participation in school-based research.

personal inquiry inquiry that employs the senses' evidence.

personal records records of private lives, such as biographies, letters, diaries, and essays.

photo elicitation a technique used to encourage dialogue between interviewer and interviewee.

photo-interviewing a data collection technique using photographs to elicit information and encourage discussion usually in conjunction with qualitative interviewing.

physical traces physical evidence left by humans in the course of their everyday lives.

pilot test a preliminary draft of a set of questions that is tested before the actual data collection.

placebo a simulated treatment of the control group that is designed to appear authentic.

population the group of elements from which a researcher samples and to which she or he might like to generalize.

positivist view of science a view that human knowledge must be based on what can be perceived.

post-positivist view of science a view that knowledge is not based on irrefutable observable grounds, that it is always somewhat speculative, but that science can provide relatively solid grounds for that speculation.

posttest the measurement of the dependent variable that occurs after the introduction of the stimulus or the independent variable.

posttest-only control group experiment an experimental design with no pretest.

predictive criterion validity a method that involves establishing how well a measure predicts future behaviors you'd expect it to be associated with.

pretest the measurement of the dependent variable that occurs before the introduction of the stimulus or the independent variable.

pretest–posttest control group experiment an experimental design with two or more randomly selected groups (an experimental and a control group) in which the researcher controls or "introduces" the independent variable and measures the dependent variable at least two times (pretest and posttest measurement).

primary data data that the same researcher collects and uses.

probability samples samples drawn in such a way as to give every member of the population a known (nonzero) chance of inclusion.

process evaluation research that monitors a program or policy to determine if it has been implemented as designed and if modifications, including alternative procedures, would be useful.

protecting study participants from harm the principle that participants in studies are not harmed physically, psychologically, emotionally, legally, socially, or financially as a result of their participation in a study.

purposive sampling a nonprobability sampling procedure that involves selecting elements based on the researcher's judgment about which elements will facilitate his or her investigation.

qualitative content analysis content analysis designed for verbal analysis.

qualitative data analysis analysis that results in the interpretation of action or representation of meanings in the researcher's own words.

qualitative interview a data collection method in which an interviewer adapts and modifies the interview for each interviewee.

qualitative research research focused on the interpretation of the action of, or representation of meaning created by, individual cases.

quantitative content analysis content analysis focused on the variable characteristics of communication.

quantitative data analysis analysis based on the statistical summary of data.

quantitative research research focused on variables, including their description and relationships.

quasi-experiment an experimental design that is missing one or more aspects of a true experiment, most frequently random assignment into experimental and control groups.

questionnaire a data collection instrument with questions and statements that are designed to solicit information from respondents.

quota sampling a nonprobability sampling procedure that involves describing the target population in terms of what are thought to be relevant criteria and then selecting sample elements to represent the "relevant" subgroups in proportion to their presence in the target population.

random assignment a technique for assigning members of the sample to experimental and control groups by chance to maximize the likelihood that the groups are similar at the beginning of the experiment.

random-digit dialing a method for selecting participants in a telephone survey that involves randomly generating telephone numbers.

range a measure of dispersion or spread designed for interval level variables. The difference between the highest and the lowest values.

rapport a sense of interpersonal harmony, connection, or compatibility between an interviewer and a respondent.

ratio level variables variables whose categories have names, whose categories may be rank-ordered in some sensible way, whose adjacent categories are a standard distance from one another, and one of whose categories is an absolute zero point—a point at which there is a complete absence of the phenomenon in question.

reliability the degree to which a measure yields consistent results.

replication a kind of elaboration in which the original relationship is replicated by all the partial relationships.

requirements for supporting causality the requirements needed to support a causal relationship include a pattern or relationship between the independent and dependent variables, determination that the independent variable occurs first, and support for the conclusion that the apparent relationship is not caused by the effect of one or more third variables.

research costs all monetary expenditures needed for planning, executing, and reporting research.

research question a question about one or more topics or concepts that can be answered through research.

research topic a concept, subject, or issue that can be studied through research.

researchable question a question that is feasible to answer through research.

respondents the participants in a survey who complete a questionnaire or interview.

response rate the percentage of the sample contacted that actually participates in a study.

sample a number of individual cases drawn from a larger population.

sampling the process of drawing a number of individual cases from a larger population.

sampling distribution the distribution of a sample statistic (such as the average) computed from many samples.

sampling error any difference between the characteristics of a sample and the characteristics of the population from which the sample is drawn.

sampling frame or study population the group of elements from which a sample is actually selected.

sampling variability the variability in sample statistics that can occur when different samples are drawn from the same population.

scale an index in which some items are given more weight than others in determining the final measure of a concept.

scientific method a way of conducting empirical research following rules that specify objectivity, logic, and communication among a community of knowledge seekers and the connection between research and theory.

screening question a question that asks for information before asking the question of interest.

secondary data research data that have been collected by someone else.

selection bias a bias in the way the experimental and control or comparison groups are selected that is responsible for preexisting differences between the groups.

self-administered questionnaires questionnaires that the respondent completes by himself or herself.

self-report method another name for questionnaires and interviews reflecting the fact that respondents are most often asked to report their own characteristics, behaviors, and attitudes.

semi-structured interviews interviews with an interview guide containing primarily open-ended questions that can be modified for each interview.

simple random sample a probability sample in which every member of a study population has been given an equal chance of selection.

skewed variable an interval level variable that has one or a few cases that fall into extreme categories.

snowball sampling a nonprobability sampling procedure that involves using members of the group of interest to identify other members of the group.

Solomon four-group design a controlled experiment with an additional experimental and control group, with each receiving a posttest only.

specification a kind of elaboration that permits the researcher to specify conditions under which the original relationship is particularly strong or weak.

spurious non-causal.

spurious relationship a non-causal relationship between two variables.

stakeholders people or groups that participate in or are affected by a program or its evaluation, such as funding agencies, policy makers, sponsors, program staff, and program participants.

standard deviation a measure of dispersion designed for interval level variables that accounts for every value's distance from the sample mean.

statistic a summary of a variable in a sample.

stimulus the experimental condition of the independent variable that is controlled or "introduced" by the researcher in an experiment.

stratified random sampling a probability sampling procedure that involves dividing the population into groups or strata defined by the presence of certain characteristics and then random sampling from each stratum.

structured interview a data collection method in which an interviewer reads a standardized list of questions to the respondent and records the respondent's answers.

study design a research strategy specifying the number of cases to be studied, the number of times data will be collected, the number of samples that will be used, and whether the researcher will try to control or manipulate the independent variable in some way.

survey a study in which the same data, usually in the form of answers to questions, are collected from one or more samples.

systematic sampling a probability sampling procedure that involves selecting every kth element from a list of population elements, after the first element has been randomly selected.

target population the population of theoretical interest.

telephone interview an interview conducted over the telephone.

test–retest method a method of checking the reliability of a test that involves comparing its results at one time with results, using the same subjects, at a later time.

testing effect the sensitizing effect on subjects of the pretest.

theoretical saturation the point where new interviewees or settings look a lot like interviewees or settings one has observed before.

theory an explanation about how and why something is as it is.

thick description reports about behavior that provide a sense of things like the intentions, motives, and meanings behind the behavior.

thin description bare-bones description of acts.

time expenditures the time it takes to complete all activities of a research project from the planning stage to the final report.

trend study a study design in which data are collected at least two times with a new sample selected from a population each time.

units of analysis the units about which information is collected, and the kind of thing a researcher wants to analyze. The term is used at the analysis stage, whereas the terms *element* or *sampling unit* can be used at the sampling stage.

units of observation the units from which information is collected.

univariate analyses analyses that tell us something about one variable.

unobtrusive measures indicators of interactions, events, or behaviors whose creation does not affect the actual data collected.

unstructured interview a data collection method in which the interviewer starts with only a general sense of the topics to be discussed and creates questions as the interaction proceeds.

validity the degree to which a measure taps what we think it's measuring.

variable a characteristic that can vary from one unit of analysis to another or for one unit of analysis over time.

video ethnography the video recording of participants and the reviewing of the resulting footage for insights into social life.

vignettes scenarios about people or situations that the researcher creates to use as part of the data collection method.

visual analysis a set of techniques used to analyze images.

visual sociology an approach to studying society and culture that employs images as a data source.

voluntary participation the principle that study participants choose to participate of their own free will.

web or Internet survey survey that is sent by e-mail or posted on a website.

REFERENCES

AAAS Science and Human Rights Coalition. 2012. Meeting Report: January 23, 2012. Retrieved on March 20, 2013 (http://srhrl.aaas.org/coalition/Meetings/2012/January/Report.pdf).

Adler, E. S. 1981. The underside of married life: Power, influence and violence. In *Women and crime in America*, 300–319, edited by L. H. Bowker. New York: Macmillan.

Adler, E. S., and J. S. Lemons. 1990. *The elect: Rhode Island's women legislators, 1922–1990.* Providence: League of Rhode Island Historical Societies.

Adler, P. A. 1985. *Wheeling and dealing.* New York: Columbia University Press.

Adler, P. A., and P. Adler. 1987. *Membership roles in field research.* Newbury Park, CA: Sage.

Adler, P. A., and P. Adler. 1994a. Observational techniques. In *Handbook of qualitative research*, 377–392, edited by N. K. Denzin and Y. S. Lincoln. Thousand Oaks, CA: Sage.

Adler, P. A., and P. Adler. 1994b. Social reproduction and the corporate other: The institutionalization of afterschool activities. *Sociological Quarterly* 35: 309–328.

Aitken, C., R. Power, and R. Dwyer. 2008. A very low response rate in an on-line survey of medical practitioners. *Australian & New Zealand Journal of Public Health* 32: 288–289.

Albig, W. 1938. The content of radio programs—1925–1935. *Social Forces* 16: 338–349.

Allen, K. R., and D. H. Demo. 1995. The families of lesbians and gay men: A new frontier in family research. *Journal of Marriage and the Family* 57: 111–128.

Allen, S. 2006, January 6. Harvard to study Katrina's long-term psychological toll. *Boston Globe* 269(6): A4.

Altman, L. E. 2005. Nobel Prize for discovering bacterium causes ulcers. *New York Times*, October 4, D3.

American National Election Studies. 2013. Our Studies. Retrieved on February 10, 2013 (http://www.electionstudies.org/OurStudies/OurStudies.htm).

American Sociological Association (ASA). 1999. Code of ethics and policies and procedures of the ASA committee on professional ethics. Washington, DC: ASA. Retrieved on September 16, 2009 (http://www.asanet.org/galleries/default-file/Code%20of%20Ethics.pdf).

American Sociological Association. 2013. Mission. Retrieved on March 23, 2013 (http://www.asanet.org/about/mission.cfm).

Anden-Papadopoulos, K. 2009. US soldiers imaging the Iraq War on YouTube. *Popular Communication* 7(1): 17–27.

Anderson, D. J. 2012. Boston College can't keep IRA materials confidential. Boston Globe, July 7, 2012 Retrieved on January 12, 2013 (http://www.bostonglobe.com/metro/2012/07/06/appeals-court-upholds-subpoenas-ruling-belfast-oral-history-project-case/LaGEiFxneQgp2OPC5KJkDP/story.html).

Anderson, E. 1999. *Code of the street.* New York: Norton.

AoIR Ethics Working Committee, 2012. Ethical Decision-Making and Internet Research: Version 2.0. Retrieved on November 20, 2012 (http://aoirethics.ijire.net/aoirethicsprintablecopy.pdf).

Applegate, B. K., and J. B. Sanborn. 2011. Public Opinion on the Harshness of Local Courts: An Experimental Test of Question Wording Effects. *Criminal Justice Review* 36: 487–497.

Aquilino, W. S. 1993. Effects of spouse presence during the interview on survey responses concerning marriage. *Public Opinion Quarterly* 57: 358–376.

Armstrong E., P. England, and A. C. K. Fogarty. 2012. Sexual Practices, Learning, and Love: Accounting for Women's Orgasm and Sexual Enjoyment in College Hookups and Relationships. *American Sociological Review* 77: 435–462.

Arthur, C., and R. Clark. 2009. Determinants of domestic violence: A cross-national study. *International Journal of Sociology of the Family* 35.

Arthur, M. M. L. 2007. Getting away from the basics: Competing explanations of curricular change. Doctoral dissertation for New York University.

Asher, R. 1992. *Polling and the public.* Washington, DC: Congressional Quarterly.

Atchley, R. C. 2003. *Social forces and aging*, 10th edition. Belmont, CA: Wadsworth.

Atkinson, R. 2001. The life story interview. In *Handbook of interview research*, 121–140, edited by J. F. Gubrium and J. A. Holstein. Thousand Oaks, CA: Sage Publications.

Attwood, F. 2009. Intimate adventures: Sex blogs, sex "blooks" and women's sexual narration. *European Journal of Cultural Studies* 12(1): 5–20.

Auerbach, J. 2000. Feminism and federally funded social science: Notes from inside. *Annals of the American Academy of Political and Social Science* 571: 30–41.

467

Babbie, E. 2012. *The practice of social research*, 13th edition. Belmont, CA: Wadsworth.

Bakalar, N. 2005. Ugly children may get parental short shrift. *New York Times*, May 3. Retrieved on September 16, 2009 (http://www.nytimes.com/2005/05/03/health/03ugly.html).

Bales, K. 1999. Popular reactions to sociological research: The case of Charles Booth. *Sociology* 33: 153–168.

Bales, R. F. 1950. *Interaction process analysis*. Reading, MA: Addison-Wesley.

Barr, A., and R. Simons. 2012. Marriage expectation among African American couples in early adulthood: A dyadic analysis. *Journal of Marriage and Family* 74(4): 726–744.

Barry, C. 1998. Choosing qualitative data analysis software: Atlas/ti and Nudist compared. *Sociological Research Online* 3. Retrieved on June 20, 2001 (www.socresonline.org.uk/socresonline/3/3/4.html).

Barry, E. 2002. After day care controversy, psychologists play nice. *Boston Globe*, September 3, E1.

Baruch, Y. 1999. Response rate in academic studies—a comparative analysis. *Human Relations* 52: 421–438.

Baruch, Y., and C. Holtom. 2008. Survey response rate levels and trends in organizational research. *Human Relations* 61: 1139–1160.

Bates, B., and M. Harmon. 1993. Do "instant polls" hit the spot? Phone-in versus random sampling of public opinion. *Journalism Quarterly* 70: 369–380.

Bates, N., J. Dahlhamer, and E. Singer. 2008. Privacy Concerns, Too Busy, or Just Not Interested: Using Doorstep Concerns to Predict Survey Nonresponse. *Journal of Official Statistics* 24: 591–612.

Baumrind, D. 1964. Some thoughts on ethics of research: After reading "Milgram's behavioral study of obedience." *American Psychologist* 19: 421–423.

Becker, B. 2011. Social disparities in children's vocabulary in early childhood. Does pre-school education help to close the gap? *British Journal of Sociology* 62: 69–88.

Becker, H. 2004. Comment on Kevin D. Haggerty, "Ethics Creep: Governing Social Science Research in the Name of Ethics." *Qualitative Sociology* 27: 415–416.

Becker, H. S. 1974. Photography and sociology. *Studies in the Anthropology of Visual Communication* 1: 3–26.

Becker, H. S., B. Geer, E. C. Hughes, and A. L. Strauss. 1961. *Boys in white: Student culture in medical school*. Chicago: University of Chicago Press.

Bell, P. 2001. Content analysis of visual images. In *Handbook of visual analysis*, 10–35, edited by T. van Leeuwan and C. Jewitt. Thousand Oaks, CA: Sage.

Berg, B. L. 1989. *Qualitative research methods of the social sciences*. Boston: Allyn & Bacon.

Berk, R. A., G. K. Smyth, and L. W. Sherman. 1988. When random assignment fails: Some lessons from the Minneapolis spouse abuse experiment. *Journal of Quantitative Criminology* 4: 209–223.

Bickman, L., and D. J. Rog. 2009. *The SAGE handbook of applied social research*, 2nd edition. Thousand Oaks, CA: SAGE Publications.

Birkeland, S., E. Murphy-Graham, and C. H. Weiss. 2005. Good reasons for ignoring good evaluation: The case of the Drug Abuse Resistance Education (DARE) program. *Evaluation and Program Planning* 28: 247–256.

Bishop, G. F. 1987. Experiments with the middle response alternative in survey questions. *Public Opinion Quarterly* 51: 220–232.

Blee, K., and A. Currier. 2011. Ethics Beyond the IRB: An introductory essay. *Qualitative Sociology* 34: 401–413.

Blumberg, S. J., and J. V. Luke. 2012. Wireless substitution: Early release of estimates from the National Health Interview Survey, July–December 2011. Center for Disease Control. Retrieved on March 23, 2013 (http://www.cdc.gov/nchs/data/nhis/earlyrelease/wireless201206.pdf).

Blumberg, S. J., J. V. Luke, G. Nadarajasundaram, M. E. Davern, and M. H. Boudreaux. 2012. Wireless Substitution: State-level Estimates from the National Health Survey, 2010–2011. Washington, DC: National Center for Health Statistics. http://www.cdc.gov/nchs/data/nhsr/nhsr061.pdf

Bock, J. 2000. Doing the right thing? Single mothers by choice and the struggle for legitimacy. *Gender & Society* 14: 62–86.

Bogenschneider, K., and L. Pallock. 2008. Responsiveness in parent–adolescent relationships: Are influences conditional? Does the reporter matter? *Journal of Marriage and Family* 70: 1015–1029.

Bohannon, J. 2012. "Social Voting" Really Does Rock the Vote. *Science NOW* (September 2012). Retrieved on February 22, 2013 (http://news.sciencemag.org/sciencenow/2012/09/social-voting-really-does-rock-t.html).

Bond, M. A. 1990. Defining the research relationship: Maximizing participation in an unequal world. In *Researching community psychology*, 183–185, edited by P. Tolan, C. Keys, F. Chertock, and L. Jason. Washington, DC: American Psychological Association.

Bond, R. M., C. J. Fariss, J. J. Jones, A. D. I. Kramer, C. Marlow, J. E. Settle, and J. H. Fowler. 2012. A 61-million-person experiment in social influence and political mobilization. *Nature* 489: 295–298.

Bornstein, D. 2012. The Dawn of the Evidence-Based Budget. *New York Times*. Retrieved on April 2, 2013 (http://opinionator.blogs.nytimes.com/2012/05/30/worthy-of-government-funding-prove-it/).

Boser, S. 2006. Ethics and power in community–campus partnerships for research. *Action Research* 4: 9–21.

Bostock, D., and J. Daley. 2007. Lifetime and current sexual assault and harassment victimization rates of active-duty United States air force women. *Violence Against Women* 13(9): 927–944.

Bourgoin, N. 1995. Suicide in prison: Some elements of a strategic analysis. *Cahiers Internationaux de Sociologie* 98: 59–105.

Bradburn, N. M., and S. Sudman. 1988. *Polls and surveys*. San Francisco: Jossey-Bass.

Brenner, P. S. 2011. Exceptional Behavior or Exceptional Identity? Overreporting of Church Attendance in the U.S. *Public Opinion Quarterly* 75: 19–41.

Brick, J. M., W. S. Edwards, and S. Lee. 2007. Sampling telephone numbers and adults, interview length, and weighting in the California Health Interview Survey Cell Phone Pilot Study. *Public Opinion Quarterly* 71: 793–813.

Brick, J. M., and D. Williams. 2013. Explaining Rising Nonresponse Rates in Cross-Sectional Surveys. *The ANNALS of the American Academy of Political and Social Science* 645: 36–59.

Brodsky, A. 2001. More than epistemology: Relationships in applied research with underserved communities. *Journal of Social Issues* 57(2): 323–335.

Brodsky, A. E., K. Rogers-Senuta, C. Leiss, C. M. Marx, C. Loomis, S. Arteaga, H. Moore, R. Benhorin, and A. Casteganera. 2004. When one plus one equals three: The role of relationships in community research. *American Journal of Community Psychology* 33: 229–241.

Browne, J. 1976. The used car game. In *The research experience*, 60–84, edited by M. P. Golden. Itasca, IL: F. E. Peacock.

Brydon-Miller, M., and P. Maguire. 2009. Participatory action research: Contributions to the development of practitioner inquiry in education. *Educational Action Research* 17: 79–93.

Budiansky, S. 1994. Blinded by the cold-war light. *U.S. News & World Report* 116: 6–7.

Burgard, S., and J. Ailshire. 2013. Gender and Time for Sleep among U.S. Adults. *American Sociological Review* 78(1): 51–69.

Burris, S., and K. Moss. 2006. U.S. health researchers review their ethics review boards: A qualitative study. *Journal of Empirical Research on Human Subjects Ethics* 1: 39–58.

Bushman, B. J., and C. A. Anderson. 2009. Comfortably numb: Desensitizing effects of violent media on helping others. *Psychological Science* 21: 273–277.

Cain, C. L. 2012. Emotions and the research interview: What hospice workers can teach us. *Health Sociology Review* 21: 396–405

Calvey, D. 2008. The art and politics of covert research: Doing "situated ethics" in the field. *Sociology* 42: 905–918.

Cammorata, N. 2009. Down at the half. *Boston Globe*, March 29, C2. Retrieved on May 26, 2009 (http://www.boston.com/bostonglobe/ideas/articles/2009/03/ 29/down_at_the_half/).

Campbell, D. T., and J. C. Stanley. 1963. *Experimental and quasi-experimental designs for research*. Chicago: Rand McNally.

Campbell, R., and A. E. Adams. 2009. Why do rape survivors volunteer for face-to-face interviews? A meta-study of victims' reasons for and concerns about research participation. *Journal of Interpersonal Violence* 24: 395–405.

CAPS. 2001. Working together: A guide to collaborative research in HIV prevention. San Francisco: University of California, Center for AIDS Prevention Studies. Retrieved on July 3, 2009 (http://www.caps.ucsf.edu/pubs/manuals/pdf/Working-Together.pdf).

Carroll-Lind, J., J. W. Chapman, J. Gregory, and G. Maxwell. 2006. The key to the gatekeppers: Passive consent an dother eithical issues surrounding the rights of children to speak on issues that concern them. *Child Abuse and Neglect* 30: 979–989.

Carroll-Lind, J., J. W. Chapman, and J. Raskauskas. 2011. Children's perceptions of violence: The nature, extent and impact of their experiences. *Social Policy Journal of New Zealand* 37: 6–18.

Carter, J. S., S. Carter, and J. Dodge. 2009. Trends in abortion attitudes by race and gender: A reassessment over a four-decade period. *Journal of Sociological Research* 1: 1–17.

Cartland, J., H. S. Ruch-Ross, M. Mason, and W. Donohue. 2008. Role sharing between evaluators and stakeholders in practice. *American Journal of Evaluation* 29: 460–477.

Center for Disease Control (CDC). 2013a. About the National Health Interview Survey. Retrieved on April 2, 2013 (http://www.cdc.gov/nchs/nhis/about_nhis.htm).

Center for Disease Control (CDC). 2013b. Behavioral Risk Factor Surveillance System 2011 Summary Data Quality Report (version 5). Washington, DC: Center for Disease Control. Retrieved on April 4, 2013 from (http://www.cdc.gov/brfss/pdf/2011_Summary_Data_Quality_Report.pdf).

Centre for Bhutan Studies (2012a). A short guide to GNH index 2012. Thimphu, Bhutan. Retrieved on March 23, 2013 from

(http://www.grossnationalhappiness.com/wp-content/uploads/2012/04/Short-GNH-Index-edited.pdf).

Centre for Bhutan Studies (2012b). Extensive Analysis of GHP Index. Retrieved on March 23, 2013 (http://www.grossnationalhappiness.com/wp-content/uploads/2012/10/An%20Extensive%20Analysis%20of%20GNH%20Index.pdf).

Chambers, D. E., K. R. Wedel, and M. K. Rodwell. 1992. *Evaluating social programs*. Boston: Allyn & Bacon.

Chang, L., and J. A. Krosnick. 2009. National Surveys Via RDD Telephone Interviewing Versus the Internet. *Public Opinion Quarterly* 73: 641–678.

Chang, L., and J. A. Krosnick. 2010. Comparing Oral Interviewing with Self-Administered Computerized Questionnaires: An Experiment. *Public Opinion Quarterly* 74: 154–167.

Charmaz, K. 1983. The grounded theory method: An explication and interpretation. In *Contemporary field research: A collection of readings*, 109–126, edited by R. M. Emerson. Boston: Little, Brown.

Chase, S. E. 1995. *Ambiguous empowerment: The work narratives of women school superintendents*. Amherst: University of Massachusetts Press.

Child Information Gateway. 2011. Online Resources for State Child Welfare Law and Policy. Retrieved on February 12, 2013 (http://www.childwelfare.gov/systemwide/laws_policies/statutes/resources.pdf).

Childs, J. H., and A. Landreth. 2006. Analyzing interviewer/respondent interactions while using a mobile computer-assisted personal interview device. *Field Methods* 18: 335–351.

Chito Childs, E. 2005. Looking behind the stereotypes of the "angry black woman": An exploration of black women's responses to interracial relationships. *Gender & Society* 19(4): 544–561.

Chiu, C. 2009. Contestation and conformity: Street and park skate-boarding in New York City public space. *Space and Culture* 12(1): 25–42.

Ciambrone, D. 2001. Illness and other assaults on self: The relative impact of HIV/AIDS on women's lives.

Sociology of Health and Illness 23(4): 517–540.

Clark, B. L. 2003. *Kiddie lit*. Baltimore: Johns Hopkins University Press.

Clark, M. C., and B. F. Sharf. 2007. The dark side of truth(s), ethical dilemmas in researching the personal. *Qualitative Inquiry* 13: 399–416.

Clark, R. 1982. Birth order and eminence: A study of elites in science, literature, sports, acting and business. *International Review of Modern Sociology* 12: 273–289.

Clark, R. 1999. Diversity in sociology: Problem or solution? *American Sociologist* 30: 22–43.

Clark, R. 2001. A curious incident: The disappearing class differences in academic achievement. *Middle School Journal* 32: 5–13.

Clark, R., and A. Lang. 2001. *Balancing yin and yang: Teaching and learning qualitative data analysis* within an undergraduate data analysis course. Presented at the Eastern Sociological Society Meetings, March, Philadelphia.

Clark, R., and A. Nunes. 2008. The face of society: Gender and race in introductory sociology books revisited. *Teaching Sociology* 36: 227–239.

Clark, R., and G. Rice. 1982. Family constellations and eminence: The birth orders of Nobel Prize winners. *Journal of Psychology* 110: 281–287.

Clark, R., and H. Kulkin. 1996. Toward a multicultural feminist perspective on fiction for young adults. *Youth and Society* 27(3): 291–312.

Clark, R., and K. Grandchamp. 2012. The methods we teach: A pitch for more emphasis on available data. *International Review of Modern Sociology* 38(1): 93–104.

Clark, R., and T. Ramsbey. 1986. Social interaction and literary creativity. *International Review of Modern Sociology* 16: 105–117.

Clark, R., J. Guilmain, P. K. Saucier, and J. Tavarez. 2003. Two steps forward, one step back: The presence of female characters and gender stereotyping in award-winning picture books between the 1930 and the 1960s. *Sex Roles* 49: 439–449.

Clark, R., J. McCoy, and P. Keller. 2008. Teach your children well: Reading lessons to and about black and white adolescent girls from black and white women authors.

International Review of Modern Sociology 34(2): 211–228.

Clark, R., P. J. Keller, A. Knights, J. A. Nabar, T. B. Ramsbey, and T. Ramsbey. 2006. Let me draw you a picture: Alternate and changing views of gender in award-winning picture books for children. Unpublished manuscript.

Clark, Roger, Rachel Lennon, and Leanna Morris. 1993. Of Caldecotts and Kings: Gendered Images in Recent American Children's Books by Black and non-Black Illustrators. *Gender & Society* 5: 227–245.

Cohen, P. 2007. As ethics panels expand grip, no field is off limits. *New York Times*, February 28. Retrieved on February 26, 2009 (http://www.nytimes.com/2007/02/28/arts/28board.html).

Collier, M. 2001. Approaches to analysis in visual anthropology. In *Handbook of visual analysis*, 35–60, edited by T. van Leeuwan and C. Jewitt. Thousand Oaks, CA: Sage.

Commission on the Measurement of Economic Performance and Social Progress. 2009.

Connolly, K., and A. Reid. 2007. Ethics review for qualitative inquiry. *Qualitative Inquiry* 13: 1013–1047.

Converse, P. D., E. W. Wolfe, X. Huang, and F. L. Oswald. 2008. Response rates for mixed-mode surveys using mail and e-mail/web. *American Journal of Evaluation* 29: 99–107.

Cook, T. 2006. Collaborative action research within developmental evaluation: Learning to see or the road to myopia? *Evaluation* 12: 418–436.

Cook, T. 2012. Where Participatory Approaches Meet Pragmatism in Funded (Health) Research: The Challenge of Finding Meaningful Spaces. *Forum: Qualitative Social Research* 13:1–22.

Couper, M. P., and S. E. Hansen. 2001. Computer-assisted interviewing. In *Handbook of interview research*, 557–575, edited by J. F. Gubrium and J. A. Holstein. Thousand Oaks, CA: Sage Publications.

Couper, M. P., and P. V. Miller. 2008. Web survey methods. *Public Opinion Quarterly* 72: 831–835.

Courser, M., and P. J. Lavrakas. 2012 Item-Nonresponse and the 10-Point Response Scale in Telephone

Surveys. *Survey Practice* 5. Retrieved on March 30, 2013 (http://surveypractice.org/index.php/SurveyPractice/article/view/28/html).

Cowan, G., and R. R. Campbell. 1994. Racism and sexism in interracial pornography: A content analysis. *Psychology of Women Quarterly* 1918: 323–338.

Coyne, L. 2010. Accessing children as research participants: Examining the role of gatekeepers. *Child: Care, Health and Development* 4: 452–454.

Crawford, S., M. Couper, and M. Lamias. 2001. Web surveys: Perceptions of burden. *Social Science Computer Review* 19: 146–162.

Croghan, R., C. Griffin, J. Hunter, and A. Phoenix. 2008. Young People's Constructions of Self: Notes on the Use and Analysis of the Photo-Elicitation Methods. *International Journal of Social Research Methodology* 11: 345–356,

Cronbach, L., and P. Meehl. 1955. Construct validity in psychological tests. *Psychological Bulletin* 52: 281–302.

Crow, G., R. Wiles, S. Heath, and V. Charles. 2006. Research methods and data quality: The implications of informed consent. *International Journal of Social Research Methodology* 9: 83–95.

Culver, L. 2004. The impact of new immigration patterns on the provision of policy services in Midwestern communities. *Journal of Criminal Justice* 32(4): 329–344.

Curtin, R., S. Presser, and E. Singer. 2000. The effects of response rate changes on the index of consumer sentiment. *Public Opinion Quarterly* 64(4): 413–428.

Curtin, R., S. Presser, and E. Singer. 2005. Changes in telephone survey nonresponse over the past quarter century. *Public Opinion Quarterly* 69: 87–98.

Davidson, J. O., and D. Layder. 1994. *Methods, sex and madness*. New York: Routledge.

Davis, J. A., and T. W. Smith. 1998. *General social surveys, 1972–1998*. Chicago: National Opinion Research Center.

Davis, J. A., T. W. Smith, and P. V. Marsden. 2007. *General social surveys, 1972–2006*. Chicago: National Opinion Research Center.

De Meyrick, J. 2005. Approval procedures and passive consent considerations in research among young people. *Health Education* 105(4): 249–258.

Denscombe, M. 2006. Web-based questionnaires and the mode effect: An evaluation based on completion rates and data contents of near-identical questionnaires delivered in different modes. *Social Science Computer Review* 24: 246–254.

Denzin, N. 1989. *The research act: A theoretical introduction to sociological methods.* Englewood Cliffs, NJ: Prentice-Hall.

Dickert, N., and J. Sugarman. 2005. Ethical goals of community consultation in research. *American Journal of Public Health* 95(7): 1123–1127.

Diener, E. 2001. Satisfaction with life scale. Retrieved on August 12, 2001 (http://www.psych.uiuc.edu /~ediener/hottopic/hottopic.html).

Diener, E., R. A. Emmons, R. J. Larsen, and S. Griffin. 1985. The satisfaction with life scale. *Journal of Personality Assessment* 49: 71–75.

DiLillo, D., S. DeGue, A. Kras, A. R. Di Loreto-Colgan, and C. Nash. 2006. Participant responses to retrospective surveys of child maltreatment: Does mode of assessment matter? *Violence and Victims* 21: 410–424.

Dillman, D., V. Lesser, R. Mason, J. Carlson, F. Milits, R. Robertson, and B. Burke. 2007. Personalization of mail surveys for general public and populations with a group identity: Results from nine studies. *Rural Sociology* 72: 632–646.

Diment, K., and S. Garrett-Jones. 2007. How demographic characteristics affect mode preference in a postal/ web mixed-mode survey of Australian researchers. *Social Science Computer Review* 25: 410–417.

Diviak, K. R., S. J. Curry, S. L. Emery, and R. Mermelstein. 2004. Human participants challenges in youth tobacco cessation research: Researchers' perspectives. *Ethics & Behavior* 14(4): 321–334.

Dixon, M., V. J. Roscigno, and R. Hodson. 2004. Unions, solidarity, and striking. *Social Forces* 83: 3–33.

Dohan, D., and M. Sanchez-Jankowski. 1998. Using computers to analyze ethnographic field data: Theoretical and practical considerations. *Annual Review of Sociology* 24: 477–499.

Donelle, L., and L. Hoffman-Goetz. 2008. Health literacy and online health discussions of North American black women. *Women & Health* 47: 71–90.

Donohue, J. J., and S. D. Levitt. 2001. The impact of legalized abortion on crime. *Quarterly Journal of Economics* 116: 379–420.

Douglas, J. D. 1976. *Investigating social research: Individual and team field research.* Beverly Hill, CA: Sage.

Dovring, K. 1973. Communication, dissenters and popular culture in eighteenth-century Europe. *Journal of Popular Culture* 7: 559–568.

Durkheim, É. [1897] 1964. *Suicide.* Reprint. Glencoe, IL: Free Press.

Durkheim, É. 1938. *The rules of sociological method.* New York: Free Press.

Duwe, G., and D. Kerschner. 2008. Removing a nail from the boot camp coffin. *Crime and Delinquency* 54: 614–643.

Economic and Social Research Council (ESRC). 2013. About ESRC. Retrieved on January 10, 2013 (http://www.esrc.ac.uk/about-esrc/).

Eliason, M., S. L. Dibble, and P. A. Robertson. 2011. Lesbian, Gay, Bisexual, and Transgender (LGBT) Physicians' Experiences in the Workplace. *Journal of Homosexuality* 58: 1355–1371.

Ellis, C., A. Bochner, N. Denzin, Y. Lincoln, J. Morse, R. Pelias, and L. Richardson. 2008. Talking and thinking about qualitative research. *Qualitative Inquiry* 14: 254–284.

Ennett, S. T., N. S. Tobler, C. L. Ringwalt, and R. L. Flewelling. 1994. How effective is drug abuse resistance education? A meta-analysis of project DARE outcome evaluations. *American Journal of Public Health* 84: 1394–1401.

Erikson, K. T. 1976. *Everything in its path: Destruction of community in the Buffalo Creek flood.* New York: Simon & Schuster.

Fallon, K., L. Swiss, and J. Viterna. 2012. Resolving the Democracy Paradox: Democratization and Women's Legislative Representation in Developing Nations, 1975 to 2009.

Feagin, J. R., and H. Vera. 2001. *Liberation sociology.* Boulder, CO: Westview.

Feagin, J. R., A. M. Orum, and G. Sjoberg. 1991. Conclusion: The present crisis in U.S. sociology. In *A case for the case study*, 269–278, edited by J. R. Feagin, A. M. Orum, and G. Sjoberg. Chapel Hill: University of North Carolina Press.

Feeley, M. 2007. Legality, social research, and the challenge of institutional review boards. *Law and Society Review* 41: 757–776.

Fernández-Peña, J. R., L. Moore, E. Goldstein, P. DeCarlo, O. Grinstead, C. Hunt, D. Bao, and H. Wilson. 2008. Making sure research is used: Community-generated recommendations for disseminating research. *Progress in Community Health Partnerships: Research, Education, and Action* 2.2. Retrieved on July 22, 2009 (http:// www.caps.ucsf.edu/CAPS/CAB/pdf /2.2.fernandez-pena.pdf).

Ferraro, F. R., E. Szigeti, K. Dawes, and S. Pan. 1999. A survey regarding the University of North Dakota institutional review board: Data, attitudes and perceptions. *Journal of Psychology* 133: 272–381.

Fetterman, D., and M. Eiler. 2000. Empowerment evaluation: A model for building evaluation and program capacity. Paper presented at the American Evaluation Association Meetings, Honolulu, Hawaii, November 1–4. Retrieved on August 2, 2001 (http://www.stanford. edu/~davidf/Capacitybld.pdf).

Fikkan, J., and E. Rothblum. 2012. Is Fat a Feminist Issue? Exploring the Gendered Nature of Weight Bias. *Sex Roles* 66: 575–592.

Finch, J., and L. Wallis. 1993. Death, inheritance and the life course. *Sociological Review Monograph* 50–68.

Fine, G. A., and K. D. Elsbach. 2000. Ethnography and experiment in social psychological theory building: Tactics for integrating qualitative field data with quantitative lab data. *Journal of Experimental Social Psychology* 36: 51–76.

Fine, M., and M. E. Torre. 2006. Intimate details, participatory action research in prison. *Action Research* 4(3): 253–269.

Fine, M., L. Weis, S. Wesson, and L. Wong. 2000. For whom? Qualitative research, representations, and social responsibilities. In *Handbook of qualitative research*, 2nd edition, 107–132, edited by N. K. Denzin and Y. S. Lincoln. Thousand Oaks, CA: Sage.

Fischman, W., B. Solomon, D. Greenspan, and H. Gardner. 2004. Making good: How young people cope with moral dilemmas at work. Cambridge, MA: Harvard University Press.

Fitzgerald, J. M. 2011. Attrition in Models of Intergenerational Links Using the PSID with Extensions to Health and to Sibling Models. *The B.E. Journal of Economic Analysis & Policy* 11 ISSN (Online) 1935–1682, DOI: 10.2202/1935-1682.2868

Flay, B. R., and L. M. Collins. 2005. Historical review of school-based randomized trials for evaluating problem behavior prevention programs. *Annals of the American Academy of Political and Social Science* 599(May): 115–146.

Fogel, C. 2007. Ethical issues in field-based criminological research in Canada. *International Journal of Criminal Justice Sciences* 2: 109–118.

Fontana, A., and J. H. Frey. 1994. Interviewing: The art of science. In *Handbook of qualitative research*, 361–376, edited by N. K. Denzin and Y. S. Lincoln. Thousand Oaks, CA: Sage.

Fontes, T. O., and M. O'Mahony. 2008. In-depth interviewing by instant messaging. *Social Research Update* 53: 1–4. Retrieved on September 16, 2009 (http://sru.soc.surrey.ac.uk/SRU53.pdf).

Foundation Center. 2012. Distribution of Foundation Grants by Subject Categories. Retrieved on March 23, 2013 (http://foundationcenter.org /findfunders/statistics/pdf/04_fund _sub/2010/10_10.pdf).

Frank, K. 2005. Exploring the motivations and fantasies of strip club customers in relations to legal regulations. *Archives of Sexual Behavior* 34: 487–506.

Frankenberg, R. 1993. *White women, race matters: The social construction of whiteness*. Minneapolis: University of Minnesota Press.

Friedan, B. 1963. *The feminine mystique*. New York: Norton.

Friedman, J., and M. Alicea. 1995. Women and heroin: The path of resistance and its consequences. *Gender & Society* 9: 432–449.

Frith, H., and C. Kitzinger. 1998. "Emotion Work" as a Participant Resource: A Feminist Analysis of Young Women's Talk-in-Interaction. *Sociology* 32: 299–320.

Fuchs, M. 2009. Gender-of-Interviewer Effects in a Video-Enhanced Web Survey: Results from a Randomized Field Experiment. *Social Psychology* 40: 37–42.

Gallagher, C. 2012. United Kingdom and the United States of America: Converging Attitudes on Immigrant Populations. Presented at the American Sociological Association Annual Meetings in August 2012 in Denver, CO.

GAO. 2007. 2010 census: Design shows progress, but managing technology acquisitions, temporary field staff, and gulf region enumeration require attention. Retrieved on June 13, 2009 (http://www.gao.gov/new. items/d07779t.pdf).

Geer, J. G. 1988. What do open-ended questions measure? *Public Opinion Quarterly* 52: 365–371.

Geer, J. G. 1991. Do open-ended questions measure "salient" issues? *Public Opinion Quarterly* 55: 360–370.

General Social Survey. 2010. Cumulative data file. SDA Archive. Retrieved on March, 2013 (http:// sda.berkeley.edu/archive.htm).

Gerich, J. 2008. Real or virtual? Response behavior in video-enhanced self-administered computer interviews. *Field Methods* 20: 356–376.

Gertz, C. 1983. *Local knowledge: Further essays in interpretative anthropology*. New York: Basic Books.

Ghose, T., D. Swendeman, G. Sheba, and D. Chowdhury. 2008. Mobilizing collective identify to reduce HIV risk among sex workers in Sonagachi, India: The boundaries, consciousness, and negotiation framework. *Social Science & Medicine* 67(2): 311–320.

Gibbs, G. 2011. Audio and Visual Analysis. Retrieved March 21, 2013 (http://onlineqda.hud.ac.uk/Intro_ CAQDAS/what_packages_are _available/audio_video_analysis .php).

Gibbs, G. 2012a. Learning Qualitative Data Analysis on the Web. Retrieved March 21, 2013 (http:// onlineqda.hud.ac.uk/Intro_QDA /what_is_qda.ph).

Gibbs, G. 2012b. Code-and-Retrieve Packages. Retrieved March 21, 2013 (http://onlineqda.hud.ac.uk /Intro_CAQDAS/what_packages _are_available/code_retrieve _packages.php).

Gillespie, W. 2008. Thirty-five years after stonewall: An exploratory study of satisfaction with police among gay, lesbian, and bisexual persons at the 34th Annual Atlanta Pride Festival. *Journal of Homosexuality* 55: 619–647.

Glaser, B. G. 2005. *The grounded theory perspective III: Theoretical coding*. Mill Valley, CA: Sociology Press.

Glaser, B. G., and A. L. Strauss. 1967. The discovery of grounded theory: Strategies for qualitative research. New York: Aldine.

Glaser, B. G., ed. 1993. *Examples of grounded theory: A reader*. Mill Valley, CA: Sociology Press.

Glasner, T., and W. Van der Vaart. 2009. Applications of calendar instruments in social surveys: A review. *Quality & Quantity* 43: 333–349.

Glazer, N. 1977. General perspectives. In *Woman in a man-made world*, 1–4, edited by N. Glazer and H. Youngelson. Chicago: Rand McNally.

Glover, T., W. Stewart, and K. Gladdys. 2008. Social ethics of landscape change: Toward community-based land-use planning. *Qualitative Inquiry* 14(3): 384–401.

Goffman, E. 1979. *Gender advertisements*. Cambridge, MA: Harvard University Press.

Goh, D. H., and Lee, C. S. 2011. An Analysis of Tweets on the Death of Michael Jackson. *Aslib Proceedings* 63(5): 432–444.

Gold, R. L. 1958. Roles in sociological field observations. *Social Forces* 36: 217–223.

Gold, R. L. 1997. The ethnographic method in sociology. *Qualitative Inquiry* 3: 388–402.

Goldstein, E., B. Freedman, A. Richards, and O. Grinstead. 2000. The Legacy Project: Lessons learned about conducting community-based research. *CAPS UC San Francisco*. Retrieved on April 1, 2013 (caps.ucsf.edu/uploads/pubs /reports/pdf/LegacyS2C.pdf).

Gonzalez-Lopez, G. 2011. Mindful ethics: Comments on informant-centered practices in sociological research. *Qualitative Sociology* 34: 447–461.

Goode, E. 1996. The ethics of deception in social research: A case study. *Qualitative Sociology* 19: 11–33.

Gorden, R. L. 1975. *Interviewing, strategy, techniques and tactics.* Homewood, IL: Dorsey.

Gorin, S., C. Hooper, C. Dyson, and C. Cabral. 2008. Ethical challenges in conducting research with hard to reach families. *Child Abuse Review* 17: 275–287.

Gould, S. J. 1995. *Dinosaur in a haystack: Reflections in natural history.* New York: Harmony.

Gould, S. J. 1997. *Full house: The spread of excellence from Plato to Darwin.* New York: Three Rivers Press.

Grady, J. 2001. Becoming a visual sociologist. *Sociological Imagination* 38 (1/2): 83–119.

Grady, J. 2007a. Advertising images as social indicators: Depictions of blacks in *Life* magazine, 1936–2000. *Visual Studies* 22(3): 211–239.

Grady, J. 2007b. Visual sociology. In *21st century sociology: A reference handbook*, 63–70, edited by C. Bryant and D. Peck. Thousand Oaks, CA: Sage Publication.

Grady, J. 2007c. Visual methods. In *The encyclopedia of sociology*, 2286–2289, edited by G. Ritzer. London: Blackwell Publications.

Graffigna, G., A. C. Bosio, and K. Olson. 2010. How do ethics assessments frame results of comparative qualitative research? A theory of technique approach. *International Journal of Social Research Methodology* 13: 341–355.

Grandchamp, K. 2009. Disney animated movies: Friend or foe of femininity? Term paper for social research methods course. Rhode Island College.

Griesler, P. C., D. B. Kandel, C. Schaf-fran, M. Hu, and M. Davies. 2008. Adolescents' inconsistency in self-reported smoking: A comparison of reports in school and household settings. *Public Opinion Quarterly* 72: 260–290.

Groves, R. 2006. Nonresponse Rates and Nonresponse Bias in Household Surveys. *Public Opinion Quarterly* 70: 646–675.

Groves, R. M. 1989. *Survey errors and survey costs.* New York: Wiley.

Groves, R. M. 2006. Nonresponse rates and nonresponse bias in households surveys. *Public Opinion Quarterly* 70(5): 646–675.

Groves, R. M., and E. Peytcheva. 2008. The impact of nonresponse rates on nonresponse bias: A meta-analysis. *Public Opinion Quarterly* 72(2): 167–189.

Gubrium, J. F., and J. A. Holstein. 2001. From the individual interview to the interview society. In *Handbook of Interview Research*, 3–32, edited by J. F. Gubrium and J. A. Holstein. Thousand Oaks, CA: Sage Publications.

Guernsey, L. 2009. What's been cut: The story of the child parent centers. Retrieved on February 26, 2009 (http://www.newamerica.net/blog/early-ed-watch/2009/whats-been-cut-story-child-parent-centers-10341).

Guillemin, M., and S. Drew. 2010. Questions of process in participant-generated visual methodologies. *Visual Studies* 25: 175–188.

Hagan, J., and A. Palloni. 1998. Immigration and crime in the United States. In *The immigration debate*, 367–387, edited by J. P. Smith and B. Edmonton. Washington, DC: National Academy Press.

Hagestad, G. O., and P. Uhlenberg. 2005. The social separation of old and young: A root of ageism. *Journal of Social Issues* 61(2): 343–360.

Haggerty, K. 2004. Ethics Creep: Governing Social Science Research in the Name of Ethics. *Qualitative Sociology* 27: 391–414.

Halse, C., and A. Honey. 2005. Unraveling ethics: Illuminating moral dilemmas of research ethics. *Signs: Journal of Women in Culture and Society* 30(4): 2149–2162.

Hamarat, E., D. Thompson, K. M. Sabrucky, D. Steele, K. B. Matheny, and F. Aysan. 2001. Perceived stress and coping resource availability as predictors of life satisfaction in young, middle-aged and older adults. *Experimental Aging Research* 27: 181–196.

Hammersley, M. 1995. *The politics of social research.* London: Sage.

Hamnett, M., D. Porter, A. Singh, and K. Kumar. 1984. *Ethics, politics, and international social science research: From critique to praxis.* Hawaii: University of Hawaii Press for East–West Center.

Hanks, R. S., and N. T. Carr. 2008. Lifelines of women in jail as self-constructed visual probes for life history research. *Marriage and Family Review* 42: 105–116.

Harding, S. 1993. Rethinking standpoint epistemology: What is "strong objectivity"? In *Feminist epistemologies*, 49–82, edited by L. Alcoff and E. Potter. New York: Routledge.

Harding, S., and K. Norberg. 2005. New feminist approaches to social science methodologies: An introduction. *Signs* 30: 2009–2015.

Hargittai. E., and E. Litt. 2012. Becoming a Tweep: How Prior Online Experiences Influence Twitter Use. Presented at the American Sociological Association Annual Meetings in August 2012 in Denver, CO.

Harkess, S., and C. A. B. Warren. 1993. The Social Relations of Intensive Interviewing Constellations of Strangeness and Science *Sociological Methods & Research* 21: 317–339.

Harris, W., L. Skogrand, and D. Hatch. 2008. Role of friendship, trust, and love in strong Latino marriages. *Marriage & Family Review* 44: 455–488.

Haskins, R., and J. Baron. 2011. The Obama Administration's Evidence-Based Social Policy Initiatives: An Overview. In *Evidence for Social Policy and Practice.* NESTA (pp. 28–35). Retrieved on April 4, 2013 (http://www.nesta.org.uk/library/documents/Expert_Essays_webv1.pdf).

Hayes, T. 2010. Labeling and the Adoption of a Deviant Status. *Deviant Behavior* 31: 274–302.

Hazell, V., and J. Clarke. 2008. Race and gender in media: A content analysis of two mainstream black magazines. *Journal of Black Studies* 39(1): 5–21.

Heise, L., M. Ellsberg, and M. Gottemoeller. 1999. Ending violence against women. *Population Reports*, Series L, No. 11. Baltimore: Johns Hopkins University School of Public Health, Population Information Program, December.

Helliwell, J., R. Layard, and J. Sachs. 2012. World Happiness Report. Columbia University: Earth Institute. Retrieved on March 23, 2013 (http://www.earth.columbia.edu/sitefiles/file/Sachs%20Writing/2012/World%20Happiness%20Report.pdf).

Hennigan, K. M., C. L. Maxson, D. Sloane, and M. Ranney. 2002. Community views on crime and policing: Survey modes effects on bias in community surveys. *Justice Quarterly* 19(3): 565–587.

Herrera, C. D. 1997. A historical interpretation of deceptive experiments in American psychology. *History of the Human Sciences* 10: 23–36.

Herzog, A. R., and W. L. Rogers. 1988. Interviewing older adults. *Public Opinion Quarterly* 52: 84–99.

Hirsch, C. E. 2009. The strength of weak enforcement: The impact of discrimination charges, legal environments, and organization conditions on workplace segregation. *American Sociological Review* 74: 245–271.

Hirschi, T. 1969. *Causes of delinquency*. Berkeley: University of California Press.

Hobbs, F. 2005. *Examining American household composition: 1990 and 2000*. U.S. Census Bureau, Census 2000 Special Reports, CENSR-24. Washington, DC: I.S. Government Printing Office.

Hobza, C., and A. B. Rochlen. 2009. Gender role conflict, drive for muscularity and the impact of ideal media portrayals on men. *Psychology of Men and Masculinity* 12: 120–130.

Hochschild, A. R., with A. Machung. 1989. *The second shift*. New York: Avon Books.

Hofferth, S. L. 2005. Secondary data analysis in family research. *Journal of Marriage and Family* 67: 891–907.

Holbrook, A. L., and J. A. Krosnick. 2010. Social desirability bias in voter turnout reports. *Public Opinion Quarterly* 74: 37–67.

Holster, K. 2004. The business of Internet egg donation: An exploration of donor and recipient experiences. Dissertation Abstracts International.

Holsti, O. R. 1969. *Content analysis for the social sciences and the humanities*. Reading, MA: Addison-Wesley.

Hoover, E. 2005. The ethics of undercover research. *Chronicle of Higher Education* 51(47): A36. Retrieved on August 7, 2005 (http://chronicle.com/free/v51/i47/47a03601.htm).

Horsfall, D., K. Noonan, and R. Leonard. 2012. Bringing our dying home: How caring for someone at end of life builds social capital and develops compassionate communities. *Health Sociology Review* 21: 373–382.

Housing and Urban Development (HUD). 2012. *Moving to Opportunity* for Fair Housing Demonstration Program - *Final* Impacts Evaluation. Retrieved on March 20, 2013 (http://www.huduser.org/portal/publications/pubasst/MTOFHD.html).

Howe, A., I. P. Cate, A. Brown, and J. Hadwin. 2008. Empathy in preschool children: The development of the Southampton test of empathy for preschoolers. *Psychological Assessment* 20(3): 305–309.

Hughes, E. C. 1960. Introduction: The place of field work in social science. In *Field work: An introduction to the social sciences*, iii–xiii, edited by B. H. Junker. Chicago: University of Chicago.

Hugman, R., E. Pittaway, and L. Bartolomei. 2011. When "Do No Harm" Is Not Enough: The Ethics of research with refugees and other vulnerable groups. *British Journal of Social Work* 41: 1271–1287.

Humphreys, L. 1975. *Tearoom trade: Impersonal sex in public places*. Chicago: Aldine.

Hurricane Katrina Community Advisory Group. 2013. Project Overview. Retrieved on March 23, 2013 (http://www.hurricanekatrina.med.harvard.edu/index.php).

Inter-Parliamentary Union. 2004. Women's suffrage. Retrieved on April 19, 2003 (http://www.ipu.org/wmn-e/suffrage.htm).

Irvine, J. 2012 Can't Ask, Can't Tell: How Institutional Review Boards Keep Sex in the Closet. *Contexts* 11: 28–33.

Jacinto, C., M. Duterte, P. Sales, and S. Murphy. 2008. "I'm not a real dealer": The identity process of ecstasy dealers. *Journal of Drug Issues* 38(2): 419–444.

Jackall, R. 1988. *Moral mazes: The world of corporate managers*. New York: Oxford University Press.

Jacobson, L., 2007. Chicago Data Bolster Case for Early-Childhood Programs. *Education Week* 26: 7.

Jaggar, E. 2005. Is thirty the new sixty? Dating, age and gender in postmodern, consumer society. *Sociology* 39: 89–106.

Johnson, J., and A. Duffet. 2012. Life After High School. New York: The Public Agenda. Retrieved on April 2, 2013 (http://www.publicagenda.org/files/life_after_high_school.pdf).

Johnston, B. R. 2010. Social Responsibility and the Anthropological Citizen. *Current Anthropology* 51: 235–247.

Johnstone, M., and C. Lee. 2009. Young Australian women's aspirations for work and family: Individual and sociocultural differences. *Gender & Society* 61(3/4): 204–220.

Jones, J. H. 1981. *Bad blood: The Tuskegee syphilis experiment*. New York: Free Press.

Jordan, S. 2003. Who stole my methodology? Co-opting PAR. *Globalisation, Societies and Education* 1: 185–200.

Jowell, R. 1998. How comparative is comparative research? *American Behavioral Scientist* 42: 168–177.

Kahneman, D., and A. Krueger. 2006. Developments in the measurement of subjective well-being. *Journal of Economic Perspectives* 20(1): 3–24.

Kaiser Family Foundation. 2005. *Sex on TV: Executive summary*. Menlo Park, CA: Henry J. Kaiser Family Foundation. Retrieved on November 20, 2009 (http://www.kff.org/entmedia/upload/Sex-on-TV-4-Full-Report.pdf).

Kaminer, A. 2012. Columbia's Gang Scholar Lives on the Edge. *New York Times*, November 30. Retrieved on March 10, 2013 (http://www.nytimes.com/2012/12/02/nyregion/sudhir-venkatesh-columbias-gang-scholar-lives-on-the-edge.html?ref=arielkaminer).

Kane, E. W., and H. Schuman. 1991. Open survey questions as measures of personal concern with issues: A reanalysis of Stouffer's *Communism, conformity and civil liberties*. *Sociological methodology* 21: 81–96.

Karney, B. R., J. Davila, C. L. Cohan, K. T. Sullivan, M. D. Johnson, and T. N. Bradbury. 1995. An empirical investigation of sampling strategies in marital research. *Journal of Marriage and the Family* 57: 909–920.

Karwalajtys, T., B. McDonough, H. Hall, M. Guirguis-Younger, L. W. Chambers, J. Kaczorowski, L. Lohfeld, and H. Brian. 2009. Development of the volunteer peer educator role in a community Cardiovascular Health Awareness Program (CHAP): A process evaluation in two communities. *Journal of Community Health* 34: 336–345.

Katz, J. 1972. *Experimentation with human beings*. New York: Russell Sage Foundation.

Keeter, S., C. Kennedy, M. Dimock, J. Best, and P. Craighill. 2006. Gauging the Impact of Growing Nonresponse on Estimates from a National RDD Telephone Survey *Public Opinion Quarterly* 70: 759–779.

Keeter, S., C. Miller, A. Kohut, R. Groves, and S. Presser. 2000. Consequences of reducing nonresponse in a national telephone survey. *Public Opinion Quarterly* 64: 125–148.

Kemmis, S., and R. McTaggart. 2003. Participatory action research. In *Strategies of qualitative inquiry*, 2nd edition, 336–396, edited by N. K. Denzin and Y. S. Lincoln. Thousand Oaks, CA: Sage.

Kempner J. 2008. The Chilling Effect: How Do Researchers React to Controversy? *PLoS Medicine* 5(11): e222. doi:10.1371/journal.pmed.0050222 Retrieved on April 3, 2013 (http://www.plosmedicine.org/article/info%3Adoi%2F10.1371%2Fjournal.pmed.0050222).

Kenschaft, L. 2005. *Reinventing marriage: The love and work of Alice Freeman Palmer and George Herbert Palmer*. Urbana and Chicago: University of Illinois Press.

Kim, J., and E. Hatfield. 2004. Love types and subjective well-being: A cross-cultural study. *Social Behavior and Personality* 32(2): 173–182.

Kim, M. S., H. W. Kim, K. H. Cha, and J. Lim. 2007. What makes Koreans happy? Exploration on the structure of happy life among Korean adults. *Social Indicators Research* 82: 265–286.

Kinsey, A. C., W. B. Pomeroy, and C. E. Martin. 1953. *Sexual behavior in the human female*. Philadelphia: Saunders.

Kirk, D. 2009. A Natural Experiment on Residential Change and Recidivism: Lessons from Hurricane Katrina. *American Sociological Review* 74: 484–504.

Kirp, D. L. 1997. Blood, sweat and tears: The Tuskegee experiment and the era of AIDS. *Tikkun* 10: 50–55.

Klofstad, C. A., S. Boulianne, and D. Basson. 2008. Matching the message to the medium: Results from an experiment on Internet survey email contacts. *Social Science Computer Review* 26: 498–509.

Knickman, J. R. 2005. When health policy is the problem. *Journal of Health Politics, Policy and Law* 30: 367–374.

Kolar, T., and I. Kolar. 2008. What respondents really expect from researchers. *Evaluation Research* 32: 363–391.

Korn, J. 1997. *Illusions of reality: A history of deception in social psychology*. Albany: State University of New York Press.

Kornrich, S., J. Brines, and K. Leupp, 2013. Egalitarianism, Housework, and Sexual Frequency in Marriage. *American Sociological Review* 78 (1): 26–50.

Kort-Butler, L, and K. Hagawen. 2011. School-Based Extracurricular Activity Involvement and Adolescent Self-Esteem: A Growth-Curve Analysis. *Journal of Youth & Adolescence* 40: 568–581.

Kotlowitz, A. 1991. *There are no children here*. New York: Doubleday.

Krafchick, J. L., T. S. Zimmerman, S. A. Haddock, and J. H. Banning. 2005. Best-selling books advising parents about gender: A feminist analysis. *Family Relations* 54: 84–100.

Kreuter, F., S. Presser, and R. Tourangeau. 2008. Social desirability bias in CATI, IVR, and web surveys: The effects of mode and question sensitivity. *Public Opinion Quarterly* 72: 847–865.

Krimerman, L. 2001. Participatory action research: Should social inquiry be conducted democratically? *Philosophy of Social Sciences* 31: 60–82.

Krippendorff, K. 2003. *Content analysis: An introduction to its methodology*, 2nd edition. Beverly Hills, CA: Sage.

Krysan, M., and M. P. Couper. 2003. Race in the live and the virtual interview: Racial deference, social desirability, and activation effects in attitude surveys. *Social Psychology Quarterly* 66: 364–383.

Krysan, M., H. Schuman, L. J. Scott, and P. Beatty. 1994. Response rates and response content in mail versus face-to-face surveys. *Public Opinion Quarterly* 58: 381–400.

Kuhn, M. H. 1962. The interview and the professional relationship. In *Human behavior and social processes*, 193–206, edited by A. M. Rose. Boston: Houghton Mifflin.

Ladd, E. C., and K. Bowman. 1998. *What's wrong: A survey of American satisfaction and complaint*. Washington, DC: AEI Press.

Landra, K., and T. Sutton. 2008. You think you know ghetto? Contemporizing the Dove "Black IQ Test." *Teaching Sociology* 36(4): 366–377.

Lang, A. 2001. Economic inequality: A quantitative study of opinions related to the income gap and government involvement in its reduction. Honors thesis at Rhode Island College.

Lasswell, H. D. 1965. Detection: Propaganda detection and the courts. In *The language of politics: Studies in quantitative semantics*, edited by H. Lasswell et al. Cambridge, MA: MIT Press.

Lauder, M. A. 2003. Covert participation observation of a deviant community: Justifying the use of deception. *Journal of Contemporary Religion* 18(2): 185–196.

Laumann, E. O., J. H. Gagnon, R. T. Michael, and S. Michaels. 1994. *The social organization of sexuality*. Chicago: University of Chicago Press.

Lavrakas, P. J. 1987. *Telephone survey methods sampling, selection and supervision*. Newbury Park, CA: Sage.

Lazar, K. 2012. Harvard study placed elderly subjects at risk. *Boston Globe* July 20, 2012. Retrieved on October 15, 2012 (http://www.bostonglobe.com/metro/2012/07/19/nursing-home-residents-receive-letter-about-risks-they-faced-from-harvard-led-study/tIk8skbZXSN35jpqU6csCM/story.html?camp=pm).

Lazarsfeld, P. F. 1944. The controversy over detailed interviews: An offer for negotiation. *Public Opinion Quarterly* 8: 38–60.

Lee, S., N. A. Mathiowetz, and R. Tourangeau. 2007. Measuring disability in surveys: Consistency over time and across respondents. *Journal of Official Statistics* 23: 163–184.

Lee, Y. S., N. Morrow-Howell, M. Johnson-Reid, and S. McCrary. 2012. The Effect of the Experience Corps Program on Student Reading Outcomes. *Education and Urban Society* 44: 97–118.

Legerski, J. P., E. Vernberg, Eric, and B. Noland. 2012. *Community Mental Health Journal* 48: 729–740.

Lemert, C. 1993. *Social theory: The multicultural and classic readings*. Boulder, CO: Westview.

Leo, R. A. 1995. Trial and tribulations: Courts, ethnography, and the need for an evidentiary privilege for academic researchers. *American Sociologist* 26: 113–134.

Leslie, P. L. 2005. The presentation of college professors in popular film. Presented at Sociological Society. Charlotte, NC.

Lester, D., and D. Lester Jr. 2009. *Writing Research Papers*. New York: Longman.

Lever, J. 1981. Multiple methods of data collection. *Urban Life* 10: 199–213.

Levitt, S. D. 2004. Understanding why crime fell in the 1990s: Four factors that explain the decline and six that do not. *Journal of Economic Perspectives* 18: 163–190.

Lewis, C. A., M. J. Dorahy, and J. F. Schumaker. 1999. Depression and life satisfaction among northern Irish adults. *Journal of Social Psychology* 139: 533–535.

Liebow, E. 1967. *Tally's corner: A study of Negro streetcorner men*. Boston: Little, Brown.

Liebow, E. 1993. *Tell them who I am: The lives of homeless women*. New York: Free Press.

Lin, J. Y., L. S. Chen, E. S. Wang, and J. M. Cheng. 2007. The relationship between extroversion and leisure motivation: Evidence from fitness center participation. *Social Behavior and Personality* 35(10): 1317–1322.

Lin, S. F., A. Beck, B. Finch, R. A. Hummer, and R.K. Master. 2012. Trends in US Older Adult Disability: Exploring Age, Period, and Cohort Effects. *American Journal of Public Health* 102: 2157–2162.

Lindsay, J. 2005. Getting the numbers: The unacknowledged work in recruiting for survey research. *Field Methods* 17(1): 119–128.

Lindsey, E. W. 1997. Feminist issues in qualitative research with formerly homeless mothers. *Affilia: Journal of Women and Social Work* 12: 57–76.

Link, M. L., and A. Mokdad. 2005. Advance letters as a means of improving respondent cooperation in random digit dial studies: A multistate experiment. *Public Opinion Quarterly* 69(4): 572–587.

Link, M. W., M. P. Battaglia, M. R. Frankel, L. Osborn, and A. H. Mokdad. 2008. A comparison of address-based sampling (ABS) versus random-digit dialing (RDD) for general population surveys. *Public Opinion Quarterly* 72: 6–27.

Lofland, J., and L. H. Lofland. 1995. *Analyzing social settings: A guide to qualitative observation and analysis*, 3rd edition. Belmont, CA: Wadsworth.

Lofland, J., D. Snow, L. Anderson, and L. H. Lofland. 2005. *Analyzing social settings: A guide to qualitative observation and analysis*, 4th edition. Belmont, CA: Wadsworth.

Logan, T. K., R. Walker, L. Shannon, and J. Cole. 2008. Combining ethical considerations with recruitment and follow-up strategies for partner violence and victimization research. *Violence Against Women* 14: 1226–1251.

Louie, M. C. Y. 2001. *Sweatshop warriors: Immigrant women workers take on the global factory*. Cambridge, MA: South End Press.

Love: Accounting for Women's Orgasm and Sexual Enjoyment in College Hookups and Relationships. *American Sociological Review* 77 (3):435–462.

Ludwig, J., G. J. Duncan, L. A. Gennetian, L. F. Katz, R. Kessler, J. R. Kling, and L. Sanbonmatsu. 2012. Neighborhood Effects on the Long-Term Well-Being of Low-Income Adults. *Science* 21: 1505–1510.

Lystra, K. 1989. *Searching the heart: Women, men and romantic love in nineteenth-century America*. New York: Oxford University Press.

MacEvoy, J. P., and S.R. Asher. 2012. When Friends Disappoint: Boys' and Girls' Responses to Transgressions of Friendship Expectations. *Child Development* 83: 104–119.

Macfarlane, B. 2009. *Researching with Integrity: The ethics of academic enquiry*. New York: Routledge.

Madge, C., and H. O'Connor. 2006. Parenting gone wired: Empowerment of new mothers on the Internet? *Social & Cultural Geography* 7: 199–220.

Madriz, E. 2000. Focus groups in feminist research. In *Handbook of qualitative research*, 2nd edition, 835–850, edited by N. K. Denzin and Y. S. Lincoln. Thousand Oaks, CA: Sage.

Maher, F. A., and M. K. Tetreault. 1994. *The feminist classroom*. New York: Basic.

Maher, K. 2005. Practicing sociology as a vocation. In *Explorations in theology and vocation*, 31–40, edited by W. E. Rogers. Greenville, NC: Furman University's Center for Theological Explorations of Vocation.

Majumdar, A. 2007. Researching South Asian women's experiences of marriage: Resisting stereotypes through an exploration of "space" and "embodiment". *Feminism & Psychology* 17: 316–322.

Malinowski, B. 1922. *Argonauts of the western pacific*. New York: Dutton.

Malter, S., L. Simich, N. Jacobson, and J. Wise. 2009. Reciprocity: An ethic for community-based participatory action research. *Action Research* 6: 305–325.

Manning, P. K. 2002. FATETHICS: Response to erich goode. *Qualitative Sociology* 25(4): 541–547.

Marecek, J., M. Fine, and L. Kidder. 1997. Working between worlds: Qualitative methods and social psychology. *Journal of Social Issues* 53: 631–645.

Marshall, B. 2002. *Heliobactor pioneers: Firsthand accounts from the scientists who discovered heliobacters 1892–1982*. Perth: University of Western Australia.

Marshall, E. 2013. Obama Lifts Ban on Funding Gun Violence Research. *ScienceInsider* January, 2013. Retrieved on April 4, 2013 (http://news.sciencemag.org/scienceinsider/2013/01/obama-lifts-ban-on-funding-gun-v.html).

Martin, L. L., T. Abend, C. Sedikides, and J. D. Green. 1997. How would I feel if…? Mood as input to a role fulfillment evaluation process. *Journal of Personality and Social Psychology* 73: 242–254.

Martineau, H. 1962. *Society in America*. New York: Anchor.

Martinez, R., and M. T. Lee. 2000. Comparing the context of immigrant homicides in Miami: Haitians, Jamaicans and Mariel. *International Migration Review* 34: 794–812.

Marx, K. [1867] 1967. *Das Kapital*. Reprint. New York: International.

Maton K. I., H. E. Wimms, S. K. Grant, M. A. Wittig, M. R. Rogers, and M. J. Vasquez. 2011. Experiences and perspectives of African American, Latina/o, Asian American, and European American psychology graduate students: A national study.

Cultural Diversity & Ethnic Minority Psychology 17: 68–78.

Maynard, G., E. Shircliff, and M. Restivo. 2012. IMF Structural Adjustment, Public Health Spending, and Tuberculosis. *International Journal of Sociology* 42(2): 5–27.

McCabe, Janice, Emily Fairchild, Liz Grauerholz, Bernice Pescosolido, and Daniel Tope. 2011. Gender in Twentieth-Century Children's Books: Patterns of Disparity in Titles and Central Characters. *Gender and Society* 25(2): 197–226.

McCall, G. J. 1984. Systematic field observation. *Annual Review of Sociology* 10: 263–282.

McClelland, S. I., and M. Fine. 2008. Embedded science: Critical analysis of abstinence-only evaluation research. *Cultural Studies, Critical Methodologies* 8: 50–81.

McCorkel, J., and K. Myers. 2003. What difference does difference make? Position and privilege in the field. *Qualitative Sociology* 26(2): 199–231.

McGraw, L., A. Zvonkovic, and A. Walker. 2000. Studying postmodern families: A feminist analysis of ethical tensions in work and family research. *Journal of Marriage and the Family* 62(February): 68–77.

McLaren, C. 2009. Analyzing qualitative data about hospitalized children: Reflections on bodily expressions. *Qualitative Report* 14, 140–154. Retrieved on June 23, 2009 (http://www.nova.edu/ssss/QR/QR14-1/mclaren.pdf).

McLaughlin, K. A., P. Berglund, M. J. Gruber, R.C. Kessler, N. A. Sampson, and A. M. Zaslavsky. 2011. Recovery from PTSD following Hurricane Katrina. *Depression and Anxiety* 28: 439–446.

Mead, M. 1933. *Sex and temperament in three primitive societies.* New York: Morrow.

Mears, A., and W. Finlay. 2005. Not just a paper doll: How models manage bodily capital and why they perform emotional labor. *Journal of Contemporary Ethnography* 34: 317–343.

Melrose, M. 2002. Labour pains: Some considerations on the difficulties of researching juvenile prostitution. *International Journal of Social Research Methodology* 5(4): 333–351.

Menard, S. 1991. *Longitudinal research.* Newbury Park, CA: Sage.

Merton, R. K., M. Fiske, and P. L. Kendall. 1956. *The focused interview.* Glencoe, IL: Free Press.

Messner, S. F. 1999. Television violence and violent crime: An aggregate analysis. In *How it's done: An invitation to social research,* 308–316, edited by E. S. Adler and R. Clark. Belmont, CA: Wadsworth.

Michael, R. T., J. H. Gagnon, E. O. Laumann, and G. Kolata. 1994. *Sex in America.* Boston: Little, Brown.

Mies, M. 1991. Women's research or feminist research? In *Beyond methodology: Feminist scholarship as lived research,* 60–84, edited by M. M. Fonow and J. A. Cook. Bloomington: Indiana University Press.

Miles, M. B., and A. M. Huberman. 1994. *Qualitative data analysis,* 2nd edition. Thousand Oaks, CA: Sage.

Milgram, S. 1974. *Obedience to authority: An experimental view.* New York: Harper & Row.

Miller, D. C., and N. J. Salkind. 2002. *Handbook of research design and social measurement.* Thousand Oaks, CA: Sage.

Miller, E. M. 1986. *Street woman.* Philadelphia: Temple University Press.

Mitchell, A. 1997. Survivors of Tuskegee study get apology from Clinton. *New York Times* 146: 9–10.

Moller, V., P. Theuns, I. Erstad, and J. Bernheim. 2008. The best and worst times of life: Narratives and assessments of subjective well-being by anamnestic comparative self assessment (ACSA) in the Easter Cape, South Africa. *Social Indicators Research* 89(1): 1–22.

Moret, M., R. Reuzel, G. van der Wilt, and J. Grin. 2007. Validity and reliability of qualitative data analysis: Interobserver agreement in reconstructing interpretative frames. *Field Methods* 19: 24–39.

Morgan, D. L. 2001. Focus group interviewing. In *Handbook of interview research,* 141–159, edited by J. F. Gubrium and J. A. Holstein. Thousand Oaks, CA: Sage Publications.

Morrow, R. 1994. *Critical theory and methodology.* Thousand Oaks, CA: Sage.

Mosteller, F. 1955. Use as evidenced by an examination of wear and tear on

selected sets of ESS. In *A study of the need for a new encyclopedic treatment of the social sciences,* edited by K. Davis. Unpublished manuscript.

Mueller, J., D. Dirks, and L. Houts Picca. 2011. Unmasking racism: Halloween costuming and engagement of the racial other. 292–303. In E. Adler and R. Clark, *An Invitation to Social Research: How It's Done.* 4th edition.

Nathan, R. 2005. *My freshman year: What a professor learned by becoming a student.* Ithaca, NY: Cornell University Press.

National Academy of Sciences. 1993. Methods and values in science. In *The "racial" economy of science,* 341–343, edited by S. Harding. Bloomington: Indiana University Press.

National Center for Education Statistics. 2013. Early Childhood Longitudinal Program: Overview. Retrieved on January 6, 2013 (http://nces.ed.gov/ecls/).

National Science Foundation. (NSF) 2012. Social Science and the Common Rule. Retrieved on April 8, 2013 (http://www.nsf.gov/bfa/dias/policy/hsfaqs.jsp).

Neimetx, C. 2011. Navigating family roles within an institutional framework: An exploration in one private Chinese orphanage. *Journal of Child & Family Studies* 20: 585–595.

Neto, F., and J. Barros. 2000. Psychosocial concomitants of loneliness among students of Cape Verde and Portugal. *Journal of Psychology* 134: 503–514.

NHS. 2013. History Nurses' Health Study. Retrieved on April 2, 2013 (http://www.channing.harvard.edu/nhs/?page_id=70).

NICHD ECCRN. 2008. Social competence with peers in third grade: Associations with earlier peer experiences in childcare. *Social Development* 17: 419–453.

Norusis, M. J. 2011. *The SPSS guide to data analysis.* Chicago: SPSS Inc.

O'Neill, K. 2008. Public sociology and participatory action research in rural sociology. *Footnotes* 36(9). Retrieved on July 2, 2009 (http://www.asanet.org/footnotes/dec08/rural.html).

OBSSR (Office Of Behavioral and Social Science Research of NIH) 2013. Current Funding Opportunities.

Retrieved on March 23, 2013 (http://obssr.od.nih.gov/funding_opportunities/foas/foas.aspx).

Office of National Statistics (ONS). 2012. Measuring National Well-being: Summary of Proposed Domains and Measures, July 2012. Retrieved on March 23, 2013 (http://www.ons.gov.uk/ons/rel/wellbeing/measuring-national-well-being/summary-of-proposed-domains-and-measures/summary-of-proposed-domains-and-measures-of-national-well-being.html#tab-About-the-ONS-Measuring-National-Well-being-Programme).

O'Halloran, S. 2006. Power and Solidarity Building in the Discourse of Alcoholics Anonymous. *Journal of Groups in Addiction & Recovery* 1: 69–95.

Olds, D. L. 2012. Improving the Life Chances of Vulnerable Children and Families with Prenatal and Infancy Support of Parents: The Nurse–Family Partnership. *Psychosocial Intervention* 21: 129–143.

Olson, J. E., B. Zdaniuk, J. Horvak, D. Pauknerová, L. A. N. Luu, M. Kovacs, J. Miluska, A. Orgocka, L. Erokhina, O. V. Mitina, L. V. Popova, N. Petkevičiūtė, M. Pejic-Bach, S. Kubušová, and M. R. Makovec. 2007. Beliefs in equality for women and men as related to economic factors in Central and Eastern Europe and the United States. *Sex Roles* 56: 297–308.

Olson, S. 1976. *Ideas and data: The process and practice of social research*. Homewood, IL: Dorsey.

Ono, H., and H.-J. Tsai. 2008. Race, parental socioeconomic status, and computer use time outside of school among young American children, 1997 to 2003. *Journal of Family Issues* 29: 1650–1672.

Oransky, M., and C. Fisher. 2009. The development and validation of meanings of adolescent masculinity scale. *Psychology of Men and Masculinity* 10(1): 57–72.

Organisation for Economic Cooperation and Development (OECD). 2012. Your Better Life Index. Retrieved on March 23, 2013 (http://www.oecdbetterlifeindex.org/).

Organisation for Economic Cooperation and Development (OECD). 2011. How's Life? Measuring Well-Being. DOI:10.1787/9789264121164-14-en. Retrieved on March 23, 2013 (http://www.oecd-ilibrary.org/economics/how-s-life/subjective-well-being_9789264121164-14-en).

Ozbay, O., and Y. Ozcan. 2008. A test of Hirschi's social bonding theory: A comparison of male and female delinquency. *International Journal of Offender Therapy and Comparative Criminology* 52(2): 134–157.

Palys, T., and J. Lowman. 2001. Social research with eyes wide shut: The limited confidentiality dilemma. *Canadian Journal of Criminology* 43: 255–267.

Palys, T., and J. Lowman. 2006. Protecting research confidentiality: Towards a research-participant shield law. *Canadian Journal of Law and Society* 21: 163–185.

Pappas, S. 2012. East-West Culture Clash? Boston, San Francisco See Happiness Differently. Retrieved on April 8, 2013 (http://www.livescience.com/23283-east-vs-west-coast-culture-differences.html).

Park, P. 1993. What is participatory research? A theoretical and methodological perspective. In *Voices of change, participatory research in the United States and Canada* 1–20, edited by P. Park, M. Brydon-Miller, B. Hall, and T. Jackson. Westport, CT: Bergin and Garvey.

Parry, M. 2012. The Neighborhood Effect 2012. *The Chronicle of Higher Education*. Retrieved on February 10, 2013 (http://chronicle.com/article/The-Neighborhood-Effect/135492/).

Pattman, R., S. Frosh, and A. Phoenix. 2005. Constructing and experiencing boyhoods in research in London. *Gender and Education* 17(5): 555–561.

Pearson, D., C. McDougall, M. Kanaan, R. Bowles, and D. Torgerson. 2011. Reducing criminal recidivism: Evaluation of citizenship, an evidence-based probation supervision. process. *Journal of Experimental Criminology* 7: 73–102.

Peirce, K., and E. Edwards. 1988. Children's construction of fantasy stories: Gender differences in conflict resolution strategies. *Sex Roles* 18: 393–404.

Perrone, K. M., C. L. Civiletto, L. K. Webb, and J. C. Fitch. 2004. Perceived barriers to and supports of the attainment of career and family goals among academically talented individuals. *International Journal of Stress Management* 11: 114–131.

Perry, C. L., K. A. Komro, S. Veblen-Mortenson, and L. M. Bosma. 2000. The Minnesota DARE PLUS Project: Creating community partnerships to prevent drug use and violence. *Journal of School Health* 70: 84–88.

Pew Research Center for People and the Press. 2013. Sampling. Retrieved March 21, 2013 (http://www.people-press.org/methodology/sampling/).

Pew Research Center, 2013. Deficit Reduction Rises on Public's Agenda for Obama's Second Term. Retrieved on April 2, 2013 (http://www.people-press.org/2013/01/24/deficit-reduction-rises-on-publics-agenda-for-obamas-second-term/).

Pew Research Forum on Religion & Public Life. 2008. US Religious Landscape Study. Retrieved on February 10, 2013 (http://religions.pewforum.org/pdf/report-religious-landscape-study-full.pdf).

Pew Research Forum on Religion & Public Life. 2012. "Nones" on the Rise: One-in-Five-Adults have no Religious Affiliation. Retrieved on February 10, 2013 (http://www.pewforum.org/uploadedFiles/Topics/Religious_Affiliation/Unaffiliated/NonesOnTheRise-full.pdf).

Plaut, C. V., H. R. Markus, J. R. Treadway, and A. S. Fu. 2012. The cultural construction of Self and well-being: A tale of two cities. *Personality and Social Psychology Bulletin* 20: 1–15.

Portes, A., J. M. Clarke, and R. Manning. 1985. After Mariel: A survey of the resettlement experiences of 1980 Cuban refugees in Miami. *Cuban Studies* 15: 37–59.

Pothier, M. 2009. Pop goes the market. The strange connection between billboard hits and stock prices. *Boston Globe*, February 22. Retrieved on February 22, 2009 (http://www.boston.com/bostonglobe/ideas/articles/2009/02/22/pop_goes_the_market/).

Ptacek, J. 1988. Why do men batter their wives? In *Feminist perspectives on wife abuse*, edited by K. Yllo and M. Bograd. Newbury Park, CA: Sage.

Public Agenda. 2002. Right to die: Major proposals. *Public Agenda Online*. Retrieved on April 18, 2002 (http://www.publicagenda.org/issues/major-proposals_details.dfm?).

Putnam, R. D. 2000. *Bowling alone: The collapse and revival of American community*. New York: Simon & Schuster.

Pyrczak, F., and R. R. Bruce. 2007. *Writing empirical research reports: A basic guide for students of the social and behavioral sciences*, 6th edition. Los Angeles, CA: Pyrczak.

Racher, F. E., J. M. Kaufert, and B. Havens. 2000. Conjoint research interviews with frail elderly couples: Methodological implications. *Journal of Nursing Research* 6: 367–379.

Radley, A., D. Hodgetts, and A. Cullen. 2005. Visualizing homelessness: A study in photography and estrangement. *Journal of Community & Applied Social Psychology* 15(4): 273–295.

Ramji, H. 2008. Exploring commonality and difference in in-depth interviewing: A case-study of researching British Asian women. *British Journal of Sociology* 59: 99–116.

Rasmussen Reports. 2005. 72% say they're willing to vote for woman president. Retrieved on June 23, 2009 (http://legacy .rasmussenreports.com/2005/ Woman% 20President.htm).

Rasmussen Reports. 2009. 44% say global warming due to planetary trends, not people. Retrieved on May 20, 2009 (http://www .rasmussenreports.com/public _content/politics/issues2/articles /44_say_global_warming_due_to _planetary_trends_not_people).

Rathje, W. L., and C. Murphy. 1992. *Rubbish! The archaeology of garbage*. New York: HarperCollins.

RealClearPolitics. 2012. General election: McCain vs. Obama. Retrieved on March 21, 2013 (http://www .realclearpolitics.com/epolls/2012).

Reger, J. 2004. Organizational "emotion work" through consciousness-raising: An analysis of a feminist organization. *Qualitative Sociology* 27(2): 205–222.

Regnerus, M., and J. Uecker. 2011. *Premarital Sex in America How Young Americans Meet, Mate, and Think about Marrying*. New York: Oxford University Press.

Reinharz, S. 1992. *Feminist methods in social research*. New York: Oxford University Press.

Report of the commission on the measurement of economic performance and social progress. Retrieved on March 23, 2013 (http://www .stiglitz-sen-fitoussi.fr/documents /rapport_anglais.pdf).

Reverby, S. M. 2011. "Normal Exposure" and inoculation syphilis: A PHS "Tuskegee" doctor in Guatemala, 1946–1949. *Journal of Policy History* 23: 6–28.

Rew, L., S. D. Horner, L. Riesch, and R. Cauvin. 2004. Computer-assisted survey interviewing of school-age children. *Advances in Nursing Science* 27(2): 129–137.

Reynolds, A. J., J. A. Temple B. A. B. White, S. R. Ou, and D. L. Robertson. 2011. Age 26 Cost-Benefit Analysis of the Child-Parent Center Early Education Program. *Child Development* 82: 379–404.

Reynolds, A. J., and D. Robertson. 2003. School-based early intervention and later child maltreatment in the Chicago Longitudinal Study. *Child Development* 74(1): 3–26.

Reynolds, A. J., L. C. Mathieson, and J. W. Topitzes. 2009. Do early childhood interventions prevent child maltreatment? A review of research. *Child Maltreatment* 14: 182–206.

Reynolds, A. J., and S. R. Ou. 2011. Path of effects from preschool to adult well-being: A confirmatory analysis of the child-parent center program. *Child Development* 82: 555–582.

Richman, W., S. Kiesler, S. Weisband, and F. Drasgow. 1999. A meta-analytic study of social desirability distortion in computer-administered questionnaires, traditional questionnaires and interviews. *Journal of Applied Psychology* 84: 754–775.

Riemer, J. W. 1977. Varieties of opportunistic research. *Urban Life* 5: 467–477.

Ringheim, K. 1995. Ethical issues in social science research within special reference to sexual behavior research. *Social Science and Medicine* 40: 1691–1697.

Risman, B. J., and R. C. Allison. 2012. A Double Standard for "Hooking Up": How Far Have we Come Toward Gender Equality? Presented at the American Sociological Association Annual Meetings in August 2012 in Denver, CO.

Rittenour, C., C. Colaner. 2012. Finding female fulfillment: Intersecting role-based and morality-based identities of motherhood, feminism, and generativity as predictors of women's self satisfaction and life satisfaction. *Sex Roles* 67(5-6): 351–362.

Robb, M. 2004. Exploring fatherhood: masculinity in the research process. *Journal of Social Work Practice* 18(3): 395–406.

Roberts, S. 2007. 51% of women are now living without spouse. *New York Times*, January 16: 1ff.

Roberts, S. 2010. New Figure for 2010 Census: $1.6 Billion under budget. *New York Times*, Retrieved September 13, 2013 (http://www. nytimes.com/2010/08/11/us /politics/11census.html?_r=0).

Robinson, J., and S. Martin. 2008. What do happy people do? *Social Indicators Research* 89: 565–571.

Rochlan, A., G. Good, and T. Carver. 2009. Predictors of gender-related barriers, work, and life satisfaction among men in nursing. *Psychology of Men & Masculinity* 10(1): 44–56.

Rollins, J. 1985. *Between women*. Philadelphia: Temple University Press.

Rosenberg, M. 1968. *The logic of survey analysis*. New York: Basic Books.

Rosenfeld, M., and R. Thomas. 2012. Searching for a mate: The rise of the internet as a social intermediary. *American Sociological Review* 77(4): 523–547.

Rosengren, K. E. 1981. Advances in Scandinavian content analysis. In *Advances in content analysis*, 9–19, edited by K. E. Rosengren. Beverly Hills, CA: Sage.

Rosenhan, D. L. 1971. On being sane in insane places. *Science* 179: 250–258.

Rossi, P. H., and H. E. Freeman. 1993. *Evaluation: A systematic approach*. Newbury Park, CA: Sage.

Rossi, P. H., and J. D. Wright. 1984. Evaluation research: An assessment. In *Annual review of sociology*, vol. 10, 332–352, edited by R. H. Turner and J. F. Short. Palo Alto, CA: Annual Reviews.

Rubin, L. B. 1983. *Intimate strangers: Men and women together*. New York: Harper & Row.

Rubin, L. B. 1994. *Families on the fault line: America's working class speaks about the family, the economy, race, and ethnicity*. New York: HarperCollins.

Runcie, J. F. 1980. *Experiencing social research*. Homewood, IL: Dorsey.

Rutherford, M. B. 2004. Authority, autonomy, and ambivalence: Moral choice in twentieth-century commencement speeches. *Sociological Forum* 19: 583–609.

Salmela-Aro, K., N. Kiuru, E. Leskinen, and J. Nurmi. 2009. School Burnout Inventory (SBI): Reliability and validity. *European Journal of Psychological Assessment* 25(1): 48–57.

Sampson, R. J. 2012. Moving and the neighborhood glass ceiling. *Science* 337: 1464–1465.

Sampson, R. J., and S. W. Raudenbush. 1999. Systematic social observation of public spaces: A new look at disorder in urban neighborhoods. *American Journal of Sociology* 105: 603–651.

Sanchez, J., C. Lomeli-Loibl, and A. A. Nelson. 2009. Sacramento's LGBTQ youth: Youth-led participatory action research for mental health justice with Youth in Focus. *Focal Point* 23: 6–8. Retrieved on July 2, 2009 (http://www.rtc.pdx.edu/PDF/fpS0902.pdf).

Satterfield, J. W., and M. G. Godfrey. 2011. The university of nebraska-lincoln football: A metaphorical, symbolic and ritualistic community event. *Forum: Qualitative Social Research* 12: 1–17.

Saucier, P. K. 2008. *We eat Cachupa, not clam Chowder: Mapping second generation Cape Verdean youth identity in the greater Boston area.* Doctoral dissertation at Northeastern University.

Scarce, R. 1995. Scholarly ethics and courtroom antics: Where researchers stand in the eyes of the law. *American Sociologist* 26(1): 87–112.

Scarce, R. 2005. *Contempt of court: A scholar's battle for free speech from behind bars.* Lanham, MD: AltaMira Press.

Schaeffer, N. C., and S. Presser. 2003. The science of asking questions. *Annual Review of Sociology* 29: 65–88.

Schieman, S. 2001. Age, education, and the sense of control: A test of the cumulative advantage hypothesis. *Research on Aging* 23: 153–178.

Scheiber, N. 2012 (November 30). The Internal Polls that made Mitt Romney think he'd win. *New Republic.* Retrieved March 21, 2013 (http://www.newrepublic.com/blog/plank/110597/exclusive-the-polls-made-mitt-romney-think-hed-win).

Schenk, K. A. Michaelis, T. Sapiano, and L. Brown. 2010. Improving the lives of vulnerable children: Implications of horizons research among orphans and other children affected by AIDS. *Public Health Reports* 125(2): 325–336.

Schieman, S. 2011. Education and the importance of religion in decision making: Do other dimensions of religiousness matter? *Journal for the Scientific Study of Religion* 50: 570–587.

Schlosser, J. A. 2008. Issues in interviewing inmates navigating the methodological landmines of prison research. *Qualitative Inquiry* 14: 1500–1525.

Schoeni, R. F., F. Stafford, K. A. Mcgonagle, and P. Andreski. 2013. Response rates in national panel surveys. *The ANNALS of the American Academy of Political and Social Science.* 645: 60–87.

Schorr, L. 2012. Broader evidence for bigger impact. *Stanford Social Innovation Review* 10: 50–55. Retrieved on April 4, 2013 (http://www.ssireview.org/articles/entry/broader_evidence_for_bigger_impact).

Schumaker, J. F., and J. D. Shea. 1993. Loneliness and life satisfaction in Japan and Australia. *Journal of Psychology* 127: 65–71.

Schuman, H., and S. Presser. 1981. *Questions and answers in attitude surveys.* New York: Academic Press.

SDA Archive. 2012. GSS cumulative datafile 1972–2008. Retrieved on April 2009 (http://sda.berkeley.edu/archive.htm).

Seale, C., J. Charteris-Black, C. Dumelow, L. Locock, and S. Ziebland. 2008. The effect of joint interviewing on the performance of gender. *Field Methods* 20: 107–128.

Seidman, I. 1991. *Interviewing as qualitative research.* New York: Teachers College, Columbia University.

Seidman, I. 1998. *Interviewing as qualitative research*, 2nd edition. New York: Teachers College, Columbia University.

Seiter, E. 1993. *Sold separately: Children and parents in consumer culture.* New Brunswick, NJ: Rutgers University Press.

Shadish, W. R., Jr., T. D. Cook, and L. C. Leviton. 1991. *Foundations of program evaluation.* Newbury Park, CA: Sage.

Shapiro, G., and P. Dawson. 1998. *Revolutionary demands: A content analysis of the Cahiers de Doleances of 1789.* Stanford, CA: Stanford University Press.

Shaw, C. R., and H. D. McKay. 1931. Social factors in juvenile delinquency. *Reports on the causes of crime*, vol. II. National Commission on Law Observance and Enforcement, Report No. 13. Washington, DC: U.S. Government Printing Office.

Shaw, R. 2011. The ethical risks of curtailing emotion in social science research: The case of organ transfer. *Health Sociology Review* 20: 58–69.

Sheets, R., and J. Mohr. 2009. Perceived social support from friends and family and psychosocial functioning in bisexual young adult college students. *Journal of Counseling Psychology* 56(1): 152–163.

Shepard III, E. M. 2001. The economic costs of D.A.R.E. Unpublished paper. Retrieved on July 3, 2009 (http://www.drugpolicy.org/docUploads/DAREfinalRP.pdf).

Sherwood, J. H. 2004. Talk about country clubs: Ideology and the reproduction of privilege. Ph.D. dissertation, North Carolina State University (http://www.lib.ncsu.edu/theses/available/etd-04062004-083555).

Shih, T., and X. Fan. 2008. Comparing response rates from web and mail surveys: A meta-analysis. *Field Methods* 20: 249–271.

Silva, J. 2012. Constructing adulthood in an age of uncertainty. *American Sociological Review* 77(4): 505–522.

Silver, H. J. 2006. Science and politics: The uneasy relationship. *Footnotes* 34(2): 1–5.

Silver, H., A. Sharpe, H. Kelly, P. Kobor, and G. Sroufe. 2013. Social and Behavioral Science Research in the FY 2012 Budget. Retrieved on February 22, 2013 (http://www.aaas.org/spp/rd/rdreport2012/12pch19.pdf).

Silverman, D. 2004. *Doing qualitative research*, 2nd edition. London: Sage Publications.

Sims, R. C. 2007. Web surveys: Applications in denominational research.

Review of Religious Research 49: 69–80.

Singer, E., and Y. Cong. 2013. The Use and Effects of Incentives in Surveys. *The ANNALS of the American Academy of Political and Social Science* 645: 112–141.

Singer, E., and F. J. Levine. 2003. Protection of human subjects and research: Recent developments and future prospects for the social sciences. *Public Opinion Quarterly* 67(1): 148–164.

Singer, P. 2005. Anti-fraud research rules take effect. *National Journal* 37(28): 2210.

Singleton, R. A., Jr., and B. C. Straits. 2001. Survey interviewing. In *Handbook of interview research*, 59–82, edited by J. F. Gubrium and J. A. Holstein. Thousand Oaks, CA: Sage Publications.

Singleton, R., Jr., B. C. Straits, M. M. Straits, and R. J. McAllister. 1988. *Approaches to social research.* New York: Oxford University Press.

Sloane, P. D., L. W. Cohen, T. R. Konrad, C. S. Williams, J. G. Schumacher, and S. Zimmerman. 2008. Physician interest in volunteer service during retirement. *Annals of Internal Medicine* 149: 317–322.

Small, S. A., S. M. Cooney, and C. O'Connor. 2009. Evidence-informed program improvement: Using principles of effectiveness to enhance the quality and impact of family-based prevention programs. *Family Relations* 58: 1–13.

Smith, A. E., J. Sim, T. Scharf, and C. Phillipson. 2004. Determinants of quality of life amongst older people in deprived neighborhoods. *Ageing and Society* 24: 793–814.

Smith, P. M., and B. B. Torrey. 1996. The future of the behavioral and social sciences. *Science* 271: 611–612.

Smith, T. W. 1989. That which we call welfare by any other name would smell sweeter: An analysis of the impact of question wording on response patterns. In *Survey research methods*, 99–107, edited by E. Singer and S. Presser. Chicago: University of Chicago Press.

Smith, T. W. 2005. Generation gaps in attitudes and values from the 1970s to the 1990s. In *On the frontier of adulthood: Theory, research and public policy*, edited by R. A. Settersen, F. F. Furstenberg Jr., and

R. G. Rumbaut. Chicago: University of Chicago Press.

Smith, T. W., J. Kim, A. Koch, and A. Park. 2005. Social-science research and the general social surveys. *ZUMA-Nachrichten* 56: 68–77.

Smyth, J. D., D. A. Dillman, L. M. Christian, and M. J. Stern. 2006. Comparing check-all and forced-choice question formats in web surveys. *Public Opinion Quarterly* 70: 66–77.

Snell, D., and D. Hodgetts. 2007. Heavy metal, identity and the social negotiation of a community of practice. *Journal of Community and Applied Social Psychology* 17: 430–445.

Sokolow, J. A. 1999. The darker side of persuasion: Stanley Milgram's experiments on obedience to authority. *Proposal Management* 31–36.

Soule, S., and B. King. 2008. Competition and resource partitioning in three social movement industries. *American Journal of Sociology* 113 (6): 1568–1610.

Southgate, D., and V. Roscigno. 2009. The impact of music on childhood and adolescent achievement. *Social Science Quarterly* 90(1): 4–21.

Spaaij, R. 2008. Violence, masculinity, and collective identity in football hooliganism. *Journal of Sports and Social Issues* 32(4): 369–392.

Srivastava, S., M. Tamir, K. M. McGonigal, O. P. John, and J. J. Gross. 2009. The social costs of emotional suppression: A prospective study of the transition to college. *Journal of Personality and Social Psychology* 96(4): 883–897.

Stake, R. E. 2003. Case studies. In *Strategies of qualitative inquiry*, 2nd edition, 134–164, edited by N. K. Denzin and Y. S. Lincoln. Thousand Oaks, CA: Sage.

Stalker, K., J. Carpenter, C. Connors, and R. Phillips. 2004. Ethical issues in social research: Difficulties encountered gaining access to children in hospital for research. *Child: Care, Health and Development* 30: 377–383.

St-Amand, O., C. Ward-Griffin, R. DeForge, A. Oudshoom, C. McWilliam, D. Forbes, M. Kloseck, and J. Hall. 2012. Making Care Decisions in Home-Based Dementia Care: Why Context Matters. *Canadian Journal of Aging* 31(4): 423–434.

Stark, L. 2007. Victims in our own minds? IRBs in myth and practice. *Law and Society Review* 41: 777–786.

Stark, L. 2012. Behind Closed Doors: IRBs and the Making of Ethical Research. Chicago: University of Chicago Press.

Starr, P. 1987. The sociology of official statistics. In *The politics of numbers*, 7–60, edited by W. Alonso and P. Starr. New York: Russell Sage Foundation.

Stenmark, C. K., A. L. Antes, L. E. Martn, Z. Bagdasarov, J. F. Johnson, L. D. Davenport, and M.D. Mumford. 2010. Ethics in the Humanities: Findings from Focus Groups. *Journal of Academic Ethics* 8: 285–300.

Stephan, Y., Caudroit, J., and Chalabaev, A. 2011. Why does a youthful subjective age lead to higher life satisfaction among older adults? The mediational role of subjective health and memory self-efficacy. *Aging and Mental Health* 15: 428–436.

Stevanovic, P., and P. A. Rupert. 2009. Work–family spillover and life satisfaction among professional psychologists. *Professional Psychology: Research an Practice* 40(1): 62–68.

Stille, A. 2001, April 14. New attention for the idea that abortion averts crime. *New York Times*. Retrieved September 13, 2013 (http//:www.nytimes.com/2001/04/14/arts/new-attention-for-the-idea-that-abortion-averts-crime.html?pagewanted=all&src=pm).

Stouffer, S. A. 1950. Some observations on study design. *American Journal of Sociology* 55: 355–361.

Strauss, A. 1987. *Qualitative analysis for social scientists.* Cambridge: Cambridge University Press.

Streb, M. J., B. Burrell, B. Frederick, and M. A. Genovese. 2008. Social desirability effects and support for a female American president. *Public Opinion Quarterly* 72: 76–89.

Subedi, B., and J. Rhee. 2008. Negotiating collaboration across differences. *Qualitative Inquiry* 14: 1070–1092.

Suchman, L., and B. Jordan. 1992. Validity and the collaborative construction of meaning in face-to-face surveys. In *Questions about questions*, 241–270, edited by J. M. Tanur. New York: Russell Sage Foundation.

Sudman, S., and N. Bradburn. 1982. *Asking questions: A practical guide to questionnaire design*. San Francisco: Jossey-Bass.

Sulik, G. A. 2007. The balancing act: Care work for the self and coping with breast cancer. *Gender & Society* 21: 857–877.

Tan, J., and Y. Ko. 2004. Using feature films to teach observation in undergraduate research methods. *Teaching Sociology* 32: 109–118.

Tang, F., V. Carr Copeland, and S. Wexler. 2012. Racial Differences in Volunteer Engagement by Older Adults: An Empowerment Perspective. *Social Work Research* 36(2): 89–100.

Taylor, L. 2005. All for him: Articles about sex in American lad magazines. *Sex Roles* 52: 152–163.

Teisl, M. F., B. Roe, and M. Vayda. 2006. Incentive effects on response rates, data quality, and survey administration costs. *International Journal of Public Opinion Research* 218: 364–373.

The Disaster Center. 2008. UCR crime statistics. Retrieved on February 28, 2009 (http://www.disaster center.com/crime/uscrime.htm).

The Economist. 1997. Don't let it happen again. 343: 27–28.

Thinley, J. Y. 2005. What is Gross National Happiness? Presented at the Second International Conference on GNH in Halifax, Canada. Retrieved on March 23, 2013 (http://www.bhutanstudies.org.bt /pubFiles/8-Rethinking.pdf).

Thurlow, M. B. 1935. An objective analysis of family life. *Family* 16: 13–19.

Tomaskovic-Devey, D., C. P. Wright, R. Czaja, and K. Miller. 2006. Self-reports of police speeding stops by race: Results from the North Carolina reverse record check survey. *Journal of Quantitative Criminology* 22: 279–297.

Tietjen, G. E. 2012. Exploring Educational Pathways: A Study of Ex-convict Reintegration. Presented at the American Sociological Association Annual Meetings in August 2012 in Denver, CO.

Topitzes, J., O. Godes, J. P. Mersky, S. Ceglarek, and A. J. Reynolds. 2009. Educational success and adult health: Findings from the Chicago longitudinal study. *Prevention Science* 10: 175–195.

Touliatos, J., B. F. Perlmutter, and M. A. Straus. 2001. *Handbook of family measurement techniques*. Thousand Oaks, CA: Sage.

Tourangeau, R., E. Singer, and S. Presser. 2003. Context effects in attitude surveys: Effects on remote items and impact on predictive validity. *Sociological Methods and Research* 31(4): 486–513.

Traver, A. E. 2007. Home(land) Décor: China adoptive parents' consumption of Chinese cultural objects for display in their homes. *Qualitative Sociology* 30: 201–220.

Trevino, A. J., 2012. Presidential Address: The Challenge of Service Sociology *Social Problems* 59: 2–20.

Trochian, W. M. K. 2006. Research methods: Knowledge base. Retrieved on January 28, 2009 (http://www.socialresearchmethods. net/kb/index.php).

Trochim, W. M., S. E. Marcus, L. C. Mâsse, R. P. Moser, and P. C. Weld. 2008. The evaluation of large research initiatives. *American Journal of Evaluation* 29: 8–28.

Turabian, K. L., revised by J. Grossman, and A. Bennett. 1996. *A manual for writers of term papers, theses, and dissertations*, 6th edition. Chicago: University of Chicago Press.

Turner, J. A. 1991. *The structure of sociological theory*, 5th edition. Belmont, CA: Wadsworth.

Tutt, D. 2008. Where the interaction is: Collisions of the situated and the mediated in living room interaction. *Qualitative Inquiry* 14(7): 1157–1179.

U.S. Census Bureau. 2000. Census bureau director says 92 percent of U.S. households accounted for. *United States Department of Commerce News*, May 31. Retrieved in October 2005 (www.census.gov /Press-Release/www/200041.html).

U.S. Census Bureau. 2009b. What is the census? US census 2010. Retrieved on June 23, 2009 (http://2010 .census.gov/2010census/).

U.S. Census Bureau. 2009c. Internet use triples in decades, Census Bureau reports. Retrieved on June 23, 2009 (http://www.census.gov/Press-Release/www/releases/archives/ communication_industries/013849. html).

U.S. Census Bureau. 2011. *Population Distribution and Change: 2000 to 2010*. Retrieved on March 21, 2013 (http://www.census.gov/prod/ cen2010/briefs/c2010br-01.pdf).

U.S. Census Bureau. 2012a. *Statistical abstract: The national data book*. Retrieved on February 13, 2009 (http://www.census.gov/compendia/ statab/).

U.S. Commission on Immigration Reform. 1994. *Restoring credibility*. Washington, DC: U.S. Commission on Immigration Reform.

U.S. Department of Justice. 2012. Annual Report to United States Sentencing Commission. Retrieved on March 23, 2013 (http://www .justice.gov/criminal/foia/docs /2012-annual-letter-to-the-us-sentencing-commission.pdf).

U.S. Government Printing Office. 1949. Trials of war criminals before the Nuremberg military tribunals under Control Council Law No. 10, Vol. 2, pp. 181–182. Washington, DC: U.S. Government Printing Office. Retrieved on March 11, 2009 (http://ohsr.od.nih.gov /guidelines/nuremberg.html).

Ulrich, L. T. 1990. *A midwife's tale: The life of Martha Ballard, based on her diary, 1785–1812*. New York: Vintage Books.

Umaña-Taylor, A. J., and M. Y. Bámaca. 2004. Conducting focus groups with Latino populations: Lessons from the field. *Family Relations* 53(3): 261–272.

United Nations Statistics Division. 2000. The world's women 2000: Trends and statistics. Table 5.D: Indicators of economic activity. Retrieved on May 17, 2004 (http://unstats.un .org/unsd/demographic/ww2000/).

Urbaniak, G. C., and S. Plous. 2009. *Research randomizer*. Retrieved on March 9, 2009 (http://randomizer .org).

University of Massachusetts. 2012. Qualitative Data Analysis Package (QDAP). Retrieved March 21, 2013 (http://www.umass.edu/qdap/).

Urbaniak, G., and S. Plaus. 2013. *Research Randomizer*. Retrieved March 21, 2013 (http://www .randomizer.org).

Van Leeuwen, T., and C. Jewitt. 2001. *Handbook of visual analysis*. Thousand Oaks, CA: Sage.

Vandell, D. L., J. Belsky, M. Burchinal, L. Steinberg N. Vandergrift, and NICHD Early Child Care Research Network. 2010. Do effects of early child care extend to age 15 years? Results From the NICHD Study of

Early Child Care and Youth Development. *Child Development*, 81, 737–756.

Venkatesh, S. 2012. Blog version of *The Blue and White*, a Columbia University undergraduate magazine retrieved on March 10, 2013 (http://bwog.com/2012/12/02/venkatesh-responds-to-the-nyt/).

Wager, N. 2011. Researching Sexual Revictimisation: Associated Ethical and Methodological Issues, and Possible Solutions. *Child Abuse Review* 20: 158–172.

Wallace, W. 1971. *The logic of science in sociology.* Chicago: Aldine.

Walton, J. 1992. Making the theoretical case. In *What is a case? Exploring the foundations of social inquiry,* 121–138, edited by C. C. Ragin and H. S. Becker. New York: Cambridge University Press.

Ward, J., and Z. Henderson. 2003. Some practical and ethical issues encountered while conducing tracking research with young people leaving the "care" system. *International Journal of Social Research Methodology* 6(3): 255–259.

Warr, M., and C. G. Ellison. 2000. Rethinking social reactions to crime: Personal and altruistic fear in family households. *American Journal of Sociology* 106: 551–578.

Warren, C. A. B., T. Barnes-Brus, H. Burgess, L. Wiebold-Lippisch, J. Hackney, G. Harkness, V. Kennedy, R. Dingwall, P. C. Rosenblatt, A. Ryen, and R. Shuy. 2003. After the interview. *Qualitative Sociology* 26(1): 93–110.

Wax, R. H. 1971. *Doing fieldwork: Warnings and advice.* Chicago: University of Chicago.

Weaver, D. 1994. The work and retirement decisions of older women: A literature review. *Social Security Bulletin* 57: 3–25.

Webb, E. J., D. T. Campbell, R. D., R.D. Schwartz, and L. Sechrest. 2000 *Unobtrusive measures,* revised edition. Thousand Oaks, CA: Sage.

Weber, M. 1947. *The theory of social and economic organization.* New York: Free Press.

Weer, C. 2006. The role of maternal employment, role-altering strategies, and gender in college students' expectation of work-family conflict. *Sex Roles* 55(7/8): 535–544.

Weisberg, H. F. 2005. *The total survey error approach.* Chicago: The University of Chicago Press.

Weisberg, H. F., J. A. Krosnick, and B. D. Bowen. 1989. *An introduction to survey research and data analysis.* Glenview, IL: Scott, Foresman.

Weiss, C. H., E. Murphy-Graham, and S. Birkeland. 2005. An alternate route to policy influence how evaluations affect D.A.R.E. *American Journal of Evaluation* 26: 12–30.

Weiss, R. S. 1994. *Learning from strangers: The art and method of qualitative interview studies.* New York: Free Press.

Weitz, R. 1991. *Life with AIDS.* New Brunswick, NJ: Rutgers University Press.

Weitzman, E., and M. B. Miles. 1995. *Computer programs for qualitative analysis.* Thousand Oaks, CA: Sage.

Weitzman, L. J. 1985. *The divorce revolution.* New York: Free Press.

Weitzman, Lenore, Deborah Eifler, E. Hokado, and C. Ross. 1972. Sex-role Socialization in Picture Books for Preschool Children. *American Journal of Sociology* 77: 1125–1150.

Westkott, M. 1979. Feminist criticism of the social sciences. *Harvard Educational Review* 49: 422–430.

Whoriskey, P. 2012. If you're happy and you know it… let the government know. *Washington Post* Retrieved on March 23, 2013 (http://www.washingtonpost.com/business/economy/if-youre-happy-and-you-know-it-let-the———government-know/2012/03/29/gIQAlSL2jS_story.html?hpid=z4).

Whyte, W. F. 1955. *Street corner society: The social structure of an Italian slum.* Chicago: University of Chicago Press.

Whyte, W. F. 1997. *Creative problem solving in the field: Reflections on a career.* Walnut Creek, CA: AltaMira Press.

Williamson, T., and A. Long. 2005. Qualitative data analysis using data displays. *Nurse Researcher* 12: 7–19.

Wilson, D. C., D. W. Moore, P. F. McKay, and D. R. Avery. 2008. Affirmative action programs for women and minorities; expressed support affected by question order. *Public Opinion Quarterly* 72: 514–522.

Wolkomir, M. 2001. Emotion work, commitment, and the authentication of the self. *Journal of Contemporary Ethnography* 30: 305–334.

Wong, J. D., and M. A. Hardy. 2009. Women's retirement expectations: How stable are they? *Journals of Gerontology Series B: Psychological Sciences and Social Sciences* 64: 77–86.

Wyman, P., C. Hendricks Brown, M. LoMurray, K. Schmeelk-Cone, M. Petrova, Y. Qin, E. Walsh, T. Xin, and W. Wang. 2010. An outcome evaluation of the sources of strength suicide prevention program delivered by adolescent peer leaders in high schools. *American Journal of Public Health* 100: 1653–1661.

Wynn, L. L. 2011. Ethnographers' experiences of institutional ethics oversight: Results from a quantitative and qualitative survey. *Journal of Policy History.*

Wynn, L. L. 2011. Ethnographers' Experiences of institutional ethics oversight: Results from a quantitative and qualitative survey. *Journal of Policy History* 23: 94–114.

Wysong, E., R. Aniskiewicz, and D. Wright. 1993. Truth and DARE: Tracking drug education to graduation and as symbolic politics. *Social Problems* 31: 448–472.

Yassour-Borochowitz, D. 2004. Reflections on the researcher-participant relationship and the ethics of dialogue. *Ethics and Behavior* 14: 175–186.

Yin, R. Y. 2009. *Case study research: Design and methods.* Thousand Oaks, CA: Sage Publications.

Zhao, S., S. Grasmuck, and J. Martin. 2008. Identity construction on Facebook: Digital empowerment in anchored relationships. *Computers in Human Behavior* 24(5): 1816–1836.

Zickuhr, K., and A. Smith, 2012. Digital divide, broadband, social networking, mobile pew research center. Retrieved on April 4, 2013 (http://pewinternet.org/Reports/2012/Digital-differences.aspx).

Zickuhr, K., and A. Smith. 2012. Internet adoption over time. Retrieved on March 21, 2013 (http://pewinternet.org/Reports/2012/Digital-differences/Main-Report/Internet-adoption-over-time.aspx).

Zients, J. D. 2012. Memorandum to the heads of executive departments and agencies; Use of evidence and evaluation in the 2014 Budget. May 18, 2012. Executive Office of the

President, Office of Management and Budget. Retrieved on April 4, 2013 (http://www.whitehouse.gov /sites/default/files/omb/memoranda /2012/m-12-14_1.pdf).

Zipp, J. F., and J. Toth. 2002. She said, he said, they said: The impact of spousal presence in survey research. *Public Opinion Quarterly* 66(2): 177–208.

Zurbriggen, E., and A. Allen. 2010. Race and Gender in the 2008 U.S. Presidential Election: A Content Analysis of Editorial Cartoons. *Analyses of Social Issues and Public Policy* 10(1): 223–247.

INDEX